THE GUINNESS
SCIENCE
FACT BOOK

GUINNESS PUBLISHING

THE GUINNESS SCIENCE FACT BOOK

Editor: Clive Carpenter
Information Systems: Alex Reid
Design: Sarah Silvé
Layout: Peters and Zabransky (UK) Ltd.
Artwork and diagrams: Peters and Zabransky (UK) Ltd., Peter Harper, Robert and Rhoda Burns, Suzanne Alexander, Pat Gibbon, Matthew Hillier, and The Maltings Partnership.

1st edition
First published 1994
Reprint 10 9 8 7 6 5 4 3 2 1 0

© **Guinness Publishing Ltd., 1994**

Published in Great Britain by Guinness Publishing Ltd,
33 London Road, Enfield, Middlesex, EN2 6DJ

All rights reserved. No part of this publication may be reproduced, stored in a retrieval system, or transmitted in any form or by any means, electronic, mechanical, photocopying, recording or otherwise, without prior permission in writing from the publisher.

Printed and bound in Great Britain by The Bath Press, Bath

British Library Cataloguing in Publication Data:
A catalogue record for this book is available from the British Library

ISBN 0-85112-765-7

THE GUINNESS SCIENCE FACT BOOK

Contents

Introduction 7-8

Biological Sciences

Evolution..10-11
Cells..12-13
Genetics...14-15
Taxonomy..16-17
Simple life forms - bacteria, viruses and protoctists..............18-19
Plants...20-23
Fungi..24
Spore-bearing plants...25-26
Conifers and their allies..27
Flowering plants..28-34
Animals - an introduction..35
Sponges..36
Coelenterates or cnidarians..37
Echinoderms...38
Worms...39
Molluscs..40
Arthropods..41
Insects...42-43
Crustaceans...44
Arachnids..45
Fishes...46-47
Amphibians...48-49
Reptiles..50-51
Birds..52-53
Mammals...54-59
The human skeleton...60-61
Human and other circulatory systems...................................62-63

Human and other respiratory systems 64-65
The human digestive system 66-67
Nervous and endocrine systems 68-69
Senses and perception 70-71
Reproduction of animals 72
Marine mammals 73

Physics

Mechanics - motion and force	74-78
Statics and forces involved in rotation	79-81
Friction and elasticity	82-83
Fluids and pressure	84-85
Machines	86
Thermodynamics	87-89
Specific heat and latent heat	90-91
Gases	92
Quantum theory and relativity	93-95
Wave types	96-97
Reflection and refraction of waves	98-99
Modulation, standing waves and diffraction	100-101
Acoustics	102-103
Optical reflection and refraction	104-105
Lens and mirrors	106-107
Lasers, holograms and fibre optics	108
Cathodes	109
Magnets and electricity	110-111
Electromagnetism	112-113
Batteries, cells and circuits	114-115
Alternating and direct current, motors and generators	116-117
Conductors	117-118
Transistors	119
Atomic and subatomic particles	120-123

Chemistry

Introduction to the principles and methods of chemistry	124-127
Elements and the periodic table	128-130
The 109 elements	131-135
Diffusion	136
States of matter	137
Noble (inert) gases	138
Chemical bonds	139-143

THE GUINNESS SCIENCE FACT BOOK

Chemical reactions 144-145
Metals 146
Alkalinity, the alkali metals and the alkaline earth metals 147-148
Halogens 149
Small molecules 150-151
Organic chemistry - natural compounds 152-155
Polymers 156-160
Synthetic products 161

Mathematics

Numbers 162-163
Fractions, decimals and percentages 164
Algebra 165
Ratio and proportion 166
Number patterns and binomials 167
Probability 168
Sets 169
Coordinates and graphs 170-171
Calculus 172-173
Mechanics 174-175
Matrices 176
Statistics 177-179
Geometry and trigonometry 180-181
Polygons 182-183
Triangles 184-185
Circles and other conic sections 186-187
Solids: areas and volumes 188-189
Polyhedra 190-191

Famous scientists

An A-Z of some of the world's most famous scientists including
nationality, birth and death dates, brief biographical details, and
their principal laws, theories, discoveries and publications. 192-217
Nobel Prize winners in Chemistry, Physics and Physiology or Medicine 218-224

Weights and measures

including Imperial/metric and Celsius/Fahrenheit conversions 225-230

Index/factfinder

.... 231-256

Advisory Editors

Biological Science	David Sweet
Chemistry	Will Hutchings
Famous Scientists	Dr Richard Weston
Mathematics	Colin Juneman
Physics	Andrew Clifford

with contributions from	
John Arblaster	Professor Peter Gahan
Ben Barkow	Professor Brian Gardiner
Dr Michael Benton	Professor Frank Glockling
Professor Michael Black	Dr Peter Hobson
E.J. Borowski	Will Hutchings
John A. Burton	Dr Gareth Jones
Gillian Carpenter	Colin Juneman
Dr Sara Churchfield	Meredith Lloyd-Evans
Dr Barry Clarke	Dr Paul Markham
Andrew Clifford	J.P. Mathias
Dr Margaret Collinson	Dr Stuart Milligan
Dr Peter V. Coveney	Dr D.M.P. Mingos
Dr John Cowan	Dr R.V. Parish
Oliver Crimmen	Dr Gillian Sales
Dr Jim Edwards	Andrew Scott
P. Ellwood	Dr J.F. Stoddart
Dr John Emsley	David Sweet
Dr Roland Emson	Dr Bryan Turner
Dr David G. Evans	David Wells
	Dr Richard Weston

Introduction

Science may be described as the systematic investigation of reality by observation, by experimentation and by induction. Galileo and Descartes were the first to insist clearly that science should employ only precise mathematical concepts in its theories. The application of such a theory to physical reality will then be a calculation in applied mathematics. This is the feature that makes physical theories so comprehensive and such precise predictors.

Physical theories are models of comprehensiveness because they explain physical processes that vary in scale from the subatomic to the astronomical, and vary in time from minute fractions of a second to billions of years. Furthermore, physical theories yield models of accurate prediction. For example, NASA was able to calculate the speed, direction and time needed to send the *Voyager 2* spacecraft, without any later course correction, on a journey of thousands of millions of kilometres passing close to each of the outer planets in turn.

Science did not exist among early civilizations. Certainly discoveries were made, but they were piecemeal. This began to change with the speculations of the early Greek philosophers, who excluded supernatural causes from their accounts of reality. Indeed, the Greeks produced theoretical models that have shaped the development of science ever since.

With the fall of Greece to the Roman Empire, science fell from grace. Few important advances were made outside medicine, and the work done was firmly within the Greek traditions and conceptual framework.

For several centuries from the fall of Rome, science was virtually unknown in Western Europe. Islamic culture alone preserved Greek knowledge, and later transmitted it back to Western Europe. Between the 13th and 15th centuries some advances were made in the fields of mechanics and optics, while scholars like Roger Bacon insisted on the importance of personal experience and observation.

The 16th century marked the coming of the so-called 'Scientific Revolution', a period of scientific progress beginning with Copernicus and culminating with Newton. Not only did science break new conceptual ground but it gained enormously in prestige as a result. Science and its trappings became highly fashionable from the later 17th century, and also attracted a great deal of royal and state patronage. The founding of the Académie des Sciences by Louis XIV in France and the Royal Society by Charles II in England were landmarks in this trend.

In the course of the 19th century science became professionalized, with clearcut career structures and hierarchies emerging, centred on universities, government departments and commercial organizations. This trend continued into the 20th century, which has seen science become highly dependent on technological advances. These have not been lacking.

Modern science is immense and extremely complex. It is virtually impossible to have an informed overview of what science as a whole is up to. This has made many people regard it with some suspicion. Nevertheless, Western civilization is fully committed to a belief in the value of scientific progress as a force for the good of humanity.

The complexity of modern science can be seen from the many branches into which scientific research and learning is divided. Some of those branches are defined here.

acoustics the branch of physics that deals with sound.
See pp. 100-03.

algebra the branch of mathematics in which unknown quantities are represented by letters or other symbols.
See p. 165.

anatomy the branch of biology that deals with the structure of animals and plants.
See pp. 20-73.

biochemistry the scientific study of the chemical composition and chemical reactions of living systems.
See pp. 12-15.

bioengineering the application of biological and engineering principles to the design and manufacture of equipment that can be used in conjunction with living organisms.

biology the science that is concerned with the study of living systems.
See pp. 10-73.

biomechanics see bioengineering.

bionics the study of living systems in order to manufacture synthetic systems that are based upon similar principles.

biophysics the scientific study that describes biological phenomena in terms of physical laws.
See pp. 12-15.

botany the branch of biology that deals with the scientific study of the classification, evolution, structure, life cycles, growth, etc., of plants.
See pp. 20-23 and 25-33.

calculus the branch of mathematics that studies continuous change in terms of the mathematical properties of the functions that represent it and these results can be interpreted in geometrical terms relating to the graph of the function.
See pp. 170-73.

cell biology see cytology.

chemistry the scientific study of matter, of the composition and properties of substances, and of the reactions by which substances are produced from, or converted from, other substances.
See pp. 124-61.

cytology the branch of biology that deals with cells.
See pp. 12-13.

ecology the scientific study of the interrelationship of organisms and their environment.

ethology the scientific study of animal behaviour.

genetics the branch of biology that deals with the scientific study of inheritance in animals and plants.
See pp. 14-15.

geometry the branch of mathematics that deals with the properties of lines, points, surfaces and solids. The rules of geometry - which have been discovered not invented - are used to derive angles, areas, distances, etc.
See pp. 180-91.

inorganic chemistry the branch of chemistry that is concerned with the study of all elements except carbon, and the compounds and interactions of those elements.
See pp. 124-51 and 156-61.

mathematics the study of numerical and spatial relationships by logic. Mathematics is concerned with quantities, magnitudes and forms which are expressed in terms of numbers and symbols. It includes algebra, calculus, geometry, statistics and trigonometry.
See pp. 162-91.

mechanics the branch of physics that describes the movement and motion of objects.
See pp. 74-83.

nuclear physics the branch of physics that deals with nuclear fission and nuclear fusion.
See pp. 120-23.

optics the branch of physics that deals with light and vision.
See pp. 104-08.

organic chemistry the branch of chemistry that deals with carbon compounds, except the oxides of carbon and the carbonates.
See pp. 152-55.

physical chemistry the branch of chemistry that deals with physical properties in relation to chemical properties.
See pp. 136-60.

physics the scientific study of the basic laws that govern matter.
See pp. 74-123.

physiology the branch of biology that deals with the functions and the processes of living organisms and of their organs and parts.
See pp. 18-73.

quantum mechanics the branch of physics that applies quantum theory to the mechanics of atomic systems.
See pp. 93-95.

statistics is sometimes treated as a branch of mathematics, but it can also be regarded as a separate science. Statistics is concerned with the collection, study and analysis of numerical data.
See pp. 177-79.

taxonomy the branch of biology that is concerned with the classification of life.
See pp. 16-17.

trigonometry the branch of mathematics that deals with the relations between the angles or the sides of triangles.
See pp. 180-81 and 182-83.

zoology the branch of biology that deals with the scientific study of the classification, evolution, structure, life cycles, growth, etc., of animals.
See pp. 34-73.

How to Use this Book

The *Guinness Science Fact Book* aims to give an accessible overview of the biological sciences, physics, chemistry and mathematics.

This book may be thought of as a completely new kind of subject dictionary. Instead of being arranged as a single A-Z listing, the main part of the book follows the thematic approach of the *Guinness Encyclopedia* and the *Guinness Encyclopedia of the Living World*. It is arranged in a series of self-contained spreads - though some subjects are dealt with in longer or shorter articles - focusing on particular areas of study and topics of interest. A topic may, therefore, be looked up in two ways: either directly in the index or under the broader subject area of which it is a part.

In each case, the scene is set with a brief introduction on an individual topic. This is followed by a glossary of the principal terms, laws, theories and matters related to that particular topic. Not only does this format give an authoritative background to a broad range of scientific topics, it also allows the user to read a scientific article and find the definitions of the relevant vocabulary in one place, without having to turn the pages of a conventional dictionary.

The section on famous scientists is, however, arranged as an A-Z listing of concise biographies.

The final section of the book consists of an A-Z index/'factfinder', providing quick access to key facts, terms, major figures, laws and many other topics, both as definitions and cross references to the thematic spreads. The references at the end of entries in the index/factfinder refer to specific mentions of the topic in the main text: where there is a string of page references, the main ones are indicated in bold type.

Biological Sciences

EVOLUTION

In the history of science there have been a handful of breakthroughs that have proved to be of fundamental importance to our understanding of ourselves and of the universe. To Newton's laws of motion and gravity, Einstein's theories of relativity and Max Planck's origination of quantum theory must be added Darwin's theory of evolution by natural selection.

The clear-cut account of creation in the Bible meant that for centuries speculation on any alternative theory was firmly discouraged. However, during the 18th century scientists began to amass increasing amounts of fossils of animals that no longer existed on earth. By the early 1800s many naturalists believed that the fossil record showed a gradual hierarchical progression from the simpler to the more complex forms of life, and this revealed the rational plan of creation through time. One of the earliest proponents of this theory of organic change was Erasmus Darwin, Charles's grandfather, in his theory of transmutation.

In 1831 Charles Darwin became the naturalist on HMS *Beagle* on her five-year voyage of discovery (1831-36). The extensive observations he made on this voyage led him to the conception of evolution by natural selection. Darwin's ideas did not appear in print until 1859 when his *On the Origin of the Species by Means of Natural Selection* was published. The effect was explosive and Darwinism was widely held to be heretical by most of the established Church. Alfred Russel Wallace, who reached similar conclusions at the same time as Darwin, has received less credit than he deserved.

adaptive evolution is a speeding up of the normal process of evolutionary divergence of species; for example, the rapid success of flowering plants over the conifers.

catastrophism the 19th-century theory that all geological changes are the result of short-lived catastrophes. This appeared to endorse the theory of Baron Georges Cuvier that all extinct and living animals had at one time coexisted and that the former had been removed by catastrophes.

coevolution is where two or more species evolve in continuous adaptation to each other; for example, in the relationship between a parasite and its host.

convergent evolution is where different organisms have evolved similar solutions to similar problems, even though they have different ancestors; for example, swallows and swifts look very similar but belong to entirely different orders.

descent with modification see natural selection.

DNA see pp. 14-15. Once the function of the DNA molecule in inheritance was established in 1953, it was discovered that it is very common for an accidental change to occur when DNA replicates itself. Such changes are known as mutations.

embryonic evidence for evolution see evidence for evolution and diagram.

evidence for evolution has been identified in embryonic development (see diagram), in vestigial structures (for example, vestigial limbs in whales and the 'hidden tail' or coccyx in humans), homologous structures (resemblances in anatomy resulting from shared ancestry), the fossil record, adaptive radiation, and biochemical similarities.

evolutionary variations see convergent evolution, parallel evolution, adaptive evolution, and coevolution.

gene see pp. 14-15. Each of the physical characteristics of an organism is dictated by a gene (a section of DNA; q.v.). At a genetic level one can regard an interbreeding population of plants or animals as a 'gene pool', an environment in which different genes compete to survive. Genes that make bodies that are better at surviving and reproducing will perpetuate themselves and will have a high frequency (i.e. be common) in the gene pool, while genes that are not so good in these respects will be rarer. As the make-up of the gene pool changes over time, so will the characteristics of the individuals of the species.

gene pool see gene.

genetic drift random shuffling of parental genes that may purely by chance perpetuate mutations.

homologous structures see evidence for evolution.

inheritance of acquired characteristics the theory proposed by Jean-Baptiste Lamarck early in the 19th century that stressed the creative power of nature to change in response to new conditions.

BIOLOGICAL SCIENCES

isolating mechanisms differences between species that may have no other function than to ensure that members of the species can recognize each other.

Mendelism the theory formulated by Gregor Mendel in 1865 (but not rediscovered until 1909) that a particular trait is either passed on or not - there is no blending.

mutation see DNA (above and pp. 14-15).

natural selection (theory of) was outlined by Darwin in his *Origin of the Species*. He showed how the process of adaptation of organisms to their environment, and hence of evolution, was due to the blind operation of everyday laws of nature, by means of a mechanism he termed 'natural selection'.

Although a major part of Darwin's achievement was to amass an overwhelming amount of evidence in support of his theory, his genius lay rather in the conclusions he was able to draw from readily available evidence. There was nothing novel in noting the considerable variations between species, nor that more offspring are produced in each generation than survive to reproduce. Yet it is an inference from these findings that provides the very cornerstone of Darwin's theory - the notion that, in the 'struggle for existence' resulting from competition among individuals, variations in attributes (some due to heredity) will affect success in survival and reproduction. In other words, a fitter individual - one that is in some way better adapted to its surroundings - will be more likely to survive and leave descendants than a less fit one, and hence will tend to perpetuate within the population those inherited differences to which it owed its success.

In this way natural selection can act over time to change or diversify the characteristics of a population. In other words there is what Darwin called 'descent with modification', dubbed by Herbert Spencer 'the survival of the fittest'. Thus many populations continuously improve their adaptations to the environment to which they are subjected, while less well-adapted populations became extinct.

parallel evolution is similar in some respects to convergent evolution, whereby unrelated plants and animals will adapt in similar ways to fill ecological niches in similar but geographically separated ecosystems; for example, the development of the marsupial 'mole', the marsupial 'wolf', etc., in isolation in Australia.

speciation the development of one or more species from an existing species.

survival of the fittest see natural selection.

transmutation (theory of) was formulated by Erasmus Darwin, (see introduction). It stated that organisms are designed to be self-improving and can develop new features and even organs in their effort to combat environmental changes.

uniformitarianism the theory popularized by the Scottish geologist Sir Charles Lyell that stated that all geological change is slow, gradual and continuous; thus rocks - and the fossils in them - are many millions of years old. Uniformitarianism was one of the bases upon which Darwin based his theory of natural selection.

vestigial structures see evidence for evolution.

COMPARATIVE EMBRYO DEVELOPMENT

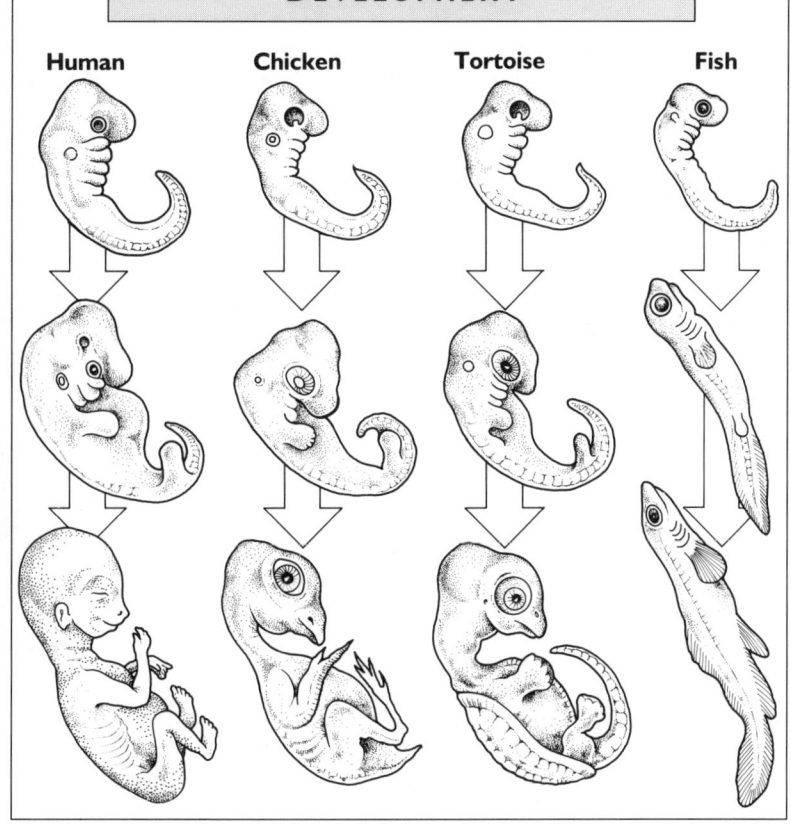

Human   **Chicken**   **Tortoise**   **Fish**

CELLS

Cells are the basic biological units of all living things, and a single cell is the smallest component of an organism that is able to function independently. Virtually every cell in an individual organism carries a complete genetic blueprint for the formation and development of that organism. Cells range in size from 0.0003mm in the case of cells of certain bacteria, to the egg - the female sex cell ovum - of the ostrich, which averages 15-20cm (6-8in) in length.

Many organisms - bacteria and protoctists (see pp. 18-19) - are unicellular. All higher animals and plants are multicellular. In multicellular organisms, groups of cells have different functions, for example, in animals there are nerve cells, blood cells, muscle cells, and so on. The lifespan of cells varies from a few days to - in the case of muscle cells - the lifetime of the organism.

At the molecular level, all cells are constructed from four groups of organic molecules: the nucleic acids (DNA and RNA; see pp. 14-15), proteins, carbohydrates (which include the sugars) and lipids. All cells are surrounded by a membrane; plant cells also have an outer cell wall, principally made of cellulose. The manner in which membranes evolved and developed has resulted in the formation of two basic types of cell: prokaryote and eukaryote.

amino acid a building block of protein, consisting of one or more carboxyl groups and one or more amino groups attached to a carbon atom. There are more than 60 naturally occurring amino acids.

carbohydrate any of a large number of organic compounds, including sugars and polysaccharides.

centrioles self-replicating organelles in animal cells, but rarely in plants. They are the focal point of production of part of cytoskeleton.

chloroplast see plastids.

chromoplast see plastids.

chromosome a coiled structure in the nucleus of cells, containing a large number of genes.

cytoplasm the solution surrounding the nucleus and the organelles, containing ions, dissolved molecules and enzymes.

endocytosis the import into a cell of materials such as nutrients for digestion or chemicals for processing.

endoplasmic reticulum a complex network of membranes; an organelle which provides an intracellular transportation system. It is connected with the Golgi apparatus (q.v.) to form an interlinked membrane system from the nucleus to the plasmamembrane. It is responsible for the synthesis of lipids and sugars. Attached ribosomes make protein for export. It also forms vacuoles.

enzyme any protein that acts as a catalyst in a biochemical reaction.

eukaryote cells evolved later than the prokaryotes, probably by the fusion of prokaryotes. They are generally larger than prokaryotes (up to 0.01mm on average, though some can be as big as 1mm), and exist as single cells (e.g. protoctists; see pp. 18-19), colonies of cells (e.g. sponges; see p. 36), or as constructs in the form of multicellular organisms (plants and animals). Their increased size and ability to form complex colonies and organisms is largely due to three features: the division of the cells into a nucleus and cytoplasm; the organelles; and the development of true sexual reproduction with egg/sperm fertilization.

exocytosis the export from a cell of such materials as polysaccharides.

gene the basic unit of inheritance, consisting of a linear section of a DNA molecule; see pp. 14-15.

Golgi apparatus an organelle; an assembly of vesicles and folded membranes which manipulate and package material from the endoplasmic reticulum for secretion. It also packages enzymes for transport to the lysosomes.

glycolysis a metabolic pathway involving the transformation of glucose into a utilizable form of energy.

glyoxysomes see microbodies.

leucoplast see plastids.

lipid any of a group of simple molecules that are essential components of cell membranes and of fats.

lysosome an organelle containing enzymes for controlled digestion of molecules, bacteria, red blood cells and worn-out organelles such as mitochondria.

membrane sheet-like tissue covering, connecting or lining cells and their organelles and structures. The membrane of all living cells shares a similar configuration, based on a dynamic molecular structure. The interaction of lipids (q.v.) forms the basic membrane, into which protein molecules are inserted and carbohydrate molecules attached. In addition to creating a membrane that separates the inside of the cell from its environment (the plasmamembrane), this basic membrane structure also provides a variety of compartments in eukaryote cells (the organelles). The mitochondria (q.v.) and chloroplasts (q.v.) have an outer membrane that

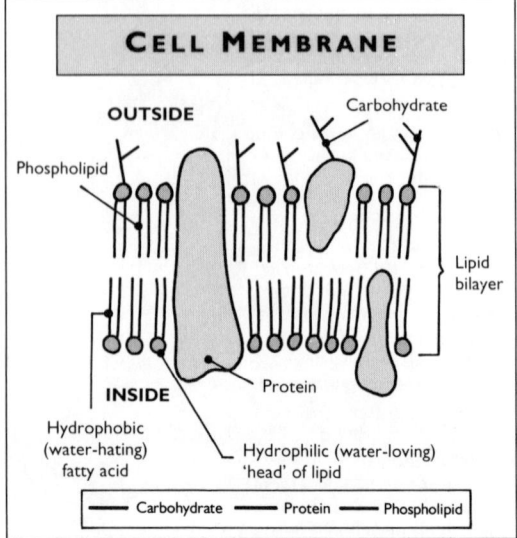

BIOLOGICAL SCIENCES

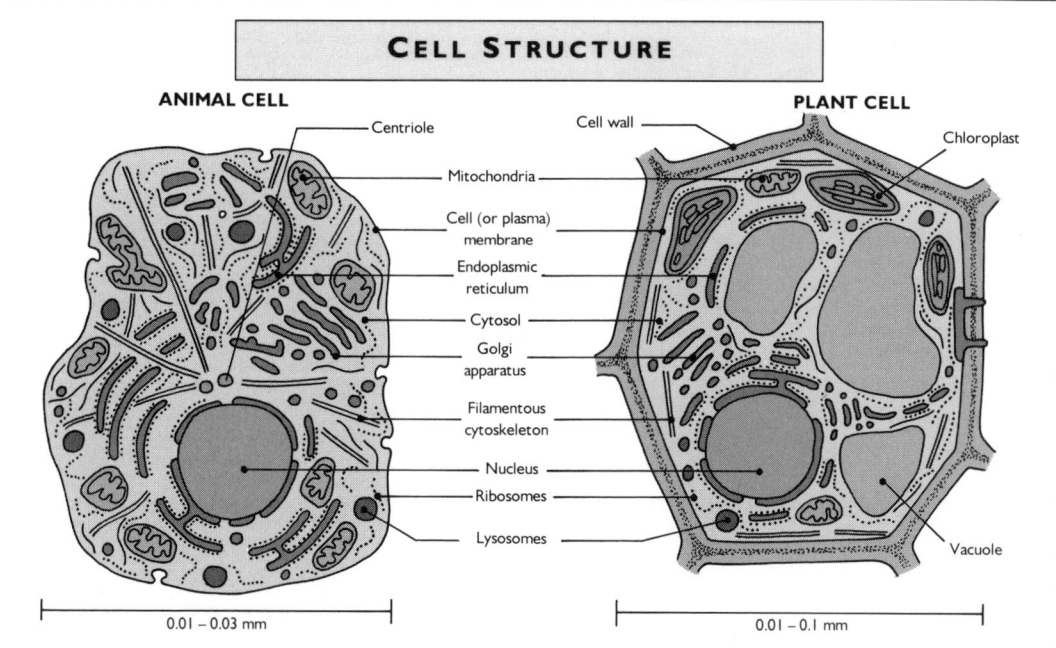

These diagrams of thin sections of generalized animal and plant cells show the many similarities of structure between the two basic types of eukaryote cell. Blue links the components of the *endomembrane system* (i.e. the principal linked membrane apart from the outer cell membrane), comprising the nuclear envelope, endoplasmic reticulum, Golgi apparatus, and transport and secretory vesicles.

separates them from the cytosol (see above). An inner membrane forms compartments within the organelles. Thus the membrane provides a large surface inside the cell on which chemical reactions can occur and permits specialized transport systems into, through and out of the cell while separating the different activities of the cell so that they do not interfere with each other. These transport systems are responsible for endocytosis and exocytosis.

mitochondria organelles which possess their own genes for partial reproduction. They are important in cell respiration for the release of energy from substrates.

multicellular consisting of an assemblage of cells.

nucleoid see prokaryote cells.

nucleus the central part of a cell. It is surrounded by a membrane and contains the genetic material in the form of chromosomes. It makes m-RNA, t-RNA and r-RNA (see pp. 14-15) and regulates cell division.

organelles miniature organs found in prokaryote cells. They have specialized functions, which permit a greater range of metabolic processes.

plasmamembrane see membrane.

plastids organelles which possess their own genes for partial reproduction. Plastids form chloroplasts in plants (which contain the chlorophyll responsible for photosynthesis), colourless leucoplasts (which store starch, protein and oil), and chromoplasts (which contain pigments).

polymer any natural or synthetic compound (e.g. starch, protein) made up of a large number of repeated molecules.

polypeptide any polymer made up of amino acids, e.g. proteins.

polysaccharides long-chain carbohydrate molecules.

prokaryote cells (which include all forms of bacteria) are the most primitive form of cell, consisting essentially of an outer membrane with some rudimentary membrane systems within it. The DNA of prokaryote cells resides in a special area, the nucleoid. Unlike the eukaryote nucleus, the nucleoid is not surrounded by a membrane, but it remains free of all structures except DNA. Again, unlike eukaryotes, prokaryotes contain no organelles.

protein any of a large number of polymers made up of polypeptide chains of amino acids.

protoplasm the material of the cell within the membrane of a prokaryote cell.

ribosome see endoplasmic reticulum.

unicellular consisting of a single cell.

vacuole a space within the cytoplasm (q.v.). It may contain food particles or water in some cellular animals. In plants it contains sap and is responsible for keeping the cell turgid.

vesicle any small, usually fluid-filled, membrane-bound sac within the cytoplasm.

GENETICS

Although theories of inheritance or heredity were put forward at least as early as the 5th and 4th centuries BC, genetics - the scientific study of inheritance - only truly began in the 18th and 19th centuries. Observations were made of how specific characteristics of plants and animals were passed from one generation to the next, to provide a rational basis for the improvement of crop plants and livestock.

The most significant breakthrough in genetics was made by the Austrian monk Gregor Mendel (1822-84). He observed specific features of the pea plant and counted the number of individuals in which each characteristic appeared through several generations. By concentrating on just a few features and determining what proportion of each generation received them, he was able to demonstrate specific patterns of inheritance. The discrete nature and independent segregation of genetic characteristics that he observed became known as Mendel's laws of inheritance, and have been shown to apply to most genetic systems.

adenine (A) one of the four bases found in the central region of the double helix of DNA. A pairs only with T; see DNA.

allele one of a pair of genes (q.v.) which together determine which of a pair of characteristics will appear in the offspring.

amino acid a building block of protein, consisting of one or more carboxyl groups and one or more amino groups attached to a carbon atom. There are over 80 naturally occurring amino acids, 20 of which occur in proteins. Each amino acid is distinguished by a different side chain group. Organisms can synthesize some amino acids.

chromosome a coiled structure in the nucleus of cells. It consists of several different types of protein wrapped tightly associated with a single DNA molecule, and each chromosome carries a large number of genes. Most eukaryote organisms (organisms with cell nuclei; see pp. 12-13) have several chromosomes, but bacteria, which do not have a nucleus, have only one.

All the cells of a particular organism have the same number of different chromosomes, but numbers vary widely between different species. There is no clear pattern to this, but plants tend to have fewer chromosomes than animals. Humans have 46 different chromosomes per cell, that is 23 pairs, of which one in each pair comes from one parent and one from the other. The pairs are referred to as homologous chromosomes.

codon (or triplet) a combination of three bases within a DNA molecule.

cytosine (C) one of the four bases found in the central region of the double helix of DNA. C pairs only with G; see DNA.

deoxyribonucleic acid see DNA.

deoxyribose a type of sugar found in a nucleotide; see nucleotide.

diploid describes normal body cells (other than sex cells) in most organisms. They have paired homologous chromosomes (q.v.). See also haploid.

DNA short for deoxyribonucleic acid.

By the start of the 20th century it was clear that organisms inherited characteristics by the reassortment and redistribution of many apparently independent factors, but the identity of the material that carried this information was unknown. To code for such a large amount of information, any type of molecule would have to be highly variable. In 1944 the American microbiologist Oswald T. Avery (1877-1955) demonstrated that the inherited characteristics of a certain bacterium could be altered by deoxyribonucleic acid (DNA) taken up from outside the cell.

To understand how genetic information was encoded required the structure of DNA to be determined. In 1953 the American James Watson (1928-) and the Englishman Francis Crick (1916-) reported that DNA is a large molecule in the shape of a double helix.

DNA is the basic genetic material of most living organisms. Although a large and apparently complex molecule, its structure is surprisingly simple. A single DNA molecule consists of two single strands wound around each other (see diagram) to form a double-helical (spiral) structure. Each strand is made up of just four chemical components

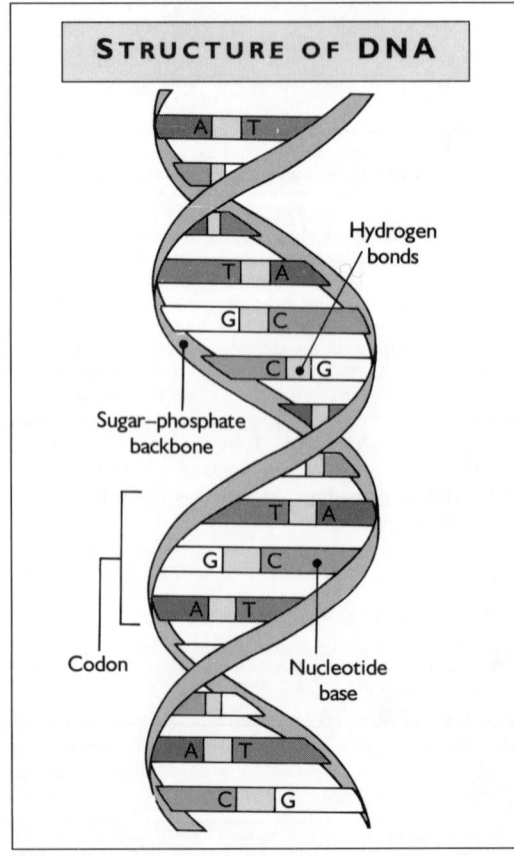

STRUCTURE OF DNA

BIOLOGICAL SCIENCES

known as nucleotides. The central region of the helix comprises four bases - adenine (A), cytosine (C), guanine (G) and thymine (T). Each base is linked by hydrogen bonds to a specific complementary base on the partner strand: A pairs only with T, and G only with C. This simple structure explains the two key properties of DNA - how it codes for the manufacture of amino acids (from which proteins are formed) and its capacity to repeat itself. Each combination of three bases (known as a triplet or codon) within a DNA molecule codes for a particular amino acid, while the specific pairing of bases explains how two identical DNA molecules can be produced by the separation and rebuilding of the two strands of the parent molecule.

DNA replication the natural process by which the DNA of a cell doubles to enable that cell to divide - as organisms grow, their body cells divide and multiply. The hydrogen bonds linking the two strands on DNA molecules break apart, and each strand uses nucleotides present in the nucleus to synthesize a new strand complementary to itself; the result is that two daughter molecules are produced, each identical to the parent molecule. This process is completed just before cells divide, so that each chromosome contains two DNA molecules instead of the usual one.

dominant gene see heterozygous gene.

double helix see DNA.

gametes the sex cells of eukaryotes. These have only half the chromosome number of somatic cells. See haploid.

gene the basic unit of inheritance, consisting of a linear section of a DNA molecule.

Mendel discovered that characteristics are passed from generation to generation in the form of discrete units. Once the structure of DNA was established, these units (genes) could be understood at the molecular level. A gene includes all the information for the structure of a particular protein or ribonucleic acid (RNA) molecule.

gene transcription the first stage in the process by which cells use genetic information stored in DNA. This involves making an RNA copy of a gene. Most of the RNA copies (known as messenger RNA) travel from the nucleus of the cell (where the genes are located) to the ribosome (particles in the cytoplasm) - where proteins are manufactured or synthesized. Ribosomes (see pp. 12-13) make proteins by joining together amino acids in the sequence dictated by the order of the triplet groups in the messenger (m-RNA).

genome the sum of all an organism's genetic information.

genotype the entire set of alleles (q.v.) in a genome.

guanine (G) one of the four bases found in the central region of the double helix of DNA. G pairs only with C; see DNA.

haploid having only one copy of each chromosome in a cell. Sex cells are haploid.

heterozygous in which a pair of alleles in a cell consists of one gene of each type. One allele may be dominant in that the product it codes for appears in preference to the other allele, which is called recessive. See homozygous.

homologous chromosome see chromosome.

homozygous the condition in which, for any pair of alleles, the cell contains both genes of the same type, i.e. both dominant or both recessive. See heterozygous.

Mendel's laws of inheritance see introduction (above).

messenger RNA (or **m-RNA**) see RNA.

mitosis the process of cell division to produce identical body cells by which each new cell receives one of the DNA molecules from every chromosome. (See DNA replication.)

mutagenesis the rate at which mutation (q.v.) occurs. Mutagenesis can be accelerated by exposure to chemicals or radiation.

mutation occurs when there is a change in the sequence of nucleotide bases in a piece of DNA. Such changes may occur naturally, as bases are added, deleted or exchanged. Accelerated mutations (see mutagenesis) may disrupt or prevent the production of proteins, thus disturbing the functioning of the organism as a whole. However, natural mutations passed on from generation to generation may also confer benefits, and are indeed essential to the process of evolution (see pp. 10-11).

nucleotide any of the four components that make up a single strand of DNA. Each of the four has the same basic composition consisting of a sugar (deoxyribose) linked to a phosphate group to form the helical backbone of DNA, and a base. Different nucleotides are distinguished only by the identity of the base. See also DNA.

nucleotide base the nitrogen-based unit that is bonded to a sugar molecule in DNA - A, T, G, C.

polymer any natural or synthetic compound made up of a large number of repeated molecular units.

protein any of a large number of polymers made up of polypeptide chains of amino acids.

recessive gene see heterozygous.

ribonucleic acid see RNA.

ribosomal RNA (or **r-RNA**) see RNA.

RNA short for ribonucleic acid. RNA occurs in different forms - m-RNA (messenger RNA), t-RNA (transfer RNA), and r-RNA (ribosomal RNA) - all of which are involved in the processes by which DNA leads to the construction of proteins. See also gene transcription (above).

somatic cells body cells (non-sex cells).

thymine (T) one of the four bases found in the central region of the double helix of DNA. T pairs only with A; see DNA.

transfer RNA (or **t-RNA**) see RNA.

triplet (or **codon**) a combination of three bases within a DNA molecule, which can act as a code for a single amino acid in a protein (q.v.).

TAXONOMY

The study of the classification of life - or taxonomy - aims to provide a rational framework in which to organize our knowledge of the great diversity of living and extinct organisms. Today, comparative biology seeks to understand the living world by searching for the underlying order that exists in nature.

The first well-authenticated system of classification goes back to Aristotle (384-322 BC), but it was not until the 17th century that the Englishman John Ray (1628-1705) proposed the first natural classification rather than an artificial scheme aiming merely to facilitate identification. In the 18th century the Swedish naturalist Carl Linnaeus (1707-78), produced a rational system of classification based on patterns of similarity between different organisms. Linnaeus developed one of the first comprehensive subordinated schemes of taxonomy.

binomial see Linnaean system.

clade in cladistics (q.v.), a group that is defined by features that are shared by all its members and that are found in members of no other group. These groups are often represented on a branching diagram (a cladogram), which shows how the various clades are linked. In evolutionary terms, a clade is a monophyletic group.

cladistics is generally believed to be the most precise and natural method of classification, and is used by most progressive taxonomists. The only groups recognized are clades.

cladogram an evolutionary tree used in cladistics; see clade.

class a classificatory group between phylum (or division) and order. For example all mammals belong to the class Mammalia. The names of all plant classes end in -opsida.

division 1) any of several high-ranking classificatory groups into which the plant kingdom is divided. The names of plant divisions all end in -phyta. The equivalent of phylum in the animal kingdom. 2) a classificatory group between phylum and class in the animal kingdom.

family a classificatory group between order and genus. Animal family names end in -idae and plant family names end in -aceae.

genus a classificatory group between family and species. Generic names are always written in italics.

kingdom the highest classificatory group. Most authorities today recognize five kingdoms: Animalia (animals), Plantae (plants), Fungi (fungi), Protoctista (formerly Protista; protists - i.e. all single-celled eukaryote organisms) and Prokaryotae (formerly Monera; bacteria and all single-celled prokaryote organisms).

Linnaean system a scheme of classification that places each organism in a group that is itself part of a larger group. The essential feature of Linnaeus's scheme is that it is binomial. He gave every distinct type or species of organism (e.g. the lion) a two-part (binomial) name (e.g. *Felis leo*) in which the second element of the name identified the individual species, while the first element placed the species in a particular genus - a group comprising all those species that showed obvious similarities with one another. For instance, he grouped together all cat-like animals in the genus *Felis*: *Felis leo* - the lion; *Felis tigris* - the tiger; *Felis pardus* - the leopard, and so on. Having thus subordinated each species to a particular genus, he went on to place groups of genera in a higher rank or category he called a family, then families in orders, and so on through classes, phyla (or divisions for plants), and finally kingdoms. Within this hierarchy each division is more embracing than the last, each containing a greater number of organisms with fewer characteristics in common.

Although Linnaeus's genera and families were for the most part natural, his higher ranks were of necessity artificial in order to deal with the vast numbers of new organisms being discovered. He used his highest taxonomic rank (kingdom) to separate plants and animals, but it is now clear that this simple division is untenable, because certain groups such as bacteria, protists and fungi fit into neither category - and are now each assigned their own kingdom. Nevertheless, the binomial system has remained unchanged to this day and every newly discovered organism is given a Latin or Latinized binomial name.

Another difficulty with the Linnaean system is that it is too simplistic to reflect the complexity of classification that is implied in Darwin's theory of evolution (see pp. 10-11). Since it is thought that every organism on this planet has arisen through a unique historical process of descent with modification, it follows that all of them should fall into uncontradicted patterns of groups within groups. Thus, although we still use such terms as order, family and class, we will need to introduce additional ranks as the analysis of the patterns of life is further elucidated.

monophyletic group a group consisting of species descending from a single ancestor and including all the most recent common ancestors of all its members.

natural classification an arrangement of living and extinct organisms into a rational framework based on presumed relationships.

order a classificatory group between class and family. For example, the order Primates in the class Mammalia contains all the primates. The names of all plant classes end in -ales, and those of several animal orders end in -iformes.

orthodox classification (or **phylogenetic classification**) provides the basis for most text book

BIOLOGICAL SCIENCES

The Classification of Man

RANK	NAME	MEMBERS	DISTINGUISHING FEATURES
KINGDOM	ANIMALIA	All multicellular organisms except plants and fungi	Nervous system
PHYLUM	CHORDATA	Lancelet, sea squirts, all vertebrates	Notochord (central nerve cord)
SUBPHYLUM	VERTEBRATA	All vertebrates, i.e. fish, amphibians, reptiles, birds, mammals	Backbone (protecting notochord)
SUPER-DIVISION			
DIVISION			
SUPERCLASS	TETRAPODA	All vertebrates except fish	Four limbs, the front pair being pentadactyl (i.e. having five digits)
CLASS	MAMMALIA	All mammals	Milk and sweat glands, hair
SUBCLASS	EUTHERIA	All mammals except monotremes and marsupials	Placenta
ORDER	PRIMATES	Lemurs, lorises, pottos, galagos, tarsier, marmosets, tamarins, monkeys, apes	Fingers with sensitive pads and nails
SUPERFAMILY	HOMINOIDEA	Apes (gibbons, orang-utan, gorilla, chimpanzees, humans)	No tail, broad chest, shoulder blades at back rather than sides
FAMILY	HOMINIDAE	Gorilla, chimpanzees, humans	Upright posture, flat face, large brain; similar blood plasma protein
SUBFAMILY	HOMININAE	Chimpanzees, humans	Similar brain shape and anatomical and protein structures; tool and weapon users
GENUS	*HOMO*	*Homo habilis*, *H. erectus* (both extinct), *H. sapiens*	Exclusively bipedal (walking on two feet), manual dexterity
SPECIES	*SAPIENS*	Neanderthal man (*H. sapiens neanderthalensis*; extinct), modern man	Double-curved spine
SUBSPECIES	*SAPIENS*	Modern man (*Homo sapiens sapiens*)	Well-developed chin

classifications. The system is based on presumed evolutionary relationships, and - as with phenetic classification - as much information as possible is taken into account. Organisms are grouped by features they share in common and that seem to reflect their common ancestry.

phenetic classification a modern classification system which aims to incorporate as much information as possible about organisms, and then groups them together on the basis of their overall similarity. All characteristics are given equal weight and no account is taken of the evolutionary relationships of the groups. This approach is amenable to numerical analysis and may be useful for large groups of simple organisms such as bacteria.

phylogenetic classification see orthodox classification.

phylum (plur. **phyla**) any of several high-ranking classificatory groups into which the animal kingdom is divided. For example, the chordates (all vertebrates plus various obscure animals such as sea squirts) are grouped in the phylum Chordata, while the vertebrates themselves are grouped into the subphylum Vertebrata. The equivalent of phylum in the plant kingdom is division.

species in general, the smallest classificatory group (although there are some subspecies). The members of a species are able to breed among themselves to produce fertile offspring: this defines the term.

subphylum see phylum.

SIMPLE LIFE FORMS - BACTERIA, VIRUSES AND PROTOCTISTS

Of the five kingdoms of living organisms, the most primitive is the kingdom Prokaryota (formerly Monera), which includes the archaebacteria, bacteria and mycoplasms. The kingdom Protoctista (formerly Protista) consists of the protozoa, the slime moulds and the Oomycota.

algae an important and diverse group of Protoctista which are oxygen producing, photosynthetic, and are mostly inhabitants of aquatic environments. They no longer include the blue-green bacteria (formerly known as blue-green algae) and are *not* plants.

amoeba any member of the rhizopod genus *Amoeba*. Members of this genus are characterized by pseudopodia (q.v.) which are used for locomotion and feeding and result in a constantly changing body shape. Most species are free-living in water or soil, but some are parasitic.

archaebacteria a group of bacteria that some scientists distinguish from other bacteria on the basis of certain distinct properties. They probably derived some 3.25 billion years ago from an ancestral prokaryote, which had itself developed from precellular forms.

autotrophic protoctists protoctist organisms that are able to convert simple inorganic molecules into complex organic molecules.

bacillus rod-like bacteria.

bacteria (sing. **bacterium**) a diverse group of prokaryote single-celled organisms.

Bacteria form the majority of prokaryote species of unicellular organisms. They vary in shape from almost spherical (coccus) to comma-shaped (vibrio) to rod-like (bacillus) to corkscrew shaped (spirochaete) and spiral (spirillium). Some bacteria, such as *Salmonella*, are motile, while others are non-motile, being grouped in packets (*Sarcina*) or chains (*Streptococcus*).

Bacteria vary in size from 0.0003mm to more than 0.02mm. Bacteria cause a wide range of diseases including cholera, food poisoning, leprosy, plague, syphilis, tetanus, tuberculosis and typhoid. Although we think of bacteria as disease-causing parasites, they are very diverse in both their habitat and biology. Although many are parasites, there are also saprophytes, free-living organisms, and bacteria that have developed a symbiotic relationship with plants.

bacteriophage a virus that infects bacteria.

blue-green algae see cyanobacteria.

chloroxybacteria green-coloured prokaryote organisms that are structurally similar to cyanobacteria. It has been suggested that they are the precursors of the chloroplasts in eukaryote cells.

cilia specialized structures used in the movement of some protoctists.

coccus spherical bacteria.

conjugation a method of reproduction used by some algae and protozoa in which individuals combine.

contractile vacuole a miniature organ (or organelle) found in some protoctists, especially those living in fresh water. They are primarily used to remove excess water.

cyanobacteria (formerly **blue-green algae**) a group of bacteria capable of photosynthesis in the same way as green plants.

Standing somewhat apart from the main group of bacteria are these blue-green algae, which occur either as small rounded cells or mobile filaments or cells, and can be free-living or present in symbiotic relationships either together with fungi to form lichens or inside the cells of higher plants.

cyst a characteristic growth on a protoctist. Different forms of cyst occur amongst the protoctists, depending upon the conditions that induce their formation. Adverse conditions may result in the formation of resistance cysts; some protoctists may form resting cysts (for example, during adverse environmental conditions); others form reproduction cysts.

diatom any member of the algal division Bacillariophyta.

DNA (or **deoxyribonucleic acid**) the basic genetic material of most living organisms; see pp. 14-15.

eukaryote any organism whose cells are eukaryotic, i.e. in which the genetic material is contained in a distinct nucleus. Protoctists and all higher animals are eukaryotes.

flagella specialized structures used in the movement of some protoctists.

heterotrophic protoctists protoctist organisms that are dependent upon the consumption of other organic matter as a source of complex molecules. Most protoctists are heterotrophic.

mesosome an area of infolded plasma membrane in a prokaryote cell.

Monera the former name of the kingdom now known as Prokaryota.

motile organism any organism that is capable of spontaneous and independent movement.

mycoplasm any of a group of minute prokaryote organisms. Mycoplasms are the smallest known free-living organisms (0.0001-0.001mm), differing from true bacteria by the absence of a cell wall and mesosomes.

myxomycetes see slime mould.

nucleocapsid see virus.

Oomycota phylum of Protoctista containing important plant parasites such as *Phytophthora infestans*, the cause of potato blight.

Paramecium a free-swimming protozoan.

plasmid any extra, small circular piece of DNA/RNA in addition to the main chromosome. Many bacteria contain one or more plasmids, which may enable their hosts to do such things as resist disinfectants or specific antibiotics.

Prokaryota one of the five kingdoms of living things, including archaebacteria, bacteria and mycoplasms. (Formerly known as Monera.)

prokaryote any unicellular organism whose cells are prokaryotic, i.e. in which there is no nucleus.

protist an obsolete name for protoctist (q.v.).

protoctist any member of the kingdom Protoctista.

The protoctists include the algae and protozoa, slime moulds and nets, all of which have developed incredible degrees of complexity and sophistication. Although remaining as single cells some of them form the first approach to multicellular organisms through the construction of colonies.

Algal forms include the major forms green, red and brown algae, also diatoms and dinoflagellates.

Protoctists are ubiquitous on land and in water (where they form a major component of plankton), as well as inside other organisms as parasites or symbionts. Free-living forms can live in the soil; slime moulds live on decaying leaves and rotting logs; other protoctists live as parasites, causing such diseases as malaria, sleeping sickness and dysentery.

Protista obsolete name for Protoctista (q.v.).

Protoctista one of the five kingdoms of living things.

protozoa obsolete name, still in common use, for any animal-like unicellular organism amongst the Protoctista.

pseudopodium (plur. **pseudopodia**) a 'false foot'; a temporary projection of an amoebic protozoan used in movement. An amoeba moves by a complex series of changes in the organism's shape.

retrovirus any of a group of RNA viruses, some of which are tumour inducing. They enter the host cell and make copies of DNA from their RNA using their own special enzyme. This DNA then integrates into the host chromosome, which is ultimately used to make viral RNA and proteins to form new viruses, and these are released non-lethally from the host cell.

RNA (or **ribonucleic acid**) an acid that occurs in different forms, all of which are involved in the processes by which DNA leads to the construction of proteins. See pp. 14-15.

Salmonella a genus of motile bacteria, including *S. typhimurium*, a cause of food poisoning.

saprophyte the commonly used name for saprotroph (q.v.).

saprotroph any organism that lives and feeds on dead matter.

Sarcina a genus of non-motile bacteria, grouped in 'packets'.

slime mould any of the myxomycetes, a group of protoctists that are formed either by individual small amoeboid cells or by the fusion of such cells.

spirillium spiral-shaped bacteria.

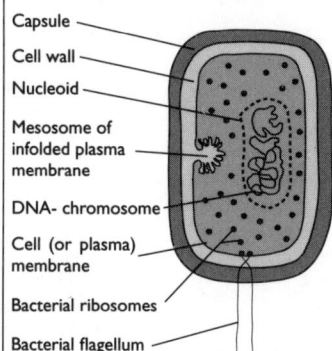

PROKARYOTE CELL STRUCTURE

Capsule
Cell wall
Nucleoid
Mesosome of infolded plasma membrane
DNA-chromosome
Cell (or plasma) membrane
Bacterial ribosomes
Bacterial flagellum

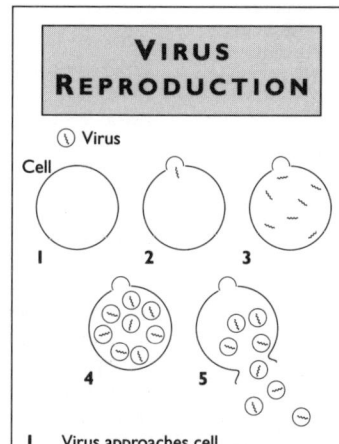

VIRUS REPRODUCTION

① Virus
Cell

1　2　3
4　5

1　Virus approaches cell
2　Virus attaches itself to cell and injects its DNA into it
3　Cell is made to replicate virus's DNA
4　New viruses are formed inside cell
5　The cell bursts and the new viruses are spread out

spirochaete corkscrew-shaped bacteria.

Streptococcus a genus of bacteria that grow in chains; they include species responsible for a variety of diseases from scarlet fever to rheumatic fever.

symbiont any organism living in symbiosis (q.v.).

symbiosis a close association between two different species from which both derive benefit.

vibrio comma-shaped bacteria.

viral replication occurs when a virus infects a host cell. On infecting such a cell, a virus can use the mechanisms of the host cell in order to make viral nucleic acids and proteins from which new virus particles can be constructed. A virus can also enter a cell, insert itself into the host chromosome, remain dormant, leave the chromosome, replicate and then leave the cell.

viroid a particle that is even smaller than a virus; e.g. potato spindle disease. Viroids are essentially small infectious single-stranded RNA molecules.

virus any of a group of minute organisms that lack a cell structure. Viruses are even simpler forms of life than bacteria. They seem to have developed directly from the plasmids present in the bacteria, with which they remain the most simple forms of life; indeed, viruses are very much a grey area between living and non-living things. The fact that they are completely dependent upon living cells as vehicles indicates that they probably evolved later than the earliest cellular organisms.

It was not until 1892 when the tobacco mosaic virus (TMV), was identified that it was shown that there are infectious agents smaller than bacteria. The bacteriophage was discovered in 1912. A large range of viruses are now known to exist. Many cause diseases - from AIDS to the common cold - and many such diseases remain incurable. Viruses have also been implicated in evolution as transmitters of blocks of genetic material from one organism to another.

Viruses range in size from 0.000018 to 0.0006mm, and it was not until the advent of the electron microscope in the 1930s that anyone actually 'saw' one, revealing variations in shape from almost spherical to rod-shaped to icosahedral (having 20 faces). At about the same time, biochemical analysis showed that viruses consist simply of a single nucleic acid - either DNA (as in the bacteriophage) or RNA (as in the TMV; see above) - surrounded by a protein layer, together forming the nucleocapsid. See also viral replication.

PLANTS

The hundreds of thousands of species of plants comprise the kingdom Plantae, one of the five kingdoms of living things. Plant species have colonized a very wide range of habitats, showing tolerance of extremes of temperature and other climatic variations. In so doing, plants have evolved an astonishing variety of adaptations, resulting in one of the most diverse - and beautiful - groups of living organisms.

The Plant kingdom has six phyla or divisions: Bryophyta (liverworts and mosses), Lycopodophtya (club mosses), Sphenophyta (horsetails), Filicinophyta (ferns), Coniferophyta (conifers) and Angiospermatophyta (flowering plants). All are eukaryotic and autotrophic.

The same basic functions occur in plants as in all other living organisms. They grow, and therefore, they assimilate materials from the environment, and this material provides the chemicals of which they are made. To maintain life they require energy; and they need water, which like the substances for growth and energy is taken up from their surroundings and transported around the plant body. Growth and the various other functions are spatially and temporally regulated and coordinated. Plants, like animals, perceive their environment and react towards it. They also reproduce. But there are unique properties of green plants - principally their structure and the fact that they are fixed in one location - that determine the way they carry out these functions.

In general, the material for energy and growth is not obtained from organic material as it is in animals. Green plants - together with the algae and some bacteria - are the only autotrophic organisms on Earth.

amino acid a building block of protein, consisting of one or more carboxyl groups and one or more amino groups attached to a carbon atom. There are over 80 naturally occurring amino acids, 20 of which occur in proteins.

angiosperm a flowering plant; any member of the phylum or division Angiospermatophyta. See pp. 28-33.

Angiospermatophyta a phylum or division of the kingdom Plantae. It comprises the flowering plants. See pp. 28-33.

Antarctic a floral region lying south of the Antarctic Circle (66° 33'S) plus the extreme southern tip of South America; see accompanying diagram.

Australian a floral region that covers not only Australia but also New Zealand, New Guinea and adjacent islands; see accompanying diagram.

autotrophic organism any organism that can make organic materials from simple inorganic sources of carbon, nitrogen, hydrogen, oxygen and other elements that are found in living things. To do this they need energy, and they derive this energy from sunlight. In so doing, they are converting simple inorganic chemicals into forms upon which the whole animal, fungal and some bacterial life depends. See photosynthesis (below).

bacteroid modified bacteria; for example, modified *Rhizobium* bacteria living in a nodule on leguminous roots. See nitrogen fixation (below).

bark tissue in woody stems and roots external to the cambium.

blade see lamina.

Boreal a floral region covering all of Europe, most of North America, part of North Africa and that part of Asia to the north of the Himalaya; see accompanying diagram.

Bryophyta a phylum or division of the kingdom Plantae. It comprises liverworts and mosses.

cambium a secondary meristem extending down the length of a root or stem. It is responsible for producing new cells, which differentiate into xylem and phloem.

carbohydrate any of a large number of organic compounds such as sugars and polysaccharides.

cellulose a polysaccharide carbohydrate made up of unbranched chains of many glucose molecules (i.e. a polymer). Its fibrils form the framework of plant cell walls. Cellulose is generally an important component in the diets of herbivorous animals.

chlorophyll the chemicals in plants that give them their green colour. The chlorophylls are particularly abundant in leaves and are contained in chloroplasts. Chlorophyll strongly absorbs the blue and red regions of the light spectrum, and this ability is used to drive the most important single chemical reaction on Earth. This reaction, which maintains both plant and animal life, is photosynthesis (q.v.).

chloroplast a miniature organ (organelle) within cells in plants.

club mosses plants of the phylum or division Lycopodophyta; see pp. 25-26.

conifer a member of the phylum or division Coniferophyta; see p. 27.

Coniferophyta a phylum or division of the kingdom Plantae. It comprises the conifers. See p. 27.

Fabaceae the legume family; see legume and nitrogen fixation.

Filicinophyta a phylum or division of the kingdom Plantae. It comprises the ferns. See pp. 25-26.

floral region (or **phytogeographic region**) a distinct region that can be defined principally on the basis of the similarity of the flowering plant inhabitants that it contains. The world can be divided into six distinct floral regions; see diagram.

glucose a simple sugar, an essential source of energy produced by plants via photosynthesis and obtained by animals from plants.

'guard cells' see stomata.

horsetails plants of the phylum or division Sphenophyta; see pp. 25-26.

lamina the blade of a leaf.

leaf a lateral organ on a plant stem, in higher plants often consisting of a petiole and lamina. Leaves play a crucial role in transpiration, respiration and photosynthesis.

leghaemoglobin a red pigment in legumes which carries oxygen; see nitrogen fixation.

legume any member of the plant family Fabaceae, including peas, beans, clover, gorse, etc., nearly all of which are symbiotic with nitrogen-fixing bacteria.

BIOLOGICAL SCIENCES

light reactions see photosynthesis.

lignin an important strengthening component of cell walls in the woody tissues of plants. It is a polymer of sugars, phenols, amino acids and alcohols.

liverworts plants of the phylum or division Bryophyta; see pp. 25-26.

Lycopodophyta a phylum or division of the kingdom Plantae. It comprises the club mosses. See pp. 25-26.

FLORAL REGIONS

- BOREAL
- NEOTROPICAL
- PALAEOTROPICAL
- SOUTH AFRICAN
- AUSTRALIAN
- ANTARCTIC

malic acid an acid produced by starch breakdown; see stomata (below).

meristem a group of cells in plants that divide by mitosis and thereby contribute to plant growth and organ formation. Meristematic activity may be generalized (as in the developing embryo) or localized (as in the apices of stems or roots). Primary meristems are those that have always been meristematic; secondary meristems develop from differentiated cells.

mitosis the mechanism by which body cells divide.

mosses plants of the phylum or division Bryophyta; see pp. 25-26.

Neotropical a floral region that covers Central and South America (except for the extreme southern tip); see accompanying diagram.

nitrogen fixation a means of obtaining nitrogen that has been developed by the legumes.

Almost all plant species in the legume family that have been examined have entered into symbiotic relationships with certain bacteria living in their roots. These bacteria, which invade the roots from the soil, can fix gaseous or molecular nitrogen. The legume family includes many important crop plants (beans, soybeans, etc.) that produce high-protein seeds, and it has been estimated that up to 50% of the nitrogen in the protein comes from the activity of the bacteria. Several species in other families also have nitrogen-fixing bacteria; these are generally plants such as bog myrtle and alder, that typically colonize soils poor in nitrates.

The bacteria infecting the leguminous plants belong to the genus *Rhizobium*, and each species of leguminous plant has a particular species of *Rhizobium* that will infect no other legume. Infection occurs on a damaged root hair, from where the bacteria will spread into the centre of the root. There, root cells are stimulated to divide to produce a nodule, inside whose cells the modified *Rhizobium* bacteria (now called bacteroids) live. The bacteroids require oxygen, and this is carried to the bacteria in leghaemoglobin, which confers a pinkish colour to the interior of the nodule. Using carbohydrate (in the form of sucrose) provided by the host plant, the bacteroids convert the nitrogen that has diffused into the nodule into ammonia, which is then changed by the nodule cells into amino acids and other nitrogenous compounds. These compounds pass into the xylem of the vascular tissue to be carried up the shoot where they are used for protein synthesis.

Nitrogen fixation in legume crops plays an extremely important part in the nitrogen economy: these plants need far less nitrogen fertilizer than other crops, and do not deplete the soil of this essential element, In addition, when ploughed in, legumes enrich the soil, and are therefore commonly used in four-crop rotation.

osmosis the movement or diffusion of water (or another solvent) from a less concentrated solution to a more concentrated solution.

Palaeotropical a floral region that covers Africa (except the northern coastal regions and the extreme south), plus southern Asia; see accompanying diagram.

petiole the stalk of a leaf.

phloem the part of the vascular tissue system (q.v.) of plants responsible for conducting substances manufactured by the plant, e.g. sugars and amino acids. The conducting cells are sieve elements, and other 'general purpose' cells are also present. Primary phloem is formed in the region of the apex of roots and shoots; secondary phloem develops from the secondary cambium, and in some cases forms part of the bark.

photoperiodism the ability of plants to judge the seasonal changes in daylight hours and, thus, to regulate the time of the year when they flower.

photosynthesis the process by which plants, algae and various bacteria convert simple inorganic molecules into complex compounds necessary for life.

Life is based upon the element carbon. Green plants obtain this from the air in the form of carbon dioxide, which they change (synthesize) into more elaborate chemicals - various carbohydrates such as sugars (glucose and sucrose) and starch. (Water is also essential as a source of hydrogen atoms in the sugars, and, as a by-product, molecules of oxygen are given off. This oxygen is essential to virtually all life on Earth.)

Chloroplasts (q.v.) from individual cells acquire carbon dioxide through the intercellular spaces after it has entered the leaf through the stomata. In photosynthesis, the carbon dioxide is then transformed into carbohydrates, initially the sugar glucose, and eventually sucrose and starch. This transformation is powered by the energy from sunlight absorbed by the chlorophylls. This is how the plant obtains carbon needed for the synthesis of materials of which it is composed. As vegetation is the primary food of all animal food chains, it is by photosynthesis that almost all of the carbon enters the living world.

Overall the chemical reaction of photosynthesis is:

$$6CO_2 + 6H_2O \rightarrow C_6H_{12}O_6 + 6O_2$$

But this simple equation hides the complex chemical nature of photosynthesis, which comprises as set of reactions involved with the absorption of light (the light reactions) and a set that do not require light. The essential feature of the light reactions is that the light energy absorbed by chlorophyll is used to split water molecules into hydrogen and oxygen. The oxygen ultimately released by plants is the source of all oxygen in the atmosphere of this planet.

In the remaining reactions carbon dioxide is converted into glucose by a complex cycle of chemical transformations, some of which use the hydrogen generated from water, and the chemical energy that has been produced from light energy.

As plants acquire their carbon dioxide, they also give off water vapour through the open stomata. As this water loss is potentially a problem in arid climates, many plants living in such environments have evolved physiological adaptations to overcome the danger. Many succulent plants, for example, keep their stomata closed during the day but open them at night when there is no drying effect of the Sun. The carbon dioxide is assimilated (or 'fixed') at night not into sugars but into certain organic acids, which later, during daylight hours, release the carbon dioxide within the leaf when the stomata are closed; the carbon dioxide then participates in photosynthesis in the normal way. In other types of plant adapted to semi-arid climates (such as maize) the carbon dioxide is again fixed very efficiently into organic acids during the day, even by leaves with partially closed stomata; these organic acids then move across to the inner cells of the leaf where the carbon dioxide is liberated and used in photosynthesis.

Because in photosynthesis carbon dioxide is used up and oxygen is given out as a waste product - the reverse of plant and animal respiration - the overall effect of plants and animals living together is to keep the atmospheric levels of these gases more or less constant.

The carbohydrates from photosynthesis provide the carbon from which nearly all the constituents of the plant body are made. Especially important among these is protein (q.v.). Protein contains nitrogen (N), an element that is taken up by plants from the soil or water, generally in the form of nitrates (NO_3- compounds). In the majority of higher plants nitrates are converted to ammonium (NH_{4+}) in the leaves; the ammonium is then combined with the carbon coming from the photosynthetic products to produce amino acids, the molecules from which proteins are produced.

phototropism the ability of plant stems to grow towards light.

phytogeographical region another name for a floral region.

PLANTS AND THE ATMOSPHERE

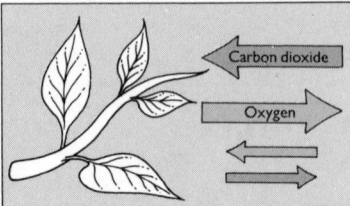

Day: plants absorb more carbon dioxide from the atmosphere by photosynthesis than they give out by respiration. They also give out more oxygen by photosynthesis than they absorb from the atmosphere for respiration.

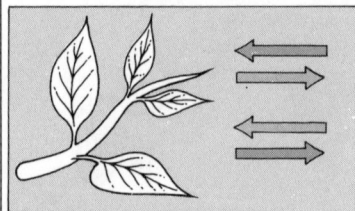

Dawn and dusk: with less light available for photosynthesis, plants give out similar amounts of oxygen and carbon dioxide as they take in.

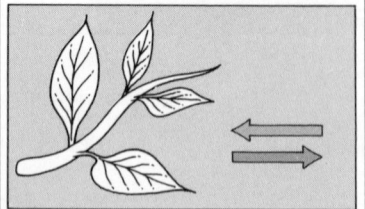

Night: with no light available, photosynthesis ceases. However, respiration continues, with oxygen being absorbed and carbon dioxide being given out.

polymer any natural or synthetic compound that is made up of repeated units; e.g. starch.

polysaccharide any of a class of large carbohydrate molecules, including cellulose, glycogen and starch.

polypeptide any polymer made up of amino acids.

primary meristem see meristem.

primary phloem see phloem.

primary xylem see xylem.

protein any of a large number of polymers made up of polypeptide chains of amino acids.

BIOLOGICAL SCIENCES

respiration (in plants) the use of oxygen to oxidize various compounds, releasing energy from them. This oxidation occurs in all living cells in plants; partly in the cytoplasm (see pp. 12-13) and partly in the mitochondria (see pp. 12-13). Respiration is used to support the synthesis of various chemical compounds, growth and the uptake and accumulation of various mineral elements.

Rhizobium a genus of bacteria found on the roots of leguminous plants; see nitrogen fixation.

secondary meristem see meristem.

secondary phloem see phloem.

secondary xylem see xylem.

South African a floral region covering the extreme southern tip of the African continent; see accompanying diagram.

Sphenophyta a phylum or division of the kingdom Plantae. It comprises the horsetails. See pp. 25-26.

stalk see petiole.

starch a polysaccharide carbohydrate into which glucose is converted for storage in plants.

stem the part of a plant bearing leaves, buds and flowers.

stomata (sing. **stoma**) the microscopic pores in the surface of leaves. They are located mostly on the lower surface of the leaf at a frequency ranging from 20 to over 1000 per square mm depending upon the species. The stomata allow carbon dioxide for photosynthesis (q.v.) into the leaf, and water vapour to escape during transpiration.

Each pore is bordered by two 'guard cells', which are typically about 0.045mm long and about 0.012mm wide. The guard cells can open and close the stomatal pore, so regulating the entry and exit of gases. They do this by taking up or losing water. When the water enters the guard cell it swells up into a crescent shape, so that when both guard cells are swollen, the pore that they border becomes bigger. Conversely, when water is lost, the cells collapse and block off the pore.

These water movements are regulated as follows. Water enters and leaves by osmosis, passing from a solution that is less concentrated to one that is stronger. Changes in potassium ion concentration occur in the guard cells: when the concentration increases, water is subsequently drawn in and the guard cell swells into its unique shape. As potassium ions enter, hydrogen ions (protons) leave: the hydrogen ions come from certain organic acids (malic acid), produced by starch breakdown.

sucrose a type of sugar into which glucose is converted in plants for transport.

sugar any of a group of carbohydrates of low molecular weight. They are a source of cellular energy, and also act as structural molecules (e.g. for plant cell walls).

symbiont a species living in symbiosis (q.v.).

symbiosis a close association between two different species from which both derive benefit.

tracheid the conducting cells in xylem.

tracheophyte a vascular plant (q.v.).

transpiration the evaporation of water from plants through the stomata (q.v.) in leaves.

vascular plant (or **tracheophyte**) any plant with a vascular system (q.v.), characteristic of all higher plants.

vascular system (in plants) the system of tissues comprising the xylem (q.v.; for conducting dissolved mineral salts and water) and the phloem (q.v.; for transporting chemicals synthesized within the plant), together with various other specialized cells (e.g. for strengthening).

vernalization the process by which a period of cold induces many plant species - such as winter wheat and barley - to form flowers.

water loss (in plants) see photosynthesis.

xylem the part of the vascular tissue of plants responsible for conducting water and dissolved mineral salts. Characteristically, the cells are elongated, with relatively thick walls strengthened by lignin. The conducting cells are the vessels and tracheids. The xylem. The xylem formed just below the apex is the primary xylem; that formed from the lateral secondary meristem is secondary xylem, making up the principal component of wood in trees and shrubs.

UPTAKE AND TRANSPORT

Uptake: Water and minerals 'sucked up' from soil via roots and transported in xylem to all parts of plant.

Translocation: Amino acids (the components of proteins) and sugars transported in phloem from leaves to all parts of plant. In some cases (e.g. some trees in springtime) sugars from the roots are transported upwards.

FUNGI

The fungi are not, as is popularly supposed, plants. They constitute an entirely separate kingdom, the Fungi. Unlike the plants, the fungi lack chlorophyll and hence cannot photosynthesize. They therefore have to obtain the carbon and energy necessary for life from other sources.

In most fungi groups there are examples of parasites, which grow on living animals, plants or other fungi, and also of saprotrophs - the function of the latter as decomposers is a very important one, assisting in the recycling of materials needed for life. Some parasitic fungi continue to live as saprotrophs on the dead remains of their hosts. Other fungi have developed a symbiotic relationship with other organisms; for example some are mycorrhizal, while others associate with algae to form lichens.

Although some fungi, such as yeasts, are unicellular, most fungi consist of a mass of filaments. Aggregates of these filaments may give rise to quite large fruiting bodies - the obvious visible parts of fungi such as toadstools. Although in some fungi the cell walls - like those of plants - contain cellulose, in most fungi the cell walls contain chitin (also the principal material in the exoskeletons of insects and other arthropods). Reproduction is by spores, which may be produced sexually or asexually. The spores are usually dispersed by the wind, but sometimes by water or insects.

There are five Phyla, of which the largest are the Ascomycota and Basidiomycota. (The names and relationships given here conform to the most recent classification.)

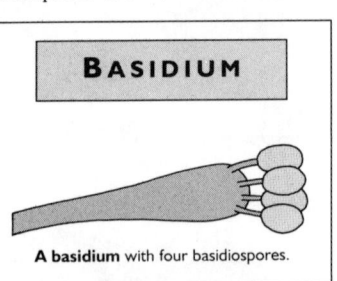

A basidium with four basidiospores.

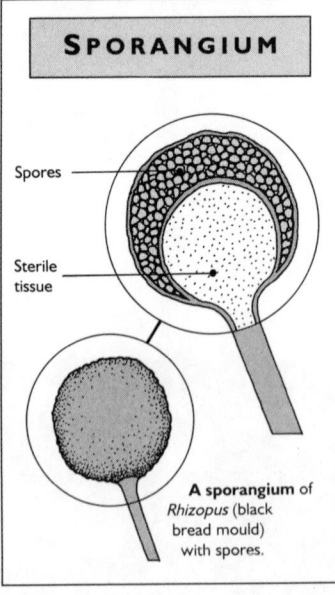

A sporangium of *Rhizopus* (black bread mould) with spores.

agaric any species of basidiomycote fungus of the orders Agaricales or Russulales. They are mushroom-shaped fungi with gills. Several are grown commercially for food.

Agaricales an order of Basidiomycota.

Ascomycota the phylum consisting of both unicellular and filamentous species. In most species the ascus contains eight spores, which may be explosively discharged several cm into the air. Microscopic forms include the unicellular yeast *Saccharomyces* (q.v.).

ascus a sac containing spores.

basidia characteristic special cells found in the fruiting bodies of basidiomycote fungi.

Basidiomycota the phylum containing most of the fungi generally thought of as mushrooms and toadstools - though it also includes microscopic plant parasites, the rusts and smuts. Many members of this class are important as decomposers, being wood-rotting or leaf-litter fungi. Others are parasitic and yet others are mycorrhizal.

basidiospore a characteristic swelling found on the basidia of basidiomycote fungi.

bolete any species of the order Boletales of basidiomycote fungi. They are mushroom-shaped fungi with pores.

budding fragmentation of the mycelium (q.v.).

Ceratocystis ulmi a filamentous member of the phylum Ascomycota. It is the cause of Dutch elm disease.

Deuteromycota a 'dustbin' group of fungi, whose true classification cannot yet be decided as they have no sexual stage, which forms the basis of classification into Ascomycota and Basidiomycota.

Fungi imperfecti see Deuteromycota.

hypha (plur. **hyphae**) an individual filament of fungi.

lichen a symbiotic relationship between an alga and a fungus. See pp. 25-26.

meiois the process by which sex cells are produced from body cells.

mycorrhizal living in a symbiotic relationship with the roots of a plant.

motile organism any organism that is capable of spontaneous and independent movement.

mushroom any species of agaric or bolete fungus. Some people would confine the term to the common mushroom and its relatives. See also Basidiomycota.

mycelium a mass of hyphae (q.v.).

Russulales an order of Basidiomycota.

rust a plant disease in which plant leaves appear rusty, caused by the microscopic species of basidiomycote fungus.

Saccharomyces a genus of the phylum Ascomycota - unicellular yeast, which is important in making bread and alcoholic beverages.

saprotroph an organism that grows on dead remains.

sporangium a body in which spores are produced.

symbiosis a close association between two different species from which both derive benefit.

toadstool a non-technical term for the visible fruiting body of certain basiodiomycote fungi. The term is sometimes restricted to poisonous varieties.

yeast see *Saccharomyces*.

Zygomycetes one of two classes of the Zygomycota. They produce asexual spores, while in sexual reproduction two special hyphae fuse to form a zygospore.

zygospore a 'resting stage' in the life cycle of zygomycete fungi. A zygospore eventually germinates to produce a sporangium containing many spores.

BIOLOGICAL SCIENCES

SPORE-BEARING PLANTS

The first known spore-bearing plants date from the later part of the Ordovician period (that is about 510-438 million years ago). By the end of the Silurian period (438-410 million years ago) the first undoubted vascular plants (q.v.), such as club mosses (see below), had established themselves on the land.

The plants covered here are characterized by the use of spores rather than seeds as their dispersal unit. These plants belong to four of the six phyla or divisions of the Plant kingdom: Bryophyta (the liverworts and mosses), Lycopodophyta (the club mosses), Sphenophyta (the horsetails) and Filicinophyta (the ferns).

Some spore-bearing plants, for example the bryophytes, become dry and are able to remain dormant during adverse conditions, but virtually all spore-bearing plants have the additional protection of a resistant outer layer - the cuticle - during adverse conditions.

alternation of generations (in spore-bearing plants) a reproductive process by which a sexually produced generation (gametophyte; see below) is followed by one (or more) asexual generations (sporophyte; see below).

In bryophytes (q.v.), it is the gametophyte stage (q.v.) that is the obvious plant; the sporophyte plant (q.v.) consists only of a 'foot', stalk and capsule. Spores are produced in the capsule. The sporophyte obtains nutrients from its gametophyte parent and is therefore never entirely independent.

In the other spore-bearing plants (the ferns, liverworts, club mosses, horsetails, etc.), it is the sporophyte stage that is the obvious plant, with the sporangia (q.v.) usually borne on the leaves. The gametophytes of these plants are independent and mostly capable of photosynthesis, but they are very small and lack cuticle and conducting tissues. They are restricted to moist sites and can easily be overwhelmed by leaf litter and other plants. In some species the gametophytes avoid this vulnerability by growing underground.

annuli (sing. **annulus**) a modified cell in the sporangial wall that enables some ferns to throw spores into the air for dispersal.

antheridia see gametophyte.

Anthocerotae the bryophyte class containing the hornworts.

archegonia see gametophyte.

Bryophyta one of the six divisions or phyla of the Plant kingdom. It contains three classes: Musci (the mosses), Hepaticae (the liverworts) and Anthocerotae (the hornworts). Although rather diminutive plants, they show remarkable diversity of form and habit. The majority are leafy, with leaves for the most part one-cell thick, so that they very easily dry out. Water transport is largely external but a few larger bryophytes have conducting tissue. Bryophytes have no roots and attach themselves by means of rhizoids.

See also alternation of generations (above).

club moss any member of the division Lycopodophyta. The club mosses are widespread and diverse, generally small, vascular, herbaceous and possess roots. Their leaves are microphyllous and some bear spores.

creeping ferns use creeping rhizomes to climb trees or twine around other objects.

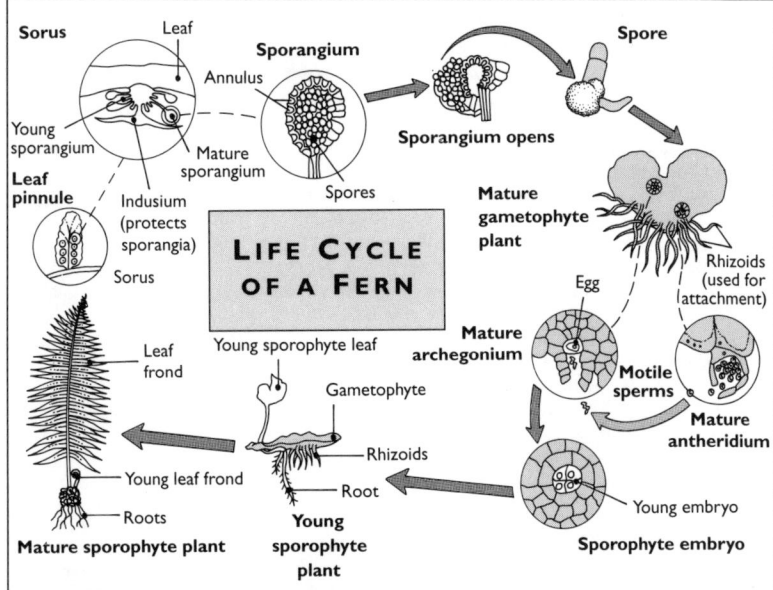

LIFE CYCLE OF A FERN

cuticle a protective layer covering the surface of spore-bearing plants. A thick cuticle forms a largely impermeable barrier which would prevent gaseous exchange (necessary for respiration and photosynthesis to occur) were it not for the presence of pores.

epiphytic ferns grow on another plant without damaging it.

fern any of some 9000 species in the Filicinophyta division. Ferns are distinguished from other spore-bearing plants by their large leaves known as fronds. These leaves have an extensive network of conducting strands and are often intricately subdivided. Sporangia are borne on leaves in groups called sori. In some species, the sporangial wall is modified into annuli (q.v.) enabling spores to be thrown into the air.

Filicinophyta one of the six divisions or phyla of the Plant kingdom. It contains the ferns.

frond the leaf of a fern.

gamete a sex cell; see gametophyte.

gametophyte the stage in a plant life cycle that produces gametes (sex cells). It alternates with an asexual sporophyte generation.

The gametophyte plant has egg-producing organs (archegonia) and sperm-producing organs (antheridia). The egg is retained on the gametophyte plant, and the flagellate sperms need water to enable them to swim to the egg for fertilization to take place. The zygote is retained by the gametophyte plant. It develops into an embryo (a diminutive young sporophyte plant).

Unlike the embryo of a seed-bearing plant, the embryo of a spore-bearing plant cannot be dispersed, and must grow wherever the gametophyte grows, however unsuitable the site.

Hepaticae the bryophyte class containing the liverworts.

heterosporous condition see sporophyte.

homosporous condition see sporophyte.

hornwort any bryophyte member of the class Anthocerotae. They are characterized by a distinctive hollow sporophyte.

horsetail any of about 15 species belonging to the division Sphenophyta. They are temperate plants which are characterized by hollow segmented stems, round which diminutive leaves and branches are borne in whorls.

liverwort any member of the bryophyte (q.v.) class Hepaticae. Some liverworts are flat and have no leaves; most, however, have leaves like mosses. These liverworts are distinct from mosses in having their leaves in two or three ranks rather than being in a spiral.

Lycopodophyta one of the six divisions or phyla of the Plant kingdom. It contains the club mosses.

macrophyllous leaf a leaf with an extensive network of vascular tissue (veins). Macrophyllous leaves are larger than microphyllous leaves (q.v.). They are characteristic of ferns (see above) and of flowering plants (see pp. 28-33).

meiosis the production of sex cells.

microphyllous leaf a leaf with only a single vascular strand. Microphyllous leaves are restricted in size. They are characteristic of the spore-bearing plants, except for the ferns. See also macrophyllous leaf.

moss any member of the bryophyte (q.v.) class Musci. All mosses are leafy and their growth habit may be either erect, often forming tufts or cushions, or creeping, forming a weft over the surface. Their leaves characteristically form a spiral.

Musci the bryophyte class comprising the mosses.

phloem the part of the vascular tissue system of plants responsible for conducting substances manufactured by the plant; see pp. 22-23.

reproduction (in spore-bearing plants) is characterized by an obvious alternation of generations, with both gametophyte and sporophyte generations being for the most part free-living. See gametophyte and sporophyte.

rhizoid a fine hair-like structure by means of which bryophytes and ferns attach themselves to a surface.

rhizome an underground stem.

root any of a number of (usually underground) leafless outgrowths of a plant. They are used for anchorage and the absorption of water and mineral nutrients.

Sphenophyta one of the six divisions or phyla of the Plant kingdom. It contains the horsetails.

sporangia (sing. **sporangium**) a structure developed by the sporophyte plant in which meiosis gives rise to spores. See sporophyte.

spore the dispersal unit of non-seed-bearing plants. Spores need protection. This is achieved by a deposition on the spore wall of sporopollenin.

sporophyte the stage in a plant life cycle that produces spores by asexual means. It alternates with a gametophyte stage.

When the sporophyte plant matures it develops sporangia (q.v.) in which meiosis (q.v.) gives rise to spores. The spores are dispersed in air currents, and if they land on a suitable site they germinate to form gametophyte plants. In most cases the spores are undifferentiated - the homosporous condition - but in a few species two spore types (male and female) are produced - the heterosporous condition.

sporopollenin the deposition on the wall of a spore that gives it protection. Sporopollenin is one of the most resistant compounds in the plant world.

LICHENS

Lichens are often linked in popular imagination with mosses although they are, in fact, not even remotely related.

Lichens are composite organisms formed by the symbiotic relationship of a fungus (see p. 24) and an alga or a cyanobacterium (see pp. 18-19). The fungus provides a protective environment for the 'green partner', which provides the fungus with the products of photosynthesis. The metabolism of lichens is suspended when they become dry or are exposed to the heat of the Sun, but they soon become active again when conditions are moist. This intermittent activity results in a very slow growth rate but enables them to colonize some relatively inhospitable habitats.

Asexual reproduction is achieved by the production of soredia, clumps of algal cells surrounded by fungus. Thus fungus and alga are disseminated together.

stomata pores in the cuticle (q.v.) of spore-bearing plants. The stomata may be opened and closed as conditions dictate.

tree ferns various species of ferns that can reach up to 25m (82ft) in height. They have a crown of leaves topping an unbranched trunk.

vascular tissue the system of tissues comprising the xylem (q.v.) and the phloem (q.v.); see pp. 22-23.

water ferns grow free-floating on water surfaces.

whisk fern any of a group of tropical or subtropical ferns that are anchored by rhizoids to other plants. Their roots are aerial. Whisk ferns are largely epiphytes.

xylem the part of the vascular tissue of plants responsible for conducting water and dissolved mineral salts. (See p. 23.)

zygote a fertilized egg.

BIOLOGICAL SCIENCES

CONIFERS AND THEIR ALLIES

Conifers - the members of the division or phyla Coniferophyta - share with the flowering plants (the angiosperms) the distinction of bearing seeds rather than spores. Seed-bearing has a number of advantages over spore-bearing in successfully spreading and increasing the species, and in ensuring survival if the parents are short-lived or suffer some catastrophe.

The conifers and their allies differ from the flowering plants in that they are all gymnosperms (q.v.).

Abies a genus of conifers; the firs.

bract scale see cone.

cone a structure containing the reproductive organs of a conifer. The mature female cone has pairs of papery or woody scales, comprising a sterile outer or bract scale and an inner ovule-carrying or ovuliferous scale. In the male cones, pollen sacs are borne on the underside of cone scales.

conifers first emerged over 300 million years ago. They are widespread and diverse with over 500 modern species. Most are tall forest trees. Their leaves are usually needle-like, usually with a single vascular strand. The xylem (see p. 23) in conifers - and in other trees and shrubs - is stiffened by lignin to form wood, and the wood is protected and insulated by bark. Resin is secreted by most conifers as an unwanted end product of metabolism. The reproductive structures are grouped in cones (see above). Most conifers have a long life cycle with several static periods - the period between pollination and fertilization often exceeds one year. Also, seedlings are very slow to establish.

In the northern hemisphere coniferous trees are a major element in the high-latitude boreal forests of North America and Eurasia. Members of the Pinaceae family dominate including *Laris* (larch), *Pinus* (pines), *Picea* (spruces), *Abies* (firs), *Pseudotsuga* (Douglas fir), and *Tsuga* (western hemlock).

Conifers are adapted to withstand harsh conditions:
- the conical shape encourages shedding of snow and optimizes interception of low-angled light;
- needle-like leaves are xeromorphic with a low ratio of surface area to volume and a thick outer layer to cope with the drying effects of frost and high winds;
- most leaves are evergreen, enabling rapid establishment of productivity as soon as conditions are suitable.

cycad any member of some 11 modern genera of cycadophyte. Most have very restricted tropical distributions. The cycads emerged about 250 million years ago and for over 100 million years they dominated land vegetation. They declined drastically upon the emergence of flowering plants. Cycads resemble small palms with a short trunk, which is covered with the bases of old leaves. A few species reach up to 18m (60ft) high.

Douglas fir see fir.

fir any species of conifer of the genera *Abies* (firs) and *Pseudotsuga* (Douglas firs).

ginkgo (or **maidenhair tree**) a single modern species of ginkgophyte, which formerly had a very broad distribution. A large, much branched tree, the ginkgo is distinguished from the conifers by its fan-shaped leaf.

gymnosperm any plant that bears a 'naked' seed, i.e. a seed that is not enclosed in an ovary.

hemlock spruce see Western hemlock.

larch any of about 10 species of conifer of the genus *Laris*.

Laris a genus of conifers; the larches.

maidenhair tree see ginkgo.

ovuliferous scale see cone.

Picea a genus of conifers; the spruces.

pine any of around 96 species of conifer of the northern hemisphere genus *Pinus*.

Pinus a genus of conifers; the pines.

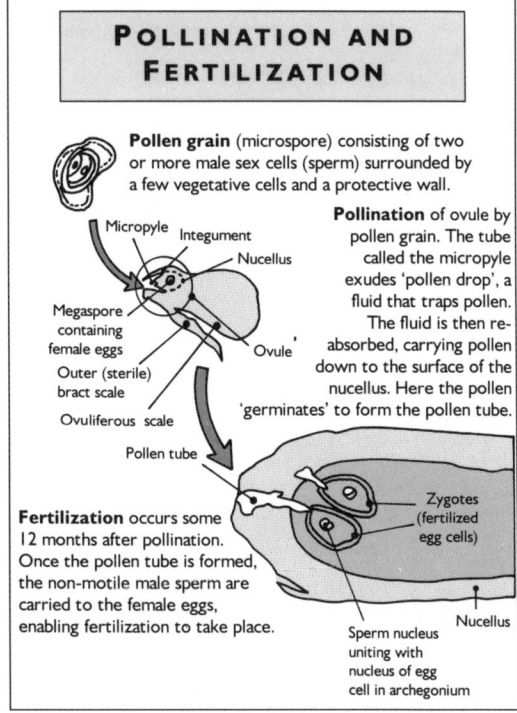

POLLINATION AND FERTILIZATION

Pollen grain (microspore) consisting of two or more male sex cells (sperm) surrounded by a few vegetative cells and a protective wall.

Pollination of ovule by pollen grain. The tube called the micropyle exudes 'pollen drop', a fluid that traps pollen. The fluid is then re-absorbed, carrying pollen down to the surface of the nucellus. Here the pollen 'germinates' to form the pollen tube.

Fertilization occurs some 12 months after pollination. Once the pollen tube is formed, the non-motile male sperm are carried to the female eggs, enabling fertilization to take place.

Labels: Micropyle, Integument, Nucellus, Megaspore containing female eggs, Outer (sterile) bract scale, Ovule, Ovuliferous scale, Pollen tube, Zygotes (fertilized egg cells), Nucellus, Sperm nucleus uniting with nucleus of egg cell in archegonium.

Pseudotsuga a genus of conifers; the Douglas firs.

spruce any of around 50 species of conifer of the northern temperate genus *Picea*.

Tsuga a genus of conifers; the western hemlocks.

western hemlock (or **hemlock spruce**) any of around 10 species of conifer of the genus *Tsuga*, which is confined to North America and Asia.

xeromorphic drought resistant.

FLOWERING PLANTS

The flowering plants belong to the phylum or division Angiospermatophyta. They are the most diverse, widespread and abundant group of multicellular plants. It is estimated that there are between 250 000 and 300 000 species - and probably many more that have not yet been collected from remote environments such as parts of the tropical rainforest. Flowering plants grow in a very wide range of habitats and to cope with this wide range the flowering plants have evolved an enormous variety of adaptations in form and in life cycle. The detailed classification of the flowering plants is based upon a variety of elements: growth form, vegetative biology, reproductive biology, wood anatomy, and chemical characteristics. There has recently been considerable reclassification and renaming of families, and this is reflected here.

acacia any of around 800 species of spiny trees and shrubs of the genus *Acacia*.

Aceraceae a dicotyledonous plant family including maples and sycamore.

Agave a genus of about 300 species of New World plants characterized by large, fleshy, tough leaves.

alder any of about 35 deciduous trees of the genus *Alnus*.

Alismataceae a monocotyledonous family; the water plantains.

Alismatidae a sub-class of monocotyledons including water plantains and pondweed.

Alliaceae a family of the sub-class Liliidae. It contains onions and garlic.

Amaryllidaceae a monocotyledonous family; daffodils and snowdrops.

anemone any of 80 species of perennial herb of the temperate genus *Anemone*.

angiosperm any seed-bearing plant in which the seed is enclosed in an ovary, i.e. a flowering plant. Compare with gymnosperm (in conifers; see p. 27).

annual any flowering plant that completes its life cycle from seed germination to death in less than one year; e.g. marigolds and petunias.

anther a pollen-containing sac found on a stamen (q.v.).

Apiaceae a dicotyledonous plant family (formerly Umbelliferae). Members include carrot, parsley, hogweeds, celery, hemlock.

apple any of around 35 species of trees and shrubs and their derivatives of the temperate genus *Malus*.

Aquifoliaceae a dicotyledonous plant family; the hollies.

Araliaceae a dicotyledonous plant family; the ivies and ginseng.

Arecaceae a monocotyledonous plant family; the palm trees.

Arencidae a sub-class of monocotyledons containing the palm trees.

ash any of around 60 species of temperate trees of the genus *Fraxinus*.

asparagus any of around 300 species of perennial herbs, shrubs and climbers of the genus *Asparagus*.

aspen the name applied to five of the 35 species of poplar.

Asteraceae a dicotyledonous plant family (formerly Compositae) in which the flowers are arranged in multiple flower heads. They include thistles, daisies, chrysanthemums, lettuce.

Asteridae a dicotyledonous sub-class.

bamboo any of around 1000 species of mainly tropical perennial woody grasses of the family Poaceae.

bark tissues in woody stems and roots external to the vascular cambium, or its corky part only. It provides a protective layer for trees and shrubs.

bean any of various leguminous plant of the family Fabaceae.

beech any of about 10 species of deciduous tree of the genus *Fagus*.

begonia any of about 350 species of tropical and subtropical herbs of the genus *Begonia*.

bellflower see campanula.

berry a fruit that usually contains a large number of small seeds; e.g. gooseberries and tomatoes.

Betulaceae a dicotyledonous family; the birches, alders and hazels.

biennial any flowering plant that completes its life cycle in more than one year but less than two years; e.g. foxgloves, carrots and cabbages. Flowering and seed production usually occur in the second year.

birch any of about 60 species of the genus *Betula*.

bluebell a member of the family Liliaceae.

borage a medical and culinary herb in the family Boraginaceae.

Brassicaceae a dicotyledonous plant family (formerly Cruciferae); broccoli, cabbage, calabrese, cauliflower, kale, mustard, rape, swede, turnip.

broadleaved tree see tree.

bromeliad any of about 40 species of tropical New World herbs of the genus *Bromelia*. Most have stiff spiny rosettes of leaves.

broom any shrub in the three leguminous genera *Cytisus*, *Genista* and *Spartium*.

bud a very compact stem with densely packed young leaves or flower parts. Buds can develop on shoots or flowers.

bulb a perennating organ; a compact underground stem bearing fleshy leaves. Examples include onions and daffodils.

bush see shrub.

buttercup any of 400 members of the genus *Ranunculus*.

cabbage a cultivar of the Brassicaeae family.

cactus any of around 800 species of plant of the family Cactaceae. They are strongly adapted to arid conditions having succulent photosynthetic stems (rather than photosynthetic leaves), which are often spiny.

calyx the collective term for the sepals, the outer whorl of leaf-like organs of a flower.

BIOLOGICAL SCIENCES

cambium a secondary meristem extending down the length of a root or stem. It is responsible for producing new cells, which differentiate into xylem (q.v.) and phloem or the periderm of the bark.

camellia any of around 84 trees of the genus *Camellia*.

campanula (or **bellflower**) any of about 300 species of usually perennial herbs of the genus *Campanula* (family Campanulaceae).

campion any of several species of the genus *Dianthus*.

carpel one of a series of four organs in a flower. Carpels consist of an ovary surmounted by a style and a stigma (q.q.v.).

Caprifoliaceae a dicotyledonous family; honeysuckle and elder.

carrot a biennial herb belonging to the family Apiaceae.

Caryophyllaceae a dicotyledonous family; pinks, campion and carnations.

Caryophyllidae a dicotyledonous sub-class containing pinks, beets, spinach, cacti, dock, knotgrass and buckwheat.

catkin an inflorescence (usually hanging) in which small, reduced, unisexual flowers are borne on a central stem, as in willow and birch.

cereals members of the family Poaceae; see grasses.

Chenopodiaceae a dicotyledonous family of plants; beet and spinach.

cherry any of a number of species of genus *Prunus* (family Rosaceae).

chrysanthemum the horticultural name for some 200 herbaceous plants of the family Asteridae.

citrus fruit the juicy fruit of two genera in the family Rutaceae; they include orange, lemon, lime, grapefruit.

clematis any of about 250 species of woody climbers of the genus *Clematis* (family Ranunculaceae).

climbing plants include both herbaceous and woody species, the latter often called lianas. Flowering plants often climb by means of touch-sensitive stems or leaves, many of which are specially modified in the form of tendrils. These organs twine around objects with which they come into contact, such as tree trunks or other climbers.

clover any of a number of annual perennial leguminous herbs of the genus *Trifolium*.

Commeliniceae a monocotyledonous family; the tradescantias.

Commelinidae a monocotyledon sub-class containing the grasses, bromeliads and tradescantias.

Compositae the former name of the family now called Asteraceae.

convolvulus a number of species of the family Convolvulaceae.

corm a perennating organ; a swollen stem base. Examples include crocuses and gladioli.

cotyledon the 'leaf' on seed embryos, sometimes modified for food storage (e.g. in peas). After germination the cotyledon can remain under ground (as in peas) or be carried above ground (e.g. French beans). The number of seed leaves provides one of the principal distinctions between the two classes of flowering plants: the Liliopsida (the monocotyledons) and the Magnoliopsida (the dicotyledons).

crocus any of 75 species belonging to the genus *Crocus* (family Iridaceae).

cross-pollination see pollination.

Cruciferae the former name of the family now known as Brassicaceae.

cultivar a variety of a cultivated plant species that has been bred for agricultural or horticultural purposes.

daffodil (or **narcissus**) any of around 60 species of bulbous herb of the family Amaryllidaceae.

daisy any of 15 species of the genus *Bellis* and other genera of the family Asteridaceae.

dandelion any of a large number of weeds of the genus *Taraxacum* (family Asteridaceae).

deciduous plants characterized by the seasonal shedding of leaves. Many broadleaved trees of the higher latitudes shed their leaves in winter, entering a dormant phase. In lower-latitude temperate zones related species may be evergreen.

dicotyledon any member of the flowering-plant class Magnoliopsida. The main distinguishing feature is a double seed leaf (cotyledon). The dicotyledons comprise about 70% of the flowering plant species.

Distinguishing features include:

- two seed leaves (cotyledons);
- leaves with branching main veins connected by a net-like venation;
- floral organs in fours or fives;
- a persistent primary root system.

Dillenidae a dicotyledonous plant subclass.

dock a number of species of perennial herb of the genus *Rumex* (family Polygonceae).

drupe an indehiscent fruit consisting of an outer layer (skin), a fleshy middle layer and a stony inner layer, within which there is a single seed. Examples include peaches and plums. An animal gathering a drupe will dispose of the stone (seed) thus dispersing it.

elder a European and North American shrub of the family Caprifoliaceae.

elm any of about 20 species of large tree of the genus *Ulmus*.

epiphyte any plant that grows on other plants without damage to them. Epiphytes rely almost entirely upon rain water as a source of minerals, taking nothing from the 'host' plant; they often have specialized cells on their leaves and roots. Flowering plant examples include many tropical orchids and bromeliads.

Ericaceae a dicotyledonous plant family including the heathers, rhododendrons and cranberry.

eucalyptus any of about 175 species of tropical tree or shrub of the mainly Australian genus *Eucalyptus*.

Euphorbiaceae a dicotyledonous plant family including the spurges.

evergreen plants characterized by the tendency of the leaves not to fall in unison. In tropical rainforests the leaves of most trees are evergreen. Conifers - see p. 27 - are also evergreen.

Fabaceae (or **Papilionaceae**) a dicotyledonous plant family (formerly Leguminosae, the legumes); it includes peas, beans, clovers, broom, lupins, peanut, etc.

Fagaceae a dicotyledonous plant family; it includes oaks and beeches.

flower the part of a plant containing the sexual organs. In true flowering plants the complete or perfect flower contains a series of four organs arranged in succession on the receptacle. From base to apex these are sepals, petals, stamens and carpels (q.q.v.). They may occur in a continuous spiral (e.g. in magnolias), sometimes with a series of gradual changes between organs (as in the sepals, petals and stamens of water lilies). Alternatively each organ occurs in a discrete, distinct whorl, often with fusion between the elements of the whorl (e.g. in the heathers, dead-nettles and primroses). Although frequently modified, the fundamental structure and function of each of the four floral organs is the same in all flowering plants.

forget-me-not any of about 50 species of the temperate genus *Myosotis*.

fruit the seed-containing structure that develops from the ovary of a flower, usually after fertilization. Some parts of the fruit may develop from other parts of the flower. A great variety of fruits have developed to attract different types of dispersers.

funicle the small stalk that emerges from an almost mature seed.

garlic a herb of the family Alliaceae.

gentian any of about 400 species of perennial herbs of the mainly alpine genus *Gentiana* (family Gentianaceae).

geranium any of about 400 species of herbs in the mainly temperate genus *Geranium* (family Geraniaceae).

ginger a tropical perennial herb belonging to the family Zingiberaceae.

gladiolus any of around 300 species of Old World bulbous herb of the genus *Gladiolus* (family Iridaceae).

gorse any of about 15 species of spiny temperate shrub of the genus *Ulex* (family Fabaceae).

Graminae the former name for the family now known as Poaceae.

grapefruit an evergreen citrus tree of the family Rutaceae.

grape vine any of several vines of the genus *Vitis*.

grasses constitute the monocotyledonous family Poaceae. This group - of which there about 9000 species - includes the cereal or grain crops that provide staple foods for much of the world's population: rice, wheat, maize, barley, rye, oats, etc. The group also includes sugar cane. Grasslands dominate large areas of both the tropical and temperate regions. In grasses typical of grassland habitats the growing points of the stems are below ground and are protected by encircling leaves.

Grossulariaceae a dicotyledonous plant family including currants and gooseberries.

Hamamelidaceae a dicotyledonous plant family including witch hazel.

Hamamelidae a dicotyledonous sub-class containing witch hazels, elms, nettles, walnuts, plane trees, birches, alders, hazels, oaks and beeches.

hardwood see tree.

hazel any of about 15 species of shrubs and trees of the genus *Corylus*.

heather a mainly European low evergreen shrub belonging to the family Ericaceae.

herb see herbaceous perennial.

herbaceous perennial any flowering plant that lacks woody cells and therefore dies back to the roots (or to other perennating organs) at the outset of frost or drought, and which produces new growth above ground upon the return of spring or rain.

herbivore any animal that feeds on plants.

hermaphrodite an individual that has both male and female reproductive organs, a common condition in plants.

Hippocastanaceae a dicotyledonous plant family; the horse chestnuts.

holly any of around 400 species of trees and shrubs of the genus *Ilex*.

honeysuckle any of about 200 shrubs and woody climbers of the temperate genus *Lonicera*.

horse chestnut see Hippocastanaceae.

hydrangea any of about 25 species of shrub of the New World genus *Hydrangea* (family Hydrangeaceae).

indehiscent (of plant structures) not opening of their own accord.

inflorescence a mass of small flowers clustered together.

Iridaceae a monocotyledonous plant family; the irises.

iris any of 300 bulbous herbs with rhizomes, belonging to the family Iridaceae.

ivy any of about 15 species of climbers and woody shrubs of the temperate genus *Hedera*.

Juglandaceae a dicotyledonous plant family; the walnut trees.

Lamiaeae a dicotyledonous plant family including the mints, thyme and basil.

Lauraceae a dicotyledonous plant family including laurel, bay and avocado.

laurel any of a number of small trees and bushes of the family Lauraceae.

legume any member of the family Fabaceae (formerly Leguminosae); see also pp. 20-23.

lemon a citrus fruit of the family Rutaceae.

lettuce any of a number of species of herb of the temperate genus *Lactuca*.

lignin an important strengthening component of cell walls in the woody tissues of plants.

Liliaceae a monocotyledonous family including lilies, tulips and bluebells.

Liliidae a monocotyledonous sub-class containing the lilies, onions, daffodils, irises, agaves, asparagus and orchids.

BIOLOGICAL SCIENCES

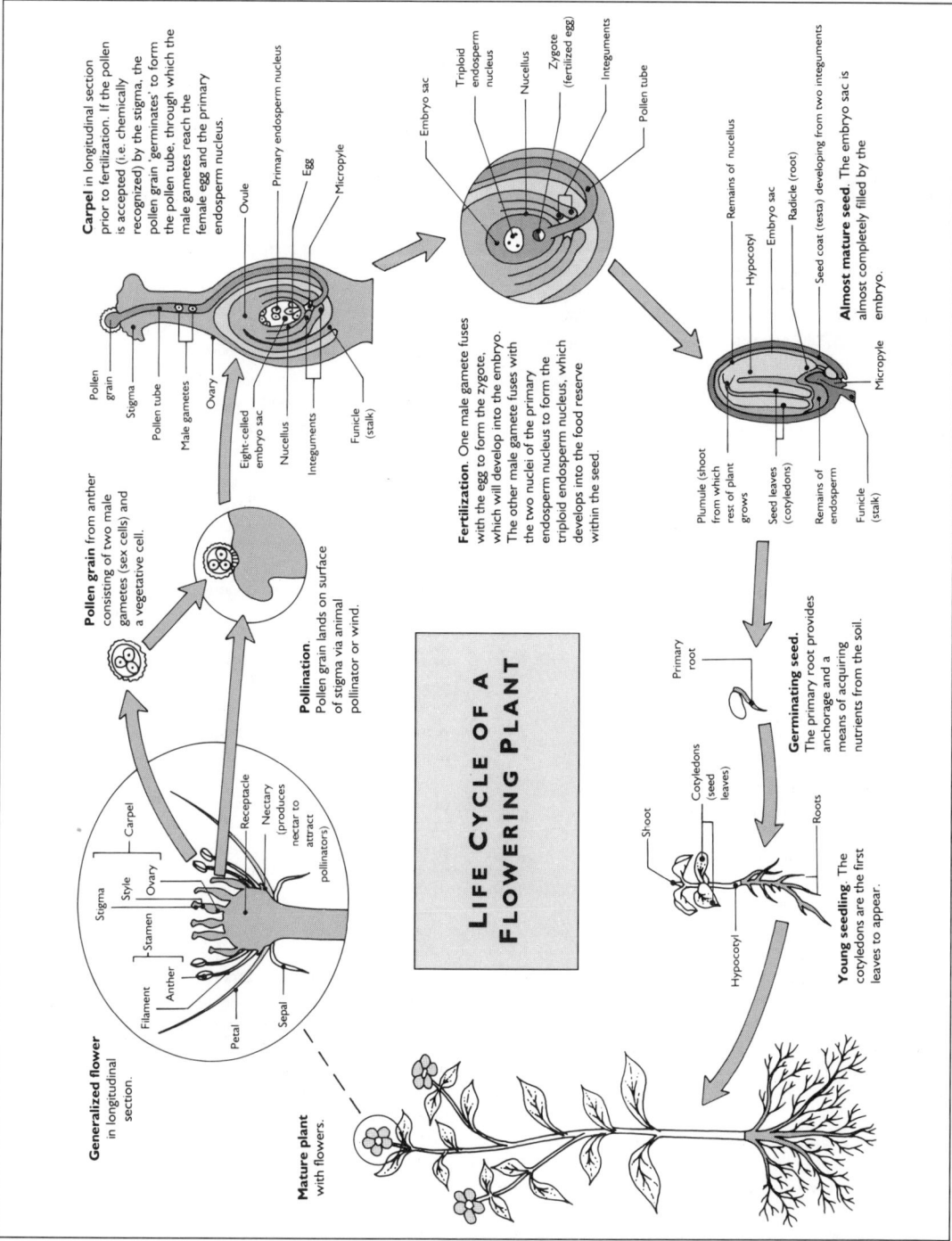

Liliopsida a plant class - the monocotyledons (q.v.).

lily any of around 90 species of bulbous herb of the family Liliaceae.

lime (or **linden**) 1. any of about 30 species of tree of the genus *Tilia* (family Tiliaceae). 2. a lemon-like citrus fruit of the family Rutaceae.

linden see lime.

lupin any of 200 species of annual and perennial herbs of the leguminous genus *Lupinus*.

magnolia any of around 85 species of trees and bushes of the Old World genus *Magnolia* (Magnoliaceae).

Magnolidiidae a dicotyledonous sub-class including magnolias, water lilies, laurels, buttercups and poppies.

Magnoliopsida a plant class - the dicotyledons (q.v.).

maple any of 200 species of evergreen and deciduous trees and shrubs of the temperate genus *Acer*.

marigold an annual herb of the family Asteraceae. There are many cultivars.

marjoram any of a number of aromatic herbs of the genus *Origanum*.

meristem a group of cells in plants that divide by mitosis and thereby contribute to plant growth and organ formation.

mimosa any of around 400 species of trees, shrubs and herbs of the leguminous genus *Mimosa* (family Mimosaceae).

mint any of about 25 species of aromatic perennial herbs of the temperate genus *Mentha*.

monocotyledon any member of the flowering-plant class Liliopsida. The main distinguishing feature is a single seed leaf (cotyledon). The monocotyledons comprise about 30% of flowering-plant species.

Most monocotyledons are herbaceous and those that are trees (e.g. palms) lack wood and therefore cannot support a branching canopy.

Distinguishing features include:

- a single seed leaf (cotyledon) in the embryo and young plant;
- parallel veins in leaves (which are usually long, narrow and pointed);
- floral organs arranged in threes;
- fibrous root systems.

mustard several herbs of the family Brassicaceae.

Myrtaceae a dicotyledonous plant family; the eucalyptuses.

nectar a sugary liquid that serves to attract the interest of pollinators.

nectar guide spots or lines that direct the pollinator to the source of nectar.

nectary that part of the flower which produces nectar to attract pollinators.

nettles any of about 50 species of temperate plants of the genus *Urtica*.

nightshade any of a number of species of the genus *Solanum*.

Nymphaeaceae a dicotyledonous family; the water lilies.

oak any of some 450 species of deciduous tree of the genus *Quercus*.

Oleaceae a dicotyledonous plant family including olive and ash.

olive a Mediterranean tree belonging to the family Oleaceae.

onion a biennial herb of the family Alliaceae.

orange a citrus tree of the family Rutaceae. There are many cultivars.

orchid any of around 18 000 members of the family Orchidaceae.

ovary (of flowering plants) a protective structure surrounding the ovule. The ovary contains one or more ovules, within each of which is a female gamete (the ovum or egg). Seeds develop within the ovary which becomes modified to form the great variety of fruit that are so important for the dispersal of seeds.

ovule see ovary.

ovum the egg (see ovary).

Paeoniaceae a dicotyledonous plant family; the peonies.

palm any member of the family Arecaceae. Most are trees. See also monocotyledon.

Palmae the former name of the family now known as Arecaceae.

pansy a number of annual and perennial herbs of the genus *Viola*.

Papaveraceae a dicotyledonous plant family; the poppies.

Papilionaceae an alternative name for the Fabaceae family.

parsley a temperate biennial herb of the family Apiaceae.

pea any of a number of leguminous plants of the family Fabaceae. There are many cultivars.

peanut an annual tropical herb of the family Fabaceae.

pear any of around 30 species of temperate trees and shrubs of the genus *Pyrus*.

peony any of 33 species of perennial herbs of the temperate genus *Paeonia*.

perennating organ an underground organ for storing food. Perennating organs - which allow biennials and herbaceous perennials to survive frost and drought - occur in a variety of forms: taproots, bulbs, corms, and tubers. In many cases these perennating organs are used for reproductive purposes.

perennial see herbaceous perennial and woody perennial.

petal one of a series of four organs in a flower. Petals are usually large, delicate and colourful. Their function is one of advertisement and attraction of pollinators.

phloem the parts of vascular tissue in plants responsible for conducting substances manufactured by the plant.

phlox any of around 65 species of annual and perennial shrubs and herbs of the North American genus *Phlox*.

pink any of several species of the genus *Dianthus*.

Platanaceae a family of the sub-class Hamamelidae. It contains the plane tree.

plane any of 10 species of the genus *Platanus*.

BIOLOGICAL SCIENCES

plum any of several species of the genus *Prunus* (family Rosaceae).

plumule the shoot from which the rest of the plant grows.

Poaceae a monocotyledonous family (formerly known as Graminae); it contains grasses, cereals and reeds.

Polemoniaceae a monocotyledonous plant family; the phloxes.

pollen male gametes (male sex cells) in flowers. The pollen grain consists of two only two gametes and one vegetative nucleus. See pollination and stamen.

pollen tube a filamentous structure extending from the germinated pollen grain. It grows to the ovule, serving to bring the male nuclei into proximity with the ovum.

pollinator any animal (usually an insect) that is an agent of pollination (q.v.).

pollination the process by which pollen is brought from the male organs to the female organs of seed-bearing plants. Most flowers are hermaphrodite so there would seem to be little problem in achieving pollination. However, although some flowers are naturally self-pollinated, there is a genetic bonus to be gained from cross-pollination, which can maintain or even increase the vigour of the population. Some plants have a genetic incompatibility system that prevents pollen functioning on stigmas of the same or closely related plants - Cox's Orange Pippin apple. An absolute barrier to self-pollination is seen in those species that have separate male and female plants, such as willow.

A flower that depends upon an animal to carry pollen to another plant needs to advertise its wares. To this end many flowers are conspicuous and may also be scented. Some have large and colourful petals or sepals (or both), while in others there are inflorescences. Colourful foliage near the flower may also assist or even take over this function. Flower colour is related to the colour vision of the pollinating animal. Bees do not distinguish red, but birds do; so red is a common colour for bird-pollinated flowers. Bees are, however, sensitive to ultraviolet light and are attracted to red poppies not by the colour red but by the ultraviolet light that is reflected off them. Flowers frequently have nectar guides. As well as acting as an advertisement, a flower must also provide - or seem to provide - a reward to the pollinator. The most common rewards are a share in the pollen itself, which is rich in protein, and nectar (q.v.). Flowers that rely on the wind for pollination naturally do not need to expend energy on attracting pollinators. However, they must produce vast quantities of pollen to saturate the air and ensure that any sticky stigma exposed to air currents will pick up pollen of the same species.

Polygonaceae a monocotyledonous plant family including dock, knotgrass and buckwheat.

pondweed any of about 100 species of aquatic herbs.

poplar any of about 35 species of trees of the northern temperate genus *Populus*.

poppy any of a large number of members of two genera of the family Papaveraceae.

Potamogetonaceae a dicotyledonous plant family; all pondweeds.

potato a South American herb belonging to the family Solanaceae.

primrose see primula.

primula any of around 600 species of temperate perennial herbs of the genus *Primula* (family Primulaceae).

Ranunculaceae a dicotyledonous plant family; it contains buttercups, anemones, clematis and delphiniums.

raspberry any of several species of the genus *Rubus* (family Rosaceae).

receptacle the reproductive axis of a flowering plant.

reed any of a number of grasses of wet places belonging to the genus *Phragmites* (family Poaceae).

rhizome a perennating organ; an underground horizontal stem (e.g. irises).

rhododendron any of over 500 species of shrubs of the genus *Rhododendron* (family Ericaceae).

root any of a number of (usually underground) leafless outgrowths of a plant, used for anchorage and the absorption of water and mineral nutrients. There are many variations of the basic root form; e.g. the tubers of dahlias, taproots of carrots and dandelions, buttress roots of some tropical trees, aerial roots in epiphytes.

Rosaceae a dicotyledonous plant family containing roses, strawberry, apple, hawthorn, cherry, etc.

rose any of about 250 species of usually prickly temperate shrub of the genus *Rosa*.

Rosidae a large dicotyledonous sub-class.

Rutaceae a dicotyledonous plant family comprising the citrus fruit.

Salicaceae a dicotyledonous plant family containing willows, poplars and aspen.

saxifrage any of about 370 species of perennial herbs of the genus *Saxifraga* (family Saxafragaceae).

scent (in flowers) see pollination.

seed the unit of dispersal containing the embryo of seed-bearing plants (both conifers - see p. 27 - and flowering plants). It develops from the fertilized ovule, and usually contains food reserves. The kernel of the seed is protected by the testa (q.v.).

seed dispersal the distribution of seeds from the parent plant. This is achieved by wind, water or animals. As seeds contain the embryo plant, and most contain a food reserve to sustain the early growth of the plant, even the smallest seed is heavy compared to a pollen grain. Modifications to the shape and surface sculpture of wind-dispersed seeds help to slow down the rate and so to increase the chances of being caught on an air current. Such features include thin, papery wings and fine hairs, which may be developed from various parts of the seed, fruit or flower - e.g. old man's beard (in which hairy appendages are derived from the style) and willowherb (in which the hairs arise from the surface of the seed).

Seed dispersal also occurs by water, for example the double coconut which floats from one island site to another. Animals disperse seeds by carrying them externally (on fur, feet or beaks) or internally (in the gut, passing out in the faeces).

Seeds that become attached to animals have hooks or spines or exude sticky substances, but generally rely upon a chance encounter with the animal. Internal dispersed seeds, on the other hand, must offer a reward to an animal in return for dispersal: seeds of this type are carried in berries and drupes (q.q.v.).

sepal one of a series of four organs in a flower. Sepals are usually small, tough and green. Their function is one of protection, especially in the bud stage, when they enclose the unopened flower. Later, in the opened flower, sepals help to prevent damage to mature ovules (q.v.) by herbivores (e.g. rabbits) or nectar robbers (e.g. bees) attacking flowers from beneath.

shrub (or **bush**) a small or medium-sized woody plant with numerous main stems.

snowdrop any of 12 species of bulbous herb of the genus *Galanthus*.

spinach any of several species of plant of the family Chenopodiaceae.

stalk the main stem of a plant. See also funicle.

stamen one of a series of four organs in a flower. Stamens consist of a stalk or filament that carries anthers. Their function is pollen production and release at the appropriate time and place.

stem the part of a plant bearing leaves, buds or flowers.

stigma the specialized surface of a carpel at which pollen is received and recognized.

strawberry any of 15 species of perennial herbs of the genus *Fragaria*.

style that part of the carpel that serves to position the stigma within the flower at an appropriate place to receive pollen from wind or animal pollination.

sycamore a large maple (q.v.) native to Europe.

taproot a perennating organ; a large single root growing vertically downwards, and from which smaller lateral roots extend (e.g. the carrot and the dandelion).

tendril a specially modified leaf or stem; see climbing plants.

testa the seed coat. It usually makes the seed resistant to harsh conditions.

Theaceae a dicotyledonous plant family including camellias and the tea plant.

thistle any prickly perennial herb of the genera *Carduus* and *Cirsium*.

thyme any of about 300 species of small aromatic shrubs of the temperate genus *Thymus*.

Tiliaceae a dicotyledonous plant family; the lime trees or lindens.

tobacco any of several species of herbs of the genus *Nicotiana*.

tomato a perennial herb of the family Solanaceae.

tradescantia any of some 40 New World herbs of the family Commelinaceae.

tree a large, woody plant having one or more main stems; distinguished from a shrub or bush by its greater size and by having fewer stems. Most forest and woodland communities are dominated by flowering-plant trees - the only exceptions being the great coniferous forests of the northern hemisphere and high altitudes. Most of these trees - such as beeches, oaks, figs, mahoganies and eucalyptuses - are dicotyledons and are often referred to as broadleaved or hardwood trees. Some are evergreen, while others are deciduous. All broadleaved trees possess an extensive growth of wood and bark, and many are highly prized for their timber. The major group of monocotyledonous trees are the palms (family Arecaceae), which grow mainly in the tropics. Palm trees do not possess wood or bark, but are supported by other tissues strengthened by lignin, and by various other structural characteristics.

tuber a perennating organ; a much-swollen underground stem (e.g. the potato).

tulip any of around 100 bulbous herb of the genus *Tulipa* (family Liliaceae).

turnip a cultivar of the Brassicaceae family.

Ulmaceae a dicotyledonous plant family including the elm.

Urticaceae a dicotyledonous plant family comprising the nettles.

vegetable any part of a plant that is used in salads or savoury dishes; note that 'vegetable' is not used as a botanical definition.

vetch any of about 140 species of annual or perennial temperate leguminous herbs of the genus *Vicia*.

violet any of about 500 species of temperate herbs of the genus *Viola* (family Violaceae).

Vitaceae a dicotyledonous plant family including grape vine and virginia creeper.

wallflower any of 10 species of annual perennial herbs and shrubs of the genus *Cheiranthus* (family Brassicaeae).

walnut any of 15 species of deciduous tree of the genus *Juglans*.

water lily any of 75 species of perennial herb of the family Nymphaeaceae.

water plantain any of several species of marsh or swamp plants belonging to the genus *Alisma* (family Alismataceae).

willow any of around 500 species of trees and shrubs of the temperate genus *Salix*.

wind pollination see pollination.

wood the hard fibrous tissues of the stems and roots of woody plants (shrubs, bushes, many climbers and trees). Wood is composed mainly of secondary xylem produced by the cambium.

woody perennial any flowering plant that takes longer than two years to complete its life cycle and does not die back to the roots; e.g. trees, shrubs and woody climbers.

xylem the part of the vascular tissue of plants responsible for conducting water and dissolved mineral salts. Characteristically, the cells are elongated, with relatively thick walls strengthened by lignin. The secondary xylem makes up the principal part of wood in trees, etc.

Zingiberaceae the monocotyledonous plant family including the ginger plant.

BIOLOGICAL SCIENCES

ANIMALS - AN INTRODUCTION

In spite of the pre-eminent importance of the animal kingdom and the almost unbounded interest in it among biologists, there is surprisingly little agreement about the precise features that characterize animals and consequently about the exact boundaries of the animal kingdom itself.

The kingdom Animalia is conventionally restricted to multicellular forms, but there is a multitude of single-celled protozoans, generally classed with other unicellular life in the kingdom Protoctista, that show all or most of the features normally regarded as defining animals and that might reasonably be classed with them. Even with this restriction, the animal kingdom is unsurpassed in size and diversity. It ranges from simple sponges (p. 36) and corals (p. 37), through a host of more complex invertebrates such as insects (pp. 42-43) and squid (p. 40), to vertebrates, which include the most advanced animals of all, the birds (pp. 52-53) and mammals (pp. 54-59).

We can only guess at the number of species contained in this enormous kingdom. Perhaps one and half million species have been scientifically described, but this must still represent only a fraction of the animals that actually exist. The majority of animals described - over one million - are insects (pp. 42-43).

Australian a faunal region covering not only Australia but also New Guinea, New Zealand and adjoining islands.

distinguishing features (of animals) include the following:

1) Animals lack the cell walls that are characteristic of plants;

2) Animals obtain their food in a quite different manner to plants. While virtually all plants are able to harness the energy of the Sun to synthesize the complex organic compounds they require from simple inorganic molecules (photosynthesis), animals rely on ready-made sources of such compounds - in other words they are dependent on plants, either by eating them directly or by feeding on the body tissues of animals that have done so.

3) Although some aquatic animals such as sea anemones and corals are more or less immobile and can rely upon their food coming to them, the majority of animals must actively find it or seize it. Unlike plants, therefore, animals generally require at least some degree of mobility.

4) Movement in multicellular animals requires a great deal of coordination between different body parts; in all but the simplest animals, such as jellyfish, this entails a reasonably sophisticated system of nervous controls.

5) Furthermore, an animal needs to move not only in a coordinated fashion but also in a particular direction; thus it requires some form of sensory apparatus tied in with its nervous system. Although plants can detect various environmental stimuli and react accordingly, animals have evolved an astonishing range of systems by which they can perceive what is going on around them.

6) Most animals have a digestive tract or cavity.

7) Most animals have a central coordinating point of the nervous system - a 'brain'.

Ethiopian a faunal region covering all of Africa except Morocco.

faunal regions the distinct regions into which the world can be divided, each of which is characterized by a distinctive range of animal species (see map).

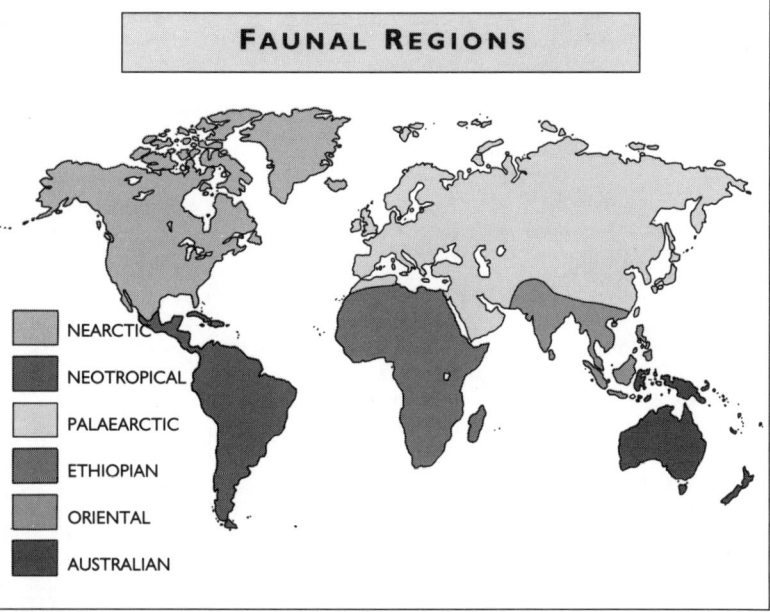

FAUNAL REGIONS

NEARCTIC
NEOTROPICAL
PALAEARCTIC
ETHIOPIAN
ORIENTAL
AUSTRALIAN

Nearctic a faunal region covering North America.

Neotropical a faunal region covering South and Central America and the Caribbean.

Oriental a faunal region covering South and Southeast Asia.

Palaearctic a faunal region covering Europe, Central, West and East Asia and Morocco.

SPONGES

Sponges, which are widespread and abundant in aquatic habitats, are among the simplest and most primitive of multi-cellular animals. Although not particularly animal-like in appearance, they actually have all the properties associated with animals. They feed on other organic matter and they reproduce sexually by means of an ovum (egg) and a spermatazoan. Movement in mature sponges is limited to slight contraction of muscle-like cells, but the larval forms are free-swimming. There is also some degree of coordination of activity, even though there is no real nervous system. Sponges are found in marine habitats, ranging from the intertidal zone to the greatest depths, and also in freshwater lakes and rivers. Some form flat, brightly coloured incrustations, others a variety of single or multiple structures in the shape of vases, chimneys or purses. Some sponges are only 1cm long, but various species in the tropics and the Antarctic can reach more than 1m (39in) in height and breadth. In the Antarctic extensive areas of the sea bed are dominated by sponges. The loose structure of sponges makes them ideal shelter for other species, and a single sponge may have thousands of other tiny animals living in it.

Calcarea the class of sponges that contains small, exclusively marine sponges with a skeleton of calcite that often protrudes through the surface of the body. The simplest sponges, they are usually purse- or vase-shaped.

coralline sponges see Sclerospongia.

covering cells see pinacocytes.

excurrent opening see osculum.

feeding (in sponges) is by extracting small particles from the surrounding water.

flagellum (plur. **flagella**) a whip-like projection of special cells in chambers of the canals; the beating of the flagellae draws water into the sponge through pore cells. These flagellate cells are also responsible for removing food particles from the water.

Demospongia the largest class of sponges and also the only one with both freshwater and marine species. These sponges have a variety of different types of spicule.

gemmules asexual reproductive bodies formed by freshwater sponges to withstand times of drought or lack of food.

glass sponges see Hexactinellida.

Hexactinellida the glass sponges; deep-water species that have complex silica skeletons often in intricate and beautiful structures.

incurrent pore pore cells in the external surface of a sponge through which water is drawn by the beating of the flagella of special cells lining chambers opening off the canals.

pinacocytes the covering cells of a sponge.

Porifera the phylum to which all sponges belong: there are four classes - Calcarea, Hexactinellida, Demospongia, and Sclerospongia.

porocyte see pore cell.

osculum the single excurrent opening at the top of a sponge through which filtered water, along with any waste products, is discharged.

reproduction (of sponges) is sexual. Eggs contained within the body of a female sponge are fertilized by sperm drawn in with the feeding currents. Free-swimming larvae develop and are then released into the sea. These disperse, settle on suitable surfaces, and metamorphose into young sponges. If two young sponges of identical genetic composition settle close to each other and come into contact as they grow, they will fuse together and become a single organism. Equally if a sponge becomes separated into pieces each will develop as an independent animal.

Sclerospongia the coralline sponges which were thought to be extinct until they were found to exist in deep water in the 1960s.

sessile the condition of adult sponges; that is they spend their lives attached to rocks or other hard surfaces.

skeleton (of sponges) may be a complex and highly specific structure, for example in the species known as Venus's flower baskets where an intricate skeleton of interwoven six-rayed spicules occurs. The skeleton makes sponges unattractive as food and few other creatures feed on them. Some species, however, have no skeleton at all.

spicules the pin-like or complex star-shaped rods of silica or calcite that form the skeleton of some sponges. Other sponges have a combination of spicules plus a meshwork of fibrous protein called spongin.

spongin a meshwork of fibrous protein that entirely or almost entirely makes up the body of a sponge (e.g. bath sponges) or which occurs in combination with spicules.

structure (of sponges) is typically a honeycomb-like structure of canals separated by cells and a hard skeleton.

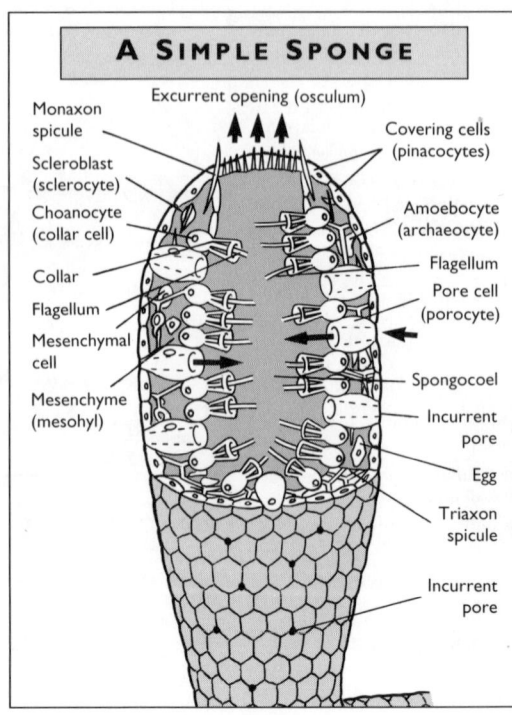

A SIMPLE SPONGE

BIOLOGICAL SCIENCES

COELENTERATES OR CNIDARIANS

The simplest of the animals with differentiated cell layers (and the only group with just two such layers) are the coelenterates, now usually called cnidarians. There are three classes within the phylum Cnidaria, with a total of some 9000 species: Scyphoza, Anthozoa and Hydrozoa. The name Cnidaria indicates the presence of stinging cells. Cnidarians are often colonial, forming flower-like structures or complex free-floating colonies. Some species build up colonies of vast proportions; amongst these are the largest living structures on earth - the coral barrier reefs of Australia. All cnidarians have a simple contractile body wall of two cell layers separated by a jelly-like layer, the mesoglea. Two basic coelenterate body plans exist: the polyp and the medusa.

alternation of generations a characteristic of some cnidarians in which one form succeeds the other, with the medusa (see below) acting as the dispersal stage and the polyp (see below) serving as a following 'fixed' generation. Most cnidarians, however, have only one form or the other.

Anthozoa the class of cnidarians that includes the sea anemones and the soft and hard corals. Anthozoans are sessile for most of their lives.

coral any of numerous marine anthozoans, the majority of which are colonial and secrete skeletons that are horny (octocorals) or calcareous (stony or reef-forming corals). The octocorals include sea fans, sea whips, sea pens and pipe corals, which are polyps bearing eight pinnate (feather-like) tentacles, but they vary very considerably in shape.

hydra any small freshwater hydrozoan of the genus *Hydra*, common in ponds and streams. The polyp phase is dominant throughout the life cycle.

Hydrozoa the class of cnidarians that includes the familiar freshwater hydra as well as many simple or branched forms found in the sea. They include the siphonophores.

jellyfish any free-swimming member of the class Scyphozoa. The medusa stage forms the principal part of their life history. Most inhabit coastal waters and are particularly prolific in summer when huge shoals may occur. Some species reach 60cm (24in) in diameter.

medusa (plur. **medusae**) a medusa is essentially an inverted free-swimming polyp, with a central downward-pointing mouth. Medusae are bell- or umbrella-shaped with long tentacles fringing the bell.

mesoglea a jelly-like layer that separates the two cell layers of all cnidarians.

nematocysts see tentacles.

octocorals see coral.

pipe coral see coral.

polyp a cnidarian that comprises a simple sac enclosing the gut cavity, with a single opening surrounded by a number of short tentacles. This opening serves both for the intake of food and for the expulsion of wastes; the end opposite the mouth is attached to an external surface.

Portuguese man-of-war see siphonophore.

reproduction (of cnidarians) is most usually asexual, by means of budding or fission. This has aided the creation of very large colonies of coral. (See also alternation of generations (above).)

Scyphozoa the class of cnidarians that includes the free-swimming jellyfish.

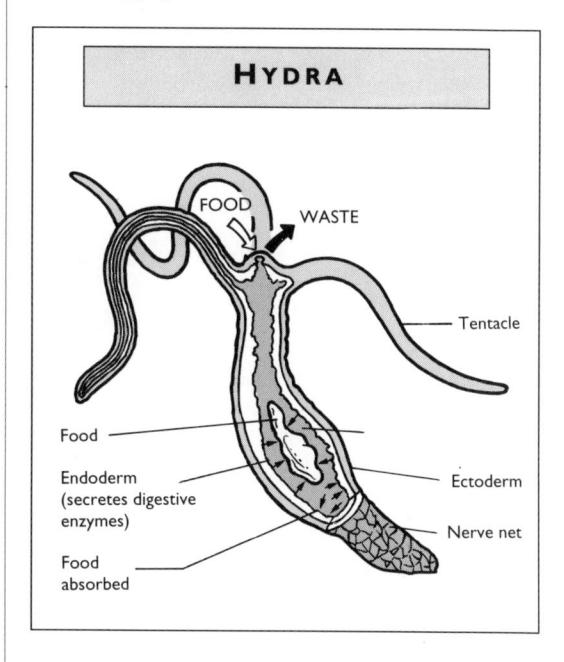

HYDRA

Labels: FOOD, WASTE, Tentacle, Food, Endoderm (secretes digestive enzymes), Food absorbed, Ectoderm, Nerve net

sea anemone any anthozoan of the order Actinaria. They are never colonial and spend their entire lives as polyps. Widely dispersed in the oceans of the world, they are normally sessile and dependent upon their food coming to them although most can crawl and a few are capable of swimming. Sea anemones can produce sexually and asexually.

sea fan see coral.

sea pen see coral.

sea whip see coral.

siphonophore any large free-swimming colonial hydrozoan of the order Siphonophora, for example the jellyfish-like Portuguese man-of-war. They display extreme polymorphism with different individuals performing different functions within the colony; some medusae pulsate to provide the propulsion for swimming, others have a purely reproductive function.

tentacles slender, elongated flexible structures of cnidarians. Members of this phylum are largely predatory and use their tentacles to capture food. While the tentacles can be contracted and may be used to grip prey, it is the batteries of unique gripping and stinging cells (nematocysts) carried on the tentacles that are the principal weapon.

ECHINODERMS

The animals of this group are extremely distinctive, since the most striking feature that they have in common - five-part radial symmetry - is virtually unparalleled in the animal kingdom. Although exclusively marine, echinoderms form an extremely successful group, often dominating the communities in which they occur, and they are found in every marine habitat, from intertidal zones to the greatest ocean depths.

There are some 5000 species in the phylum Echinodermata, which is subdivided into six classes. As well as five-part symmetry, the common features of the group are an internal skeleton based on calcite (a mineral composed of calcium carbonate) and tube feet. Frequently echinoderms are very spiny (their scientific name means spiny skinned).

Asteroidea the class of echinoderms that includes the starfish.

brittlestar (or **serpent star**) any member of the class Ophiuroidea. Brittlestars have long, thin, spiny arms that are flexible and linked to a compact central disc. They are very mobile and found in huge numbers in all marine habitats. Their common name reflects their readiness to cast off arms and regrow them.

calyx the huge crown formed by the arms of a sea lily.

Crinoidea the class of echinoderms that includes sea lilies.

Echinoidea the class of echinoderms that includes the sea urchins.

feather star another name for sea lily (q.v.).

Holothuroidea the class of echinoderms that includes the sea cucumbers.

Ophiuroidea the class of echinoderms that includes the brittlestars.

pedicellarium (plur. **pedicellaria**) small pincer-like organs found on echinoderms. The pedicellaria are used in defence.

reproduction (of echinoderms) is mainly sexual. Echinoderms disperse through the free-swimming larvae, but they are also able to multiply their numbers by asexual means. Remarkable powers of regeneration are characteristic of the group: individuals that split in half or are severely damaged can replace missing parts without difficulty.

sea cucumber any member of the class Holothuroidea. Worm- or sausage-like in appearance sea cucumbers have little internal skeleton and lack spines. Although they can become rigid by altering the state of their internal tissues, they are normally soft and extremely flexible; this allows a range of movement and some can swim actively. They live on the sea bottom, feeding by means of tentacles that surround the mouth. Some are sedentary.

sea lily any member of the class Crinoidea. The earliest forms of echinoderm are thought to have resembled sea lilies or feather stars, the most primitive living echinoderms. Two groups of sea lilies exist today: stalked, sedentary forms that are found exclusively in deep water and mobile, stalkless forms occurring principally in shallow water. Both feed by sieving small particles from surrounding water. As a large surface is required for effective sieving, sea lilies have evolved numerous branched arms and may have 10, 20 or 40 arms to create a huge calyx.

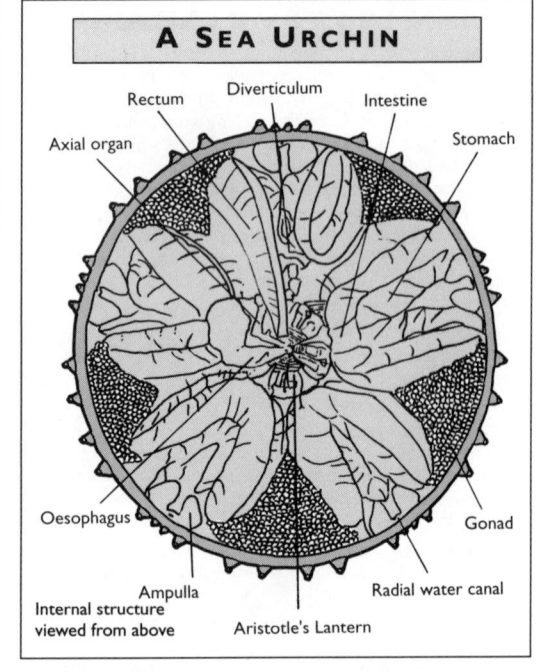

A SEA URCHIN

Internal structure viewed from above. Labels: Rectum, Diverticulum, Intestine, Stomach, Axial organ, Oesophagus, Gonad, Ampulla, Radial water canal, Aristotle's Lantern.

sea urchin any member of the class Echinoidea. Sea urchins come in a variety of sizes and shapes - globular, oval and disc-shaped. They have a complete skeleton (or test) of strong linked plates and are densely covered with mobile spines used in locomotion and defence, and with small pedicellaria. Both spines and pedicellaria may be poisonous. Their tube feet are arranged in five rows around the test and variously used in movement, food collection, attachment and sensory perception. Some live on hard surfaces and graze on small plants which they consume using a complex jaw arrangement, called Aristotle's lantern.

serpent star see brittlestar.

starfish any member of the class Asteroidea. The familiar starfish usually have five hollow arms linked to an ill-defined central disc. Some, however, can have six or more arms, while sunstars can have as many as 40. A few reach 1m (39in) in diameter, but most are very much smaller. Small spines and pedicellaria cover the upper surface, while the arm edges are often set with large defensive plates. On the underside are five rows of tube feet stretching from the central mouth to the arm tips. The tube feet are used for walking, digging and grasping prey, and those near the arm tips are sensory. Starfish are mostly scavengers or predators, finding their food by smell.

sunstar a starfish (q.v.) with many arms.

test the complete skeleton of a sea urchin.

tube feet small, hollow walking and feeding organs that are characteristic of all echinoderms. Tube feet are often equipped with suckers and are connected to a unique system of internal body cavities.

BIOLOGICAL SCIENCES

WORMS

The many invertebrate animals known as 'worms' are, in fact, of very diverse origins; what they share is a typically slender elongated body form, not a common ancestry. Worms are successful animals, and they are extremely numerous in a wide range of habitats. Many are parasitic, but there are also a number of free-living forms occurring in the soil and in aquatic habitats. Worms (and all 'higher animals') are composed of three cell layers, which allows greater structural sophistication. The digestive system of most worms is a continuous tube with a mouth and an anus.

annelid any member of the phylum Annelida, the segmented worms. Annelids have long, thin bodies with distinct front and hind ends. The body is made up of a series of separate segments, each of which can act semi-independently and usually bears parapodia and/or chaetae. Internally an extensive body cavity separates the gut from the muscular body wall, and the nervous, muscular and circulatory systems are well developed. Annelids occur in three highly specialized classes.

bristleworms the polychaetes (q.v.).

Cestoda the flatworm class that includes the tapeworms.

chaetae hairlike protrusions found on annelids.

cilia tiny hair-like projections found on the sides of worms.

earthworm any member of the annelid class Oligochaeta. Earthworms and their freshwater relatives burrow their way through earth or detritus, feeding on decaying vegetable matter. They are long, thin worms without external structures except retractable chaetae (used in burrowing) and a reproductive organ (the clitellum).

eelworm a nematode (q.v.).

fan worms see polychaetes.

flatworm any member of the phylum Platyhelminthes. Flatworms are simple, bilaterally symmetrical animals with an upper and a lower surface. They have a distinct head with a simple brain and simple organ systems. Food is taken in and waste is expelled through a single opening. The simple structure of these worms means that they can regenerate lost parts with ease.

fluke see trematode.

Hirudinea the annelid class that includes the leeches.

hookworm a nematode (q.v.).

leech any member of the annelid class Hirudinea. Leeches are the most specialized annelids. Predatory or parasitic, they are common in aquatic habitats and in moist terrestrial habitats. Suckers at each end of the body are used to grasp prey and in the characteristic movement pattern of these animals. They have a well-developed sense of smell. Parasitic leeches have saliva that contains an anaesthetic to prevent detection and an anticoagulant so that blood remains fluid in the gut.

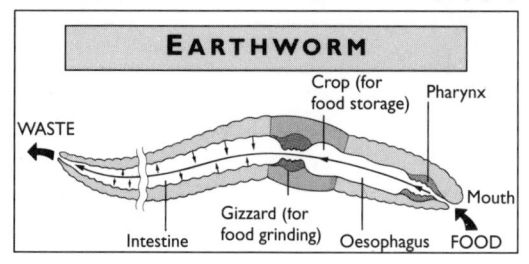

lugworms see polychaetes.

nematode any member of the phylum Nematoda. The phylum includes eelworms, threadworms and hookworms. These worms have long, thin bodies that are sharply-pointed at one or both ends and a thick, resistant cuticle ('skin'), which is often transparent. They range in size from the microscopic to 1m (39in) long. Unlike flatworms, nematodes have a second opening to the gut (an anus) and internal fluid-filled vacuoles separating the body wall and the gut, allowing both to function more efficiently. This fluid makes the body stiff and they can only perform writhing movements. About 15 000 species have been described but very many more are thought to exist. Many are parasitic to plants or animals.

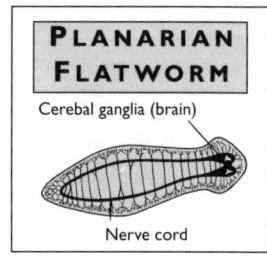

Oligochaeta the annelid class that includes the earthworms.

paddleworms see polychaetes.

parapodia limb-like protrusions found on annelids.

planarians the largest group of turbellarians, characterized by a central mouth on the underside.

Platyhelminthes the phylum that contains the flatworms (q.v.).

polychaetes (or **bristleworms**) members of the annelid class Polychaeta which includes paddleworms, lugworms, ragworms and fan worms. They are characterized by a pair of crawling limbs (parapodia) on each segment. Most live on or near the sea bed.

ragworms see polychaetes.

roundworms see nematodes.

segmented worm see annelid.

tapeworm any member of the flatworm class Cestoda. Tapeworms are ribbon-like worms living a parasitic life in the guts of vertebrates. Their bodies are made up of many repeated, egg-producing segments, and some may reach several metres long. They are unusual in having no gut, all food being absorbed through the body wall.

threadworm a nematode (q.v.).

trematode (or **fluke**) any member of the flatworm class Trematoda. Similar in shape to the turbellarians, trematodes live as external and internal parasites equipped with suckers to grip their hosts. They have complex life cycles, usually involving an aquatic snail and at least one other species. Often hermaphrodites, they produce a series of larvae, each of which may reproduce asexually.

turbellarian any member of the flatworm class Turbellaria. They are free-living, small, flattened, leaf-like worms found in aquatic and damp terrestrial habitats, where they live as predators or scavengers. They move by beating cilia in a fine mucus. They are hermaphrodite -each individual has both male and female organs - and reproduction is usually sexual.

PLANARIAN FLATWORM
- Cerebral ganglia (brain)
- Nerve cord

EARTHWORM
- WASTE
- Crop (for food storage)
- Pharynx
- Mouth
- FOOD
- Oesophagus
- Gizzard (for food grinding)
- Intestine

MOLLUSCS

The phylum Mollusca is a highly diverse group that encompasses seven classes and contains over 50 000 species. Although each class of molluscs has become specialized for particular ways of life, there are a number of common characteristics, some or all of which are found in all molluscs. These include a broad locomotory foot, a protective shell, a radula, and a ctenidia.

bivalve any member of the class Bivalvia, which comprises molluscs whose shell is made of two parts (or valves) connected by an elastic hinge. The group contains about 14 000 species including clams, mussels, oysters and cockles. Bivalves are essentially living filter pumps, with no radula and no distinct head. Large volumes of water are drawn into the shell through a siphon by the rhythmic beating of thousands of tiny projections (cilia) on the massive gills, and small floating particles are sieved off as the water passes over the gills. Few bivalves can swim; most live permanently attached to rocks or some other surface.

byssi the strong threads by means of which some bivalves attach themselves to rocks.

cephalopod any member of the gastropod class Cephalopoda, which includes squid, cuttlefish and octopuses. These animals have evolved into highly specialized predators - they are fast-moving, alert and more intelligent than any other invertebrates. They grow quickly but reproduce only once before dying. All cephalopods have the ability to change colour.

clam any burrowing marine or freshwater bivalve of the genus *Venus*.

cockle the name given to some 250 species of bivalve mollusc; they are found in estuarine conditions.

ctenidia the special respiratory gills of molluscs.

cuttlefish any marine cephalopod mollusc of the genus *Sepia*; predatory animals, they are often found in shallow inshore waters. Cuttlefish are similar to squid but are inshore species that swim near the bottom. Unlike squid, cuttlefish have an internal shell, which acts as a flotation device.

gastropod any member of the class Gastropoda, the largest group of molluscs. Gastropods include such diverse creatures as slugs, snails, periwinkles, topshells, and limpets. Members of this group have exploited not only marine and freshwater habitats but also terrestrial ones. Their diversity also extends to feeding habits and structure. Among the 35 000 or so species there are herbivores, carnivores, parasites and particle-feeders; they range in size from tiny forms to giant conchs that can be up to 70cm (28in). All gastropods have a muscular foot for walking or swimming, and almost all have a radula. In carnivorous species the radula has sharp pointed teeth for tearing, while in the specialized cone shells it is modified into a single harpoon-like weapon armed with a poison gland. Most gastropods have external shells, which are strong and versatile. Usually there is a hard, outer protein layer, a mixed protein-chalk layer, and an inner chalk layer. Some shells are simple cones, but most are complex spirals. Most gastropods reproduce sexually, producing free-swimming veligers (q.v.), but some lay large eggs protected by tough capsules, which develop directly into miniature adults.

mussel any bivalve mollusc of the family Mytilidae. They attach themselves to rocks by means of byssi. Principally intertidal, mussels - an important source of food - are found worldwide.

Mytilidae the order that contains the mussels.

octopus any marine cephalopod mollusc belonging to the order Octopoda. Mainly bottom-living predators, octopuses differ from squid and cuttlefish in having eight tentacles. Most octopuses are relatively small although some reach up to 1.8m (6ft) in length. They have extraordinarily sophisticated nervous systems.

oyster the name given to a number of sedentary bivalve molluscs especially those of the family Ostreidae. Oysters live firmly attached to hard surfaces such as rocks. They are cultivated for food and for pearls.

Pulmonata the gastropod subclass that contains the snails.

radula the tongue-like feeding organ of molluscs.

sea slug (or **nudibranch**) any predatory marine gastropod of the order Nudibranchia. They lack an external shell and are brightly coloured.

slug any terrestrial gastropod mollusc in which the shell is absent or greatly reduced.

snail any of the numerous gastropod molluscs, most of which belong to the subclass Pulmonata.

squid a marine cephalopod mollusc. Squid have long, streamlined bodies and no shell. At one end is the head, which bears two large eyes and ten tentacles - developed from the modified foot. Squid - which have well-developed sensory systems - are fast-moving predatory animals, found in most seas.

valve either of the two parts of the shell of a bivalve (see above).

veliger the free-swimming larva produced by gastropods.

Two kinds of bivalve molluscs showing internal structure — Protobranch (Ligament, Palp lamellae, Posterior adductor, Pedal retractors, Mouth, Anus, Foot, Protobranch gill, Palp proboscis, Siphons); Lamellibranch (Lamellibranch gill, Pedal retractor, Palp, Mouth, Foot, Anterior adductor).

BIOLOGICAL SCIENCES

ARTHROPODS

Judged by the number and variety of species, the arthropods are the most successful of all living creatures. Crustaceans, such as crabs and shrimps, are the dominant arthropods in the sea (see p. 44), while on land insects (see pp. 42-43) can be found in most available habitats. Albeit to a lesser degree than insects, the other important arthropod groups - arachnids (spiders, mites and scorpions; see p. 45), and myriapods (millipedes and centipedes) - are highly successful animals.

Whatever their evolutionary origins, the arthropods represent a highly diverse collection of animals. As a group they are characterized by having external skeletons and pairs of jointed limbs.

centipede any of some 2800 predatory myriapods, characterized by a body comprising 15 or so segments, each bearing a pair of legs.

chitin a complex molecule that is the base of the cuticle. Chitin is based on long-chain molecules of the sugar acetyl glucosamine; these become cross-linked to give tough fibres, which are embedded in a protein matrix.

cuticle a protective layer of the exoskeleton. The cuticle covers the animal's soft underlying parts. The cuticle is based on chitin.

ecdysis an alternative name for moulting in insects.

evolution (of arthropods) has been vigorously debated. The fact that all arthropods have segmented exoskeletons and jointed limbs was once thought sufficient to show that they had evolved from a single ancestral stock in which these features occurred. However, recent comparative analyses of the chemical composition of arthropod skeletons and of the detailed structure of joints, mouthparts and legs suggest that this may not be so - that several distinct evolutionary lines may have branched off from a segmented worm-like ancestor and that the basic arthropod features evolved several times over.

exoskeleton the external skeleton of all arthropods. The success of the arthropods lies in the external skeleton, which both supports and protects the animal. The exoskeleton is formed by the cuticle. When mixed with differing proportions of proteins and inorganic salts, such as calcium carbonate, the arthropod skeleton shows remarkable versatility, varying from the soft elastic bag that surrounds a caterpillar to the immensely hard claws of a crab. The exoskeleton is further strengthened by tanning.

Movement is only possible where the exoskeleton is not hardened, hence the need for joints (q.v.).

Despite its success, the exoskeleton has some obvious drawbacks. The most important of these is that the hard, external skeleton limits the size of the animal (see growth).

growth (of arthropods) can only take place if the old hard exoskeleton is discarded and a new larger one is grown in its place. This process is known as moulting. Moulting has to occur a number of times in all arthropods as they grow to full adult size. At each moult the animal is vulnerable to attack by predators, parasites or disease and - in the case of terrestrial species - the danger of dehydration. Much of the old exoskeleton is absorbed and re-used in the new one, but even so, at each moult there is a loss of old cast 'skin'. The need to moult may also influence the absolute size that arthropods can reach, since in the course of moulting an arthropod must be able to hold its shape until the new cuticle has hardened.

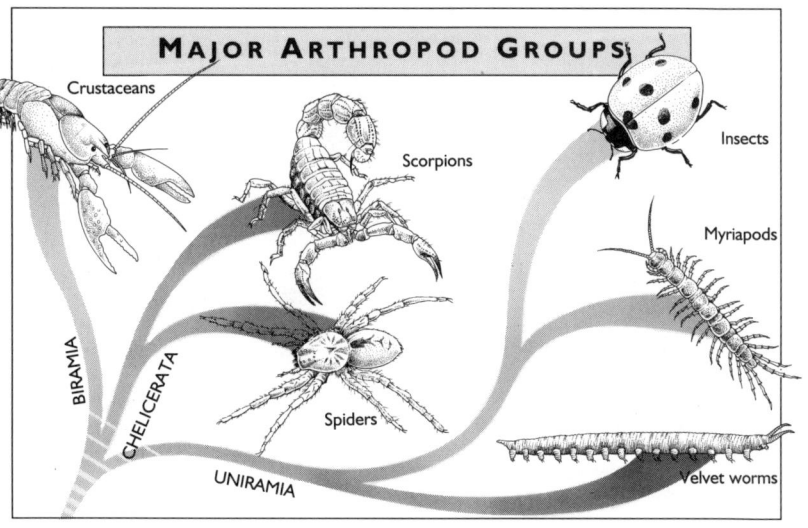

joints (of arthropods) are composed of flexible cuticle and allow as varied a set of movements as possible, as varied, in fact, as those found in animals with internal skeletons.

millipede any of some 8000 herbivorous myriapods, characterized by a body comprising numerous segments, each bearing a pair of legs.

moulting (or **ecdysis**) see growth (of arthropods).

myriapod name given to several groups of arthropods characterized by a long segmented body and many legs, the most important of which are centipedes and millipedes.

tanning a process that hardens the exoskeleton of arthropods. The darkening of the exoskeleton that commonly accompanies tanning is due to an excess of the organic compound quinone, which turns into the pigment melanin after hardening of the proteins is complete.

velvet worm any arthropod of the class Onchychopora, small, worm-like animals with short joint-less limbs.

INSECTS

An insect is any arthropod of the class Insecta, an enormously successful group comprising several million species (only a proportion of which have so far been described). Conventionally the class is divided into wingless insects (apterygotes, of which there are four orders) and winged insects (pterygotes, of which there are 25 orders), but this is not universally accepted. The body form of the insect has proved to be the most successful of any living organism. According to some estimates, over 80% of all plant and animal species are insects and they have colonized almost every habitat except the sea where crustaceans are the dominant arthropods. The success of insects stems from the light, strong, wax-covered cuticle, which is highly protective, especially against dehydration. The small size of insects has meant that the amount of living space available to them is almost infinite. In general, adult insects are easily recognizable by their three body sections - head, thorax and abdomen.

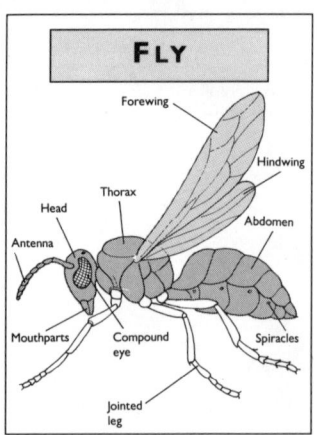

abdomen (of insects) the hindmost section of the body behind the thorax. The digestive and reproductive systems of insects occupy the abdomen. In most insects the abdomen is clearly segmented, and there are sometimes small paired appendages at the end, which are used in reproduction.

antenna (plur. **antennae**) one of a pair of appendages projecting from the head of insects. They are generally highly mobile and have a sensory function, but they are sometimes modified for other purposes such as swimming and defence.

apterygote any insect of the group Apterygota, the 'wingless insects'. They include four orders: thysanurans (bristletails), diplurans, collembolans (springtails) and proturans.

arthropod any member of the phylum Arthropoda, the largest and most important invertebrate group. As well as insects they include crustaceans, arachnids, centipedes and millipedes. See p. 41.

blood system (of insects) a relatively simple system whereby nutrients, waste products, etc. are dissolved and circulated around an insect's body. The organs float in a blood-filled cavity called the haemocoel. The blood (or haemolymph) is stirred by a tubular heart, which lies along the dorsal (back) side of the abdomen. A single blood vessel extends forwards from the heart to the head. The blood is pumped to the head and then percolates back through the thorax to the abdomen. The blood carries nutrients, waste products and carbon dioxide, and also contains corpuscles that counter invading bacteria, but (in contrast to vertebrates) it does not carry oxygen.

caterpillar the larva of a butterfly or moth.

clasper a male copulatory structure in some insects.

cuticle (in arthropods) a protective hard layer covering the epidermis of arthropods, including insects. It is also known as the epidermis.

digestion (of insects) the process whereby complex food molecules are broken down by insects into simpler compounds that can be absorbed by them and used as a source of energy. The insect gut consists of three sections. The foregut starts at the mouth, passes through the thorax as a narrow oesophagus and then swells in the abdomen to form a crop for food storage. The midgut, which often has pockets lined with secretory cells, is the place where digestion and secretion of food occur.

exoskeleton a tough outer covering made of chitin (a polysaccharide) consisting of the cuticle and the epidermis, and used to prevent water loss and to attach muscles internally.

endopterygote any pterygote insect of the group Endopterygota, comprising nine orders and over 85% of all known insect species. Familiar endopterygotes include beetles, butterflies and moths, flies, and bees, ants and wasps. The group is characterized by complete metamorphosis, i.e. juveniles bear no similarity to adults, and the adult form is attained via larval and pupal stages.

epidermis see cuticle.

excretion (of insects) the process whereby waste products are removed from an insect's body fluids and expelled from the body. An insect's excretory organs, equivalent to our kidneys, are the Malpighian tubules, which are connected to the alimentary canal at the junction of the mid- and hindguts. In the hindgut water is extracted from waste food and excretory products. The resulting mixture of dry faecal and excretory material is expelled through the anus.

exopterygote any pterygote insect of the group Exopterygota, comprising 16 orders and roughly 12% of all known insect species. Familiar exopterygotes include grasshoppers, dragonflies, earwigs and aphids. The group is characterized by incomplete metamorphosis, i.e. juveniles resemble adults and there is no pupal stage.

ganglia (sing. **ganglion**) an encapsulated group of nerve-cell bodies allowing coordination between pathways of the nervous system.

head (of insects) the front section of the body of an insect. It usually has a pair of compound eyes for vision and often two or three small single eyes (ocelli), which respond to varying light levels. Two antennae and two pairs of palps at the mouth provide information on both touch and taste. The mouthparts are typically adapted for biting, but in many species they are modified for other functions. Within the head is a bilobed brain, which closely connects with the optic lobes of the compound eyes; two nerve cords pass from the brain to the first of a series of segmentally arranged ganglia that extend along the ventral (front) side of the insect.

imago an insect in its final mature winged state.

larva (plur. **larvae**) an insect in an immature but active state, often markedly different in structure and life style from the adult. Larvae typically undergo a dramatic metamorphosis to become adult.

legs (of insects) insects have three pairs of legs attached to the lower part of the thorax. Some of the leg muscles are in the limbs themselves, but there are others within the thorax; indeed in some insects muscles are shared by the wings and the legs. Insects' legs are highly varied in form, ranging from the strong digging forelegs of the mole cricket to the long hindlegs of grasshoppers, which are adapted for jumping. Aquatic insects have legs adapted for swimming - diving beetles and water boatmen, for instance, have flattened hindlegs fringed with long hairs, which are used like oars.

Malpighian tubules the excretory organs of insects.

metamorphosis a radical structural transformation in an animal. In insects two kinds of metamorphosis are recognized, endopterygotes and exopterygotes. In insects the adult form is not reached by progressive growth (as it is, for example, in mammals). Instead, intermediate larval stages occur, which differ radically from the adult in both structure and life style. In such cases the adult stage is attained by means of a major transformation of structure (metamorphosis). A number of larval stages may follow each other before the adult stage is reached. A caterpillar is a larva that metamorphoses during the pupal stage to a butterfly.

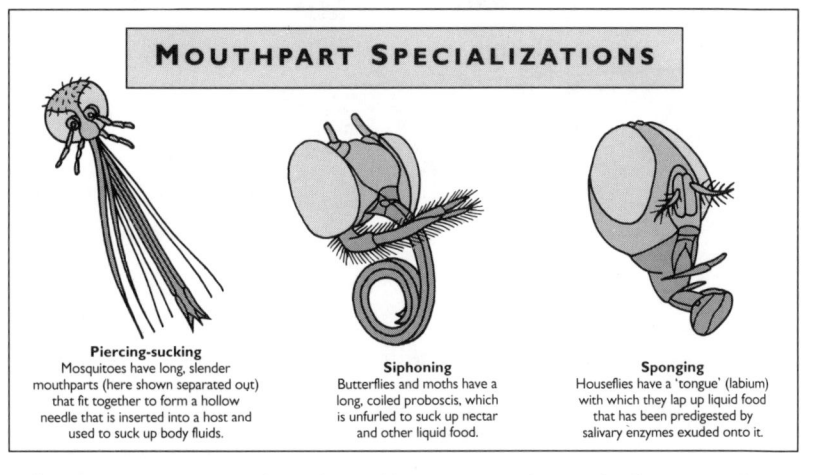

ocellus (plur. **ocelli**) a simple eye found in some insects.

ovipositor the egg-laying structure of female insects, at the tip of the abdomen. In some insects (bees, wasps, etc.) it is modified into a sting.

palp a segmented sense organ of touch and taste, paired and comprising some of the mouth parts of insects.

pterygote any insect of the group Pterygota, the 'winged insects', accounting for the vast majority of living insect species. The group is subdivided into the exopterygotes (16 orders), such as grasshoppers, dragonflies and aphids, which undergo incomplete metamorphosis; and the endopterygotes (nine orders), such as butterflies and bees, which undergo complete metamorphosis. Although belonging to fewer orders, the endopterygotes are far more numerous in terms of species.

pupa the intermediate, usually inactive, form of an insect between the larval and imago (adult) stages.

reproductive systems (of insects) the system of associated organs (genitalia) by means of which insects are able to reproduce. The reproductive systems of insects are located in the abdomen. Males have claspers with which to hold the female during sperm transfer, while females often have a structure called the ovipositor, which is used to place fertilized eggs in chosen sites, from crevices in bark to the tissues of another insect.

respiration (of insects) see spiracle.

spiracle any of a series of small, paired openings in the cuticle of an insect by which air enters and leaves the tracheae. While this system is clearly successful in insects, it would be unsuitable for larger animals since the spiracles and tracheae would have to be unmanageably large.

sting a sharp organ in insects such as bees and wasps which is adapted to wound a predator or prey by piercing and injecting poison. The sting is formed from the ovipositor.

thorax the section of an insect's body between the head and the abdomen to which the legs and wings are attached. The thorax is the locomotory centre. It consists of a rigid box containing the muscles that operate the two pairs of wings. Three pairs of legs are attached to the lower part of the thorax.

trachea any of the numerous tiny tubes by which air is conveyed from the spiracles, via even smaller tubes (tracheoles) to the tissues.

winged insects see pterygote.

wingless insect see apterygote.

wings (of insects) the flight structure of insects. The wings are usually thin yet strong, with the wing membrane stiffened by a network of veins. Some insects move their wings by contraction of muscles attached directly to them, while others use an indirect 'click' mechanism. Wings may have other functions besides flight. In some insects, such as beetles and grasshoppers, the forewings are hardened, covering and protecting the large hindwings underneath. Crickets use their wings to produce sound, while in many insects the wings are coloured or patterned and used in courtship display, or as camouflage.

MOUTHPART SPECIALIZATIONS

Piercing-sucking
Mosquitoes have long, slender mouthparts (here shown separated out) that fit together to form a hollow needle that is inserted into a host and used to suck up body fluids.

Siphoning
Butterflies and moths have a long, coiled proboscis, which is unfurled to suck up nectar and other liquid food.

Sponging
Houseflies have a 'tongue' (labium) with which they lap up liquid food that has been predigested by salivary enzymes exuded onto it.

CRUSTACEANS

Crustaceans have been called 'the insects of the sea', for their success as marine invertebrates is comparable to that of insects on land. They are the major component of the marine zooplankton - the vast mass of small animals that drift through the upper waters of the sea - and they are extremely abundant elsewhere. The freshwater species, though numerous, suffer competition from insects and are less abundant. A very few - land crabs and woodlice - have successfully adapted to life on land. The crustaceans (subphylum Crustacea) form a very varied group of arthropods with some 42 000 species divided into 10 classes. Of these the most important are the fairy shrimps (Branchiopods), copepods (Copepoda), crabs, shrimps and prawns (Decapoda),

ANATOMY OF A LOBSTER

Labels: Stomach, Anterior aorta, Testis, Heart, Posterior aorta, Gut, Compound eye, Anus, Telson, Antenna, Antenna exopodite, Mandible, Nerve cord

sand hoppers (Amphipoda), and woodlice (Isopoda). Crustaceans are typically elongated animals with segmented bodies clearly divided into a cephalothorax and a tail or abdomen. Each segment bears a pair of limbs.

amphipods small shrimp-like crustaceans found in enormous numbers in marine and freshwater habitats.

barnacle any marine crustacean of the class Cirripedia. Most live as sedentary filter-feeders, settling in groups glued down by means of head glands. Some are parasitic on crabs and jellyfish.

branchiopod see fairy shrimp.

carapace a protective shell plate, which is made of chitin strengthened with calcium carbonate to form a hard shell.

cephalothorax the head/thorax region of a crustacean. It is often protected by a carapace. The head segments bear sensory antennae, compound eyes and mouthparts. Advanced crustaceans, such as crabs, have strong mouthparts well suited to chewing and biting and many are predatory.

chitin a complex sugar molecule, the principal component of arthropod exoskeletons.

copepod any minute marine or freshwater crustacean of the class Copepoda, a major constituent of the zooplankton and the most numerous of crustaceans.

crab any of more than 4500 decapod crustaceans of the genus *Cancer*. They have a reduced abdomen that is concealed beneath a short broad cephalothorax, and the first pair of limbs is modified as pincers.

crayfish any freshwater decapod crustacean of the genera *Astacus* and *Cambarus*. They are omnivorous, are found worldwide and are widely farmed.

cuticle a protective hard layer.

decapod any member of the class Decapoda, the largest and most important group of crustaceans with around 10 000 species. The group includes crabs, spider crabs, shrimps, lobsters, prawns and crayfish.

fairy shrimp (brine shrimp) any small, primitive crustacean of the class Branchiopoda, found in abundance in temporary pools. They are classic primitive crustaceans with rows of similar limbs used for swimming and feeding. They sieve particles from water using hairs on the limbs.

growth (of crustaceans) is by moulting. In this hormonally controlled process a new, soft cuticle is laid down beneath the old one, which by this time has weakened by reabsorption of most of its valuable material. As the animal breaks out of the old cuticle, its body swells by intake of air or water to the new size before the new cuticle begins to harden. Moulting of this type occurs periodically throughout a crustacean's life.

isopod any crustacean of the class Isopoda; see woodlouse.

krill any small shrimp-like crustacean of the order Euphausiacea, the principal food of baleen whales.

limbs (of crustaceans) occur in primitive forms as a series of similar jointed structures behind the cephalothorax. In more advanced crustaceans the limbs have different functions and are structurally modified. The limbs on the thorax are generally used for walking, digging or swimming, while the abdominal limbs are usually adapted for respiratory or reproductive functions. The front limbs of crustaceans often take the form of claws specialized for food capture and defence.

lobster any large decapod crustacean of the genus *Homarus*. All lobsters have the first pair of limbs modified into pincers.

prawn the name given to numerous decapod crustaceans, especially those of the genus *Palaemon*. They typically have flattened bodies and two pairs of pincers. See also shrimp.

shrimp the common name given to numerous small decapod crustaceans, especially those of the genus *Crangon*. Shrimps typically have flattened transparent bodies and a fan-like tail. The difference between shrimps and prawns is imprecise.

spider crab any decapod crustacean of the family Maiidae, with a small triangular body and extremely long legs.

water fleas a group of tiny freshwater crustaceans, closely related to the fairy shrimp.

woodlouse any crustacean of the class Isopoda, the principal group of terrestrial crustaceans. They have flattened bodies, simple antennae and seven pairs of similar legs. They live as scavengers. The closely related species found on rocky shores are usually known as slaters.

BIOLOGICAL SCIENCES

ARACHNIDS

With over 60 000 species, the arachnids (class Arachnida) are the largest group of arthropods after the insects (see pp. 42-43). Scorpions, spiders and ticks are the most important members of this group, which also includes harvestmen and pseudoscorpions. Although far less numerous than insects, many arachnids are common and familiar terrestrial animals, and - unlike the insects - several members of the group have colonized marine habitats. Arachnids also have considerable economic importance: spiders play a significant role in controlling insect numbers, while ticks and mites are involved in the transmission of a number of diseases in humans and animals.

cephalothorax a combined head and abdomen in arachnids.

harvestman any long-legged spider-like arachnid of the order Opiliones, found in grassland and other vegetation in tropical and temperate areas. They are often abundant in late summer - hence their name.

mite any tiny parasitic arachnid of the order Acari, including gall mites (family Eriophydae), chicken mites (Dermanyssidae), itch mites (Sarcoptidae), harvest mites (Trombidiidae), spider mites (see below) and water mites (Hydrachnidae).

scorpion any of about 800 predatory arachnids of the order Xiphosura, characterized by large pincers, a squat eight-legged body, and a long abdomen with a prominent pointed sting. Scorpions were among the earliest terrestrial animals, first appearing about 410-355 million years ago. The modern forms are secretive and nocturnal. They sense their prey by means of highly sensitive vibration and odour receptors on the underside. They then wait for it to move into range, seize it with their pincers and inject a paralysing venom, which in some cases may cause death to humans. The female broods the young, usually carrying them on her back until they are large enough to care for themselves.

spider any predatory arachnid of the order Araneae, characterized by eight legs, a large abdomen, and a combined head and thorax (cephalothorax). Of the 30 000 described species most are terrestrial. As well as bearing the four pairs of legs the cephalothorax has two other pairs of appendages - one is a pair of hollow fangs used to inject venom into prey; the other is leg-like in females, and a complicated reproductive structure in males. As many as eight large eyes may be seen at the front end. At the tip of the abdomen there are a number of glands (developed only in females) that are used to produce silks. Fluids secreted by these glands dry as threads as they are forced out through three pairs of external nozzles called spinnerets. Different types of thread are produced - trapping thread and dry threads used to anchor webs. Spiders have evolved many techniques to trap prey - including jumping, ambushing, the use of 'trapdoors' and wrapping in thread.

spider mite any herbivorous mite of the family Tetranychidae, such as the red spider mite, which is a pest in orchards.

tick any tiny parasitic arachnid of the families Ixodidae (hard ticks) and Argasidae (soft ticks) in the order Acari. Many thousands of species have been described and many others await discovery.

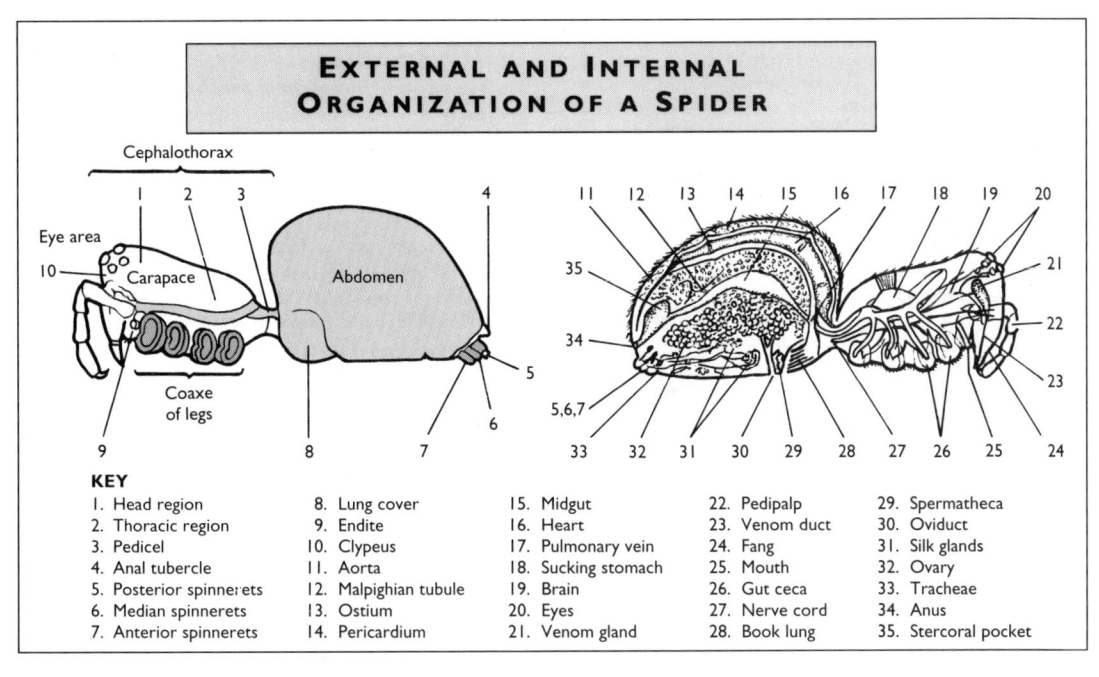

EXTERNAL AND INTERNAL ORGANIZATION OF A SPIDER

KEY

1. Head region
2. Thoracic region
3. Pedicel
4. Anal tubercle
5. Posterior spinnerets
6. Median spinnerets
7. Anterior spinnerets
8. Lung cover
9. Endite
10. Clypeus
11. Aorta
12. Malpighian tubule
13. Ostium
14. Pericardium
15. Midgut
16. Heart
17. Pulmonary vein
18. Sucking stomach
19. Brain
20. Eyes
21. Venom gland
22. Pedipalp
23. Venom duct
24. Fang
25. Mouth
26. Gut ceca
27. Nerve cord
28. Book lung
29. Spermatheca
30. Oviduct
31. Silk glands
32. Ovary
33. Tracheae
34. Anus
35. Stercoral pocket

FISHES

Fishes are cold-blooded vertebrates that belong to either of the two classes Chondrichthyes or Osteichthyes. The number of known fish species is roughly equivalent to the combined total for all the other vertebrate groups put together. The characteristic adaptations of fishes are related to the need to propel themselves through water - a much denser medium than air - and to extract oxygen from water. An elongated, streamlined body shape, a well-muscled hind end of the body and a powerful tail fin are all adaptations to propulsion through water, while the gills absorb oxygen with great efficiency.

actinopterygians the ray-finned fishes (see below).

agnathans the jawless fish; any fish of a primitive group characterized by the absence of (true) jaws, sometimes assigned to the class Agnatha but probably in fact of more diverse origins. All the earliest fossil fishes were jawless, but the group is represented today only by the hagfishes and lampreys.

air bladder another name for the swim bladder.

anal fin the single fin that in most fishes is on the underside of a fish and in front of the tail.

bony fish (or **osteichthyans**) any fish of the class Osteichthyes, one of the major groups of living fishes. Bony fishes are distinguished from jawless fishes and cartilaginous fishes by the possession of an internal skeleton of endochondral bones, that is bone that replaces cartilage during the course of development. The bony fishes are divided into two principal subclasses: the fleshy-finned fishes (lungfishes and coelacanths) and the ray-finned fishes, the latter group comprising the great majority of living species.

cartilaginous fish (or **chondrichthyans**) any fish belonging to the class Chondrichthyes, which includes sharks, rays and chimaeras. They are so named because their entire skeleton is made from cartilage, in contrast to bony fishes. The great majority of species are marine predators, often favouring the relatively shallow waters of the continental margins. Fertilization is internal; some species lay eggs, but many give birth to live young. The cartilaginous fishes are of limited economic importance.

cartilage a firm, translucent tissue formed by special cells called chondrocytes. These cells make up the bulk of the skeleton of young vertebrates but most is converted to bone in adults. Cartilaginous fishes retain a cartilaginous skeleton throughout life.

caudal fin the tail fin of a fish.

Chondrichthyes the class comprising cartilaginous fishes.

cod common name given to many members of the family Gadidae.

dorsal fin the fin that runs down the centre of the back of a fish.

fin any of the projecting structures by which fishes swim, steer, maintain balance, etc. Fins are supported by rays (cartilaginous, bony or horny), which may be soft or hard. The dorsal, caudal (tail) and anal fins are single, while the pectoral and pelvic fins are in pairs.

flatfish any of about 538 (almost exclusively) marine fishes of the order Pleuronectiformes, mainly occurring in relatively shallow waters (under 200m/656ft) of tropical and temperate seas. Young flatfishes are symmetrical, but during development one eye migrates over the head to lie next to the one on the other side; the flattened adult then swims with its eyeless side facing downwards. Usually this is the left side. Most flatfishes lie on the bottom, and many have the ability to change colour to merge with their background. Many species are important food fishes.

gills the respiratory organs used for gas exchange in water by molluscs, amphibians and fish. Since water contains around 30 times less oxygen than air, fishes' gills have to be very efficient to obtain this gas from water. Although a number of shark species have six or seven pairs of gills and some hagfishes have 16, most fishes have four gills on each side of the head, which look like a row of V-shaped bars. Red filaments on the hind edge of the bars extract oxygen from the water flowing over them. Blood passes through the gill filaments in the opposite direction to the flow of water, thus meeting successively 'fresher' water (containing more and more oxygen) as it does so. Thanks to this arrangement the gills are able to absorb up to 80% of the oxygen passing over them. Water for respiration is generally taken in through the mouth and expelled once it has passed over the gills.

gill cavity a cavity which in most teleosts produces a sucking motion. Combined with compression of the water by the mouth, this sucking motion ensures a pressurized flow of water to the gills, so increasing the oxygen supply.

gill raker a structure of the gills of fishes such as mackerel and herring hat sieves plankton from water.

gill slit one or more openings in the pharynx region of fishes, containing the gills.

hering either of two species of the genus *Clupea*.

jawless fishes see agnathans.

lateral-line system a sensory system in fishes involved in the detection of pressure, vibration, etc.

Osteichthyes the class that comprises bony fish.

pectoral fin in most fishes, the pair of fins just behind the head. When turned to face the water, pectoral fins and pelvic fins are used in 'braking'.

pelvic fin in most fishes, the pair of fins centrally-placed on either side of the underside of a fish.

Pleuronectiformes the order that comprises flatfish (see above).

poikilothermic (or **ectothermic**) relating to an animal whose body temperature varies with and is largely dependent on the environmental temperature. Fishes - as well as amphibians and reptiles - are poikilothermic, that is 'cold-blooded'. Fishes do not alter their body temperature to compensate for the temperature fluctuations in their environment as birds and mammals do.

ray any of 250 members of the subclass Selachii.

ray-finned fishes (or **actinopterygians**) any bony fish of the subclass Actinopterygii, the dominant group of living fishes. Apart from a few primitive orders (sturgeon, paddlefish, gar, bowfin, bichir), all modern ray-finned fishes belong to the subdivision Teleostei, the teleosts.

scales disc-like protective overlapping plates that form a covering in the majority of bony fishes. The forward edge of each scale is embedded in the dermis (the inner layer of skin) and all the scales are covered by a continuous layer of epidermis (the outermost layer). Although this is the most common type of scale, highly modified forms are found, for example, in sturgeons and swordfishes. Some groups, such as mackerel, tunnies and eels, have very small scales, while others, including lampreys, some catfishes and very deep-sea fishes, have none at all.

shark any of 250 cartilaginous fish of the subclass Selachii.

swim bladder (or **air bladder**) a gas-filled sac in the upper part of the body cavity of teleost fish. The bladder is either filled with air taken in at the surface of the water via the mouth or charged with gases from the blood by means of a special gas gland. The swim bladder enables fish to achieve buoyancy in the water, thus all locomotory effort can be used to produce and control directional movement.

swimming (in fish) movement through water principally by means of segmental muscle contractions. In most fishes forward propulsion is produced by a series of contractions that cause undulations of the lateral curvature of the body. As these waves pass backwards down the body, the effect is to present a series of moving inclined planes pushing outwards and backwards against the water. These waves are of very small amplitude at the head end, which oscillates only slightly from side to side as a fish swims, but they become more pronounced towards the tail. The tail fin may contribute up to 90% of the total thrust, but some fishes, such as eels, have very small tails and it is the undulations of the body alone that provide the propulsive force. Many fishes depart quite widely from the usual swimming method. Flatfishes, for example, undulate the whole body and/or the vertical fins (which are in the horizontal plane) to produce the waves that generate thrust. Fins can contribute towards propulsion but, in most fish, are generally used for steering and balance.

teleost any ray-finned fish belonging to the subdivision Teleostei, comprising over 20 000 species, or about 95% of living fish species. Teleosts typically have swim bladders, and tails in which the upper and lower halves are equal.

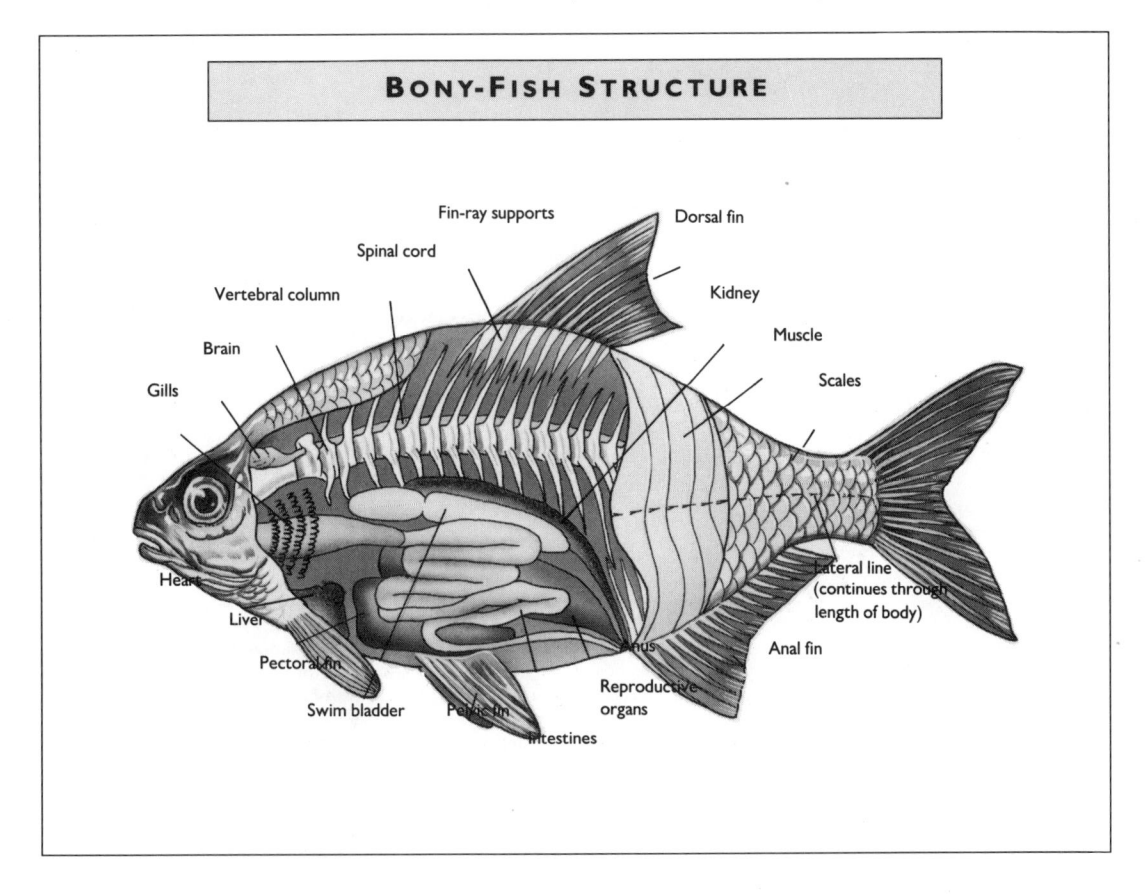

BONY-FISH STRUCTURE

Labels: Fin-ray supports, Spinal cord, Vertebral column, Brain, Gills, Dorsal fin, Kidney, Muscle, Scales, Heart, Liver, Pectoral fin, Swim bladder, Pelvic fin, Anus, Intestines, Reproductive organs, Anal fin, Lateral line (continues through length of body)

AMPHIBIANS

The amphibians played a central role in the evolution of life on earth, since they were the first of the vertebrates to leave the water to live part of their lives on land. This meant that they and those that followed them - reptiles, birds and mammals - were able to colonize a huge array of terrestrial habitats. Modern amphibians are less diverse than their predecessors. They belong to three orders: frogs and toads to the order Anura; newts and salamanders to the order Urodela; and caecilians to the order Gymnophiona.

The basic body plan of modern amphibians reflects their adaptation to life on land and in water. In order to spend at least part if their lives on land, amphibians must minimize the danger of drying out when exposed to air. At the same time, they must be able to obtain and use gaseous oxygen from the air for respiration. Finally, they have had to develop means of moving on land under the full force of gravity, without the buoyancy provided by water. Amphibians are a diverse group, but most share several common features.

Anura the amphibian order containing frogs and toads.

anuran any amphibian of the order Anura, comprising about 24 families and approximately 3500 species of frog and toad (the 'tailless amphibians'). Although anurans are structurally rather uniform, they exhibit a great diversity of habitat, and terrestrial, arboreal, burrowing and totally aquatic forms are known. Most anurans have a free-swimming tadpole stage, but many, including the midwife toads and Darwin's frog, show some form of parental care.

axolotl a species of Mexican salamander which reproduces by neoteny (q.v.).

Bufo bufo the common European toad.

Bufonidae the toad family.

caecilian any amphibian of the order Gymnophiona, comprising some 150 species in 5 families. Caecilians are limbless amphibians occurring in tropical habitats. Most are burrowing, but one group is aquatic. Caecilians have developed streamlined heads and their eyes have become reduced in size and in function.

Caudata an alternative name for the order Urodela.

eggs (of amphibians) are relatively large and yolky. They have gelatinous outer capsules but lack the protective membranes seen in higher animals; see also box.

frog the common name of many Anuran amphibians (the group that also includes toads). The term is often used to distinguish smooth-skinned, short-legged jumping species from warty-skinned, short-legged forms, which are generally called 'toads', but the distinction is imprecise and has no taxonomic significance; frequently loosely related species belonging to the same family are indifferently labelled frogs or toads. In a restricted sense, the name 'frog' may be used for the 'true' frogs of the family Ranidae, such as the common European frog *Rana temporaria*; in a broader sense it is sometimes used to refer to any anuran.

Typical frogs and toads live on or near the ground and take fast-moving prey such as insects and spiders. The large keen eyes, long hind legs to assist the forward thrust of the body, and a large head with a broad mouth and a wide gape have all increased the likelihood of capturing prey.

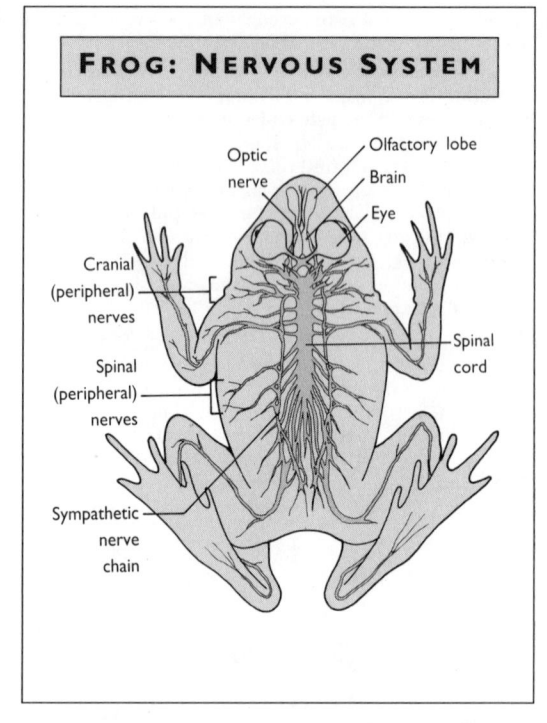

Gymnophiona the order comprising the caecilians.

locomotion (of amphibians) is by swimming in the larval stage and again in the adult stage when amphibians return to the water to breed. Tadpoles swim by lashing their tails from side to side, while aquatic caecilians, newts and salamanders swim in a fish-like manner with an S-shaped wave passing down the body from the head. Most frogs and toads swim by simultaneously thrusting the hind legs against the water. Extensive webbing between the toes assists in this movement. Most frogs and toads have well-developed hind limbs that can be folded and used to leap or hop.

lungless salamander any salamander of the family Plethodontidae; see lungs (below).

lungs (of amphibians) are simple sac-like organs, which are well supplied with blood vessels, permitting gas exchange while limiting water loss. They are usually paired, but in caecilians the left lung is greatly reduced. In some salamanders the lungs are lost altogether - they breathe through the skin and the mouth cavity.

metamorphosis the change from the larval to adult stage; see box. The change is far more dramatic in frogs than in salamanders or caecilians. See also neoteny (below).

BIOLOGICAL SCIENCES

migration (of amphibians) occurs because many frogs, toads and newts show great fidelity to certain breeding sites. They often migrate some kilometres, passing other 'suitable' sites to reach their 'home' pond.

neoteny the ability of some newts and salamanders to breed while still in the tadpole stage. The best known of these sexually mature larval forms is the Mexican axolotl.

newt any of 22 species of highly aquatic salamander, included in the 'true' salamander family Salamandridae. Newts are notable for their highly ritualized courtship behaviour.

nictitating membrane a third eyelid present in all amphibians except caecilians. It keeps the corneal surface clean and moist.

pelvic patch (or **pelvic seat**) an area of baggy skin on the underside of the thighs of toads. Through this patch - which is in contact with the ground when the toad is sitting - a toad may take up to 70% of the water that passes through the skin.

Plethodontidae the family containing the lungless salamanders, the largest salamander family.

Rana temporaria the common European frog.

Ranidae the frog family.

salamander (or **urodele**) any amphibian of the order Urodela (or Caudata), comprising some 350 species of tailed amphibians in 9 families. Salamanders are typically lizard-like in appearance, but they do not have a scaly skin. The term 'salamander' is sometimes used in a restricted sense to refer to members of the family Salamandridae ('true' salamanders and newts). Salamanders live on or near the ground in less open situations than frogs and toads. They take slow-moving invertebrates such as worms, slugs and snails.

Salamandridae the 'true' salamander family; salamanders and newts.

siren any salamander of the family Sirenidae, comprising three species from southern USA and Mexico. Permanently aquatic; they are eel-like, they have external gills and tiny fore limbs but no hind limbs. They are highly distinctive and are sometimes placed in an order of their own.

Sirenidae the siren family.

skin (of amphibians) is hairless and kept moist by mucus secreted from special glands embedded in the skin; it is also permeable and capable of some control over water uptake and loss.

tadpole the aquatic larva of an amphibian; see box.

tails (of amphibians) are universal in the tadpole stage. However, adult frogs and toads lack tails, having lost the s-shaped movement of the larval stage.

toad the common name given to many of the anuran amphibians, generally to warty-skinned species, but the distinction between frogs and toads is often arbitrary. In a restricted sense, the name 'toad' may be used for the 'true' toads of the family Bufonidae, such as the common European toad *Bufo bufo*. See also frog (above).

LIFE CYCLE OF A TYPICAL FROG

1. In a 'typical' frog the fertilized eggs, or spawn, initially have a spherical yolk, which assumes an oval shape after about a week. After ten days the position of the head, body and tail are apparent, and external gills soon appear. At about two weeks the tadpole hatches out from the egg mass and attaches itself to waterweed or some convenient surface by means of an adhesive organ; there is as yet no mouth – it feeds off its yolk supply.

2. Within a few days a mouth is formed and the tadpole feeds on algae, plant material and pond microorganisms. Soon after the mouth appears, internal gills are formed and the external ones are absorbed.

3. At about five weeks hind-limb buds appear, at seven weeks toes are discernible, and by ten weeks proper feet are formed.

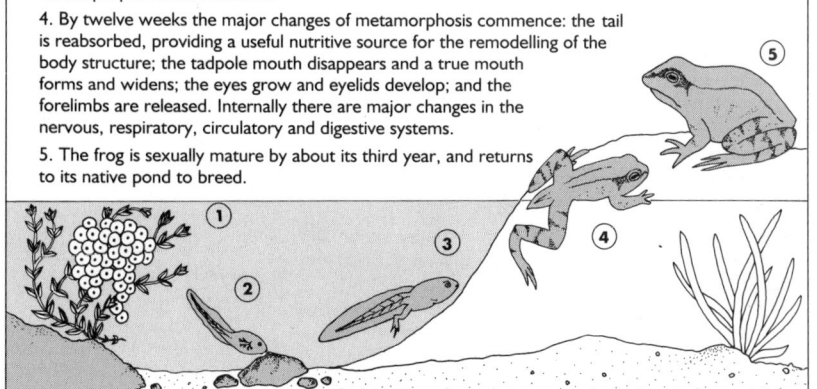

4. By twelve weeks the major changes of metamorphosis commence: the tail is reabsorbed, providing a useful nutritive source for the remodelling of the body structure; the tadpole mouth disappears and a true mouth forms and widens; the eyes grow and eyelids develop; and the forelimbs are released. Internally there are major changes in the nervous, respiratory, circulatory and digestive systems.

5. The frog is sexually mature by about its third year, and returns to its native pond to breed.

---

tongue (of amphibians) is generally long and can be extended in capturing and controlling prey.

Urodela (or **Caudata**) the amphibian order that contains newts, salamanders and sirens. The sirens are sometimes assigned to a separate order.

urodele another name for a salamander.

REPTILES

The Carboniferous period was characterized by swamps which were dominated by amphibians, but, as hotter and drier conditions emerged about 300 million years ago, the reptiles - which were not so reliant on water - found themselves at an enormous advantage. From this time their conquest of the land began and the 'age of the reptiles' had begun. The extinction of the dinosaurs 65 million years ago remains an unsolved mystery.

Reptiles have succeeded in making a complete break from water by developing two features lacking in living amphibians - scaled waterproof skins and shelled yolk-bearing eggs. It is particularly due to these developments that reptiles have succeeded in adapting to a far wider range of habitats than amphibians. Of the once mighty reptiles there are now only four orders: Chelonia (turtles, tortoises and terrapins), Crocodilia (crocodiles), Squamata (snakes, lizards and amphisbaenians), and Rhynchocephalia (tuatara).

agamid any of some 300 species of Old World lizard of the family Agamidae.

albumen see egg.

alligator either of two species of aquatic reptile belonging to the genus *Alligator*, the American and the Chinese alligators.

amnion see egg.

amphisbaenian any of about 100 small, virtually blind, limbless, burrowing lizards belonging to the suborder Amphisbaenia. They feed principally on worms and other invertebrates.

basilisk any small American iguanid of the genus *Basiliscus*.

boa see constrictor.

carapace the upper part, or dome, of the shell of a turtle.

chameleon any of around 85 species of mainly tree-dwelling lizard of the family Chamaeleontidae, famous for their capacity to change colour rapidly.

Chelonia the reptilian order that includes terrapins, tortoises and turtles.

Chelonidae the family containing the turtles.

chorion see egg.

cloaca the single passage through which eggs, sperm and body wastes are released. (Birds, fish and amphibians also share this feature.)

cobra any highly venomous elapid snake of the genus *Naja* or *Ophiophagus*. When alarmed, they give a threat display by expanding the ribs in the neck region (the 'hood').

cold-blooded see poikilothermic.

colubrid any snake of the family Colubridae, the largest family of snake with over 1000 species, mostly harmless.

constrictor any of around 80 species of non-venomous snake in the family Boidae that kill their prey by constriction, including pythons and boas.

coral snake any of several small, venomous, brightly coloured, New World elapid snakes.

cranial kinesis see lizard.

crocodile any of 13 species of large predatory reptile of the family Crocodilidae. All crocodiles share a body armoured with bony plates set in the skin of the back, a long deep-sided tail used in swimming and a long-snouted skull. Both crocodiles and alligators have a large pair of teeth near the front of the lower jaw, which in crocodiles are visible when the mouth is closed. They have a secondary palate that allows them to breathe under water.

Crocodilia the reptilian order that includes the crocodiles.

Crotalidae the family comprising the pit vipers.

dinosaur any reptile of the orders Saurischia and Ornithischia that dominated life on earth for over 160 million years; see introduction.

ectotherm any animal which is dependent on its surroundings to maintain its normal body temperature. Reptiles are ectotherms; see also poikilothermic.

egg (reptilian) is well protected by a chalky or leathery shell. It contains an embryo enclosed in its private 'pond' by a membrane called the amnion. A yolk enclosed in a yolk sac provides food; the albumen (egg white) provides extra cushioning and serves as a reservoir of water and proteins. The entire structure is surrounded by a protective membrane called the chorion. Many species of lizard and snake are ovoviviparous - they retain their eggs in the oviduct during the development of the young, which hatch as the eggs are laid. However, most reptiles abandon their eggs.

elapid any of some 200 species of venomous snake belonging to the family Elapidae. They include cobras, coral snakes, mambas and kraits.

Emydidae the chelonian family that includes the terrapins.

European wall lizard see lacertid.

fang see venom.

gecko any of over 600 small, mainly insectivorous, lizards belonging to the family Gekkonidae.

Gekkonidae the lizard family comprising the geckos.

hood see cobra.

Hydrophiidae the family containing the sea snakes.

iguanadon see iguanid.

iguanid any of around 600 species of lizard of the mainly New World family Iguanidae. Most are agile tree- or ground-dwellers. They include iguanas and basilisks.

Komodo dragon see monitor.

krait a brightly coloured south Asian elapid snake.

lacertid any of over 150 species of lizard belonging to the family Lacertidae. Widespread, they vary considerably in size. The family includes the European wall lizard (*Lacerta muralis*).

lizard any reptile of the suborder Sauria, containing some 3000 species and constituting (with snakes and amphisbaenians) the most successful group of living reptiles. The major families include agamids, chameleons, geckos, iguanids, lacertids, monitors, skinks, and teiids.

The skull of the lizard is made up of several separate mobile elements, a form of modification known as cranial kinesis. The kinetic skull of lizards improves their ability to grasp food items.

BIOLOGICAL SCIENCES

locomotion (of snakes) lacking limbs, snakes have developed a number of highly characteristic modes of movement: serpentine (the most characteristic), concertina, straight line (by stretching and contracting), and sidewinding.

mamba any tropical African elapid snake. They are notoriously venomous and aggressive.

monitor (or **vanarid**) any of about 30 species of lizard of the family Vanaridae. All are large predatory carnivores. They include the Komodo dragon, the largest of all lizards.

moulting (in reptiles) see sloughing.

Ophidia another name for the suborder Serpentes.

Ornithischia see dinosaur.

ovoviviparous see egg.

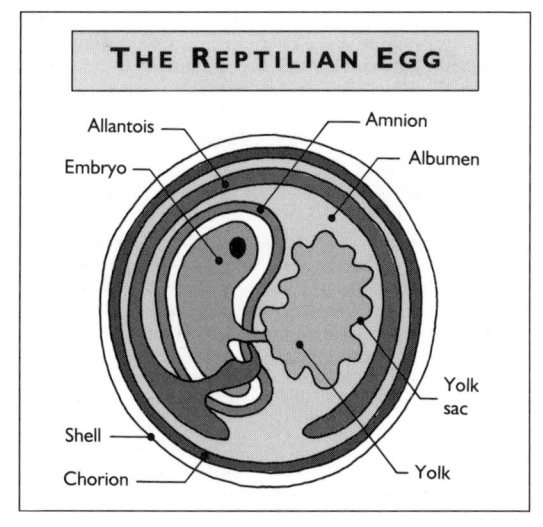

THE REPTILIAN EGG

Labels: Allantois, Amnion, Embryo, Albumen, Yolk sac, Shell, Chorion, Yolk

pit viper any of around 120 highly venomous species of snake belonging to the largely New World family Crotalidae.

plastron the lower, or softer belly, part of the shell of a turtle.

poikilothermic (or cold-blooded) relating to an animal whose body temperature varies with and is largely dependent upon the environmental temperature. Reptiles are poikilothermic, as are all animals except birds and mammals. See also ectotherm.

python see constrictor.

Rhynchocephalia the reptilian order that comprises the tuatara.

Saurischia see dinosaur.

Scincidae the family comprising the skinks.

sea snake any of around 50 species of snake of the family Hydrophiidae. Living in the Indian and Pacific Oceans, all sea snakes are highly venomous.

Serpentes the suborder comprising the snakes.

shell (of turtles) is made of between 20 and 30 bony plates, which are covered by keratin or horn; see carapace and plastron.

skink any lizard of the family Scincidae, the largest family of lizards with around 800 species.

sloughing the process of shedding the skin.

snake any reptile of the suborder Serpentes (or Ophidia), containing over 2000 species and constituting (with the lizards and amphisbaenians) the order Squamata. The major families include colubrids, constrictors, elapids, pit vipers, sea snakes, and vipers.

Snakes have two main characteristics: the loss of their limbs and extreme cranial kinesis (see lizard). Evolving from their lizard ancestors 120 million years ago, snakes have become highly successful hunters, killing their prey by suffocation, by biting or by venom. The smallest snakes are 20cm (8in) long and feed on termites; the longest constrictors are 10m (33ft) long. The skull in all forms is an extremely loose structure, with numerous joints that allow the snake to dislocate its skull in order to swallow huge prey animals that may be several times the normal diameter of the mouth. See also locomotion (above).

Squamata the reptilian order that includes snakes, lizards and amphisbaenians.

teiid any of over 200 species of lizard of the family Teiidae, mainly tropical and subtropical in distribution.

terrapin name given to a number of freshwater turtles of the family Emydidae. They have flattened, webbed limbs and mainly feed on small aquatic animals.

Testudinidae the chelonian family that contains the tortoises.

tortoise any of several slow-moving terrestrial chelonian reptiles, especially of the family Testudinidae. They are typically herbivorous, with a high-domed shell and clawed feet.

tuatara the only surviving species of the reptilian order Rhynchocephalia; a medium-sized lizard-like animal with a primitive non-kinetic skull. It feeds on invertebrates such as snails and worms. Its closest relatives are known from the Triassic and Jurassic periods, some 220-150 million years ago.

turtle any of several aquatic chelonian reptiles, especially of the marine family Chelonidae. In US and scientific usage, the term may apply to any chelonian reptile, including tortoises and terrapins. Turtles - which are characterized by their shells (q.v.) have no teeth; instead the bony edges of the jaws are lined with a horny beak, with which they can handle most food. Their limbs are short and strong, and may be adapted for walking on land or (if expanded into a broad paddle) for swimming. All chelonians can pull in their limbs and short tail, and retract the head, if in danger.

Vanaridae the family containing the monitors.

venom snake poison. The salivary glands of many snakes produce venom. The chemical composition of snake venom varies from species to species. In front-fanged snakes, two very large teeth swing down as the mouth opens wide, and the snake stabs its prey and injects venom.

vertebral column (of reptiles) is well developed, with snakes having up to 450 vertebrae, each with a pair of ribs.

viper any venomous snake of the Old World family Viperidae, containing some 40 species.

BIRDS

Birds (class Aves) probably evolved from small, lightly built, flesh-eating dinosaurs. A recent reclassification - with only 23 orders - based upon DNA analysis has been proposed. Most of the 8600 or so species of bird alive today are able to fly. The stresses imposed by flight are very great, and a high degree of structural development has been necessary to make it possible. The three features that conspicuously distinguish birds from other animals are a beak, a pair of wings, and feathers.

albatross any of 14 large gliding birds of the family Procellariidae.
Anseriformes (order) ducks, geese and swans.
Apodiformes (order) swifts.
beak (or **bill**) the projecting jaws of a bird. The beak is covered in hardened keratin but is not dead tissue. Over the jaws there is a beak-generating organ, producing the beak's epidermal tissues. Beaks are highly adapted to meet the needs of particular species.
bee-eater any of 26 slender birds of the family Meropidae.
bird of prey common name given to birds that hunt and kill other animals for food. They include hawks, eagles, kites, buzzards, falcons, ospreys and secretary birds (order Ciconiiformes) and the nocturnal birds of prey (order Strigiformes).
Bucerotiformes (order) hornbills.
Ciconiiformes (order) most birds of prey (except owls), sandgrouse, gulls, waders, grebes, cormorants, gannets, herons, flamingos, ibises, spoonbills, pelicans, storks, frigate birds, divers, penguins, shearwaters, petrels and albatrosses.
claws solid hooked nails at the end of the hind limbs of birds. Ground birds use their claws to dig and gather food; perching birds use them to grip branches. Three digits face forwards for propulsion, and one usually faces backwards for support. In zygodactylous birds - such as owls and woodpeckers - the third toe also faces backwards to increase the strength of their grip.
Coliiformes (order) colies.
Columbiformes (order) pigeons and doves.
contour feathers feathers which serve to streamline the bird's body and act as an outer insulating layer by trapping warm air.
Coraciiformes (order) kingfishers, bee-eaters and rollers.
cormorant any of about 35 large water birds of the family Phalacrocoracidae.
Craciformes (order) curassows.
crane any of 15 species of the family Gruidae, characterized by long, slender necks and bills.
crop a small storage compartment for food preceding the gizzard.
cuckoos any of about 100 species of the order Cuculiformes, about one half of which are characterized by laying their eggs in the nests of other birds.
Cuculiformes (order) cuckoos.
digestion (in birds) see crop and gizzard.

dove any of approximately 155 of the smaller, more slender species of the order Columbiformes.
down a layer of soft, fine feathers which form the principal insulating covering.
duck any of about 130 species of waterfowl belonging to the family Anatidae.
eagle any of about 30 large birds of prey of the family Accipitridae.
feather any of the keratin-based epidermal structures of a bird that together make up its plumage. Feathers are made of the protein keratin. Closest to the skin is the down; beyond this are the contour feathers. Tail feathers and flight feathers on the wings have the strongest shafts (quills), in order to withstand the tremendous stresses of flight and steering. With the exception of down, birds' feathers have a complex microstructure that allows the vanes of each side of the quill to 'knit' together to produce a smooth surface. Feathers may be adapted to different purposes, other than flight and body insulation. Birds spend much time preening feathers.
finch name given to a variety of small, seed-eating passerine birds.
flamingo any of 5 species of slender birds with long legs (family Phoenicopteridae).
flight the great power needed for flight is provided by two pairs of massive pectoral muscles anchored to the large, keeled sternum. These muscles work in the same direction, but one set pulls directly on the wings to bring them down and in, while the other is carried over the shoulder by a tendon-pulley arrangement to pull the wings up and back. A bird's wing has a characteristic profile with a convex upper surface and a concave lower surface. Air moving over the wing has to travel further, and thus faster, over the upper surface than the lower; this causes a reduction in air pressure above the wing, and hence creates lift. Although this profile is shared by all flying birds, there is a great variety in styles of flight.
furcula the wishbone, by which wings are attached to the skeleton. The furcula increases the spring of the wing beat and damps the unequal and jarring stresses imposed when a bird wheels or changes speed suddenly.
Galbuliformes (order) jacamars and puffbirds.
Galliformes (order) partridges, pheasants, grouse, etc.
gannet any of 3 species of large sea bird (family Sulidae).
gizzard a special compartment early in the digestive tract with thick muscular walls. As birds have no teeth they often swallow small stones to assist the grinding process.
grebe any of about 20 swimming and diving birds of the family Podicipedidae.
goose the name given to various birds of the family Anatidae. The distinction between geese and ducks is not clear-cut.
grouse any of 11 gamebirds of the family Phasianidae.
Gruiformes (order) cranes and rails.
gull any of about 50 species of sea bird with webbed feet (family Laridae).
hawk common name for many smaller birds of prey.

BIOLOGICAL SCIENCES

heron any of 38 species of long-legged wading birds (family Ardeidae; includes egrets and bitterns).

hornbill any of 56 species of the order Bucerotiformes, characterized by large downwardly curved bills.

hummingbird any of 319 species of tiny birds (family Trochilidae) characterized by long bills. Hummingbirds hover by means of a figure-of-eight 'stirring' motion.

ibis any of 25 species of wading bird of the family Treshkiornithidae.

kingfisher any of 94 compact-bodied birds of the order Coraciiformes.

kite any of 21 birds of prey of the family Accipitridae.

Musophagiformes (order) turacos.

nightjar any of 68 cryptically coloured birds of the family Caprimulgidae.

ostrich the world's largest bird (family Struthionidae), which can reach 2.75m tall.

owl any of about 175 birds of prey belonging to the order Strigiformes. They are characterized by flattened faces that allow the large eyes to point forwards.

parrot any of about 350 often brightly coloured species of the order Psittaciformes.

partridge any of about 60 gamebirds of the family Phasianidae.

passerine birds the Passeriformes.

Passeriformes (order) the songbirds and 'perching birds' - over 5 500 species (about 60% of living birds).

pelican any of 7 large, heavily built water birds of the family Pelecanidae.

penguin any of 17 species of the family Spheniscidae, characterized by short stubby wings used as underwater flippers.

pheasant any of 50 large gamebirds of the family Phasianidae.

Piciformes (order) woodpeckers and barbets.

pigeon any of approximately 140 of the larger species of the order Columbiformes.

plumage see feather.

preening the careful cleaning, oiling and arranging of feathers.

Psittaciformes (order) parrots.

quill the shaft of a feather.

rhea either of two large flightless birds of the family Rheidae.

stork any of 17 species of long-legged wading birds of the family Ciconiidae.

Strigiformes (order) owls and nightjars.

Struthioniformes (order) ostriches and rheas.

swan any of 7 heavily built birds belonging to the family Anatidae.

swift any of about 100 small birds of the order Apodiformes, which are characterized by spending much of their time flying fast.

syrinx the singing organ of birds. Located at the base of the trachea, this is particularly well developed in song birds. As air vibrates the membranes of the syrinx, its muscles alter their tension, allowing differences in timbre and pitch to be produced.

tail see feather.

thrush any of about 180 species of passerine birds.

Tinaformes (order) tinamous.

Trochiliformes (order) hummingbirds.

Trogoniformes (order) trogons.

Turniciformes (order) buttonquails.

Upupiformes (order) hoopoe.

vulture any of 18 large scavenging birds of the families Ciconiidae and Accipitridae.

wing the flight structure of birds. Birds' wings are highly modified fore limbs: the digits are reduced in size, and the wrist bones are elongated and fused, to protect the supporting structure for the flight feathers. The wings are attached to the skeleton by the mobile shoulder-joints and by the furcula.

woodpecker any of over 300 species of the family Picidae, characterized by their long strong bills and their ability to chisel nests in trees.

zygodactylous birds see claws.

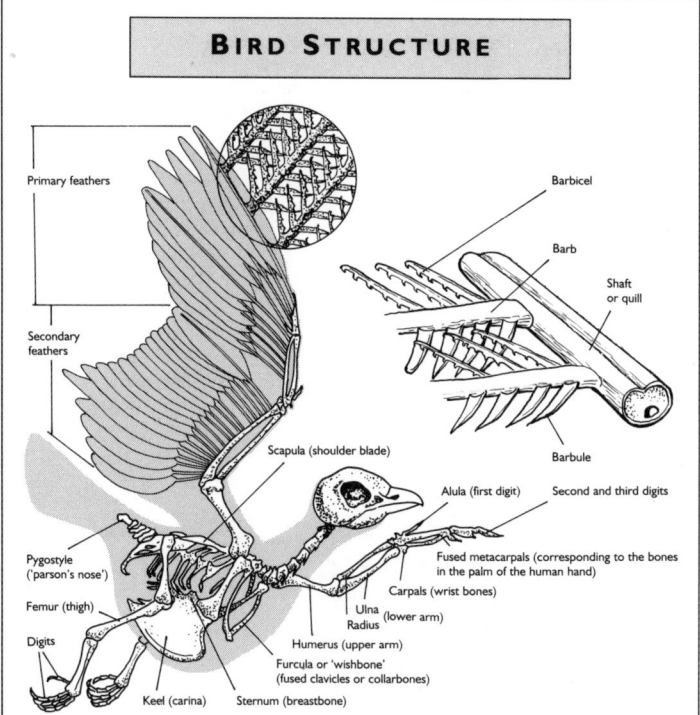

BIRD STRUCTURE

Labels: Primary feathers; Barbicel; Barb; Shaft or quill; Secondary feathers; Barbule; Scapula (shoulder blade); Alula (first digit); Second and third digits; Fused metacarpals (corresponding to the bones in the palm of the human hand); Carpals (wrist bones); Pygostyle ('parson's nose'); Ulna; Radius (lower arm); Femur (thigh); Humerus (upper arm); Digits; Furcula or 'wishbone' (fused clavicles or collarbones); Keel (carina); Sternum (breastbone)

MAMMALS

The class Mammalia has fewer than 5000 species, of an extraordinary variety of size and form. Characteristics of mammals include the following (although not all are found in all mammals):

1) The production of milk from mammary glands in the female. (The monotremes (q.v.) produce milk but do not have teats with which to suckle the young.)
2) Mammals are homoiothermic - they maintain a constant internal temperature by means of a high metabolic rate. Birds are also homoiothermic but their external 'lagging' is feathers rather than the insulating covering of hair or fur that is unique to mammals.
3) Only mammals have three sound-conducting bones (or ossicles) in the middle ear - the malleus, incus and stapes.
4) Mammals have larger and are far fewer teeth than reptiles. Mammalian teeth are specialized in different parts of the jaws for different functions. At the front, the incisors have developed for biting and canines for stabbing, while at the back cusped teeth (molars and premolars) are adapted for chewing and grinding.
5) Only mammals have milk teeth, which are replaced by a permanent set of teeth in the adult.
6) The roof of the mammalian mouth is characteristically arched and almost all mammals also have a secondary palate.
7) The way in which the head articulates with the rest of the body is unique to mammals. The joint system of the neck involving a pivotal joint between the first and second vertebrae allows the head to be turned from side to side, but not to the same degree as birds, which can rotate the head through 180°.
8) Intelligence is a most significant characteristic of mammals - although the most difficult to quantify. The cerebral cortex is better developed in higher mammals than in any other group of vertebrates.
9) The chest and abdominal cavities of mammals are separated by a muscular diaphragm.

aardvark an ant- and termite-feeding mammal of sub-Saharan Africa with a long tubular snout, long ears and stocky body. It has its own order (Tubulidentata).

aardwolf see hyena.

agouti any of 10 species of large cavy-like South American rodents.

alpaca a domesticated South American camelid (q.v.).

anteater any of four species of toothless edentate (q.v.) mammal of the family Myrmecophagidae found in Central and South America.

antelope common name given to many bovid (q.v.) mammals, typically graceful, long-legged grazers and browsers - they include oryxes, gazelles, impala, waterbuck and wildebeest. (For larger antelopes see Bovinae.)

ape any higher primate of the superfamily Hominoidea, an exclusively Old World group (except for *Homo sapiens*).

armadillo any of 20 species of edentate (q.v.) mammal of the family Dasypodidae found in Central and South America. Distinguished by their body armour of plates, armadillos are omnivorous.

artiodactyl any member of the order Artiodactyla - the even-toed ungulates - with some 200 species including pigs, sheep, cattle, deer and antelope. Artiodactyls are typically big mammals with large bellies to carry their bulky food of grasses and leaves. They have large heads with long jaws and a battery of grinding teeth. The third and fourth toes form the typical semicircular cloven hoof. Except for the pigs, peccaries and hippopotamuses, all artiodactyls are ruminants. Artiodactyls have a four-chambered stomach (but see camel) incorporating a large fermentation chamber (the rumen) containing bacteria to help break down the food. Food is swallowed with little chewing, passes to the rumen where it is moistened and softened and is then regurgitated to the mouth and chewed as cud. Swallowed for a second time, it bypasses the rumen and passes to the other chambers of the stomach. This system allows ruminants to eat quickly and to digest slowly and safely.

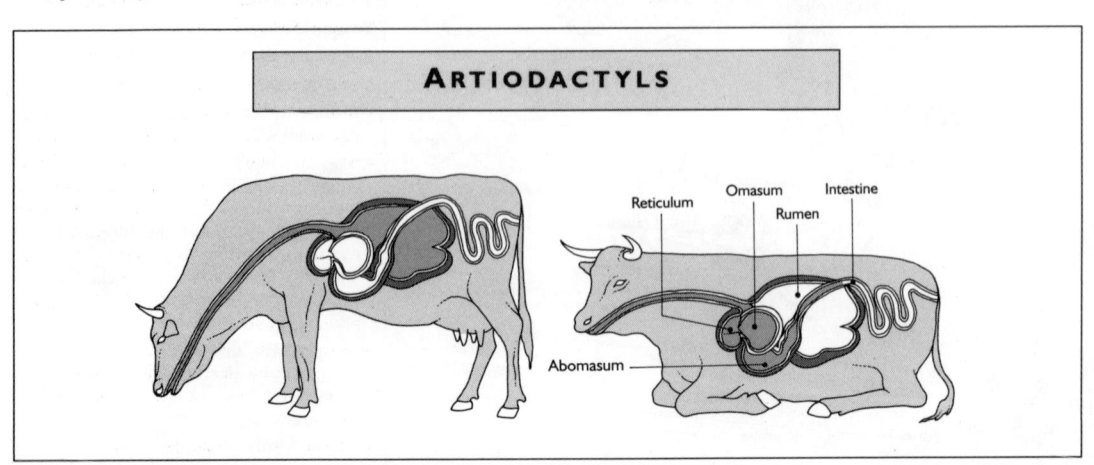

ARTIODACTYLS

BIOLOGICAL SCIENCES

ass either of two perissodactyl (q.v.) members of the family Equidae - the African wild ass is the ancestor of the domesticated donkey.

aye-aye a lemur (q.v.); see also primate.

baboon any large, often brightly coloured, Old World monkey of the genus *Papio*.

badger any of several stockily built mustelids (q.v.), typically with black and white head markings.

bandicoot any of 15 species of marsupial (q.v.) of the Australasian family Peramelidae.

Barbary ape see macaque.

bat any mammal of the order Chiroptera, one of the largest groups of mammals and the only one to have achieved true flight. There are two suborders: the Megachiroptera (with a single family of about 170 species of fruit bat) and the Microchiroptera (with about 800 other species of bat), most of which are capable of echolocation (q.v.). The flying membrane of bats consists of skin stretched between the four extremely elongated fingers of each hand; only the thumb remains free.

bear any of eight species of the omnivorous mammalian family Ursidae. Grizzly and polar bears are the largest of all carnivores. They are characterized by large heads with relatively small rounded ears, dog-like faces with long snouts and heavily built bodies with small tails.

beaver either of two large swimming rodents of the Castoridae, characterized by building dams to form ponds.

bison either of two massively built cattle-like bovids (q.v.) of the genus *Bison*: the North American bison or plains buffalo and the European bison or wisent.

bobcat a stocky, reddish-brown North American cat (q.v.).

bongo a chestnut brown African antelope with short horns; see Bovinae.

bovid any ruminating artiodactyl (q.v.) mammal belonging to the family Bovidae, which includes cattle, sheep, goats, antelopes and duikers. Bovids show considerable variation in size and form, but all are characterized by horns in the adult male (and in the females of some species). Bovid horns are unbranched and never shed.

Bovinae a ruminant mammalian subfamily that includes cattle, bison, buffalo and the spiral-horned antelopes such as the eland, bongo and bushbuck.

Bradypodidae an edentate (q.v.) mammalian family; the three-toed sloths.

browser any mammal that forages leaves, shoots and other vegetation raised above the ground.

buffalo either of two massively built species of cattle - the dark-brown, herd-dwelling African buffalo and the larger, black Asiatic water buffalo.

bushbuck an African antelope with spiral horns; see Bovinae.

camel either of two large ruminating camelids (q.v.), both adapted to life in arid desert conditions: the domesticated one-humped African camel (or dromedary) and the two-humped Bactrian camel of Central Asia.

camelid any six species of the family Camelidae, including the Old World camels and the New World llamas, guanacos, etc. Camelids differ from other ruminants (q.v.) in that their stomachs have three rather than four chambers. They have characteristic long thin necks, small heads, and two toes on each foot. All are adapted to arid conditions.

canid any member of the mammalian family Canidae comprising 35 species including wolves, foxes, jackals, dogs and the coyote. Canids are remarkably similar in appearance: they have slim, muscular, deep chested bodies. The sense of smell is particularly strongly developed.

capuchin any lively New World monkey of the genus *Cebus*, whose members have a 'hood' of thick hair on the back of the head.

capybara the largest living rodent; a stocky semi-aquatic South American mammal.

caribou (or **reindeer**) a large deer with branched antlers in both sexes. It is native to the tundra of Eurasia (the reindeer) and of North America (the caribou), and has been domesticated.

carnivore a carnivorous animal, i.e. one that feeds primarily or solely on other animals. Carnivores are distinguished by their canine teeth, which are usually large, curved and dagger-like (for stabbing and tearing flesh) and small pointed incisors for holding prey and nipping bones. They have four or five claws on each foot.

Castoridae a rodent family; the beavers.

cat any of 34 species of carnivore of the family Felidae. They are distinguished by shorter skulls and jaws than the other carnivores.

cattle common name given to large bovid (q.v.) mammals of the genus *Bos* and related genera in the subfamily Bovinae, such as gaur, banteng, yak, kouprey, bison and buffalo. Domestic cattle are descended from the aurochs.

cavy any of several small, stockily built South American rodents of the genus *Cavia*. The guinea pig is a domesticated cavy.

Cervidae the mammalian family of ruminating artiodactyls (q.v.); the deer family (q.v.).

cheetah a slim, lithe African cat (q.v.) - reputedly the fastest land animal.

chimpanzee either of two species of ape of the genus *Pan* (superfamily Hominoidea).

chinchilla either of two soft-furred, agile Andean rodents with large ears and a long floppy tail.

chipmunk see squirrel.

Chiroptera a mammalian order; the bats.

civet any cat-like viverrid (q.v.), typically omnivorous, solitary and nocturnal. They are found in Africa and southern Asia.

cloven hoof the distinctive foot of artiodactyls (q.v.).

coati any small, social member of the family Procyonidae, native to Central and South America.

colobus any slender, tree-dwelling, African monkey of the genera *Colobus* and *Procolobus*.

coyote a North American canid (q.v.) resembling a small wolf with a grey coat.

dasurid any of 50 species of carnivorous marsupial (q.v.) of the family Dasuridae.

Dedelphidae a marsupial (q.v.) family; the opossums (q.v.).

deer any of 38 species of the family Cervidae (q.v.), which includes red, roe, sika, and fallow deer, wapiti, moose, and caribou (or reindeer). Compared with other hoofed mammals, deer are finely built with long legs and slim bodies. Their most distinctive feature is branched antlers. Except for reindeer, these are borne only by the males and are shed annually.

dog the common name given to many canids (q.v.), especially the larger more social species.

donkey see ass.

dromedary see camel.

duck-billed platypus see platypus.

duiker a small antelope-like African bovid (q.v.) of the subfamily Cephalophinae.

echidna (or **spiny anteater**) either of two species of mainly insectivorous, Australasian monotremes (q.v.) which are covered in long protective spines.

echolocation non-visual orientation and detection involving the emission of high-frequency sounds and monitoring their echoes from intervening objects. The system is used by bats and toothed whales.

edentates any mammal belonging to the order Edentata which includes the anteaters, armadillos and sloths. They are distinguished by the strange and unique additional articulations between the vertebrae at the base of the spine. These bony projections give extra reinforcement particularly when digging.

eland a large stocky grey or brown African antelope; see Bovinae.

elephant either of two trunked herbivorous mammals of the family Elephantidae, the largest living land animals: the African and the Indian elephant. The most characteristic feature is the long, flexible, muscular trunk, which is an elongation of the nose and the upper lip, with nostrils at the end. Elephants' teeth are also distinctive - they possess neither canines nor lower incisors; the tusks are much enlarged upper incisors.

elephant shrew any of 15 species of tiny African shrew-like insectivores (q.v.) of the suborder Macroscelidae.

elk see moose. Elk is also the North American name for the wapiti (q.v.).

Equidae the perissodactyl (q.v.) family containing horses, asses and zebras.

Erinaceidae a mammalian family; the hedgehogs and moonrats.

Felidae the cat family (q.v.).

fox common name given to many canids (q.v.), especially the smaller more solitary ones.

gazelle any of 18 species of small, elegant, agile antelopes of the genus *Gazella*.

genet any small cat-like viverrid (q.v.), typically striped or spotted.

gerbil any of some 81 species of mouse-like, burrowing rodents of the family Muridae.

gibbon any of nine species of Southeast Asian ape (q.v.) of the family Hylobatidae.

giraffe a sub-Saharan artiodactyl of the family Giraffidae. It is characterized by its long neck (which contains the normal mammalian seven vertebrae).

gnu see wildebeest.

goat common name for various sure-footed, agile bovid (q.v.) mammals of the genus *Capra*, such as the ibex.

gorilla a herbivorous great ape of the superfamily Hominoidea. A native of central Africa, the gorilla is, after the chimpanzee, man's closest relative.

grazer any mammal that crops grass and other ground vegetation.

guanaco a South America camelid (q.v.), the ancestor of the domesticated llama and alpaca.

hair (or **fur**) see characteristics of mammals.

hare any lagomorph (q.v.) mammal of the genus *Lepus*. In contrast to rabbits, hares are typically solitary, produced fully furred young and live in shallow scrapes rather than burrows.

hedgehog the common name given to 12 species of the mammalian family Erinaceidae. Distinguished by their covering of spines, hedgehogs are insectivorous.

herbivore a herbivorous animal, i.e. one that feeds primarily or solely on plants.

hippopotamus either of two non-ruminating artiodactyl (q.v.) mammals of the family Hippopotamidae: the massively built common hippopotamus of sub-Saharan Africa and the pygmy hippopotamus of West Africa.

Hominidae a subfamily of apes (q.v.); man, the chimpanzee and the gorilla.

homoiothermic 'warm-blooded', i.e. able to maintain a steady body temperature above that of the surroundings.

Homo sapiens sapiens the human being; see pp. 60-72.

horse the common name given to two perissodactyls (q.v.). of the family Equidae: the domesticated horse, probably descended from the now extinct tarpan, and Przewalski's horse, the only true surviving wild horse. 'Horse' also refers to other members of the family, i.e. asses and zebras.

howler any large New World monkey of the genus *Alouatta*, characterized by a booming call.

hyena any of three species of carnivore of the family Hyaenidae, which also includes the unusual aardwolf (which feeds on termites).

Hylobatidae a family of smaller apes (q.v.); the gibbons.

hyrax any of 11 species of small, rodent-like herbivorous ungulates (q.v.) belonging to the African and Middle Eastern order Hyracoidea.

ibex a nimble large goat (q.v.) with massive horns. It inhabits alpine Eurasia.

impala a common gregarious African antelope, known for its leaps when frightened.

BIOLOGICAL SCIENCES

insectivore any member of the order Insectivora, a diverse group that is divided into three major groups: the hedgehog types, the shrew types and the tenrec types.

insectivorous animal one that feeds primarily or solely on insects; see also insectivore.

jackal any of four small fox-like canids (q.v.) of the genus *Canis*.

kangaroo any Australian herbivorous marsupial (q.v.) of the genus *Macropodus* and related genera in the family Macropodidae (smaller wallabies, etc.). Kangaroos typically have powerful hind limbs adapted for jumping.

kinkajou a fruit-eating tree-dweller of the family Procyonidae.

koala a marsupial (q.v.) of eastern Australia, a highly specialized tree-dweller.

lagomorph any member of the order Lagomorpha including rabbits, hares and pikas. Lagomorphs are distinguished by long, exceptionally soft fur, furred feet, eyes high set on the sides of the head and narrow nostrils, with a fold of skin allowing them to be closed.

langur any slender, tree-dwelling, herbivorous Old World monkey of the genus *Presbytis*.

lemming any of several species of small vole-like rodents of the genus *Lemmus*.

lemur any of 21 species of lower primate, mainly restricted to Madagascar.

leopard a solitary, versatile African and Asian big cat (q.v.), noted for its spotted coat. The black leopard is called the panther.

Leporidae a lagomorph (q.v.) family; the rabbits and hares.

lion a large, social big cat (q.v.) now largely confined to sub-Saharan Africa.

llama a domesticated South American camelid (q.v.).

loris any of several slow-moving, nocturnal lower primates of the Southeast Asian family Lorisidae.

lynx a pale brown spotted cat (q.v.) distinguished by its short tail and found in Eurasia and North America.

macaque any Old World monkey of the genus *Macaca*, which includes the rhesus monkey and Barbary ape.

Macropodus a marsupial genus; the kangaroos.

mara a large, long-legged South American cavy (q.v.).

marine mammals see p. 73.

marmoset any of several tree-dwelling New World monkeys (q.v.).

marmot any large, squirrel-like, burrowing, colonial, Alpine rodent of the genus *Marmota*.

marsupial any of 266 species of the Australasian and New World mammalian order Marsupiala including the possums and opossums, bandicoots, wombats, kangaroos and wallabies, the numbat and the koala. Marsupials are distinguished from other mammals by various features of their reproduction. In its form and development, the marsupial egg more closely resembles that of birds and reptiles than that of mammals. Marsupial young are born in an undeveloped state, tiny and blind, and make their own way unassisted from the birth canal to the mother's teats which - except in some small marsupials - are situated in a pouch.

Megalonychidae an edentate (q.v.) mammalian order; the two-toed sloths.

Metatheria a mammalian infraclass which includes a single order, comprising the marsupials.

mole the common name given to 27 species of insectivores of the family Talpidae. Moles are highly adapted as burrowers, feeding mainly on earthworms.

mole rat a small, burrowing, solitary cavy-like rodent of the family Bathyergidae.

mongoose any small, agile viverrid (q.v.) of the genus *Herpestes*, widespread in Africa and Asia.

monkey common name given to all the higher primates except the tarsiers, apes and man. Two major groups are recognized: New World monkeys (marmosets, tamarinds, capuchins and howlers) and Old World monkeys (macaques, baboons, langurs, colobus, etc.).

monotreme any member of the Australasian mammalian order Monotremata. (q.v.). There are only three monotreme species - the duck-billed platypus and the echidnas. Even though they are furry and feed their young on milk, they lay eggs that are structurally very similar to those of birds, and - like birds - they incubate them in a nest or burrow. Monotremes do not have breasts or teats but have special glands that ooze milk, which the young lap up from the fur.

moonrat any of five species of 'spineless hedgehogs' of the family Erinaceidae.

moose (or **elk**) a stocky forest-dwelling deer with a broad muzzle and shoulder hump. In North America it is called the moose, in Eurasia the elk.

Moschidae a family of ruminating artiodactyls (q.v.); the musk deer.

mouse common name of many species of small rodents of the family Muridae. (The difference between a mouse and a rat is largely one of size.)

Muridae a rodent family containing many mouse-like mammals.

musk deer three species of small deer-like artiodactyls lacking antlers. The males have long pointed protruding upper canines.

mustelid any of 67 species of the mammalian carnivore family Mustelidae, including weasels, badgers, otters and skunks. They are distinguished by strong carnassial teeth (long cutting teeth) and by anal scent glands.

Myremecobiidae a marsupial (q.v.) family; the numbat.

Myremecophagidae an edentate (q.v.) family; the anteaters.

numbat a pouchless Australian marsupial (q.v.), the sole member of the family Myremecobiidae.

ocelot a yellowish cat (q.v.) with black markings found in Central and South America.

Ochotonidae a lagomorph (q.v.) family of carnivores; the pikas.

okapi a shy artiodactyl (q.v.) native to central Africa. It is, with the much larger giraffe, the only member of the family Giraffidae.

omnivore an omnivorous animal; any animal that has unspecified feeding habits, i.e. it feeds both on plants and other animals.

opossum any of about 70 species of marsupial (q.v.) mammal of the family Dedelphidae, widely distributed in North, Central and South America.

orang-utan a hominoid great ape, the most distantly related to man; a large primate restricted to Sumatra and Borneo.

oryx any of three species of elegant, long-horned antelope of arid Africa and Arabia.

otter any of several aquatic mustelids (q.v.), typically with streamlined bodies and webbed feet.

panda (giant) a large black and white mammal usually placed in the bear family (q.v.) inhabiting bamboo forests in China.

pangolin (or **scaly anteater**) any of seven species of the Old World family Manidae (order Pholidota). Their bodies are covered in overlapping horny scales.

panther see leopard.

peccary any of three non-ruminating, pig-like, New World artiodactyl (q.v.) mammals of the family Tayassuidae.

Peramelidae a marsupial (q.v.) family; the bandicoots.

perissodactyl any herbivorous hoofed mammal of the order Perissodactyla (the odd-toed ungulates), comprising 16 species in three families: Equidae (horses, asses and zebras), Rhinocerotidae (rhinoceroses) and Tapiridae (tapirs). The weight of the body is borne on the central toes, with the main axis of the limb passing through the third (central) toe, which is the longest. In horses only the single central toe is functional; in rhinoceroses three hoofed toes are present and in tapirs four. All other toes are absent or present only as vestiges. Perissodactyls do not ruminate; their digestive processes are assisted by bacterial fermentation which takes place not in the stomach but in the enlarged caecum and in the hindgut.

Petrogale a marsupial (q.v.) genus; the wallabies.

Pholidota a mammalian order; the pangolins.

pig any of eight stockily built, non-ruminating, Old World artiodactyl (q.v.) mammals of the family Suidae. Pigs are distinguished by their flexible, muscular snouts, and males by their curved tusks. The domestic pig is descended from the wild boar, the wild pig of Eurasia.

pika any of 14 lagomorphs belonging to the family Ochotonidae; hare-like herbivores found in Asia and North America.

platypus a small, semi-aquatic Australian monotreme (q.v.) mammal that forages underwater for freshwater invertebrates.

porcupine any large, mainly nocturnal, cavy-like rodent that is typically covered in stiff spines.

possum the common name for a large number of small, arboreal, nocturnal marsupials (q.v.) from Australia - they include phalangers and gliders. See also opossum.

prairie dog see squirrel.

primate any of 180 species of mammal including lemurs, lorises, monkeys, apes and *Homo sapiens*. Primates are characterized by having four digits on each hand and foot, with (except for the aye-aye) the claws of other mammals being replaced by flattened nails. Hands and feet are highly modified to allow these animals to grasp and manipulate objects with a greater dexterity than any other mammals. Primates are also distinguished by their relatively poor sense of smell and by their relatively large brain size.

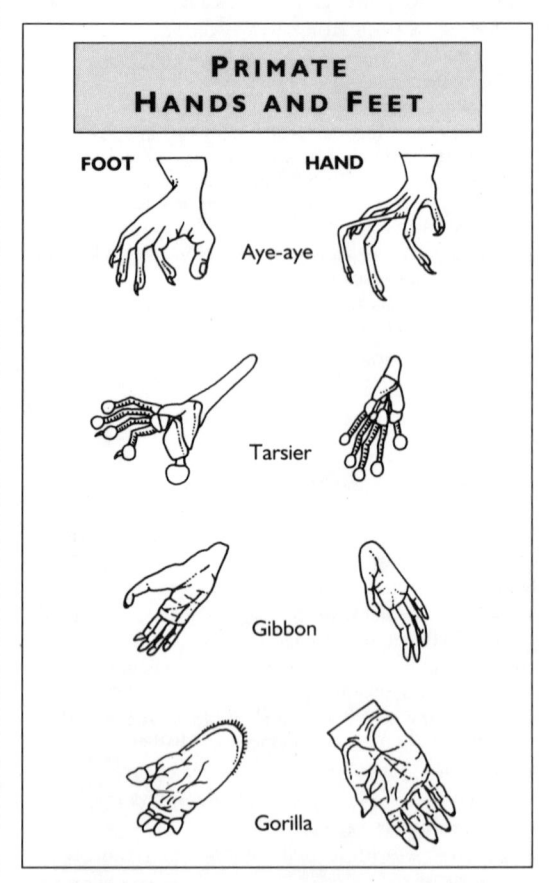

PRIMATE HANDS AND FEET

FOOT — HAND

Aye-aye

Tarsier

Gibbon

Gorilla

primitive ungulates comprise four distinct orders: the proboscides, the hyraxes, the sirens and the order Tubulidentata (the aardvark). They differ dramatically in form, but they share certain anatomical and biochemical features - elephants and hyraxes, for example, both have toes bearing short, flattened nails rather than well-developed hooves, and both have grinding cheek teeth.

proboscidean any trunked mammal of the order Proboscidea, now represented only by two species of elephant.

BIOLOGICAL SCIENCES

Procyonidae a carnivorous mammalian family; the racoons, coatis and kinkajous.

pronghorn a ruminating artiodactyl which is the New World equivalent to the antelope.

prototherian any member of the small subclass Prototheria distinguished by the fact that its members lay eggs. The subclass comprises a single order the monotremes (q.v.).

rabbit common name given to those members of the family Leporidae that live socially in burrows and give birth to very immature young. Rabbits are lagomorphs (q.v.).

racoon any small New World mammal of the genus *Procyon*.

rat common name given to numerous larger species of the rodent family Muridae; see mouse.

rhesus monkey see macaque.

rhinoceros any massively built, horned perissodactyl (q.v.) mammal (family Rhinocerotidae) of sub-Saharan Africa (two species) and Southeast Asia (three species).

rodent any of almost 1700 species of the family Rodentia, which represents 40% of all living mammalian species. Three distinct types occur: squirrel-like, cavy-like and mouse-like. Rodents are characterized by two pairs of chisel-like incisors that continue to grow through the animal's life.

ruminant see artiodactyl.

Sciuridae a rodent family; the squirrels.

sheep common name given to various bovid (q.v.) mammals of the genus *Ovis* and related genera in the subfamily Caprinae (which also includes goats).

shrew any of over 200 species of insectivore (q.v.) of the family Soricidae. Shrews are small, agile predators with characteristic long snouts.

skunk any of several New World mustelids (q.v.), typically with black and white markings. They eject a foul-smelling liquid at attackers.

sloth any slow-moving, herbivorous New World edentate (q.v.) mammal of the families Megalonychidae (2 species of two-toed sloth) and Bradypodidae (3 species of three-toed sloth). Adapted for life in the trees - they rarely come to the ground.

Soricidae a mammalian family of insectivores, including nearly 270 species of shrews and moles.

spiny anteater see echidna.

squirrel any of some 270 species of the rodent family Sciuridae. Many are arboreal; others are ground-dwelling including the chipmunk and prairie dog.

stoat a weasel-like mustelid (q.v.) with an elongated body.

Suidae a non-ruminant artiodactyl family; the pigs.

Tachyglossidae a monotreme mammalian family; the echidnas.

Talpidae a family of mammalian insectivores (q.v.); the moles.

tapir any sturdily-built, forest-dwelling perissodactyl (q.v.) mammal of Central and South America (three species) and Southeast Asia (one species), with a distinctive, short, flexible trunk.

tarpan see horse.

Tasmanian devil a squat, powerfully built dasurid (q.v.), confined to Tasmania.

Tayassuidae a non-ruminating artiodactyl (q.v.) family; the peccaries.

tenrec any of 31 species of shrew-like insectivores (q.v.) of the mainly Madagascan family Tenrecidae.

therian a mammal that bears live young, i.e. all mammals except the monotremes.

tiger the largest of the cats (q.v.); a solitary predator now largely confined to southern Asia.

tree shrew any of 18 species of the family Tupalidae, which comprises small squirrel-like animals whose status as rodents or primates is debatable.

Tubulidentata a primitive ungulate (q.v.) represented today only by the aardvark.

Tupalidae a mammalian family; the tree shrews (q.v.).

ungulate any mammal that has hooves. The group of mammals so defined is divided into two sub-groups (orders): the perissodactyls and the artiodactyls (q.q.v.).

Ursidae the family of carnivores comprising the bears (q.v.).

vicuna a South American camelid (q.v.).

viverrid any carnivore of the Old World order Viverridae, comprising 66 species of civets, mongooses, genets, etc. They are typically small, lithe and long-bodied, with long bushy tails.

vole any small, burrowing, herbivorous rodent of the genus *Microtus*.

Vombatidae a marsupial (q.v.) family; the wombats.

wallaby any herbivorous marsupial (q.v.) of the genus *Petrogale*. Wallabies are similar to kangaroos (q.v.) but smaller.

wapiti a large deer (q.v.) resembling the red deer. It is found in Eurasia and North America (where it is called the elk).

waterbuck a large stocky, shaggy antelope of the African savannah.

weasel common name given to many small Eurasian mustelid (q.v.) carnivores.

wildebeest (or **gnu**) either of two stocky, gregarious antelopes of the genus *Connochaetes*, both inhabiting the African savannah.

wisent see bison.

wolf either of two dog-like canids (q.v.) of the genus *Canis* - the grey or timber wolf and the red wolf.

wombat any of three short-legged, stocky, burrowing marsupials (q.v.) of the family Vombatidae found in Australia and Tasmania. They are nocturnal and herbivorous.

yak a massively built dark-brown species of cattle with shaggy fur. It inhabits mountainous Central Asia.

zebra any of three surviving species of distinctively striped African perissodactyl (q.v.) mammals of the family Equidae.

THE HUMAN SKELETON

The human skeleton is a hard structure that supports the body, protects the soft inner organs, and provides anchorage for muscles. As in all vertebrates the human skeleton is internal. (In insects and other arthropods the skeleton is external, an exoskeleton. The human skeleton comprises 206 bones (see below and diagram).

ankle see tarsus.

anvil see incus.

arm either of the human forelimbs. Each arm comprises 30 bones: humerus 1 bone; radius 1 bone; ulna 1 bone;

carpus (comprising scaphoid 1 bone, lunate 1 bone, triquetral 1 bone, pisiform 1 bone, trapezium 1 bone, trapezoid 1 bone, capitate 1 bone, hamate 1 bone);

metacarpals 5 bones;

phalanges (comprising first digit or thumb 2 bones, second digit 3 bones, third digit 3 bones, fourth digit 3 bones, fifth digit 3 bones),

breastbone see sternum.

carpals the bones that make up the carpus or wrist.

carpus the wrist comprising a number of smaller bones (carpals); see also arm.

cervical vertebrae the vertebrae (q.v.) of the neck supporting the head. Two of these provide articulating surfaces against which the head can move in relation to the backbone.

clavicle either of the collar bones attached to the sternum.

coccyx the lowermost element of the back bone; a vestigial human tail.

collar bone see clavicle.

digit a finger or toe.

ear the sense organ of vertebrates that is specialized to detect sound and to help maintain balance. The two human ears each contain 3 bones: the malleus, the incus and the stapes.

femur the thigh bone; it articulates at one end with the pelvis and at the other with the tibia. See leg.

fibula the smaller of two bones in the lower leg.

finger see digit.

foot see metatarsus.

hammer another name for the malleus.

hip see pelvic girdle.

humerus the long bone of the upper arm (q.v.).

hyoid a single bone in the throat supporting the tongue.

ilium the largest of the three bones that combine to make up one half of the pelvic girdle. The ilium has a flattened 'wing' attached by ligaments to the sacrum.

incus (or '**anvil**') one of the three auditory ossicles in the mammalian inner ear.

ischium the most posterior of the three bones that combine to make up one half of the pelvic girdle.

joint a point or junction between two body parts at which movement is possible. Various types occur. A ball-and-socket joint, as in the hip and shoulder, allows movement in all directions, including rotation, but is susceptible to dislocation. A saddle joint allows versatile movement in several directions, and very significantly - in the case of the primitive thumb joint - the 'opposition' of the thumb to the fingers that is characteristic of precise movements such as grasping. A hinge joint, as in the elbow and knee, allows swinging movement, mostly in a single plane. A pivotal joint is mainly restricted to rotational movement - movement of the head from side to side is primarily due to a pivotal joint between the first and second neck vertebrae. A condyloid joint, such as the wrist joint, allows both rotation and backward and forward movement. A plane joint, such as that between the pelvis and the base of the spine, allows only very limited movement, except in pregnancy when the pelvis expands to accommodate the growing foetus.

lacrimal near the glands that secrete tears.

leg either of the human hind or lower limbs. Each leg contains 29 bones: femur 1 bone; tibia 1 bone; fibula 1 bone;

tarsus (comprising talus 1 bone, calcaneus 1 bone, navicular 1 bone, medial cuneiform 1 bone, intermediate cuneiform 1 bone, lateral cuneiform 1 bone, cuboid 1 bone);

metatarsals 5 bones;

phalanges (comprising first digit or big toe 2 bones, second digit 3 bones, third digit 3 bones, fourth digit 3 bones, fifth digit 3 bones).

malleus (or '**hammer**') one of the three auditory ossicles in the mammalian inner ear.

mandible the lower jaw of vertebrates; see skull.

metatarsus the foot comprising a series of small rod-shaped bones (metatarsals).

occiput the back part of the skull (q.v.).

ossicle a small bone, especially (in mammals) any one of the three ossicles of the ear (q.v.).

parietal bone either of two bones forming part of the top and sides of the skull.

patella either of the two knee caps.

pectoral girdle (or **shoulder girdle**) the structure to which the arms are attached. The human shoulder girdle comprises 4 bones:

clavicle (a pair) 2 bones;

scapula (including the coracoid; a pair) 2 bones.

pelvic girdle (or **pelvis**) the structure to which the legs are attached. Also known as the hip girdle, it articulates dorsally with the backbone. The human pelvic girdle comprises 2 bones - a pair of hip bones that represent the fusion of the ilium, ischium and pubis.

phalanx any of the bones that make up the digits of either the hand or the foot; see arm.

pubis the foremost of the three bones that combine to make up one half of the pelvic girdle.

radius the smaller of the two bones in the lower section of the arm (q.v.); see also ulna.

BIOLOGICAL SCIENCES

ribs see vertebral ribs.

sacral vertebrae five vertebrae fused together to form the sacrum. They articulate securely with the pelvic girdle (q.v.) and, as they are fused to form a single bone, they provide firm support.

sacrum see sacral vertebrae.

scapula either of the shoulder blades attached to the backbone.

shoulder blade see scapula.

shoulder girdle see pectoral girdle.

skull the part of the skeleton enclosing the brain. The human skull comprises 22 bones: occipital 1 bone; parietal (a pair) 2 bones; sphenoid 1 bone; ethmoid 1 bone; inferior nasal conchae (a pair) 2 bones; frontal (a pair - fused) 1 bone; nasal (a pair) 2 bones; lacrimal (a pair) 2 bones; temporal (a pair) 2 bones; maxilla (a pair) 2 bones; zygomatic (a pair) 2 bones; vomer 1 bone; palatine (a pair) 2 bones; mandible (a pair -fused) 1 bone.

spinal column see vertebra.

stapes (or '**stirrup**') one of three auditory ossicles in the mammalian inner ear.

sternum (or **breastbone**) the breastbone in vertebrates. The human sternum comprises three bones:

manubrium, sternebrae, and xiphisternum.

stirrup see stapes.

tarsals see tarsus.

tarsus the ankle comprising a number of small bones (called tarsals), see leg.

temporal bone either of a pair of compound bones that form the sides of the skull (q.v.).

thoracic vertebrae the vertebrae (q.v.) of the upper back. They articulate with the ribs and are characterized by a number of articulating facets for the attachment of ribs.

thumb the first digit of the human hand.

tibia the larger of two bones in the lower leg.

toe see digit.

ulna the larger of the two bones in the lower section of the arm (q.v.); see also radius.

vertebra (plur. **vertebrae**) any of the independent bony segments of the spinal column, through which the spinal cord passes. The human body has 26 vertebrae: cervical 7 bones; thoracic 12 bones; lumbar 5 bones; sacral (five vertebrae fused together to form the sacrum) 1 bone; coccyx 1 bone.

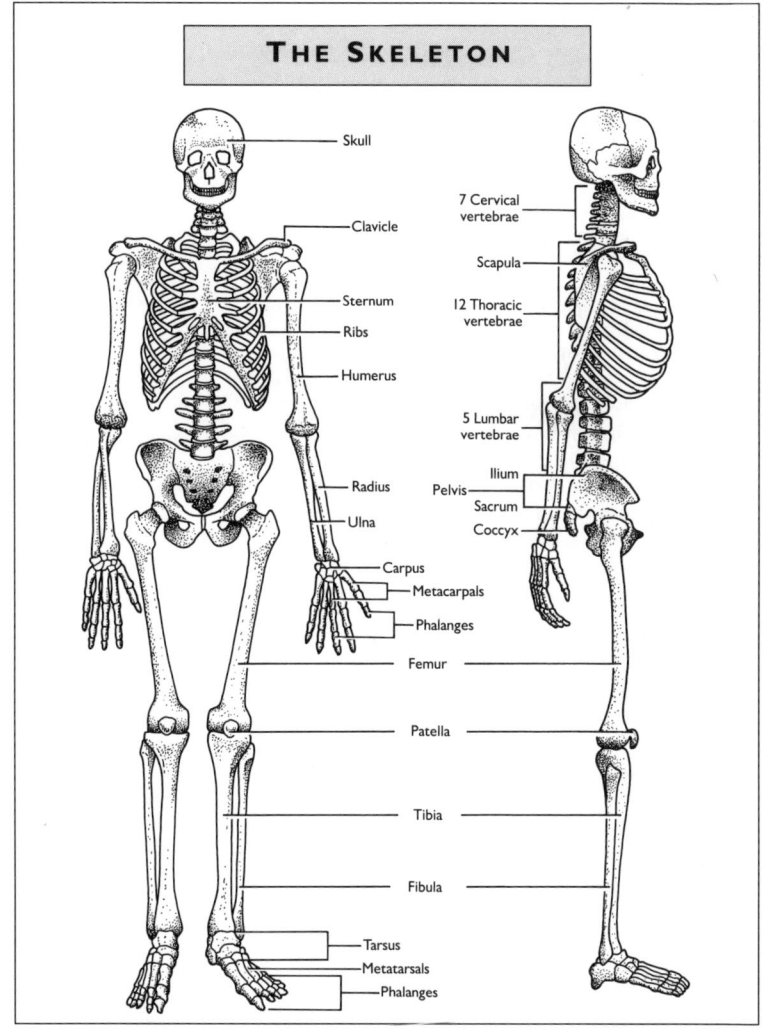

vertebral rib any of a series of slender curved bones forming a cage that encloses, supports and protects the lungs and heart. Pairs of ribs articulate with the thoracic vertebrae at the back and the sternum at the front. The human body has 24 vertebral ribs:

ribs 'true' (7 pairs) 14 bones, which articulate directly with the sternum;

ribs 'false' (5 pairs, of which 2 pairs are 'floating' and 3 pairs articulate with the sternum via elastic cartilage) 10 bones.

wrist see carpus.

HUMAN AND OTHER CIRCULATORY SYSTEMS

Circulatory systems are responsible for bringing oxygen and nutrients to cells and carrying away carbon dioxide and other waste products.

Each animal cell requires an uninterrupted supply of oxygen and nutrients, and waste products must be continuously removed. Small, simple animals that live in a watery environment, such as sponges and flatworms, have no special circulatory systems. Their bodies are only a few cells thick and diffusion alone is sufficient. In larger animals, diffusion is not sufficient to meet the needs of all the cells in every part of the body, and circulatory systems have evolved to transport materials both between the cells and between cells and the outside environment.

See also respiratory systems (pp. 64-65).

Circulatory systems typically consist of three components:

1) A fluid transport medium (blood) in which nutrients, gases and waste products are dissolved.

2) A mechanism by which the blood is pumped around the body (the heart).

3) A system of spaces through which blood circulates.

Arthropods (see pp. 41-45) and many molluscs (see p. 40) have open circulatory systems. Vertebrates (see pp. 46-59), echinoderms (see p. 38) and most molluscs (see p. 40) have closed circulatory systems.

artery a blood vessel by which blood is carried away from the heart.

atrium (plur. **atria**) a chamber in the heart in which blood is collected from the veins and passed to a ventricle (q.v.).

blood the fluid in which nutrients, oxygen and waste products (including carbon dioxide) are dissolved and carried around the body. In vertebrates, the blood consists of a fluid medium (plasma; q.v.) in which cells are suspended. The white blood cells (leucocytes) are important in combating invading organisms, while the red blood cells (erythrocytes) carry the respiratory pigment (q.v.).

capillary a minute, thin-walled blood vessel.

closed circulatory system the circulatory system found in vertebrates including humans (see pp. 46-59), echinoderms (see p. 38) and most molluscs (see p. 40).

In closed circulatory systems the blood passes through a continuous network of flexible blood vessels, and can be pumped more quickly. Arteries branch into smaller and smaller vessels, carrying blood away from the heart to all parts of the body.

The smallest vessels (capillaries) are quite permeable and allow fluid to leak out to form the tissue fluid (interstitial fluid) that surrounds all cells. The exchange of gases, nutrients and waste materials between cells and the blood takes place by diffusion through the interstitial fluid. Vertebrates have a separate system of small vessels, the lymphatic system, by which excess interstitial fluid is collected and returned to the blood. After passing through the capillaries, blood returns to the heart through a network of veins.

Pressure is necessary to force blood through vessels and around circulatory systems. This pressure may simply be generated by movements of the surrounding muscles in the body. Return of blood through the veins in the human leg is assisted by contractions of the leg muscles during walking.

In more advanced animals, part or parts of the circulatory system are specialized as one or more hearts to pump blood. In the simplest cases, rhythmical contractions (peristalsis) of the muscles in the vessel walls force the blood along. (Annelid worms - see p. 39 - have a number of vessels that are specialized to act as hearts.)

Valves are used in hearts and in veins to ensure that the flow of blood proceeds in the right direction.

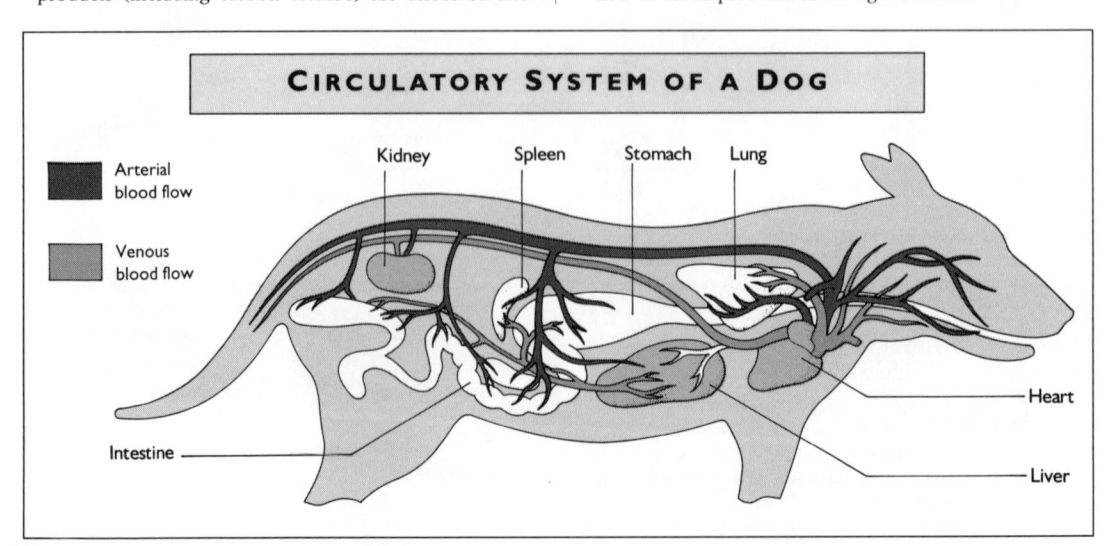

CIRCULATORY SYSTEM OF A DOG

BIOLOGICAL SCIENCES

erythrocyte a red blood cell.

double circulatory system see heart (below).

haemocoel a blood-filled space in an open circulatory system (q.v.).

heart any specialized structure by which blood is pumped around the body.

More specialized hearts consist of strong, muscular chambers. Although these have evolved from contractile blood vessels, they are able to pump much larger volumes of blood with greater force. Such structures range from the relatively simple heart of the fish to the high-pressure systems that are found in higher vertebrates such as birds and mammals.

Lower vertebrates, such as fish, have a single circulatory system. In a fish such as a shark, deoxygenated blood from the veins is collected in a thin-walled chamber (atrium) and then passed to a muscular, thick-walled chamber (ventricle), which contracts forcibly to send it to the gills, where it gains oxygen, and then straight to the rest of the body. Because the passage of blood from the heart to the organs and tissues takes place via the gills, the pressure in such a system is rather low.

As vertebrate life changed from aquatic to terrestrial and air-breathing, the circulation becomes more complex.

In birds and mammals, the heart has two atria and two ventricles, which serve as a double circulatory system (see diagram). Deoxygenated blood from the organs and tissues returns to the right side of the heart, which pumps it to the lungs. The oxygenated blood then returns to the left side of the heart, which pumps it out with great force to the rest of the body. This high-pressure flow of oxygenated blood to active cells allows them to maintain the high metabolic rate characteristic of birds and mammals.

interstitial fluid see closed circulatory system (above) and lymphatic system (below).

leucocyte a white blood cell.

lymphatic system in vertebrates, a network of small vessels by which the interstitial fluid that surrounds cells and through which gas exchange takes place is collected and returned to the blood.

open circulatory system a circulatory system that is found in arthropods (see pp. 41-45) and many molluscs (p. 40). In such a system the heart pumps blood into vessels that are open-ended. Blood flows out of the open vessels to bathe all the cells directly. The blood-filled spaces are called haemocoels. Blood returns to the heart via other open-ended vessels or through holes in the heart. Blood flow in such open systems is relatively slow.

peristalsis rhythmical contractions of the muscles in the arterial walls that force the blood along.

plasma (of blood) the fluid portion of blood.

VERTEBRATE CIRCULATION

- Deoxygenated blood to right lung
- Right atrium
- Right ventricle
- Deoxygenated blood from tissues
- Oxygenated blood to tissues
- Deoxygenated blood to left lung
- Oxygenated blood from lungs
- Left atrium
- Left ventricle

red blood cell see erythrocyte.

respiratory pigment a special molecule in blood that has a high-oxygen carrying capacity and which combines with oxygen at the respiratory surfaces and releases it to tissues in the body.

single circulatory system a circulatory system typical of lower vertebrates, e.g. fish. See heart (above).

valve a structure (especially in the heart and veins) that closes temporarily to prevent the back-flow of blood.

vein a blood vessel by which deoxygenated blood is carried to the heart.

ventricle a chamber in the heart that receives blood from an atrium (q.v.) and pumps it through the arteries.

white blood cell see leucocyte.

HUMAN AND OTHER RESPIRATORY SYSTEMS

All animal cells require a constant supply of oxygen, so that food can be oxidized ('burnt') to release energy. At the same time, nutrients must be carried to each cell, and the various waste products of cellular activity - including carbon dioxide, the by-product of respiration - must be transported away.

In active, multicellular animals, special systems have developed to meet these needs. Respiratory systems allow sufficient oxygen to be obtained from the environment and carbon dioxide to be removed.

The way an animal obtains oxygen from the environment is dictated by its basic body plan, its life style, and whether it lives in water or in air. The amount of oxygen entering an animal and the amount of carbon dioxide lost depends upon the surface area available for gas exchange. In small, simple animals with low energy requirements, such as most worms, sponges and hydras, gas exchange across the body surface is sufficient. These animals tend to be small, with a low metabolic rate and a low demand for oxygen. Movements of the animals or of the surrounding medium are sufficient to keep relatively fresh, oxygen-rich air or water in contact with the body surface. Most amphibians (see pp. 48-49) are also very efficient at 'skin-breathing', although it is not their only means of respiration.

In larger animals, special respiratory systems have developed. Although these differ in structure, all are designed to provide a large surface area for the exchange of gases. A system of ventilation is also needed, whereby the oxygen-carrying air or water in contact with the respiratory surfaces is replaced at regular intervals. The more active an animal is, the greater its energy demand and the more oxygen it requires. Active, warm-blooded (homoiothermic) animals such as birds and mammals, which have greater energy requirements, have respiratory systems with very large surface areas and efficient ventilation.

air sac see alveoli.

alveoli (sing. **alveolus**; or **air sac**) numerous tiny swellings at the termini of the bronchioles in the lungs that increase the surface area for gas exchange with the blood.

amphibians, respiratory system of see lung.

birds, respiratory system of see lung.

bronchi (sing. **bronchus**) in air-breathing vertebrates, two large branches by which air is conveyed from the trachea to the two lungs.

bronchiole any of numerous tiny branching tubes in the lungs.

diaphragm a muscular partition that separates the chest from the abdomen in mammals.

gills are specialized respiratory structures found in some aquatic animals. They are thin and often folded to provide a large surface area for gas exchange. While their outer surfaces are in contact with water, internally they usually have a network of blood vessels. Gills may be external, extending out into the water as in the tadpoles of amphibians, but in most vertebrates they are internal. In fish, the gills are usually located inside a series of slits that perforate the wall of the pharynx, and are ventilated by water passing in through the mouth and out through the slits. In bony fish, each gill consists of many thin filaments. Efficiency of gas exchange is maximized by passing blood through the gill filaments in the opposite direction to the flow of water.

haemocyanin the respiratory pigment (q.v.) in many invertebrates, some molluscs and arthropods. See also oxygen (below).

haemoglobin the respiratory pigment (q.v.) in vertebrate blood. See also oxygen (below).

insects, respiratory system of see trachea (below).

lung a large respiratory structure used for exchange of gases with air.

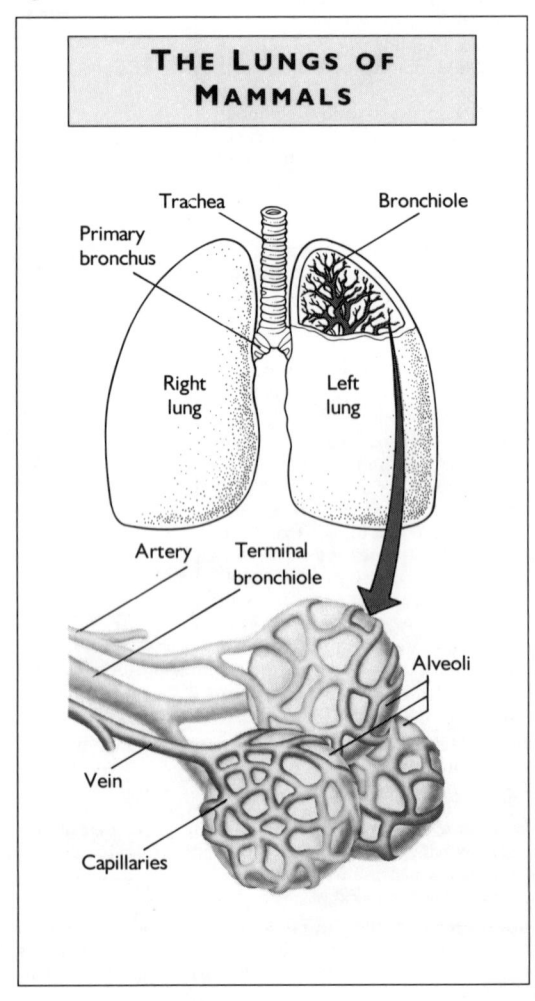

THE LUNGS OF MAMMALS

Some invertebrates have lungs. Spiders have lung books (q.v.). In lower vertebrates, such as frogs and reptiles, the lungs are very simple sacs that are ventilated by movements of the body or mouth.

In mammals (including humans), which have very high metabolic rates, the lungs are complex and efficiently ventilated. Regular ventilation is achieved by contractions of the diaphragm. The enormous surface area of mammalian lungs is due to millions of tiny air-sacs known as alveoli. Air is drawn through the nose and down the trachea and passes into the lungs through a system of ever-narrowing, branching tubes (bronchi and bronchioles), which finally terminate in millions of alveoli. The wall of each alveolus is lined with a network of minute, thin-walled blood vessels called capillaries. As deoxygenated blood from the heart passes through the capillaries, it loses the carbon dioxide it is carrying, which diffuses into the alveoli to be breathed out from the lungs, and picks up oxygen; the oxygenated blood is then returned to the heart, to be pumped around the body once again.

The specialized respiratory system of birds is arranged so that there is a one-way flow of air through the lungs. Air passes into a number of air sacs and through tiny thin-walled tubes (parabronchi), where exchange of gases between air and blood takes place. The flow of air is maintained by movements of the chest and by the squeezing effect of muscles on the air sacs during flight.

lung books the lungs of spiders (see p. 45). Lung books are air-filled cavities in the abdomen into which air enters via a spiracle. A series of parallel, blood-filled plates hang in each cavity to exchange gases.

oxygen a highly reactive element (symbol O; see p. 131), and the most abundant in the Earth's crust. In its molecular form (O_2) it is a gas which comprises 21% of the atmosphere. In this form it is essential to the respiration of plants (see pp. 20-23) and animals - as described here, and in the energy-producing processes (cell respiration) of virtually all cells apart from certain bacteria (see pp. 18-19). It is also crucial to all life as a constituent of water (H_2O) and many organic compounds.

Oxygen does not diffuse very far in tissues, penetrating only about 1mm (1/25in) into tissues in sufficient amounts to maintain life. To ensure that sufficient oxygen reaches all cells, the respiratory surfaces are usually well supplied with blood. They are also very thin, to allow oxygen to diffuse quickly.

Only small amounts of oxygen can be carried in simple solution, so many animals have special molecules called respiratory pigments in their blood. These molecules are able to combine with oxygen, picking it up at respiratory surfaces and liberating it to the cells inside the body. Haemoglobin - the respiratory pigment in vertebrates - contains iron atoms and is responsible for its red colour. It increases the capacity of the blood to carry oxygen by about 75 times. Although haemoglobin is also present in many invertebrates, some molluscs and arthropods have a blue, copper-containing pigment called haemocyanin. The metal atoms in both these molecules are important in binding oxygen. In vertebrates the respiratory pigment is contained within blood cells, but in many invertebrates the pigments are free in the blood fluid or plasma.

parabronchi (sing. **parabronchus**) minute respiratory tubes in birds.

pharynx the cavity formed by the upper part of the alimentary canal (see pp. 66-67), lying behind the mouth and in front of the larynx and oesophagus. In fish gill slits develop in this area.

reptiles, respiratory system of see lung.

respiration in air is the method of obtaining oxygen for all land animals. Air contains far more oxygen than water - about 30 times as much for a given volume. Getting oxygen from air is therefore much easier than from water. Another advantage of air breathing is that air has a low heat capacity and relatively little heat is lost despite the continuous ventilation of large respiratory surfaces. It is no coincidence that the only truly warm-blooded animals - birds and mammals - are air-breathers. Breathing air does have some disadvantages, however. The large respiratory surfaces have to be kept moist, so that gases can dissolve in the surface layers before they diffuse across. Evaporation of water from these moist surfaces can be an important cause of water loss in terrestrial animals. (Water lost in this way can easily be seen on cold days, when water vapour in the breath condenses in the cold air.)

respiration in water is the method of obtaining oxygen for most fish and aquatic invertebrates. Getting oxygen from water is more difficult than from air. A fish may use up to 20% of its total energy in just ventilating its gills, while a mammal may use only 1-2% of its energy in breathing air.

respiratory pigment a special molecule in blood that has a high oxygen-carrying capacity and which combines with oxygen at the respiratory surfaces and releases it to tissues within the body. The respiratory pigment in vertebrates is haemoglobin. See also oxygen (above).

spiracle any of a series of small, paired openings in the cuticle of an insect by which air enters and leaves the tracheae (q.v.).

trachea (plur. **tracheae**)

1) the windpipe; in air-breathing vertebrates, the tube by which air is conveyed from the throat to the lungs.

2) in insects, any of the numerous tiny tubes by which air is conveyed from the spiracles. Insects (see pp. 42-43) are mostly small, active terrestrial animals. To avoid excessive water loss, they are covered by a hard, impermeable cuticle. Air enters the body through a unique system of narrow, branching tubes (tracheae) that carry air deep within the body. The size of the openings to the exterior (spiracles) can be adjusted according to external conditions, and in active insects movements of the body help to move air in and out through the spiracles.

ventilation (in respiration) the means by which oxygen-carrying air or water is brought into contact with the respiratory surfaces; see introduction (above).

vertebrates, respiratory system of see lung (above) respiration in air and respiration in water.

windpipe see trachea.

THE HUMAN DIGESTIVE SYSTEM

Unlike plants, which can synthesize everything they require using energy from the Sun, animals, including human beings, must obtain their nutrients and energy from food. Digestion is the process in which the energy and nutrients contained in food are broken down into a suitable form to be absorbed by the body and utilized as a source of energy, or to synthesize substances such as proteins, enzymes and hormones that are required for the normal functioning of the body.

The nutrients required by the body are proteins, carbohydrates, fats, mineral salts and vitamins. Water is not a nutrient but an adequate intake is essential to replace the water that is lost each day through the skin and lungs, and in urine and faeces.

alimentary canal another name for the gastrointestinal tract.

amino acid an organic compound containing one or more amino groups. Eight amino acids must be provided by the diet - these are called the essential amino acids. The others can be synthesized from other amino acids.

anus the opening at the end of the digestive tract through which waste products are discharged.

appendix (or **vermiform appendix**) a small narrow pouch extending from the lower part of the caecum.

ascorbic acid see vitamin C.

bile a secretion of the liver formed by the breakdown of haemoglobin. It helps break down fats in the small intestine. Bile ducts conduct bile to the duodenum.

biotin a B vitamin essential for metabolism of fat. It is found in egg yolks, liver, tomatoes, raspberries, artichokes.

caecum a pouch which forms the first part of the colon.

carbohydrate any of a large number of organic compounds. Carbohydrates contain carbon, hydrogen and oxygen and provide energy. The most importasnt carbohydrate is starch, a polysaccharide obtainable mainly from cereals. The simple carbohydrates are the monosaccharides (glucose, fructose and galactose) and the disaccharides (sucrose, lactose and maltose). A disaccharide consists of two molecules of monosaccharide. Good sources of simple carbohydrates are fruits, honey, milk and table sugar.

cellulose a polysaccharide carbohydrate made up of unbranched chains of many glucose molecules.

chyme the semi-liquid state into which food is reduced in the human stomach.

colon the large intestine.

cyanocobalamin see vitamin B12.

digestive tract see gastrointestinal tract.

disaccharides see carbohydrates.

duodenum the first part of the small intestine extending from the stomach to the jejunum. On average the duodenum is about 25cm long.

enzyme any protein that acts as a catalyst in certain biochemical reactions.

essential amino acids see amino acids.

faeces the waste residue of food, dead and live bacteria, and water, expelled through the rectum.

fats provide twice the amount of energy as carbohydrates and protein. Fatty acids may be either saturated or unsaturated. A diet high in fat, particularly 'saturated fat' (as found, for example, in red meat and dairy products), has been linked to the development of heart disease.

folic acid a B vitamin essential for maturing red blood cells in bone marrow. It is found in spinach, liver, broccoli, peanuts.

fructose a monosaccharide carbohydrate.

galactose a monosaccharide carbohydrate.

gall bladder a membranous muscular sac associated with the liver and in which bile made in the liver is stored.

gastrointestinal tract a long tube about 9m long through the human body from the mouth to the anus. Here the complex structures present in food are mixed with enzymes and broken down into their simple constituents. These are small enough to be absorbed through the wall of the intestine into the bloodstream.

Food is chewed in the mouth and mixed with saliva which begins starch digestion. It passes through the oesophagus into the stomach which produces the enzyme pepsin and hydrochloric acid to start protein digestion, and mixes the food until it is in a semi-liquid state called chyme. This is then released slowly into the duodenum.

Most digestion takes place in the duodenum. Enzymes, secreted by the pancreas and the duodenum lining, convert fats into fatty acids and glycerol, and polysaccharides into glucose and fructose. These are then absorbed through the wall of the ileum. Glucose, fructose and amino acids are absorbed into the bloodstream and carried to the liver. Fatty acids and glycerol are absorbed into the lymphatic system and enter the bloodstream.

Substances that cannot be digested pass into the colon. Some compounds are fermented by the bacteria there and others are egested as waste products in the faeces via the rectum.

glucose a monosaccharide carbohydrate.

glycogen the form in which carbohydrates are stored in the liver and muscles. It is a polysaccharide carbohydrate consisting of glucose units and is readily converted to glucose.

gullet see oesophagus.

ileum the lower part of the small intestine, on average about 4m long.

intestine the part of the alimentary canal between the stomach and the rectum.

jejunum the middle part of the small intestine, on average about 2.5m long.

lactose a disaccharide carbohydrate.

large intestine (or **colon**) the latter part of the intestine between the small intestine and the rectum. Its main function is to reabsorb water from food wastes into the bloodstream. See gastrointestinal tract.

BIOLOGICAL SCIENCES

liver a large organ well-supplied with blood vessels. It has four functions, two of which are associated with the digestive system:
- the production of bile (see above);
- the reception of all the products of food absorption.

maltose a disaccharide carbohydrate.

monosaccharides see carbohydrates.

mineral salts soluble ions essential in the diet, such as calcium, iron, phosphate, sodium and chloride.

mouth the first part of the digestive tract; see gastrointestinal tract (above).

niacin see nicotinic acid.

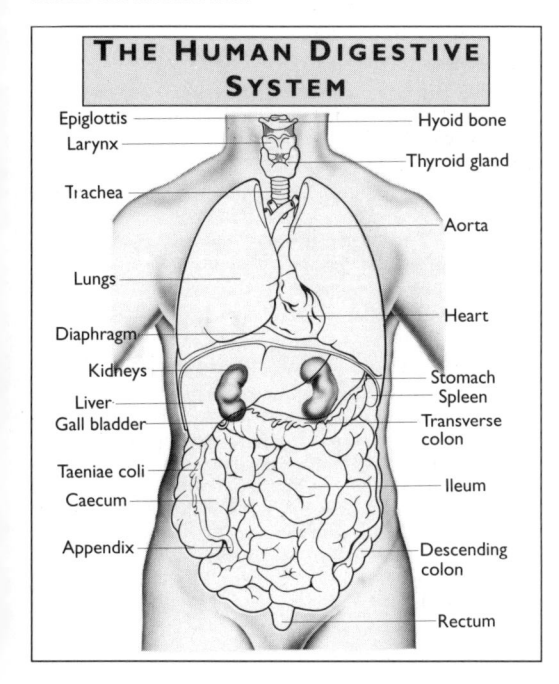

THE HUMAN DIGESTIVE SYSTEM

Epiglottis, Larynx, Trachea, Lungs, Diaphragm, Kidneys, Liver, Gall bladder, Taeniae coli, Caecum, Appendix, Hyoid bone, Thyroid gland, Aorta, Heart, Stomach, Spleen, Transverse colon, Ileum, Descending colon, Rectum

nicotinic acid (niacin) a B vitamin essential for metabolism of carbohydrates and the functioning of the digestive tract and nervous system. It is found in yeast, fish, meat, cereals, peas and beans.

oesophagus (or gullet) a 23cm muscular tube by which food passes from the mouth to the gastrointestinal tract (q.v.).

pancreas a large compound gland near the stomach. One of its functions is to secrete digestive enzymes.

pepsin an enzyme produced in the stomach. When activated by acid it starts the digestion of polypeptide chains (proteins).

peptide the bond linking the amino acids in a polypeptide.

polysaccharide any of a large number of polymers made of peptide chains of amino acids.

proteins are made up of large numbers of amino acids. There are about 20 different amino acids and they can be arranged in any order to produce a larger number of different proteins. Proteins provide cell structure, help fight infections, transport substances around the body and form enzymes and hormones. Milk, eggs, meat, fish and pulses are rich in proteins.

pyloric sphincter the small circular opening at the base of the stomach through which partly digested food passes to the duodenum.

pyridoxine see vitamin B6.

rectum the last section of the gastrointestinal tract. (q.v.).

retinol see vitamin A.

riboflavin see vitamin B2.

small bowel see small intestine.

small intestine the longest part of the gastrointestinal tract (q.v.). It comprises the duodenum, jejunum and ileum.

starch a polysaccharide carbohydrate into which glucose is converted for storage in plants, and which forms the main source of carbohydrate for humans.

stomach a muscular bag between the oesophagus and duodenum; see gastrointestinal tract.

sucrose a disaccharide carbohydrate. Sucrose (table sugar) is formed from a molecule of glucose and a molecule of fructose.

thiamin see vitamin B1.

vitamins complex chemical compounds that are essential in small quantities for many chemical reactions. If a vitamin is lacking in the diet a deficiency disease arises. An excess of certain vitamins can also be dangerous.

vitamin A (retinol) a vitamin essential for growth, vision of poor light, health of the cornea and resistance to infection. It is found in dairy products, fish-liver oils, egg yolks.

vitamin B1 (thiamin) a vitamin essential for carbohydrate metabolism and nervous system functioning. It is found in yeast, egg yolks, liver, wheatgerm, peas and beans.

vitamin B2 (riboflavin) a vitamin essential for tissue respiration. It is found in yeast, meat extracts, milk, liver, kidneys, cheese, eggs, green vegetables.

vitamin B6 (pyridoxine) a vitamin essential for metabolism of fat and protein. It is found in liver, eggs, meat, peas and beans.

vitamin B12 (cyanocobalamin) a vitamin essential for maturing of red blood cells in bone marrow. It is found in liver, fish, eggs, meat.

vitamin C (ascorbic acid) a vitamin essential for formation of red blood cells, antibodies and connective tissue, the formation and maintenance of bones, maintenance of blood capillaries. It is found in blackcurrants, citrus fruits, green vegetables, potatoes.

vitamin D a vitamin essential for absorption of calcium and phosphorus. It is found in fish-liver oil, eggs, butter, cheese. Humans can synthesize vitamin D by exposing the skin to sunlight.

vitamin E is thought to prevent oxidation of unsaturated fatty acids in cells. It is found in vegetable oils, cereals, green vegetables, eggs, butter.

vitamin K is associated with the clotting mechanism of blood. It is found in green vegetables and liver, and can be synthesized in the human gut.

NERVOUS AND ENDOCRINE SYSTEMS

In a multicellular animal, the activity of all the cells must be coordinated if the body is to work as a single functional unit. An animal must be able to respond as a whole to specific sensory stimuli such as heat and touch that are received at a single point on its body. At the same time, the development and functioning of the various tissues and organs involved in physiological processes such as digestion, reproduction and respiration must be precisely coordinated. (Communication occurs even within single-cell living systems (see pp. 18-19) in which chemical signals pass backwards and forwards between the cytoplasm and the nucleus to coordinate the cell's internal activities.)

The coordination system of animal cells - which may be widely separated in the body - depends upon two basic mechanisms. Nervous systems based upon vast networks of intercellular connections provide a means of extremely rapid communication between cells, while endocrine systems allow communication between cells in different tissues by means of chemicals (hormones) in the blood.

For the body to work as an integrated unit, the nervous and endocrine systems must be exactly coordinated. Some nerves produce hormones, so providing the link between the neural and endocrine systems - these neurones are neurosecretory cells (q.v.). In vertebrates, the main link between the nervous and endocrine systems is provided by the pituitary gland (q.v.).

acetylcholine see neurotransmitters.

afferent nerves carry information from sensory receptors in the outer regions of the body to the central nervous system (q.v.).

autonomic function an involuntary function.

axon a single long extension of a nerve cell by which impulses are conducted from the cell body.

brain a concentration of nerve cells acting as the central coordinating point of a nervous system, progressively complex in more complex animals. Different regions of the brain may become specialized to perform different functions. This specialization is most pronounced in vertebrates.

In simple vertebrates, the brain is divided into three main regions: the forebrain, the midbrain and the hindbrain. The forebrain receives information from the nose. The midbrain receives information from the eyes, while the hindbrain receives information from the ears, the organs associated with balance, and the skin. During embryonic development, the roof of the hindbrain enlarges to become the cerebellum (q.v.). The floor of the hindbrain thickens into the medulla oblongata (q.v.).

As the life style and behaviour of vertebrates became more complex, so did their brain. In mammals, two cerebral hemispheres (together forming the cerebrum) develop from the forebrain, and in more advanced species, the surface area of the brain is increased by numerous folds or convolutions. In humans, this area of the brain is associated with intelligent behaviour, language, memory and consciousness. See also central nervous system.

central nervous system (or **CNS**) in vertebrates, comprises the brain and spinal cord. Within the central nervous system, linking the afferent and efferent nerves, are large numbers of special nerves known as interneurones. These form a complex but highly organized network of interconnecting nerve cells and are responsible for integrating the information coming from all sources. All the patterns of movement and behaviour in animals are the result of the way interneurones process and pass on the incoming signals. To integrate all the incoming information, large concentrations of nerve cells in the head region develop into a distinct brain (q.v.).

cerebellum in vertebrates, the part of the brain associated with voluntary movement and balance.

cerebrum see brain (above).

convolutions (of the brain) see brain (above).

dendrite any of a number of branching projections from a nerve cell by which impulses are conducted to the cell body. See nerve cell (above).

dopamine see neurotransmitters.

efferent nerves carry signals from the central nervous system to outlying systems: stimulation of muscle systems, by means of motor nerves, causes contraction, while stimulation of glandular systems causes secretion.

endocrine cell any cell that has been modified to produce secretions that act as chemical signals over much greater distances.

endocrine gland see endocrine system and gland.

endocrine system a system of glands that produce chemicals (hormones, q.v.) that are transported in the blood and carried to distant tissues, whose activity they modify. The system is principally involved in the control and regulation of growth, reproduction, the internal environment, and energy production. Chemical signals are used by all multicellular animals. In vertebrates, many different hormones are known, affecting almost every part of the body. Examples include insulin, growth hormone, thyroxine, oestradiol and testosterone.

forebrain see brain (above).

ganglion (plur. **ganglia**) an encapsulated group of nerve-ending cells.

gland any organ producing a secretion that is either carried by the blood (an endocrine gland) or passed via a duct to where it is needed (e.g. salivary and mammary glands).

growth hormone a hormone produced by the pituitary gland. It controls general body growth and metabolism.

hindbrain see brain (above).

hormone a chemical signal produced by a particular type of cell and carried in the blood. The structure of each hormone is highly specific, and the 'target cells' that respond to it have special receptor molecules that recognize and bind with the hormone concerned. The binding of the hormone to the receptor causes changes in the metabolic or genetic machinery of the target cell in such a way as to alter its activity. Endocrine cells producing particular hormones are often grouped together in endocrine glands. See also endocrine system (above).

hypothalamus part of the vertebrate brain, controlling the pituitary gland (q.v.) and important as an integrating centre for many autonomic functions.

impulse a tiny electrical pulse acting as a signal between nerve cells (q.v.).

insulin a hormone produced from the pancreas. It controls the levels of blood sugar.

interneurone see central nervous system (above).

medulla oblongata in vertebrates, part of the brain associated with involuntary actions such as breathing and heart beat.

midbrain see brain (above).

motile systems systems involved with movement.

motor nerve see efferent nerves.

nerve cell (or **neurone**) the functional unit of the nervous system. Many millions of these work together to form an intricate network that far outstrips in complexity the circuitry of the most advanced computer. Neurones vary greatly in shape, but typically they consist of a cell body from which there are a variable number of very thin, branching projections (dendrites) and a single much longer extension (the axon).

Communication along nerves depends upon tiny electrical signals. These are sent as discrete electrical pulses (impulses) that can be propagated along the axons. Each impulse is continuously regenerated as it passes along the axon and can therefore be transmitted over great distances. Information about the strength of the original signal is encoded in the frequency of impulses in each nerve axon and in the number of nerves that are carrying such impulses. Axons normally terminate at synapses, where they link up with the cell bodies or dendrites of other neurones, or with secretory cells or muscles. At the synapse, the electrical impulses passing along the axon cause the release of tiny amounts of neurotransmitters, which diffuse across to the next cell. See also nervous system.

nerve net the simplest form of nervous system found. It occurs in cnidarian animals such as the hydra. The nerve cells are scattered throughout the body and there is no central coordinating point. From each nerve cell there is a number of long projections, and an irregular, interconnecting network of these nerve fibres surround the body. Signals are transmitted in all directions through this diffuse network, becoming less intense as they spread from the source. The stronger the initial stimulus, the more nerves in the net will be affected. In animals that have more complex motile systems, some of the nerves are linked together in distinct tracts.

nervous system a network of cells adapted for very rapid communication over long distances within an animal body.

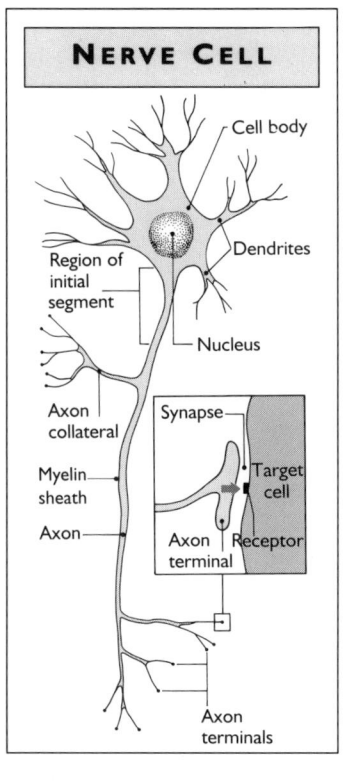

NERVE CELL

- Cell body
- Dendrites
- Region of initial segment
- Nucleus
- Axon collateral
- Synapse
- Myelin sheath
- Target cell
- Axon
- Axon terminal
- Receptor
- Axon terminals

The simplest nervous system found is the nerve net (q.v.). In more complex animals, nerve cells increase in number and greater organization develops. Not only do collections of nerve fibres occur together in nerve tracts but the cell bodies of neurones are grouped together in ganglia. With this greater organization, the nerves develop specific functions - see afferent nerves and efferent nerves.

neuromodulator any chemical substance that acts beyond the vicinity from which it is released. Neuromodulators produce much more diffuse effects and change the levels of excitability in whole groups of neurones.

neurone an alternative name for a nerve cell (q.v.).

neurosecretory cell a hormone-secreting nerve cell. The hormones produced pass down the axons to be stored at the nerve terminals and are released during neural activity.

neurotransmitters tiny amounts of chemicals which are released at the synapse (q.v.) when electrical impulses pass along the axon (q.v.). Many different chemicals are known to act as neurotransmitters: acetylocholine, noradrenaline, dopamine, glycine and serotonin are common examples. The neurotransmitters bind to specific receptors and cause new electrical or chemical signals to be initiated in the neighbouring cell. See also nerve cell and neuromodulators.

noradrenaline see neurotransmitters.

oestradiol in female vertebrates, a hormone produced from the ovaries. It influences sexual characteristics.

pancreas in vertebrates, an endocrine gland; see insulin.

peripheral nervous system (or **PNS**) the combination of nerves, ganglia and the brain monitoring the involuntary functions of the body.

pituitary gland in vertebrates, an endocrine gland. This gland is under the control of the hypothalamus (q.v.). See also growth hormone.

reflex action a quick, automatic muscular response to a sudden stimulus involving only sensory and motor nerves and the spinal cord.

serotonin see neurotransmitters.

synapse a specialized junction between nerve cells. See nerve cell.

testosterone in male vertebrates, a hormone produced by the testes. It influences sexual characteristics.

thyroid gland in vertebrates, an endocrine gland; see thyroxine.

thyroxine a hormone associated with body growth and metabolism. It is produced by the thyroid gland.

SENSES AND PERCEPTION

Animals gain information about what is going on in the world around them and within their own bodies through their senses. This information is received by various receptor cells as sensory stimuli and is transformed into nerve impulses that are transmitted to the central nervous system (see pp. 68-69). Sensory information passed to the brain may result in an awareness of the stimulus (sensation), which can be studied in animals by observing changes in their behaviour. Alternatively, it may result in the formation of an internal image of the stimulus (sensory perception), which is difficult to study in non-human animals.

The various senses are based on the sensitivity of different types of sensory cell to different forms of energy. Sensations of taste and smell, for example, are due to receptors responsive to chemical energy, while vision depends upon receptors sensitive to the electromagnetic energy of light. When a stimulus reaches a sensory cell, it causes a change in the electrical potential varying with the strength of the stimulus. The potential change is then transmitted across the cell or passed on to a neighbouring cell, where nerve impulses are produced. Impulses may be produced for the full duration of the stimulus or only at the beginning (a phasic response). Some sensory cells are phasitonic (q.v.). Impulses from the peripheral sensory cells may excite or inhibit activity in subsequent sensory neurones (nerve cells), and these may interact with each other so that information is processed before or after reaching the brain. More central sensory neurones - see feature detectors - may respond to complex aspects of the stimulus, such as curved edges, the direction of movement, or a particular type of sound.

adaptation see phasic response.

ampullae of Lorenzini see temperature reception (below).

balance a sense whose receptors are located in the semicircular canals of the ear.

chemoreception sensitivity to chemical stimuli.

cold receptor a mammalian temperature receptor capable of responding to a decrease in temperature over the range 12-35°C (54-95°F). See also warm receptor and temperature reception.

compound eye a feature of insects; see lens (below).

cornea a transparent membrane at the front of the vertebrate eye.

ear the organ of hearing and balance (q.v.). Sound waves stroking the ear drum set up vibrations in the three auditory ossicles. These stimulate receptor cells in the cochlea, which communicate with the brain.

eye, of vertebrates the organ of sight. In vertebrates, the retina, a sheet of photoreceptors (see photoreception) and associated cells is arranged on the inside of a sphere, the eyeball. Light from the visual field enters through a curved outer cornea and passes through a lens, both of which bend light rays to focus them on the retina, where an inverted image is formed (the brain re-inverts this image). Each photoreceptor detects light from a minute part of this image. The lens can be moved or changed in shape by muscles in such a way that the eye can adjust to both near and far objects. Between the cornea and the lens lies the iris, an opaque area with a central aperture (the pupil) which regulates the amount of light entering the eye. In bright light, muscles in the iris contract reducing the size of the pupil.

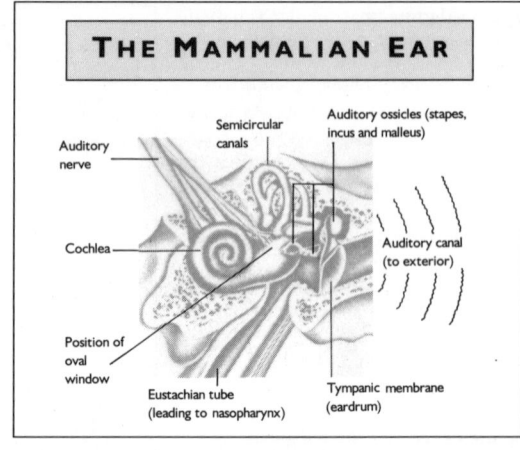

feature detector a cell that responds to a particular aspect of a stimulus; see introduction (above).

gustation the sense of taste (q.v.).

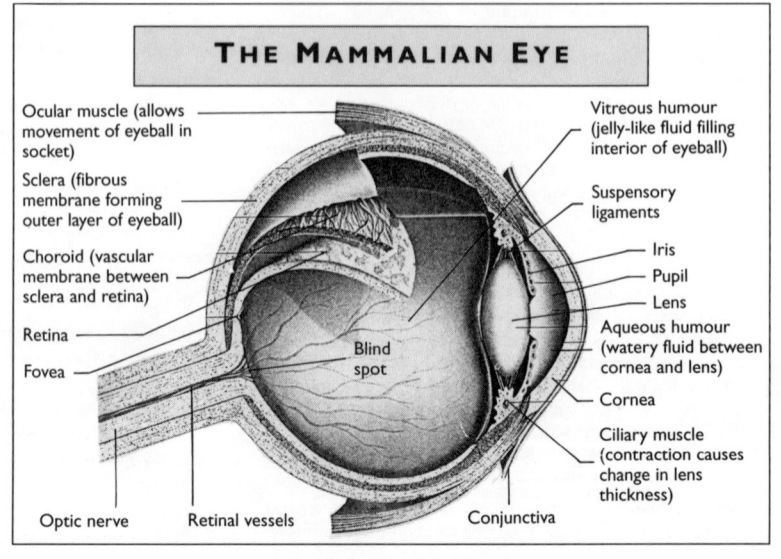

BIOLOGICAL SCIENCES

iris an opaque muscular diaphragm in the vertebrate eye that surrounds the pupil (q.v.) and alters its size and, in some animals, its shape.

Jacobsen's organ see smell (below).

lens the part of the vertebrate eye responsible for focusing incoming light on the retina (q.v.). In the compound eyes of insects there are many lenses.

ocellus a group of photoreceptive cells in a shallow pit. Such organs detect light intensity and sometimes direction, and occur in a wide range of invertebrates from jellyfish to insects.

olfaction the sense of smell (q.v.).

phasic response an impulse produced only at the beginning of a stimulus. For example, the sensation of our clothes fades soon after we put them on as a result of this process (known as adaptation), so preventing the nervous system from being overloaded with information.

phasitonic response a response at more than one level. For example, temperature receptors signal a change in temperature with a rapid burst of impulses and then signal the level of temperature at a lower but consistent rate of impulses related to the actual temperature.

pheromone a chemical signal released by one animal that has a specific behavioural or physiological effect on another member of the same species.

photopigment see photoreception.

photoreception the ability to detect light - an ability which depends upon the presence of light-sensitive receptor cells of special molecules called photopigments, which absorb light energy and change their shape, ultimately causing a change in electrical potential across the cell membrane. Different pigments absorb a different range of light wavelengths from 400 nanometres (red) to 700nm (violet), which is therefore the spectrum of visible light. Insects, however, are able to detect wavelengths in the ultraviolet region. The most familiar pigment is rhodopsin, which occurs in the rod cells (q.v.) of vertebrates and absorbs light across a wide spectrum. Light reception is widespread in the animal kingdom. The simplest type of light sensitive organ is the ocellus (q.v.). True vision is the ability to form an image of the light source. There are various ways of doing this. In one system, similar in principle to a pinhole camera, a narrow beam of light is allowed to pass through a tiny aperture onto a photoreceptive sheet, which is made up of an array of photoreceptors arranged on a curved surface behind the aperture. See eye (above).

pupil the central, usually circular aperture in the iris of vertebrate eyes, the size of which dictates the amount of light entering the eye.

retina a light-sensitive membrane at the back of the vertebrate eye.

sensation an awareness of an external stimulus.

sense organ a group of sensory cells of the same type.

sensilla invertebrate taste organs; see taste (below).

sensory perception an internal image of a stimulus.

sensory signals see introduction (above).

sight see photoreception (above).

smell one of the basic senses. The detection of distant objects, including food, predators, mates, migratory land and the home base, often involves smell. Insects may have 40 000 to 200 000 olfactory receptors on their antennae. The olfactory receptors of vertebrates lie in cavities. In many fish, for instance, there are blind-ending nasal sacs in the head. In mammals the olfactory membrane lies at the top of the nose. In keen-scented mammals such as rabbits and dogs, the membrane is greatly folded to increase its surface area and hence the number of receptor cells.

In humans the olfactory membrane covers 5cm^2 ($\frac{3}{4}$ sq in) and contains about 5 million olfactory cells, but in the German shepherd dog in covers 150cm^2 (23 sq in) and comprises nearer 220 million olfactory cells. Many terrestrial vertebrates have a second olfactory area, the vomeronasal organ or Jacobsen's organ, which lies in the ventral region of the nose or in the roof of the mouth. In mammals such as guinea pigs, mice and cats, it appears to be associated with the detection of sex pheromones. In snakes odours are picked up on the tongue, which is then inserted into the openings of the vomeronasal organ on the roof of the mouth.

spectrum (of visible light) see photoreception.

taste one of the basic senses. Sensitivity to chemical stimuli is found in all organisms, including bacteria. The sense of taste, which is generally concerned with detecting food, requires contact with the source of the chemical. If the source is distant, the sense of smell is involved. The gustatory cells of invertebrates may be found in any part of the body and often grouped together into organs called sensilla. In vertebrates, gustatory receptors are grouped together into taste buds (q.v.). Humans can distinguish four basic taste qualities - sweet, salt, sour and bitter - each of which is most strongly detected by a particular area of the tongue.

taste bud a grouping of gustatory receptors in vertebrates. These are commonly found on the tongue, but may be on the roof of the mouth and in the throat. (In some bottom-dwelling fish they are on projections known as barbels.)

temperature reception (or **thermoreception**) a basic sense of many living systems. The mechanism of temperature reception is not well understood. Insects, especially parasites, can detect small changes in temperature. Some fish have temperature receptors on the head at the bases of jelly-filled ducts called the ampullae of Lorenzini. Mammals and birds have thermoreceptors in the skin, tongue, central nervous system and deep organs. Mammals have two kinds of temperature receptor: warm receptors and cold receptors.

thermoreception an alternative name for temperature reception.

threshold level the lowest intensity of stimulation that any cell responds to.

tonic response an impulse that is produced for the full duration of a stimulus.

vomeronasal organ see smell (above).

warm receptor a mammalian temperature receptor capable of responding to an increase in temperature over the range 22-45°C (72-113°F). See also cold receptor and temperature reception.

REPRODUCTION OF ANIMALS

One of the fundamental characteristics of living systems is the ability to reproduce. This has ensured the survival of many different and diverse patterns of life over thousands of millions of years, even though particular individuals live for only a short time. The end result of reproduction is a generation of new individuals, which are in turn capable of reproducing. To accomplish this, reproduction involves many different processes: not only must new individuals be produced, but they must also survive and grow, developing the special structural and behavioural characteristics of their species as they do so. There are two types of reproduction: asexual reproduction and sexual reproduction.

acrosome see sperm.

asexual reproduction involves just a single parent, which produces offspring by splitting or dividing the parent body. All the offspring produced in this way are identical both to the parent and to each other. A form of asexual reproduction is budding (q.v.).

budding a form of asexual reproduction (q.v.), in which a small part of the body grows and separates from the parent to develop into a new individual. This is common in cnidaria (q.v.). The ability to reproduce asexually allows animals to reproduce very rapidly under favourable conditions.

cervix the necklike part of the uterus.

copulation close physical contact between male and female in order to transfer sperm, occurring in many animals practising internal fertilization including birds, mammals and most insects.

egg (or **ovum**) the gamete produced by the female parent. The eggs produced by the female are typically large because they contain a considerable amount of stored material to support the early development of the embryo. This represents a considerable drain upon the resources of the mother and sets a limit to the number of eggs that can be produced.

external fertilization see fertilization.

Fallopian tube either of two slender tubes that carry ova from the ovary to the uterus.

fertilization in sexual reproduction, the fusion of sex cells (gametes) to form a zygote (q.v.). In animals fertilization may take place externally, as in most fish and amphibians such as frogs, or internally, in insects, birds and mammals.

flagellum the tail of the sperm (q.v.).

gamete a sex cell.

genitalia see sexual reproduction.

gonad in animals, an organ in which gametes (q.v.) are produced; see testes and ovary.

hermaphrodite an individual that has both male and female reproductive organs, a common condition in invertebrate animals.

internal fertilization see fertilization.

intromittent organ an organ of the male that is inserted into the female genital tract to transfer sperm, e.g. the mammalian penis.

ovary in animals, a female organ (gonad), usually paired, in which egg cells or ova (see ovum) and various hormones (e.g. oestrogen) are produced.

ovum (plur. **ova**) the egg (q.v.).

parthenogenesis a specialized form of reproduction in which eggs develop into new individuals without the need for fertilization. This occurs sporadically throughout the animal kingdom, for example in some minute aquatic invertebrates and in some lizards and insects - in honeybees, unfertilized eggs develop into females (workers and queens) while fertilized eggs develop into males.

penis an intromittent organ (q.v.).

sexual reproduction the method by which most animals reproduce. New individuals are formed not by simple division from a single parent but as the result of the fusion of two sex cells (gametes), each from a different parent. The parents are usually sexually differentiated (male and female) each producing a different type of gamete. The internal ducts that store and deliver the gametes differ between the sexes, and the areas surrounding the external openings of the tracts (the external genitalia) are often specialized to meet the needs of mating and fertilization. The two cells that fuse to form the new individual are highly specialized - eggs and sperm. Fusion between the two gametes at fertilization produces a new, single-celled individual (the zygote; q.v.).

sperm (short for **spermatozoa**) the gamete produced by the male parent. Sperm are much smaller than eggs. They are motile cells which are adapted to find, penetrate and fuse with eggs. The small head of the sperm contains the nucleus, and a bag of enzymes (the acrosome) at the very front of the sperm is important in helping the sperm to penetrate the various protective layers of the egg. A long flagellum provides the means for swimming. The small size of sperm means that only limited resources are required to produce millions of them. Once released, the motility of sperm increases the chances that eggs and sperm will meet. Complex molecular mechanisms allow the sperm to fuse with an egg of the appropriate species and prevent other sperm fertilizing the same egg.

spermatozoa see sperm.

spermatophore a package or capsule of sperm picked from the ground by the female or transferred to her directly, a feature of various animal groups practising internal fertilization, including many crustaceans, insects and salamanders.

testis (plur. **testes**) in male animals, an organ (gonad), usually paired, in which sperm are produced. In most animals the testes are internal, but in mammals they are generally contained in a scrotal sac.

uterus (or **womb**) in female mammals, a hollow muscular cavity in the pelvic region in which the foetus develops.

vagina the elastic, muscular tube of female mammals extending from the cervix to the vulva, that receives the penis during intercourse and serves as the birth canal.

vulva the external genital opening of female mammals.

womb see uterus.

zygote the single-celled individual that is produced as the result of the fusion of a male and a female gamete. The zygote is diploid, that is it has a full set of chromosomes.

BIOLOGICAL SCIENCES

MARINE MAMMALS

The mammals first evolved on land but at least three separate groups have returned to the seas, where their fish ancestors had lived millions of years before, and have become adapted to life in the waters. The cetaceans (whales and dolphins) and the sirenians (the dugong) live their entire lives in water, and have lost the hind limbs and the ability to move on land. The pinnipeds (seals and walruses) come ashore to breed and retain their hind limbs, but move awkwardly on land.

The pressures of adapting to life in water have led to a degree of uniformity in the appearance of marine mammals, but their origins are in fact very diverse. Cetaceans are thought to have evolved from the same ancestral stock as the ungulates or hoofed mammals, diverging from their common ancestor some 650 million years ago. The sirenians are related to the proboscideans - which include modern elephants. The pinnipeds are thought to be very closely related to carnivorous animals such as bears and weasels, and indeed are sometimes included in the order Carnivora.

adaptations for life in water are achieved in a number of ways. A major problem for a warm-blooded animal in water is reducing loss of body heat. All marine animals tend to be large in size, since a low surface area to volume ratio reduces heat loss. Sirenians reach a mass of 900kg (1984lb), while the largest seals - elephant seals - can approach 4 tonnes. The cetaceans include the largest animals that have ever lived: the blue whale can grow to lengths of 35m (115ft) and have an estimated mass of 130 tonnes. Loss of heat is also reduced by an insulating layer beneath the skin known as blubber. In whales this can be up to 1m (3ft 3in) thick.

Like fishes, marine mammals have streamlined bodies to reduce drag as they move through water. Movement is hindered by parts of the body that protrude, so external features such as limbs, ears and hair are reduced. In the cetaceans and sirenians all external signs of the hind limbs have disappeared, the pelvic girdle is vestigial, and hair is virtually or totally absent. The seals and walruses need to move around on land to breed, and so they retain hair for warmth and also have hind limbs, but these are reduced in length and are not very efficient on land.

The fore limbs of all marine mammals, and the hind limbs where present, are modified to form webbed paddles. The hands are long and flat, and in the cetaceans there are extra finger bones to give increased length. Only the parts of a limb below the elbow or knee protrude and movement is limited, generally to an up and down movement. In cetaceans and sirenians horizontally flattened tail flukes stiffened with cartilage provide the propulsion for swimming.

Many marine mammals can remain submerged for up to 30 minutes (although shorter dives are more common), and some for as long as two hours. They can close off air passages and survive on low levels of oxygen by using anaerobic respiration in the muscles, and a system for storing oxygenated blood called the retia mirabilia.

baleen (or **whalebone**) a horny, flexible material arranged in a series of thin sheets and hanging from the upper jaw of baleen whales, used in filtering plankton from water.

blubber see adaptation to life in water.

cetacean any aquatic mammal of the order Cetacea, which is divided into two main groups: the toothed whales (suborder Odontoceti, 72 species), which include dolphins, porpoises and most of the smaller whales, and the baleen whales (suborder Mysticeti, 10 species), which include the large to very large filter-feeding whales.

Delphinidae a cetacean family; the dolphins.

dolphin any small marine toothed cetacean (q.v.) of the family Delphinidae (32 species), found in all oceans. They are known for their intelligence.

dugong a herbivorous marine mammal, the only species in the family Dugongidae (order Sirenia). They are found in tropical and subtropical coastal waters in the Indian and western Pacific Oceans.

manatee any of three species of herbivorous marine mammal of the family Trichechidae (order Sirenia). They are found in tropical and subtropical coastal waters of the Atlantic Ocean.

Mysticeti a cetacean suborder.

Odobenidae a pinniped family; the walrus (q.v.).

Odontoceti a cetacean suborder.

Otariidae a seal family (q.v.).

Phocidae a seal family (q.v.).

Phocoenidae a cetacean family; porpoises.

pinniped any carnivorous marine mammal of the order Pinnipedia, which comprises three families - two families of seals and the walrus.

porpoise any small toothed cetacean (q.v.) of the family Phocoenidae (32 species), inhabiting coastal waters, mainly in the northern hemisphere.

seal any carnivorous marine mammal of the families Phocidae ('true seals', 18 species) and Otariidae (eared seals, including fur seals and sea lions, 14 species).

sea cow see sirenian.

sea lion see seal.

sirenian (or **sea cow**) any herbivorous marine mammal of the order Sirenia; see manatee and dugong.

walrus a large carnivorous pinniped (q.v.) mammal found in Arctic coastal waters, the sole species of the family Odobenidae.

whale common name for most of the larger cetaceans (q.v.). See also cetacean and introduction (above), and adaptations for life in water.

Physics

MECHANICS - MOTION AND FORCE

Physics is the study of the basic laws that govern matter, Mechanics is the branch of physics that describes the movement or motion of objects, ranging in scale from a planet to the smallest particle within an atom. Sir Isaac Newton developed a theory of mechanics that has proved highly successful in describing most types of motion, and his work has been acclaimed as one of the greatest advances in the history of science.

The Newtonian approach, although valid for velocities and dimensions within normal experience, has been shown to fail for velocities approaching the speed of light and for dimensions on a subatomic scale. Newton's discoveries are considered to be a special case within a more general theory.

Newtonian mechanics were so successful that a mechanistic belief developed in which it was thought that with the knowledge of Newton's laws (and later those of electromagnetism; see pp. 112-13) it would be possible to predict the future of the Universe if the positions, velocities and accelerations of all particles at any one instant were known. Later the quantum theory and the Heisenberg Uncertainty Principle confounded the belief by predicting the fundamental impossibility of making simultaneous measurements of the position and velocity of a particle with infinite accuracy.

acceleration the rate of change of the velocity, if a body moves with a changing velocity.

Acceleration may be defined as the change in velocity in a given time interval. Its dimensions are velocity divided by time, and are (usually) given in metres per second (ms^{-2}).

At its simplest, the equation may be given as:

$$\text{Acceleration} = \frac{\text{change in velocity}}{\text{time taken for this change.}}$$

Note that since velocity is speed in a given direction a change in velocity may involve a change in direction rather than speed or perhaps a combination of speed and direction. See circular motion (below) and centripetal acceleration (below).

When a body moves with uniform acceleration, the displacement, velocity and acceleration are related - these relationships are described in the kinematic equations.

acceleration due to gravity see gravitational acceleration.

acceleration of free fall see gravitational acceleration.

centrifugal force a 'centre-fleeing' force.

In an accelerating or non-inertial frame of reference, Newton's second law of motion will not work unless some fictitious force is introduced. For example, passengers on a fair ground merry-go-round feel as if they are being forced outward when the machine is operating. This is ascribed to a 'centre-fleeing' or centrifugal force.

The passengers experience this 'centre-fleeing' force because they are moving within the system; they are within an accelerating frame of reference (see circular motion, below). To an observer on the ground it appears that the passengers on the merry-go-round should fly off at a tangent to the circular motion unless there were a force keeping them aboard. This is the centripetal force and is experienced as the friction between each passenger and the seat. If a passenger were to fall off, it would be because the centripetal force was not strong enough, not because the centrifugal force was too great.

centripetal acceleration literally means 'centre-seeking' acceleration. It is the acceleration of a body moving in a circular path (see circular motion; below). Centripetal acceleration is directed inward, towards the centre of the circle (see diagram).

centripetal force a force acting on a body which causes it to move in a circular path. See centrifugal force; above.

circular motion occurs when a body moves in a circular path at constant speed - in this case its direction of motion (and therefore its velocity) will be changing continuously. Since the velocity is changing, the body must have acceleration, which is also changing continuously. Thus the laws of uniformly accelerated motion do not apply. The acceleration that occurs in this case is centripetal acceleration (q.v.).

CENTRIPETAL ACCELERATION

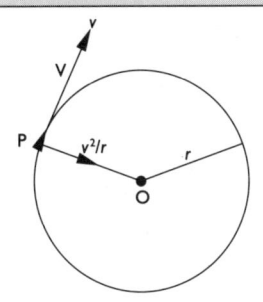

At point P the body is moving with instantaneous velocity v. The centripetal force along PO is v^2/r (where r is the radius of the circle), and this force prevents the body from moving in a straight line along PV. The magnitude of V is the speed which is constant.

conservation of momentum, principle of states that, when two bodies interact, the total momentum before impact is the same as the total momentum after impact. Thus the total of the components of momentum in any direction before and after the interaction are equal.

This can be stated as:

$$m_1 u_1 + m_2 u_2 = m_1 v_1 + m_2 v_2$$

where

m_1, m_2 are the masses

u_1, u_2 are the initial velocities

v_1, v_2 are the resultant velocities of the bodies.

decreasing velocity means that the velocity has different values at different times; it is decreasing with time. See velocity (below) and diagram.

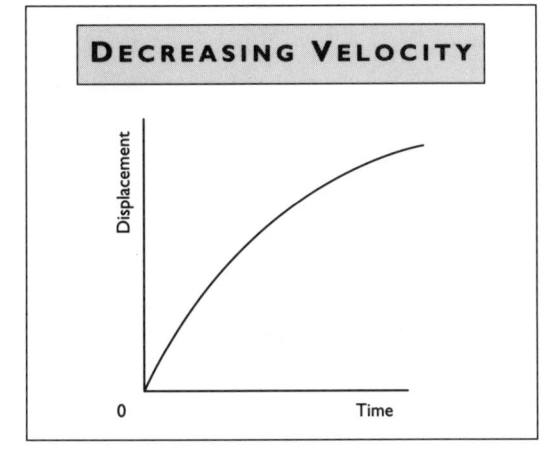

displacement see motion (below).

electro-weak force see fundamental forces.

force (in physics) something that causes a change in the velocity of an object.

fundamental forces (in physics) are the four forces that occur in nature - they are gravitational force (see below), the electromagnetic force (see pp. 112-13) and the strong and weak nuclear forces (see pp. 120-21).

The electromagnetic and weak forces have recently been shown to be part of an electro-weak force.

gravitation see gravitational force (below).

gravitational acceleration (or **acceleration due to**

gravity or (**acceleration of free fall**) the downward acceleration of objects falling towards the Earth.

The 16th-century Italian physicist and astronomer Galileo Galilei investigated the motion of objects falling freely in air. he believed that all objects falling freely towards the Earth have the same downward acceleration. This is gravitational acceleration. Near the surface of the Earth it is $9.80 m s^{-2}$, but there are small variations in its value depending upon latitude and elevation.

In the idealized situation, air resistance is neglected, although in a practical experiment it would have to be considered. In a demonstration on the Moon in August 1971, an American astronaut showed that, under conditions where air resistance is negligible, a feather and a hammer, released at the same moment from the same height, would fall side by side. They landed on the lunar surface together.

gravitational constant (symbol G) is the constant of proportionality implicit in Newton's law of gravitation (q.v.).

gravitational force or **gravity** see box p. 76.

horizontal component of velocity see instantaneous velocity.

increasing velocity means that the velocity has a different value at different times; it is increasing with time. See velocity (below) and diagram.

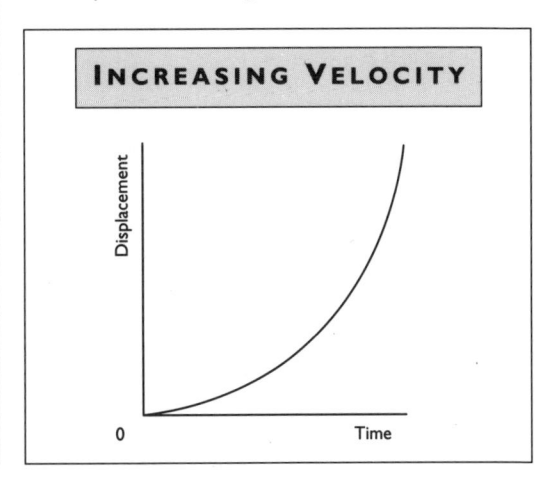

DECREASING VELOCITY

INCREASING VELOCITY

inertia see Newton's first law of motion (below).

instantaneous velocity the velocity of any instant.

When real motion is considered, both the magnitude and the direction of the velocity have to be investigated. A golf ball, hit upwards, will return to the ground. During flight its velocity will change in both magnitude and direction. In this case, instead of average velocity, the instantaneous velocities have to be evaluated. The velocity can, at any instant, be considered to be acting in two directions, vertical and horizontal. Then the velocity at that instant can be split into a vertical component and a horizontal component.

Each component can be considered as being uniformly accelerated rectilinear motion, so the kinetic equations (q.v.) can be applied in each direction. Then the instantaneous velocity and position at any point of the flight can be calculated.

kilogram (symbol kg) the SI unit of mass. The kilogram is defined as a mass equal to that of the international platinum-iridium prototype which is kept by the International Bureau of Weights and Measures at Sèvres, near Paris.

kinematic equations (or **laws of uniformly accelerated motion**) see box.

kinematics the study of bodies in motion, ignoring masses and forces.

linear motion see motion.

mass (of a body) is often confused with its weight. The mass is the amount of matter in the body, and is a measure of its inertia (see Newton's first law of motion). See also weight (below).

mechanics see introduction.

momentum (of a body) is defined as the product of its mass and velocity. Newton called momentum the 'quantity of motion'.

GRAVITATIONAL FORCE

Gravitational force (or **gravity**) is one of the four fundamental forces (q.v.) that occur in nature.

Gravitational force is the mutual force of attraction between masses. The gravitational force is much weaker than the other three fundamental forces (see above). However, this long-range force should not be thought of as weak force.

An object resting on a table is acted on by the gravitational force of the whole Earth - a significant force. The almost equal force exerted by the table on that object is the result of short-range forces exerted by molecules on its surface.

Newton's law of gravitation was first described in his *Philosophiae Naturalis Principia Mathematica* (The Mathematical Principles of Natural Philosophy), which he wrote in 1687. Newton used the notion of a particle, by which he meant a body so small that its dimensions are negligible compared to other distances.

Newton's law of gravitation states that every particle in the universe attracts every other particle with a force that is directly proportional to the product of their masses and inversely proportional to the square of the distance between them.

This may be given as

$$F = \frac{Gm_1 m_2}{x^2}$$

where

G is the gravitational constant (q.v.),

F is the force,

m_1, m_2 are the masses,

x is the distance between the particles.

The law is an 'inverse-square law', since the magnitude of the force is inversely proportional to the square of the distance between the masses. (A similar inverse-square law applies for the force between two electric charges - Coulomb's law.)

If Newton's law of gravity is combined with his second law of motion (q.v.) it follows that:

$$g = \frac{GM}{d^2}$$

where

g is the acceleration due to gravity,

G is the gravitational constant (q.v.),

M is the mass of the Earth, and

d is the distance of the body from the centre of the Earth.

(For a body on the surface of the Earth g (that is gravity) = $9.80665 \, ms^{-2}$)

Gravity also exists on other planets and their satellites (their moons), but because gravity depends upon the mass of those bodies and their diameters, the strength of the gravitational force is not the same as it is on Earth.

PHYSICS

DISPLACEMENT

Anna walks from A to X, then to Y, then to B. AB is her displacement, and the arrow shows the direction of the displacement

motion (of a body) occurs when a body is moving in time and space. If the body moves from one position to another, the straight line joining its starting point to its finishing point is its displacement, see diagram.

Displacement has both magnitude and direction, and is therefore said to be a vector quantity. The motion is linear.

newton (symbol **N**) the SI unit of force. The newton is defined as the resultant force that, acting on a mass of 1kg, produces an acceleration of $1 ms^{-2}$.

Newton's laws of motion see box p. 78.

Newton's first law of motion see box p. 78.

Newton's second law of motion see box p. 78.

Newton's third law of motion see box p. 78.

Newton's law of gravitation see gravitational force (above).

particle (Newtonian) was defined by Newton as being a body so small that its dimensions are negligible compared to other distances. See gravitational force (above).

principle of the conservation of momentum see conservation of momentum, principle of.

quantity of motion see momentum (above) and Newton's second law of motion.

scalar quantity a quantity in which direction is either not applicable or not specified.

speed the ratio of a distance covered by a body in a given amount of time to that time.

At its simplest, the equation may be given as:

$$\text{Average speed} = \frac{\text{distance moved}}{\text{time taken}}$$

Speed - compared with velocity - has magnitude, but is not considered to be in any particular direction. Speed is a scalar quantity rather than a vector quantity.

uniform acceleration uniformly accelerated motion.

uniformly accelerated motion, laws of see kinematic equations.

uniform velocity means that the speed and the direction taken remain the same. See velocity (below) and diagram.

In the accompanying diagram, *AB* represents the distance that is travelled in the period of time that is represented by *OB*. Thus:

$$\text{Velocity} = \frac{AB}{OB}$$

velocity the rate at which a body moves in a straight line or rectilinearly.

At its simplest the equation may be given as:

$$\text{Velocity} = \frac{\text{distance moved in a particular direction}}{\text{time taken}}$$

Like displacement, velocity has both magnitude and direction and is a vector quantity. (See also speed.)

The average velocity during this rectilinear motion is defined as the change in

KINEMATIC EQUATIONS

Kinematic equations (or **laws of uniformly accelerated motion**) equations that describe the relationships of displacement, velocity and acceleration.

For a body moving in a straight line with uniformly accelerated motion:

1. $v = u + at$
2. $s = \frac{1}{2} at^2$
3. $v^2 = u^2 + 2as$
4. $s = \frac{1}{2} t (u + v)$

where

s = displacement
t = time
u = initial or starting velocity
v = velocity after time t
a = acceleration.

To use the kinematic equations to solve a problem in kinematics it is necessary to identify the information given in the problem, then to identify which of the four equations can be manipulated to give the answer required.

displacement divided by the total time taken. Its dimensions are therefore length divided by time, and are given in metres per second (ms^{-1}). The instantaneous velocity at any point is the rate of change of velocity at that point.

vector a quantity in which both the magnitude and the direction must be stated.

vertical component of velocity see instantaneous velocity.

weight the gravitational force acting on the body. Weight, therefore, varies with location. Thus a body can have the same mass on the Moon as on Earth, but its weight on the Moon will be less than on Earth since the gravitational force on the Moon is approximately one sixth of that on Earth. The same person, stepping on a set of compression scales at the bottom of a mountain and then at the top, would weigh less at the top because of the slight decrease in the gravitational force, which results from the slight increase in distance from the centre of Earth.

Weight is sometimes confused with mass (q.v.).

The unit of weight is the newton (q.v.).

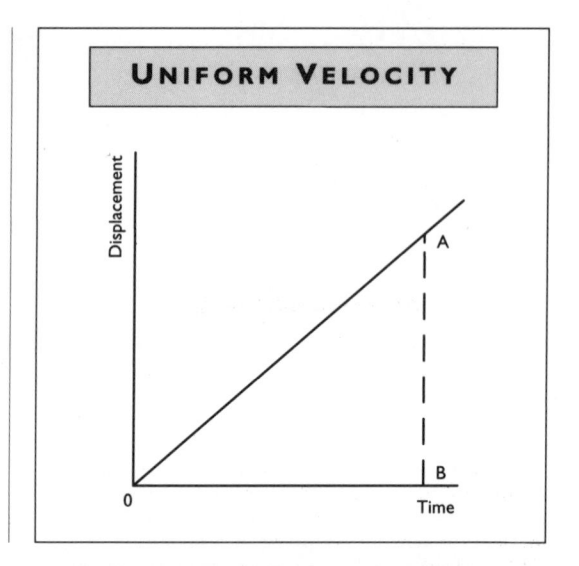

UNIFORM VELOCITY

NEWTON'S LAWS OF MOTION

Newton's laws of motion state relationships between the acceleration of a body and the forces acting on it.

Newton's first law of motion states that a body will remain at rest or travelling in a straight line at constant speed unless it is acted upon by an external force.

Notice that the force has to be an external one. In general, a body does not exert a force upon itself.

The tendency of a body to remain at rest or moving with constant velocity is called the inertia of the body. The inertia is related to the mass, which is the amount of substance in the body. The unit of mass is the kilogram (kg).

Newton's second law of motion states that the resultant force exerted on a body is directly proportional to the acceleration produced by the force. (See weight and force.)

The unit of force is known as the newton (N) - see above.

Newton expressed his second law by stating that the force acting on a body is equal to the rate of change in its 'quantity of motion', which is now called momentum (q.v.).

$$F = ma$$

or

$$F = \frac{mv_2 - mv_1}{t}$$

where
- F is the force exerted,
- m is the mass of the body,
- a is the acceleration,
- v_1 is the initial velocity,
- v_2 is the final velocity, and
- t is the time for which the force acts.

Newton's third law of motion states that a single isolated force cannot exist on its own; there is always a resulting 'mirror-image' force. In Newton's words:

'To every action there is always opposed an equal reaction'.

(It is important that the equal and opposite forces do not act on the same body.)

This means that, because any two masses exert on each other a mutual gravitational attraction, the Earth is always attracted towards a ball as much as the ball is attracted towards the Earth. Because of the huge difference in their sizes, however, the observable result is the downward acceleration of the ball.

The principle of the conservation of momentum (q.v.) follows from the third law.

STATICS AND FORCES INVOLVED IN ROTATION

In addition to the fundamental forces (see pp. 74-78), other forces may be encountered. Because of their different natures, solids and fluids appear in some ways to react differently to similar applied forces. When forces are applied to solids they tend to resist.

Newton's first law of motion (see pp. 74-78), stated for a single particle, can also apply to real bodies that have definite shapes and sizes and consist of many particles. Such a body may be at equilibrium (q.v.). This means that it is acted on by net zero force, and that it has no tendency to rotate.

In the simple action of opening a door, or using a spanner, a turning force is exerted. The size of the turning effect depends upon two factors: firstly, the magnitude of the force and, secondly, the distance of the line of action of the force from the pivot. A large turning effect can be produced even if only a small force is exerted if the distance from the pivot is large - this is, for example, the effect that can be produced with certain spanners.

centre of mass (of a body) is a point, normally within the body itself, such that the net resultant force produces an acceleration at this point, as though all the mass of the body were concentrated at that point.

The weight of a body is the force with which the Earth attracts it (see pp. 74-78). The many particles that go to make up the body are each attracted to the Earth with the same force. Thus, the Earth's pull on a body comprises a very large number of equal parallel forces. These parallel forces can be replaced by the concept of a single force that acts through the point known as the centre of mass.

The centre of gravity of a body is the point of application of the resultant force due to the attraction of the Earth on the body.

If a uniform gravitational field is present, the centre of gravity coincides with the centre of mass. Thus all the weight can be considered to act at this single point. The stability of an object is helped by keeping the centre of gravity as low as possible. A racing car is low-slung to improve stability (see diagrams, below).

CENTRE OF MASS

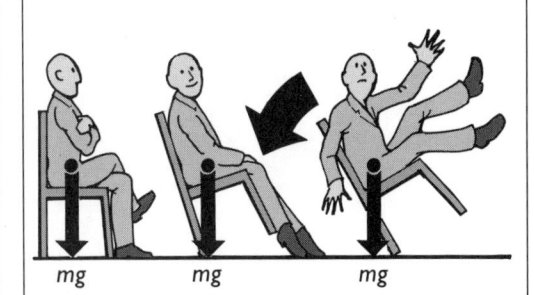

Centre of mass. In the left-hand picture the person is in a stable state because the perpendicular through his centre of mass falls between the legs of the chair. In the centre picture, although the chair is tipped, the perpendicular still falls within the safe area between the legs of the chair. In the right-hand picture the perpendicular falls outside the legs of the chair. The person sitting on the chair is in trouble...

CENTRE OF GRAVITY

Centre of gravity. A racing car has a very low centre of gravity and will remain stable even on a slope. A loaded truck will have a high centre of gravity and will topple over if driven on too steep a slope.

THE GUINNESS SCIENCE FACT BOOK

The centre of mass of a body may be located by means of a simple experiment. If a thin sheet of material is balanced on a straight edge (see diagram) in a number of positions the centre of mass (marked cm on the accompanying figure) may be determined.

with a plumbline from each hole. If the position of the plumbline is marked on the sheet (as in the accompanying diagram), the point of intersection of the three lines so drawn may be determined. This point of intersection is the centre of mass - marked cm.

LOCATING THE CENTRE OF MASS

Straight edge

Balance positions

Check

cm

LOCATING THE CENTRE OF MASS

cm

Plumbline

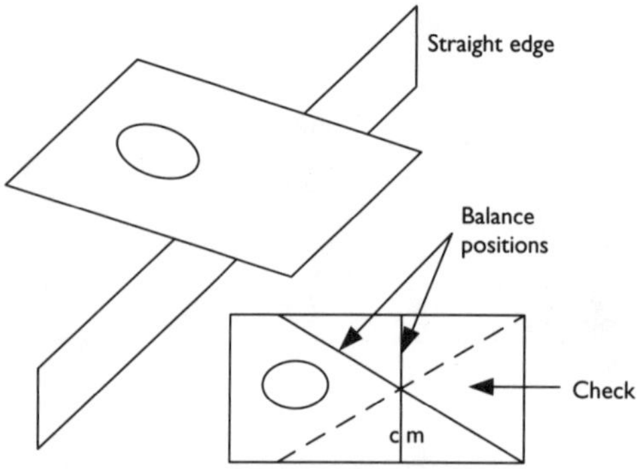

The centre of mass of an irregular sheet may be determined by making three holes near to the edge of the sheet and then, in turn, suspending the sheet along

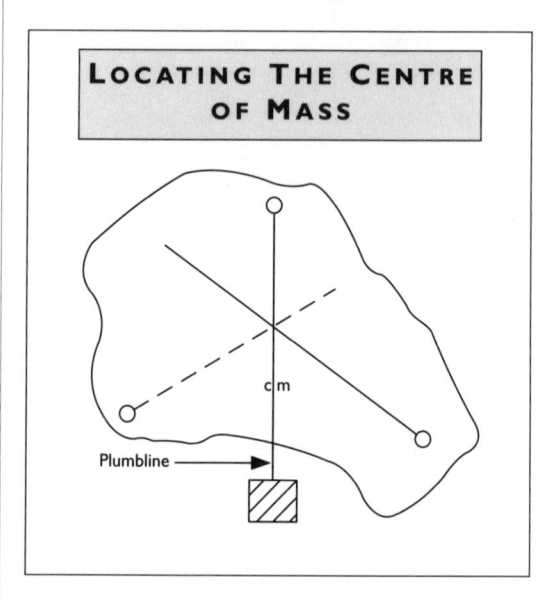

couple two equal and opposite forces applied to the same body where the two forces do not act in the same line; see moment (below).

EQUILIBRIUM

Equilibrium. The net force on the body is zero. The mass is balanced by the reaction force:
$R = mg$
(mg is the gravitational force acting on the mass).
The men are pulling with equal force:
$F = F$.
The body will not move.

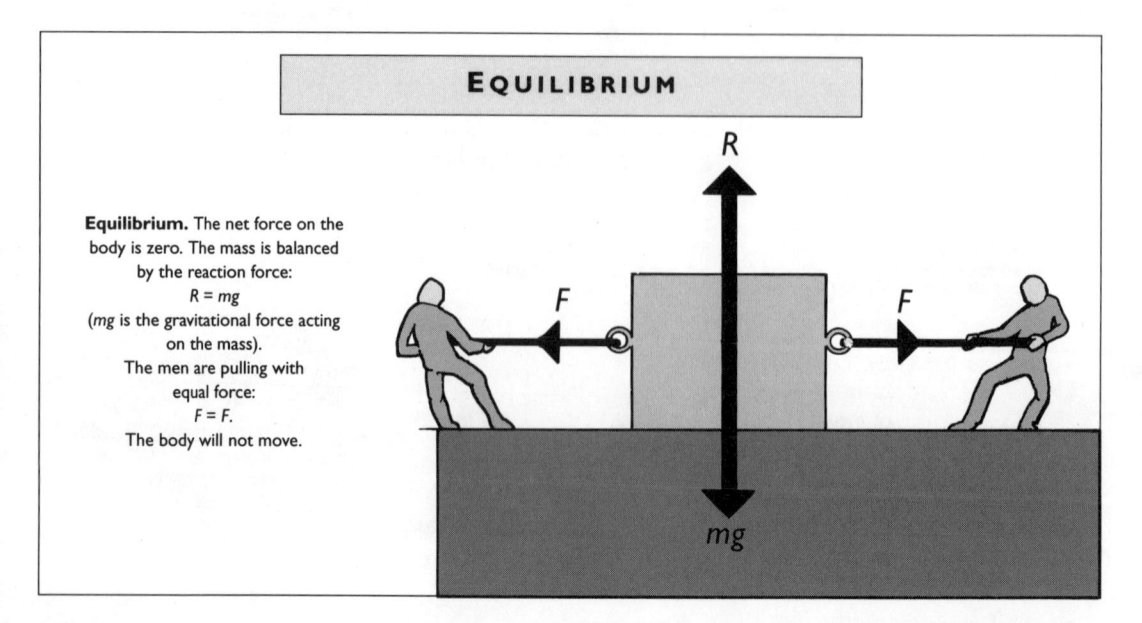

R

F   F

mg

A COUPLE

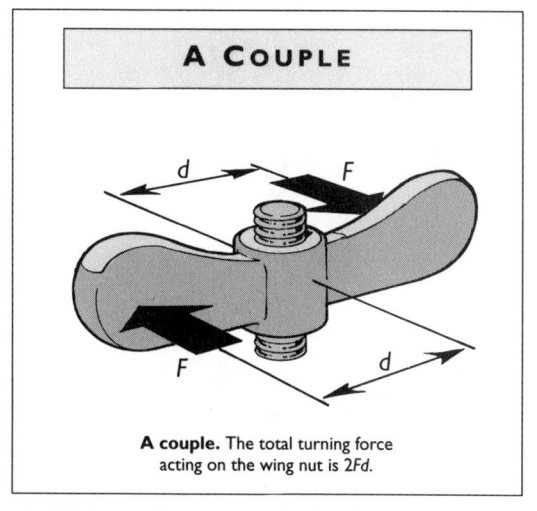

A couple. The total turning force acting on the wing nut is $2Fd$.

TORQUE

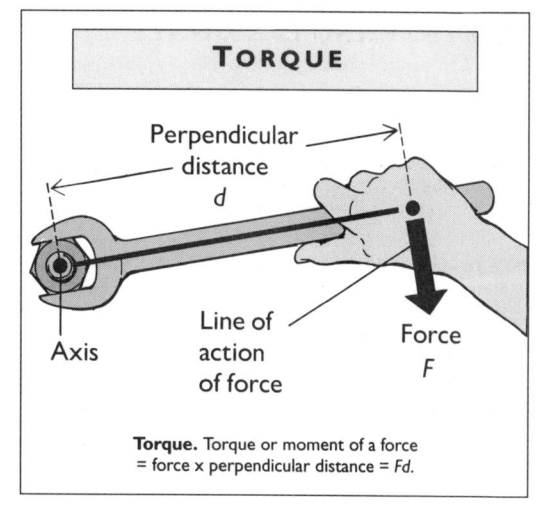

Torque. Torque or moment of a force = force × perpendicular distance = Fd.

equilibrium the state in which a body is at rest or moving with constant velocity.
fulcrum an alternative term used to describe a pivot.
moment (or **torque**) see box p. 81.
static equilibrium see zero net force (below).
torque see moment of force (above).

zero net force the condition when the total or resultant of all the forces acting on a body are zero - i.e. all the forces cancel each other out; see diagram.
If the body is at rest it is in static equilibrium. (Studies of such conditions are important in the design of bridges, dams and buildings.) See also introduction (above).

MOMENT

moment (or **torque**) the size of the turning effect.

The moment (or torque) of a force about a point is the product of the force acting on a body and the perpendicular distance from the axis of the rotation of the body to the line of action of the force - see diagram (below).

Torque measures the tendency of a force to cause the body to rotate. In this case torque causes angular acceleration, which is the rate of change of angular velocity of the body.

Torque has units of force × distance, usually expressed as newton metres (Nm).

Torque is increased if either the force or the perpendicular distance is increased. If a wedge is used to keep a door open, it has maximum effect if it is placed on the floor as far away from the hinge as possible.

When a body is acted upon by two equal and opposite forces, not in the same line, then the result is a couple, which has a constant turning moment about any axis perpendicular to the plane in which they act - see diagram.

Moments can be observed in a simple experiment using a ruler, a fulcrum and several small weights. A weight is tied to one end of the ruler and a second weight is tied to the other end of the ruler. The ruler is then placed on a fulcrum and is adjusted so that a balance is achieved. When the body is in equilibrium, the sum of the anti-clockwise moments about the fulcrum is equal to the sum of the anti-clockwise moments about the same point - thus the sum of the clockwise moments by the weighted ruler on one side of the fulcrum is equal to the sum of the anti-clockwise moments by the other end of the weighted ruler on the side of the fulcrum.

FRICTION AND ELASTICITY

In addition to the fundamental forces (see pp. 74-78), other forces, such as friction and elasticity, may be encountered. Because of their different natures, solids and fluids appear in some ways to react differently to similar applied forces.

When forces are applied to solids they tend to resist. Friction inhibits displacement, but is overcome after a certain limit. Bodies may be deformed by tensions.

bulk modulus of elasticity one of the three types of elastic modulus (q.v.). This modulus characterizes the behaviour of a substance subject to a uniform volume comparison. The bulk modulus is the ratio of the pressure on a body to its fractional decrease in volume.
See also Young's modulus and shear modulus.

coefficient of friction (symbol μ) is the ratio of the frictional force (F) to the normal reaction (N) hence

$$\mu = \frac{F}{N}$$

The coefficients are different for limiting and sliding friction.

deformation (in physics) the act of impairing the form or shape of a body.

elasticity deals with deformations that disappear when the external applied forces are removed. Most bodies may be deformed by the action of external forces and behave elastically for small deformations.
See strain and stress (below).

elastic modulus the constant of proportionality in stress (q.v.) and strain (q.v.); in other words, the ratio of the stress applied to a body to the strain produced.
The elastic modulus varies according to the material and the type of deformation.
See also Young's modulus, shear modulus, bulk modulus and Hooke's law.

friction the force that resists the motion of one surface relative to another with which it is in contact. See sliding friction, static friction, and rolling friction.

Friction is caused because any and every body, no matter how smooth it may look, has bumps, lumps and cracks on it that can be seen microscopically. Because of this the number of points of contact between two surfaces are few; they are represented by these irregularities and not by the entire area of the adjoining surfaces. Consequently, very high pressure results in local pressure welding of the two surfaces. When one, or both, of these surfaces move, the welds are broken and are remade time after time and, thus, friction occurs.

A body resting on a horizontal surface has a normal contact force (R) between itself and the surface. The normal contact force acts perpendicularly to the surface. A horizontal force (B) is then introduced with the intention of moving the body towards the right. An equal horizontal force (F), that is resisting the force to move will be experienced to the left of the body. As the horizontal force B is increased, F also increases up to the limiting frictional force, at which point the body will just move.
Once the body is moving the value of F falls.

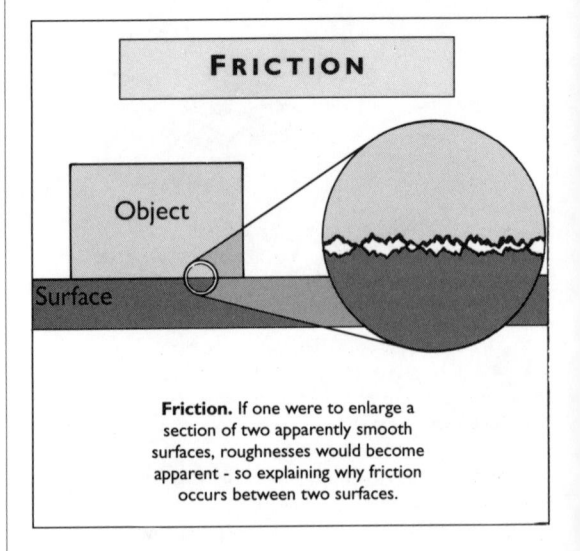

Friction. If one were to enlarge a section of two apparently smooth surfaces, roughnesses would become apparent - so explaining why friction occurs between two surfaces.

Hooke's law applies to a special example of deformation - the extension or elongation of a spring by an applied force.
Hooke's law, which was formulated by the English scientist Robert Hooke (1635-1703), states that, for small forces, the extension is proportional to the applied force. Thus a spring balance will have a uniform scale for the measurement of various weights.
Rubber does not obey Hooke's law because the ratio of applied force to extension depends upon the particular force at which it is measured; i.e. it is 'easy' to stretch at first but becomes very stiff as the force increases.
See also perfect elasticity and imperfect elasticity.

imperfect elasticity occurs when a substance does not readily regain its initial state. It can also be demonstrated in the case of a soft rubber ball which - when dropped on hard ground - bounces to only about one half of its initial height.
See also perfect elasticity and Hooke's law.

kinetic frictional force see sliding friction.

limiting frictional force see friction (above).

normal contact force see friction (above).

perfect elasticity the state where a substance returns to its initial state readily. This is found, for instance, in the case of spring steel, which in scientific terms, is almost perfectly elastic. See also imperfect elasticity and Hooke's law.

perfect plasticity is the opposite of perfect elasticity: the material shows *no* tendency to return to its original size and shape.

plastic flow occurs when a substance behaves in a viscous manner.

Some bodies behave elastically for low values of stress, but above a critical level they behave in a perfectly viscous manner and 'flow' like thick treacle with irreversible deformation. See diagram.

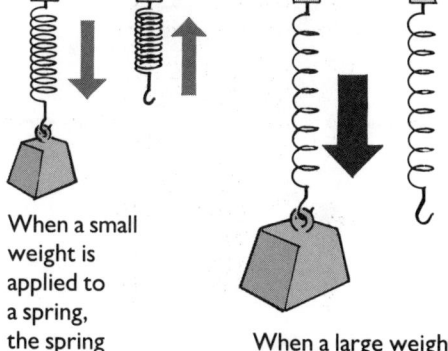

ELASTICITY AND PLASTIC FLOW

When a small weight is applied to a spring, the spring remains elastic and will revert to its unloaded shape

When a large weight is applied to a spring, the spring becomes plastic and will not revert to its unloaded shape

EXTENSION

Plastic flow

Critical point

Elasticity

FORCE

Elasticity and plastic flow. Some bodies behave elastically when small forces are applied, but above a critical point they experience plastic flow, i.e. they are irreversibly extended.

radian the angle subtended at the centre of a circle by an arc of equal length to the radius of the circle.

rigidity modulus an alternative name for the shear modulus (q.v.).

rolling friction occurs when a wheel rolls. Energy is dissipated through the system, because of imperfect elasticity (q.v.). This effect does not depend upon surfaces and is unaffected by lubrication.

shear modulus of elasticity (or **rigidity modulus of elasticity**) one of the three types of elastic modulus (q.v.). The shear modulus relates to the type of deformation where planes in a solid slide past each other.

The shear modulus of elasticity is the tangential force per unit area which is divided by the angular deformation of in radians.

See also Young's modulus and bulk modulus.

sliding friction occurs when a solid body slides on a rough surface. Its progress is hindered by an interaction of the surface of the solid with the surface it is moving on. This is called the kinetic frictional force. See also static friction (below).

static friction occurs as described in the following situation.

Before the object moves, the resultant force acting on it must be zero. The frictional force acting between the object and the surface on which it rests cannot exceed its limiting value. Thus, when the other forces acting on the object, against friction, exceed this value the object is caused to accelerate. The limiting or maximum value of the frictional force occurs when the stationary object acted on by the resultant force is just about to slip.

Both sliding friction (see above) and static friction involve interaction with a solid surface. The frictional forces depend on the two contacting surfaces and in particular on the presence of any surface contaminants. The friction between metal surfaces is largely due to adhesion, shearing and deformation within and around the regions of real contact.

Energy is dissipated in sliding friction and appears as internal energy, which can be observed as heat. Thus car brakes heat up when used to slow a vehicle.

The results of friction may be reduced by the use of lubricants between the surfaces of contact. This is one function of the oil used in car engines.

strain is a measure of the amount of deformation expressed as a fraction of the original length or volume. See elasticity.

stress is a quantity proportional to the force causing the deformation. (See elasticity; above.) Its value at any point is given by the magnitude of the force acting at that point divided by the area over which it acts.

It is found that for small stresses the stress is proportional to the strain. The constant of proportionality is called the elastic modulus (q.v.).

Young's modulus of elasticity one of the three types of elastic modulus (q.v.). Young's modulus refers to longitudinal stress (q.v.) and strain (q.v.). It refers to changes in the length of a material under the action of an applied force.

See also shear modulus and bulk modulus.

FLUIDS AND PRESSURE

Various forces affecting solids are described on pp. 79-81 and pp. 82-83. Here forces, including pressure, affecting fluids are dealt with.

Fluids, although lacking definite shape, are held together by internal forces. They exert pressure on the walls of the containing vessel. Fluids - by definition - have a tendency to flow; this may be greater in some substances than in others and is governed by the viscosity of the fluid.

aneroid barometer a barometer containing no liquid (see barometer below). This device is a flat cylindrical metal box that has been partially evacuated of air and then sealed.

Archimedes' principle see buoyancy force (below).

atmospheric pressure results from the 'weight' of the large volume of air that surrounds the planet Earth. As a result of this weight, air exerts a pressure on the Earth and everything upon Earth. This atmospheric pressure - which is measured in newtons per square metre - may be measured using a barometer (q.v.). At sea level, it is equivalent to the weight of a column of mercury about 0.76m high, which is about 1.01×10^6. It varies by up to about 5%, depending on the weather systems overhead.

barometer a device used for measuring atmospheric pressure (see above).

A simple barometer, as shown in the diagram here, comprises a thick-walled glass tube about a metre long, closed one end and filled almost to the top with mercury - it is necessary to have a vacuum at the top of the tube as atmospheric pressure is to be measured. The tube is inverted several times and a large air bubble will form; this bubble will collect any smaller bubbles as it goes up and down the tube upon each inversion. More mercury is gradually added so that the tube eventually becomes full. The open end is then covered as the the tube is inserted - open end down - into mercury contained in a reservoir. When this end is opened again, the level of the mercury in the tube will fall until the vertical difference between the level of the mercury in the two containers is about 76cm. Daily small rises and falls in the level of the mercury will chart changes in atmospheric pressure.

buoyancy force was described by the Greek mathematician and physicist

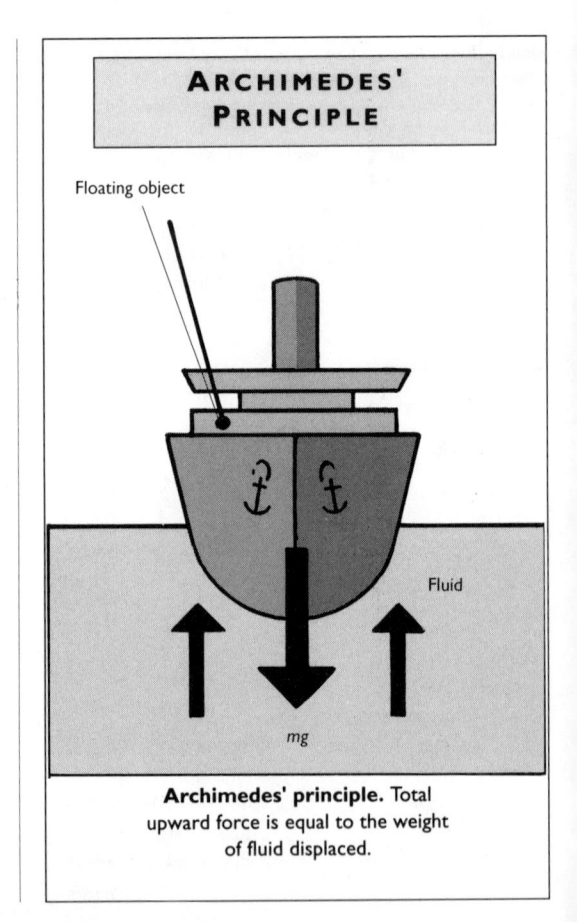

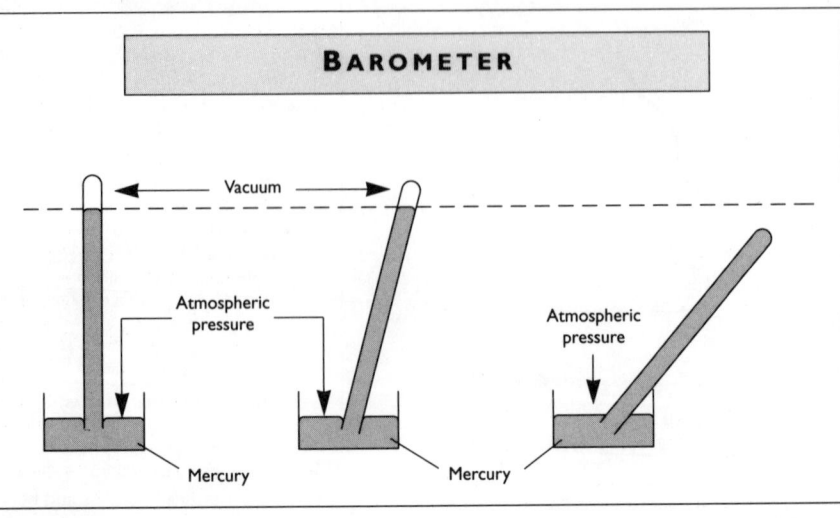

ARCHIMEDES' PRINCIPLE

Archimedes' principle. Total upward force is equal to the weight of fluid displaced.

BAROMETER

Archimedes (287-212 BC). Archimedes' principle states that an object placed in a fluid is buoyed up by a force equal to the weight of fluid displaced by the body. See the accompanying diagram.

A body with density greater than that of the fluid will sink, because the fluid it displaces weighs less than it does itself. A body with density less than that of the fluid will float. The following equation can be given:

density × volume = mass.

A submarine varies its density by flooding ballast tanks with sea water or emptying them; this enables it to dive or rise to the surface.

density is sometimes described as the 'lightness' or 'heaviness' of a material. More precisely density is the mass of a substance per unit of volume. In SI units density is measured in kgm^{-3}.

$$\text{Density} = \frac{\text{mass}}{\text{volume}}.$$

Equal volumes of different substances clearly have very considerably different masses. The aluminium alloys that are used in the manufacture of airliners, for example, have a very different density from the steel that is used in the manufacture of a railway locomotive.

hydrostatics the study of fluids at rest.

laminar flow the steady state when adjacent layers of a fluid flow smoothly past each other.

manometer an instrument used for measuring the pressure of a gas. It is basically a U-shaped tube that contains water.

newtonian fluid see viscosity.

pascal the SI unit of pressure. It is equivalent to 1 newton per square metre Nm^{-2}.

pressure is defined as the perpendicular or normal force per unit area of a plane surface in, for example, a fluid. Its unit is the pascal (symbol Pa; q.v.).

At all points in the fluid at the same depth the pressure is the same. The pressure depends only upon depth in an enclosed fluid, and is independent of cross-sectional area. (In the hydraulic brakes of a car - see the accompanying diagram - a force is applied by the foot pedal to a small piston. The pressure is transmitted via the hydraulic fluid to a larger piston connected to the brake. In this way the force applied to the brake is magnified by comparison with the force applied to the pedal.)

Reynold's number see turbulence.

turbulence the state when - if the flow velocity of a fluid is increased - the flow becomes disordered with irregular and random motions.

Turbulence may also be seen in the smoke rising from a cigarette. When smoke rises from a cigarette it starts with smooth laminar flow (q.v.) but soon breaks into turbulent flow with the formation of eddies.

Reynold's number is used to predict the onset of turbulence. It is defined as:

$$Re = \frac{(\text{speed} \times \text{density} \times \text{dimension})}{\text{viscous force}}.$$

or, alternatively, as a ratio of the inertial force to the viscous force:

$$Re = \frac{\text{inertial force}}{\text{viscous force}}.$$

This is a pure ratio so has no units. It is a characteristic of the system and the dimension may be the diameter of a pipe or the radius of a ballbearing. Viscosity is relevant for small values of Reynold's number. Above certain values turbulence is likely to break out. Thus, for the fall of a very small raindrop, resistance is viscous and is proportional to the product of the viscosity of air, the radius of the raindrop and its speed. For a large raindrop, the resistance is proportional to the density of air, the square of the radius of the raindrop and the square of its speed. Sometimes when special smooth surfaces are involved the onset of turbulence is delayed.

viscosity a measure of the resistance to flow that a fluid offers when it is subjected to shear stress (see pp. 82-83).

Viscosity relates to the internal friction in the flow of a fluid - how adjacent layers in the fluid exert retarding forces on each other. This arises from cohesion of the molecules in the fluid.

In a solid deformation of adjacent layers is usually elastic. In a fluid, however, there is no permanent resistance to change of shape; the layers can slide past each other, with continuous displacement of these layers.

Fluids are described as newtonian if they obey Newton's law that the ratio of the applied rate of shearing has a constant value. This is not true for many fluid substances. Some paints, for example, do not have constant values for the coefficient of viscosity; as the paint is stirred it flows more easily and the coefficient is diminished. Molten lava is another non-newtonian fluid.

See also laminar flow and turbulence.

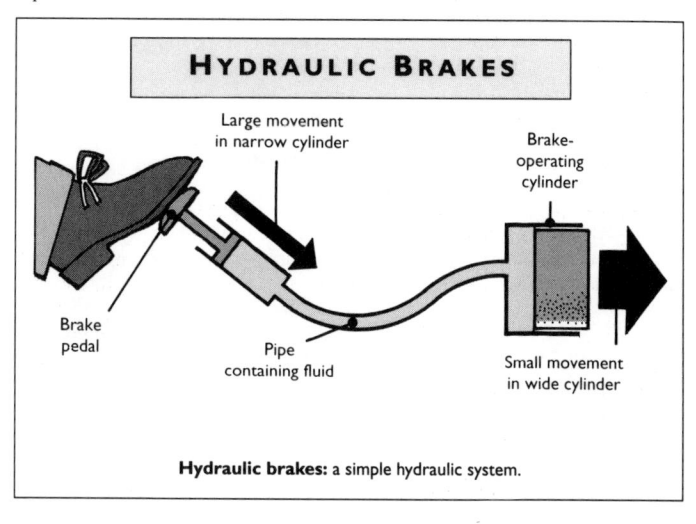

HYDRAULIC BRAKES

Large movement in narrow cylinder

Brake-operating cylinder

Brake pedal

Pipe containing fluid

Small movement in wide cylinder

Hydraulic brakes: a simple hydraulic system.

MACHINES

A machine may be defined as any device by means of which a force (the effort) that is applied at one point can be used to overcome another force (the load) at another point in a system. In general, a machine is designed so that the effort is less than the load: such a machine may be defined as a 'force multiplier'. In the system that is the machine the force that the effort exerts is multiplied by the machine so that it becomes equal to the force that the load exerts. Until the forces are equal the load may not be moved; see mechanical advantage (below). However, some machines may be defined as 'distance multipliers' (e.g. a bicycle) where the load moves further than the effort; to achieve this the effort is bigger than the load.

efficiency (of a machine) the ratio of the useful work that is done by a machine to the total work that is put into it.

This is expressed as a percentage:

$$\text{efficiency} = \frac{\text{work output} \times 100\%}{\text{work input}}$$

Work done = force × distance, thus:

$$\text{efficiency} = \frac{\text{load} \times \text{distance load moves} \times 100\%}{\text{effort} \times \text{distance load moves}}$$

and

$$\text{efficiency} = \frac{\text{mechanical advantage} \times 100\%}{\text{velocity rate}}$$

In an ideal machine, no energy would be wasted. In such a case efficiency would be 100% and the mechanical advantage and the velocity ratio would be equal. An ideal machine does not, however, exist as, in practice, energy has to be used in, for example, overcoming friction.

force (symbol F) changes an object's motion, making it move either more quickly or less quickly, or making it change direction. The unit of force is the newton.

'force multiplier' see introduction, above.

fulcrum the pivot or support about which a lever turns in raising something.

lever a simple machine consisting of any rigid body which is pivoted about a fulcrum.

The work of a lever is based upon the principle of moments (q.v.). When a force (effort) is applied to one end or point of the lever, this overcomes the other force (the load) at the other end or at another point of the lever. Simple examples of this include turning a nut with pincers or removing a nail with the back of the head of a hammer.

mechanical advantage of a machine (see introduction) may be given in the following equation:

$$\text{mechanical advantage} = \frac{\text{load}}{\text{effort}}$$

Example: If an effort of 10N is applied to a lever (q.v.) to overcome a load of 60N, the mechanical advantage is said to be six.

NB. Mechanical advantage has no dimensions; in other words, it has no units.

moment the size of a turning effect.

pulley a simple machine comprising one or more wheels with a grooved rim through which a rope or chain is passed. Several pulleys are often mounted together so as form a block.

In the block and tackle arrangement in the accompanying diagram:

E = effort, and
L = load.

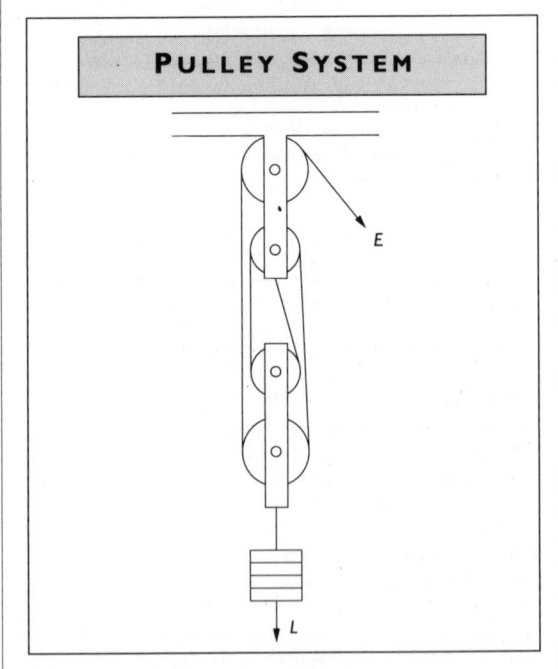

PULLEY SYSTEM

The pulleys at the top and at the base are both shown to be on separate axles. This has been done for the sake of clarity only - in normal circumstances the top pulleys would be mounted on a single axle and the bottom pulleys would be mounted on another single axle.

To raise the load by 0.5m each of the four ropes that are supporting the lower block has to be shortened by 0.5m. This means that effort has to be applied through a distance of 2m. Thus the velocity ratio (q.v.) =

$$\frac{2m}{0.5m} = 4 \text{ (also the number of pulleys)}$$

turning effect a force depending upon the magnitude of the force exerted and the distance of the line of action of that force from the fulcrum.

velocity ratio the ratio of the distance moved by the effort to the distance moved by the load in the same time.

This can be given in the following equation:

$$\text{velocity ratio} = \frac{\text{distance moved by the effort}}{\text{distance moved by the load at the same time}}$$

NB. Velocity ratio has no dimensions; in other words, it has no units.

PHYSICS

THERMODYNAMICS

Thermodynamics is the study of heat and temperature.

absolute scale an alternative name for the kelvin scale, see the box on Thermodynamic temperature scale or kelvin scale (p. 88).

absolute zero see thermodynamic temperature scale.

calorie a former unit of internal energy (q.v. and kelvin). A calorie is equivalent to 4.2 joules and is defined as the heat required to raise the temperature of 1 gram of water from 14.5°C (58.1°F) to 15.5°C (58.9°F). (The unit used by nutritionists is actually the kilocalorie - which is commonly written Calorie but is equal to 1000 calories.

Carnot engine see second law of thermodynamics.

closed system a system in which no external forces are experienced.

conservation of energy a fundamental principle of thermodynamics that states that the total magnitude of a certain physical property of a system - its mass, its charge or its energy - will remain unchanged even though there may be exchanges of that property between the various components of that system. This may be more simply expressed that energy is neither created nor destroyed.

This theory was developed in the late 19th century by about a dozen scientists including James Joule and Baron Herman von Helmholtz. Although there seemed to be plenty of evidence that energy was not conserved, this important principle was eventually established.

Much of the energy that seems to be lost in typical interactions - such as a box sliding across a floor - is converted into internal energy; in the case of the sliding box, this the kinetic energy (see below) gained by the atoms and molecules within the box and the floor as they interact and are pulled from their equilibrium position. This 'hidden energy' is thermal energy. Strictly speaking, heat is transferred between two bodies as a result of a change in temperature, although the term 'heat' is commonly used for the thermal energy as well.

Processes that turn kinetic energy into thermal energy include viscosity and friction. In a steam engine heat is turned into work.

energy is the capacity of a body to do work.

The total energy stored in a closed system (q.v.) remains constant, however it may be transformed. This is the principle of conservation of energy (see above). It may take the form of mechanical energy (kinetic or potential; q.q.v.), electrical energy, chemical energy, or heat energy. There are other forms of energy including gravitational energy, magnetic energy, the energy of electromagnetic radiation, and the energy of matter.

equilibrium temperature see internal energy.

first law of thermodynamics see box p. 89.

flow of heat a transfer of energy resulting from differences in temperature.

heat a form of energy; the term 'heat' is commonly used for thermal energy as well.

ideal gas is one that would obey the ideal gas law perfectly (see pp. 92-93). In fact, no gas is ideal but most behave sufficiently closely that the ideal gas law can be used in calculation provided the pressures are low and temperatures well above those at which they liquefy. At ordinary temperatures, dry air can be considered as a very good approximation to an ideal gas.

ideal gas scale see thermodynamic temperature scale.

internal energy the molecular energy (kinetic and potential) within a body. When this energy is transferred from a place of high energy to one of lower energy, it is described as a flow of heat.

If two bodies of different temperatures are placed in thermal contact with each other, after a time they are found both to be at the same temperature. Energy is transferred from the warmer to the colder body, until both are at a new equilibrium temperature.

joule (J) the SI unit of work or internal energy. A joule is defined as the work done on a body when it is displaced 1 metre as the result of the action of a force of 1 newton acting in the direction of motion. This is given by the following equation:

$$1J = 1Nm.$$

The result can be expressed more generally as in the accompanying diagram.

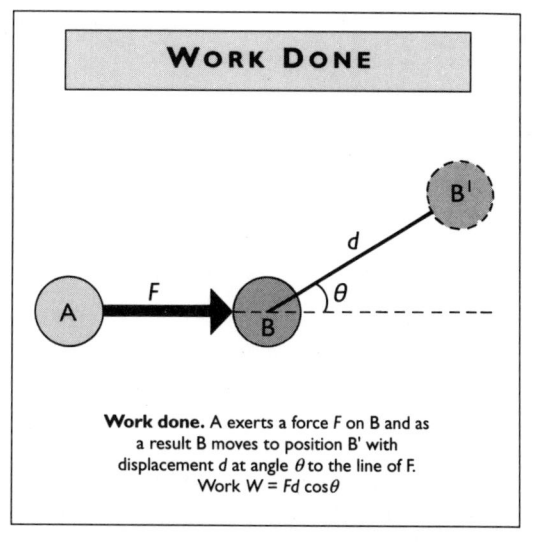

WORK DONE

Work done. A exerts a force F on B and as a result B moves to position B' with displacement d at angle θ to the line of F.
Work $W = Fd \cos\theta$

Units of internal energy used previously include the calorie, see above.

kelvin (K) the SI unit of thermodynamic temperature equal to the fraction $\frac{1}{273.15}$ of the thermodynamic temperature of the triple point of water.

The kelvin is named after the Scottish physicist William Thompson, later Lord Kelvin, who did important work in thermodynamics and electricity.

KELVIN SCALE

thermodynamic temperature scale (or **kelvin scale** or **ideal gas scale**) is based on the kelvin. The thermodynamic temperature scale is used in both practical and theoretical physics.

On the thermodynamic temperature or kelvin scale the freezing point of water is 273.15K (0°C or 32°F) and its boiling point is 373.15K (100°C or 212°F): thus one degree kelvin is equal in magnitude to one degree in the Celsius scale. The temperature of 0 (zero) K is known as absolute zero. At this temperature, for an ideal gas, the volume would be infinitely large and the molecular kinetic energy zero.

Whereas the Celsius scale is positive above the melting point of ice (0°C) and negative below it, the kelvin scale (or absolute scale) does not have any negative values since its zero is absolute zero.

A conversion chart for the kelvin and Celsius scales follows:

K	°C
273.15K	0°C
278.15K	5°C
283.15K	10°C
288.15K	15°C
293.15K	20°C
298.15K	25°C
303.15K	30°C
308.15K	35°C
313.15K	40°C
318.15K	45°C
323.15K	50°C
328.15K	55°C
333.15K	60°C
338.15K	65°C
343.15K	70°C
348.15K	75°C
353.15K	80°C
358.15K	85°C
363.15K	90°C
368.15K	95°C
373.15K	100°C

kelvin scale see box.

kinetic energy the organized energy of a moving body. The kinetic energy of a body is the energy that it has because it is moving. It is equal to half the product of the mass and the square of the velocity; see diagram.

KINETIC ENERGY

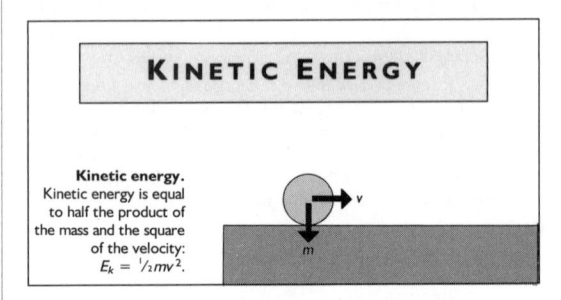

Kinetic energy.
Kinetic energy is equal to half the product of the mass and the square of the velocity: $E_k = \frac{1}{2}mv^2$.

laws of thermodynamics see first law of thermodynamics and second law of thermodynamics.

newton metre see joule.

potential energy acts as a store of energy. It can be converted into kinetic energy or it can be used to do work. In contrast to kinetic energy, which is dependent upon velocity, potential energy is dependent upon position.

The gravitational potential energy of a body of mass m at height h above the ground is mgh, where g is the acceleration due to gravity (see diagram). This gravitational potential energy is equal to the work that the Earth's gravitational field will do on the body as it moves to ground level.

If a body moves upward against the gravitational force, work is done on it and there is an increase in gravitational potential energy.

POTENTIAL ENERGY

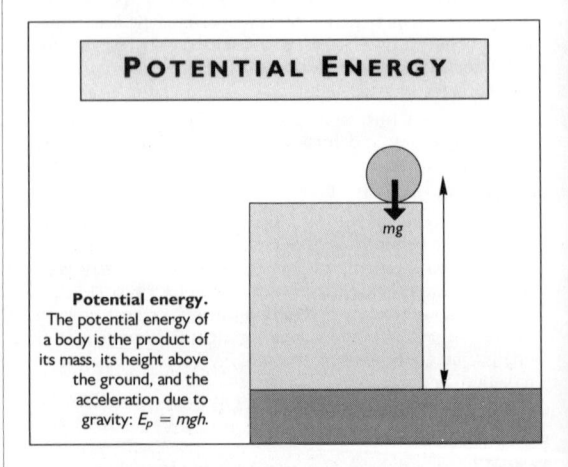

Potential energy.
The potential energy of a body is the product of its mass, its height above the ground, and the acceleration due to gravity: $E_p = mgh$.

second law of thermodynamics see box.

sink (of an engine) the place where energy is removed from the system.

temperature (of a substance) a measure of its internal energy or 'hotness' of a body, *not* the heat of the body. Thermometers are used to measure temperature. They may be based on the change in volume of a liquid (as in a mercury thermometer), the change in length of a strip of metal (as used in many thermometers), or the change in electrical resistance of a conductor. Other parameters may also be involved in measuring temperature.

thermal energy disorganized energy due to the motion of atoms and molecules.

thermodynamic temperature scale (or **kelvin scale** or **ideal gas scale**) see box p. 88.

thermometer see temperature.

triple point the temperature at which the vapour, liquid and solid phases of a substance are at equilibrium.

work (done on a body by a constant force) is defined as the product of the magnitude of the force and the consequent displacement of the body in the direction of the force (see diagram). (When a force acts on a body, causing acceleration in the direction of the force, work is done.)

The unit of work is the joule, sometimes referred to as the newton metre.

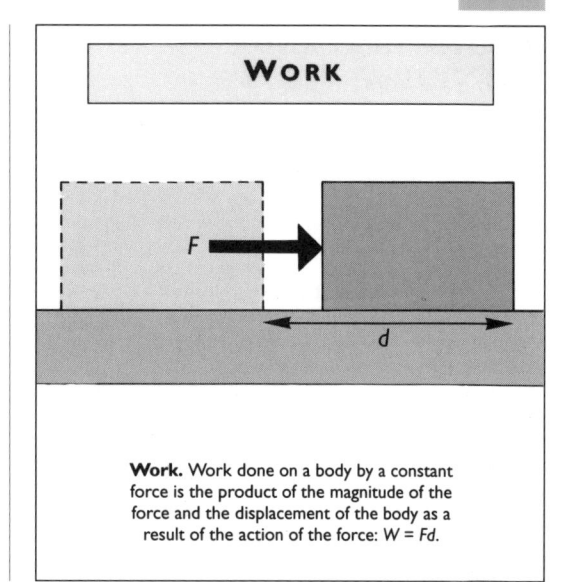

WORK

Work. Work done on a body by a constant force is the product of the magnitude of the force and the displacement of the body as a result of the action of the force: $W = Fd$.

LAWS OF THERMODYNAMICS

first law of thermodynamics states that if, during an interaction, a quantity of heat is absorbed by a body, it is equal to the sum of the increase in internal energy of the body and any external work done by the body. This law is a development of the law of conservation of energy.

The increase in internal energy will be made up of an increase in the kinetic energy of the molecules in the body and an increase in their potential energy, since work will have to be done against intermolecular forces as the body expands.

The change in internal energy of a body thus depends only on its initial and final states. The change may be the result of an increase in energy in any form - thermal, mechanical, gravitational, etc. Another statement of this law is that is possible to convert work totally into heat.

second law of thermodynamics states the converse to the situation described above under the entry for the first law of thermodynamics. There are several ways in which the second law may be stated but, essentially, it means that heat cannot itself flow from a cold object to a hot object. Thus the law shows that certain processes may only operate in one direction. It implies heat cannot be completely converted into work.

The second law was established after work by a French engineer Sadi Carnot, who was trying to build the most efficient engine. His ideal engine - the Carnot engine - established an upper limit for the efficiency with which thermal energy could be converted into mechanical energy. Real engines fall short of this ideal efficiency because of losses due to friction and heat conduction. As the temperature of the sink (q.v.) in a working engine is near room temperature, the amount of work that can be done is restricted by the relatively small temperature difference. This limits the efficiency of most steam engines to about 30-40%. Thus it makes sense to use the vast amounts of waste heat from electrical power stations for heating purposes rather than allow it to be lost in cooling towers.

SPECIFIC HEAT AND LATENT HEAT

Heat is a form of energy (see Thermodynamics; pp. 87-89). Like any other form of energy, heat is measured in joules (see below).

Archimedes' principle states that the weight of liquid displaced by a floating body is equal to the weight of the body or, more generally, the weight of fluid displaced equals the upthrust (apparent weight loss) of a body when completely or partially immersed in the fluid.

conduction (of heat) occurs when kinetic and molecular energy is passed from one molecule to another.

Metals are good conductors of heat because of electrons that transport energy through the material. Air is a poor conductor in comparison. (Thus a string vest keeps the wearer warm by trapping air and so preventing the conduction of heat outwards from the body.)

convection (of heat) results from the motion of the heated substance.

Warm air is less dense than cold air and so, according to Archimedes' principle, it rises. Convection is the main mechanism for mixing the atmosphere and diluting pollutants emitted into the air.

fusion (in physics) melting.

In the same way that latent heat is taken in when water changes to steam at the same temperature, so the same thing happens when ice melts to form water.

The specific latent heat of fusion of a substance is the quantity of heat that is required to convert unit mass of that substance from a solid state into a liquid state without a change of temperature.

The value of specific latent heat of fusion for water ice at 0°C is 3.34×10^5 J/kg.

Thus if m kilograms of a particular substance, whose specific heat is L, undergo a phase change from one state to another without any change in temperature, then the heat that is either absorbed or given out (depending upon the nature of the phase change) will be mL joules. Thus

$$\text{energy change} = mL \text{ joules.}$$

heat transfer see conduction, convection and radiation.

hidden heat an uncommon alternative name for latent heat.

joule (symbol J) the SI unit of work and energy equal to the work done when the point of application of a force of one newton moves - in the direction of the force - a distance of one metre. 1 joule = 0.2388 calorie. (The joule is named after James Prescott Joule.)

kelvin (symbol K) the SI unit of thermodynamic temperature. One kelvin is equal to the fraction $\frac{1}{273.16}$ of the thermodynamic temperature of the triple point (see below) of water. (The kelvin is name after Lord Kelvin.)

latent heat literally means 'hidden' or 'concealed' heat.

When heat flows between a body and its surroundings there is usually a change in the temperature of the body, as well as changes in internal energies. This is not so when a change of state occurs, as from solid to liquid or from liquid to gas (a phase change). Such a change involves a change in the internal energy of the body only.

The amount of heat needed to make the change of phase is called the hidden heat or latent heat. To change water at 100°C (212°F) to water vapour requires nearly seven times as much heat (latent heat of vaporization; see vaporization below) as to change ice to water (latent heat of fusion; see fusion below). This varies for water at different temperatures - more heat is required to change it to water vapour at 80°C (176°F), less at 110°C (230°F). In each case the attractive forces binding the water molecules together must be loosened or broken.

Latent heat has an important effect on climate (see diagram). A similar cycle takes place in a heat pump or a refrigerator.

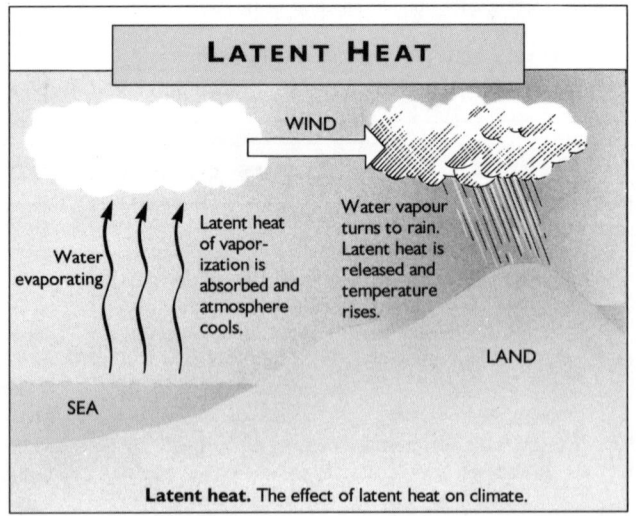

Latent heat. The effect of latent heat on climate.

phase change a change of form or state, for example from solid to liquid or from liquid to gas. A phase change involves a change only in the internal energy of the body.

radiation (of heat) is the third process of heat transfer.

All bodies radiate energy in the form of electromagnetic waves (see pp. 112-13). This radiation may pass across a vacuum, and thus the Earth receives energy radiated from the Sun. A body remains at a constant temperature when it both radiates and receives energy at the same rate.

specific heat heat expressed per unit mass.

If equal masses of two substances - water and oil - were to be heated in separate containers for the same period of time and in identical conditions, the temperature of one of those substances - water - would rise more rapidly than the other substance - oil. These two substances thus have different specific heat capacities.

The specific heat capacity of a substance may be defined as the quantity of heat that is required to raise the temperature of one kilogram of that substance by 1 K. It is thus expressed in units of joules per kilogram per kelvin.

If m kilograms of a substance, of specific heat capacity c, are raised in temperature (T) by K, then the heat required would be mcT joules.

$$\text{Heat required} = mcT \text{ joules.}$$

Water requires 4200 joules of heat to raise the temperature of one kilogram by 1K. The specific heat capacity of water is thus 4200J/kg K. (This is considerably higher than the figure for many other substances. Thus a much greater amount of energy is needed to raise the temperature of water by a particular amount than would be required to raise the same mass of certain other substances to the same temperature. For this reason, water is a very suitable substance to be used in radiators.)

The temperature of a substance can be raised by supplying energy to the substance. This can be done in a number of ways:
- by obtaining the energy from another hot body;
- by supplying energy by means of an electric current;
- by the object to fall by the necessary amount - this supplies mechanical energy.

The specific heat capacity of a solid can be measured using this latter mechanical method in the following manner. A lead shot - whose temperature has been recorded - is dropped into a wide container which is then sealed at both ends with a rubber bung. The container is then rapidly inverted about 100 times and the temperature of the lead shot is taken again. The specific heat may be calculated from the following equation (where n is the number of times the container has been inverted, h is the height of the container, mgh is the energy supplied, c is the specific heat capacity and T is the temperature rise of the lead).

$$mcT = nmgh.$$

Hence $$c = \frac{ngh}{T}.$$

When two substances at different temperatures are mixed - and then come to a common temperature - the heat that is lost by cooling in one substance will be equal to the heat that is gained by the other substance.

A more accurate method of measuring the specific heat capacity of a solid is described below.

A 12V immersion heater H (ideally, with a power of either 24W or 36W) is placed into a hole which has been made in a solid - for example, an aluminium block. A second hole in the same block is made for the insertion of a thermometer T. (A few drops of oil may be placed between the heater and the aluminium block and the thermometer and the aluminium block in order to ensure a good thermal contact. An insulating jacket placed around the block will also reduce any heat loss.)

The temperature of the aluminium block is taken. The apparatus is connected and then switched on for - say 30 minutes - and both the voltage and the current are kept as constant as possible (by making any necessary adjustments to the variable resistor R) and the values of both are noted. At the end of 30 minutes the temperature of the aluminium block is taken again. The specific heat of the block may be worked out according to the following equation:

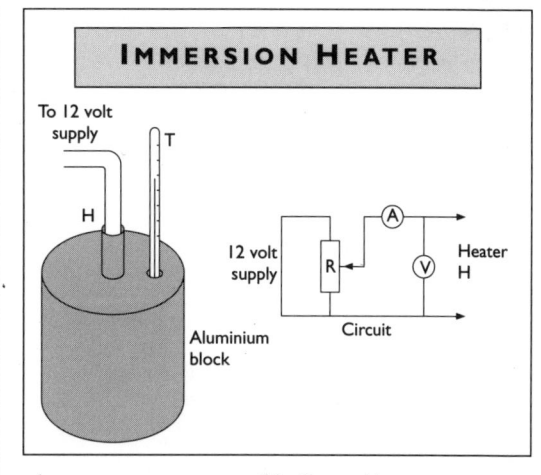

$$\text{energy supplied} = \text{heat gained by the block}$$
$$VIt = mc(\theta_2 - \theta_1)$$

where V is the potential difference, I is the electric current, t is the time in seconds, m is the mass, c is the specific heat capacity, and $\theta_2 - \theta_1$ is the temperature rise K.

(θ_1 is the first temperature taken of the block; θ_2 is the final temperature of the block.)

specific latent heat of fusion see fusion (above).

specific latent heat of vaporization see vaporization (below).

triple point the temperature and pressure at which the vapour, liquid and solid states of a substance are in equilibrium. For water, the triple point occurs at 273,16K and 611.2Pa. This forms the basis of the definition of the kelvin (q.v.).

vaporization the conversion of water, or any solution, to a vapour state.

If water is heated it begins to boil at 100°C (212°F) at atmospheric pressure. Once the water reaches boiling point its temperature will remain at 100°C even if heating continues. In heating, the water steadily absorbs energy but the temperature does not rise further. The energy is used instead to convert the water from its liquid state into the vapour state. This involves 'freeing' the molecules from the influence of other molecules and thus allowing them to move more independently. This energy is latent heat (q.v.).

The specific latent heat of vaporization of a substance is the quantity of heat required to change unit mass of the substance from the liquid state into the vapour state without additional change in temperature.

Its value for water at 100°C is 2.26×10^6 J/kg.

GASES

Three important laws - Boyle's law, Charles' law and the pressure law - relate to the behaviour of gases.

The relation between volume and pressure at a constant temperature is expressed in Boyle's law.

The relation between volume and temperature at a constant pressure is expressed in Charles' law.

The relation between pressure and temperature at a constant volume is expressed in the pressure law.

The three quantities of a gas - volume, temperature and pressure - are thus related and can be further related in the equations of the ideal or universal gas law.

Boyle's law states that the volume of a fixed mass of gas is inversely proportional to the pressure, provided the temperature remains constant. Thus, the pressure multiplied by the volume is constant. This may be given as:

$$pV = \text{constant.}$$

Charles' law states that the volume of a fixed mass of gas is directly proportional to its absolute temperature provided the pressure remains constant. Thus, the volume divided by the absolute temperature is a constant. This may be give as:

$$\frac{V}{T} = \text{constant.}$$

ideal gas law combines the equations derived from Boyle's law, Charles' law and the pressure law. The more general equation may be given - for a fixed mass of gas - as:

$$\frac{pV}{T} = \text{constant.}$$

or

$$\frac{p_1 V_1}{T_1} = \frac{p_2 V_2}{T_2}$$

where

P = pressure of the gas,
T = temperature of the gas on the Kelvin scale, and
V = volume the gas occupies.

If one mole of gas is used the equation is given as:

$$PV = RT$$

where

R = the universal gas constant.

See also diagram and kinetic theory of gases.

internal energy (of a gas) see kinetic theory of gases.

kinetic the result of movement.

kinetic theory of gases takes Newton's laws and applies them to a group of molecules. It treats a gas as if it were made up of extremely small - dimensionless - particles, all in constant random motion. It is based on an ideal gas.

One conclusion is that the pressure and the volume of that gas are related to the average kinetic energy for each molecule. The kinetic theory explains that pressure in a gas is due to the impact and elastic rebound of the molecules on the containing walls around the gas. There is an equation that relates the pressure, temperature and volume of an ideal gas - see diagram.

The temperature of an ideal gas is a measure of the average molecular kinetic energies. At a higher temperature the mean speed of the molecules is increased. For air at room temperature and atmospheric pressure the mean speed is about $500 ms^{-1}$ (about 1800km/h or 1100mph, the velocity of a bullet).

The internal energy of a gas is associated with the motion of its molecules and their potential energy. For a gas that is more complex than one with monatomic molecules, account has to be taken of the energies associated with the rotation and vibration of its molecules as well as their speed.

monatomic pertaining to a single atom.

pressure law (of gas) states that the pressure of a fixed mass of gas is directly proportional to its absolute temperature provided its volume remains constant. Thus, the pressure divided by the temperature is a constant. This may be given as:

$$\frac{P}{T} = \text{constant.}$$

universal gas law see ideal gas law.

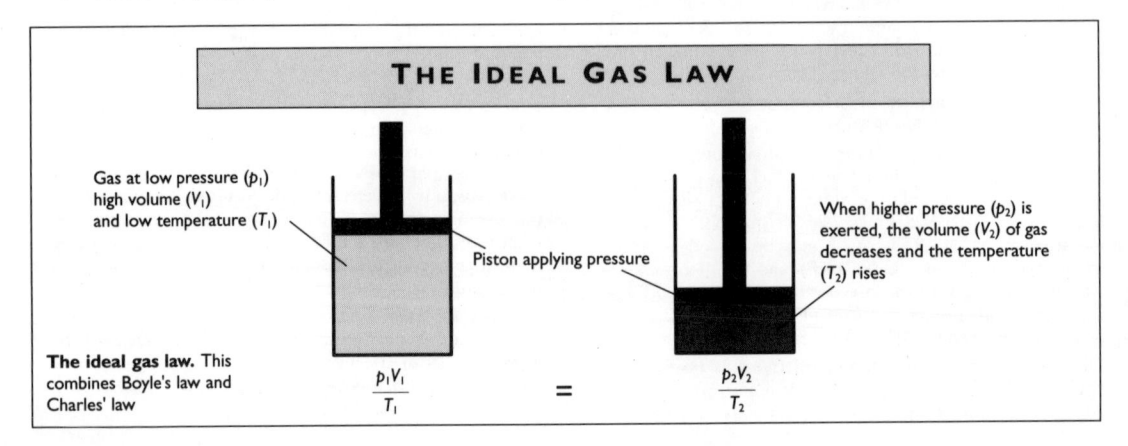

THE IDEAL GAS LAW

Gas at low pressure (p_1) high volume (V_1) and low temperature (T_1)

Piston applying pressure

When higher pressure (p_2) is exerted, the volume (V_2) of gas decreases and the temperature (T_2) rises

The ideal gas law. This combines Boyle's law and Charles' law

$$\frac{p_1 V_1}{T_1} = \frac{p_2 V_2}{T_2}$$

QUANTUM THEORY AND RELATIVITY

The physical world is not as simple as the theories of Newton supposed, although such views are appropriate simplifications for large objects moving relatively slowly with respect to the observer. Three of the most important theories of the 20th century are the quantum theory and the theories of special and general relativity.

curved space-time see general relativity theory.

dual nature of light (theory of) states that light behaves as a wave during interference experiments (see diagram) but as a stream of particles during the photoelectric effect (q.v.). Further work on this phenomenon has led to the acceptance of wave-particle duality (see below).

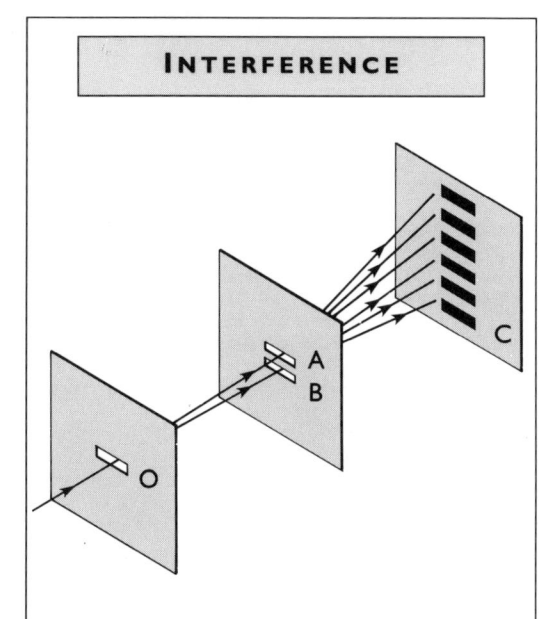

INTERFERENCE

Interference. The waves passing through slits A and B and reaching the screen C will be either in phase or out of phase and will either reinforce or cancel each other. The result is a series of light and dark bands on the screen. Reinforcement occurs when the path difference is a whole number of wavelengths.

frame of reference the position and motion of an object can be measured with respect to various spatial axes, or 'frames of reference', centred on the most convenient points, usually chosen to make the description appear simple.

general relativity theory is an extension of special relativity to include gravitational fields and accelerating reference frames. Space-time, mass and gravity are interdependent. The concept of curved space-time was proposed by Einstein in his general theory of relativity. The motion of astronomical bodies is controlled by this curvature of space and time close to large masses.

Heisenberg uncertainty principle see uncertainty principle.

inertial frame a frame of reference (q.v.) that is unaccelerated and does not contain a gravitational field.

indeterminism (principle of) see uncertainty principle.

particles (behaviour of) cannot be described by the theories of classical physics, since there is no equivalence to subatomic particles in everyday mechanics. Thus it is not helpful to discuss the behaviour of electrons in atoms in terms of tiny 'planets' orbiting a 'sun'. The French physicist de Broglie (1892-1987) suggested that if light waves can behave like particles, then particles might in certain circumstances behave like waves. Later experiments confirmed that under appropriate conditions particles can exhibit wave phenomena.

The de Broglie equation is given as:

$$\lambda = \frac{h}{p}$$

where

λ is the wavelength of the particle of momentum p,

h is a constant called Planck's constant (value 6.626×10^{-34} Js).

photoelectric effect the effect when - if light (usually visible or ultraviolet) is directed onto a piece of metal in a vacuum - electrons are knocked from the surface of the metal. For light of a given wavelength, the number of electrons emitted per second increases with the intensity of the light, although the energies of the electrons are independent of the intensity. Increasing the frequency of the incident radiation increases the energies of the emitted electrons, providing the frequency is above a certain 'threshold frequency' that depends on the metal.

The discovery of the photoelectric effect led the German physicist Einstein to deduce that the energy in a light beam exists in small discrete packets called photons or quanta.

photons the name given to light quanta, or 'packets' of energy. They both have particle and wave behaviour. Photons can be detected in experiments in which light is allowed to fall on a detector, usually photographic film. This has led to the theory of the dual nature of light (q.v.).

The way a system is described depends upon the apparatus with which it is interacting: light behaves as a wave when it passes through slits in an interference experiment, but as streams of particles when it hits a detector. The photon nature of light is revealed by experiments - such as the two-slit interference experiment shown in the accompanying diagram.

Planck's constant see particles (above).

quantum (plur. **quanta**) a separate packet of electromagnetic energy. The energy E of a quantum is given by:

$$E = hf$$

where

　　h is Planck's constant (see particles, above),

　　f is the frequency of radiation.

quantum mechanics is the study of the observable behaviour of particles. This includes electromagnetic radiation in all its details (see pp. 112-13). In particular, it is the only appropriate theory for describing the effects that occur on an atomic scale.

Quantum mechanics deals exclusively with what can be observed, and does not attempt to describe what is happening in between measurements. This is not true of classical theories, which are essentially complete descriptions of what is occurring whether or not attempts are made to measure it. In quantum mechanics the experimenter is directly included within the theory. Quantum mechanics predicts all the possible results of making a measurement, but it does not say which one will occur when an experiment is actually carried out. All that can be known is the probability of something being seen. In some experiments one event is very much more likely than any other, therefore most of the time this is what will be found, but sometimes one of the less probable events will occur. It is impossible to predict which will occur; the only way to find out is by making the appropriate measurement.

quantum theory states that nothing can be measured or observed without disturbing it: the observer can affect the outcome of the effect of being measured. Quantum theory is the only correct description of effects on an atomic scale, and special relativity must be used when speeds approaching the speed of light, with respect to the observer, is involved.

At the beginning of the 20th century scientists such as the German physicist Planck discovered that the theories of classical physics were not sufficient to explain certain phenomena on the subatomic scale, particularly in the field of electromagnetic radiation and the study of light waves. Their work resulted in the development of the quantum theory. The Scottish physicist Maxwell had developed a theory about the electromagnetic wave nature of light (see pp. 112-13), and this was crucial to the development of quantum theory. Maxwell showed that at any point on a beam of light there is a magnetic field and an electric field that are perpendicular to each other and to the direction of the light beam. The fields oscillate millions of times every second, forming a wave pattern (see diagram).

special relativity theory was proposed by Einstein in 1905. Einstein stated that all inertial frames are equally good for carrying out experiments. This assumption, coupled with the evidence that the speed of light is the same in all frames, led Einstein to develop the theory of special relativity. This theory has been extensively tested using particles accelerators, where electrons or protons travel at speeds within a fraction of 1% of the speed of light. The masses of such particles measured by an observer in the laboratory in which particles are travelling are higher than the masses measured by an observer at rest with respect to the particles.

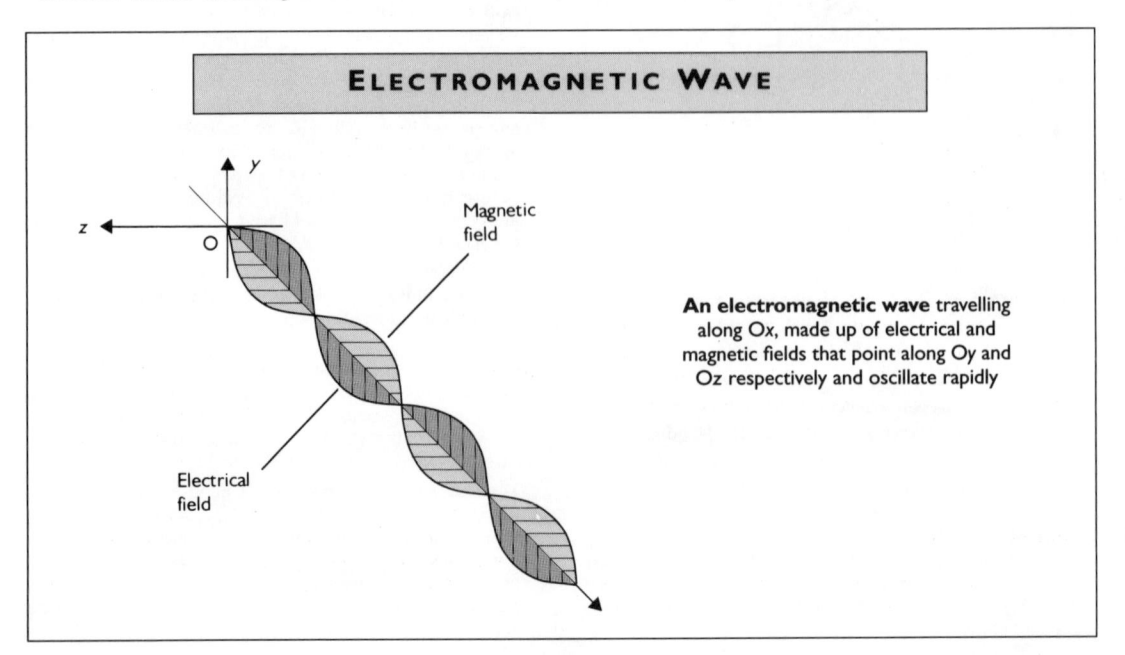

ELECTROMAGNETIC WAVE

An electromagnetic wave travelling along Ox, made up of electrical and magnetic fields that point along Oy and Oz respectively and oscillate rapidly

The classical view of time is that if two events take place simultaneously with reference to one frame then they must also occur simultaneously within another frame. In terms of special relativity, however, two events that occur simultaneously in one frame may not be seen as simultaneous in another frame moving relative to the first; see diagram. The sequence of cause and effect in related events is not, however, affected. Also, in special relativity each observer has an individual time scale. In special relativity time and space have to be considered as unified and not as two separate things. This means that time is related to the frame of reference in which it is being measured.

The equations of special relativity lead to the very simple prediction that the length of a moving body in the direction of its motion measured in another frame is reduced by a factor dependent on its velocity with respect to the observer. A clock may be moving with a uniform velocity in one frame can be measured as running slow in another frame. Its fastest rate is in its own frame, and at speeds - relative to the observer - approaching the speed of light the clock rate approaches zero. This effect is a time dilation and may appear to be a paradox although its veracity can be tested. See also general relativity.

time dilation see special relativity theory.

uncertainty principle (or **Heisenberg uncertainty principle** or **principle of indeterminism**) states that it is not possible to know with unlimited accuracy both the position and the momentum of a particle. The German physicist Heisenberg interpreted wave-particle duality differently to Einstein (see photon; above). He proposed that when a beam of light is directed at a screen with two slits, the interference pattern formed exists only if we do not know which slit the photon passed through. If we make an additional measurement and determine which slit was traversed, we destroy the interference pattern. Heisenberg showed that it is impossible to measure position and momentum simultaneously with infinite accuracy; he expressed his findings in the uncertainty principle named after him. This changed the thinking about the precision with which simultaneous measurement of two physical quantities can be made.

wave-particle duality is a fundamental principle in quantum physics; see dual nature of light (above).

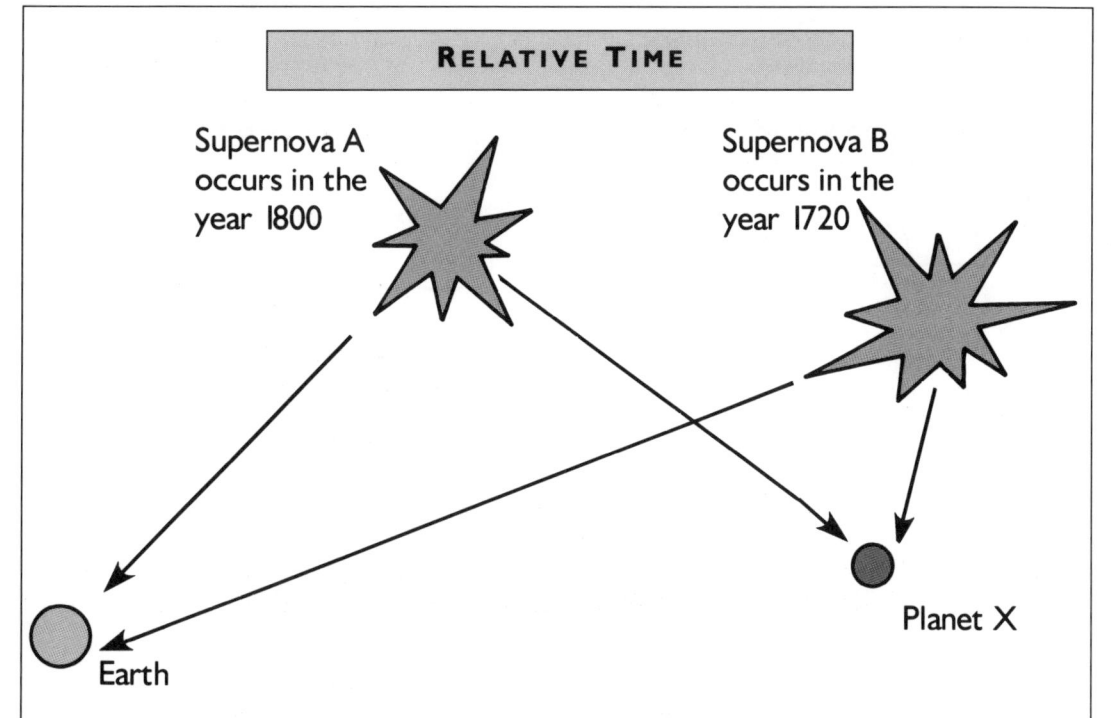

RELATIVE TIME

Supernova A occurs in the year 1800

Supernova B occurs in the year 1720

Earth

Planet X

Relative time. Supernova A occurs 120 light years from Earth (i.e. light takes 120 years to travel from Supernova A to Earth), and Supernova B is 200 light years from Earth. From Earth the two events are observed simultaneously in 1920, even though Supernova B occured 80 years before Supernova A. From planet X, in contrast, the two events are not observed simultaneously. Planet X is only 40 light years from Supernova B, but 150 light years from Supernova A. Therefore, on planet X, Supernova B is obseved in 1760, while Supernova is not seen until 1950.

WAVE TYPES

Water waves are phenomena that can be seen, and the effects of sound waves are sensed directly by the ear. Some of the waves in the electromagnetic spectrum can also be sensed by the body: light waves by the eye, and the heating effect of infrared waves. However, other electromagnetic waves cannot be experienced directly through any of the human senses, and even infrared can generally only be observed using specialized detectors.

Wave phenomena are found in all areas of physics, and similar mathematical equations may be used in each application. Some of the general principles of wave types are explored here.

amplitude the maximum displacement from the equilibrium position.

frequency (symbol f) the rate of repetition of a regular event. The frequency of wave motion is defined as the number of complete oscillations or cycles per second; see diagram. The unit of frequency is the hertz (Hz).

AMPLITUDE

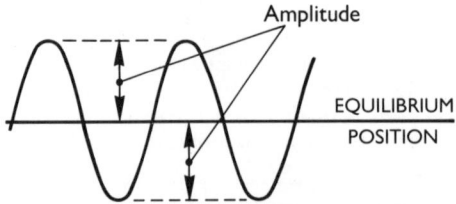

Amplitude is the maximum displacement from the equilibrium position.

PERIOD

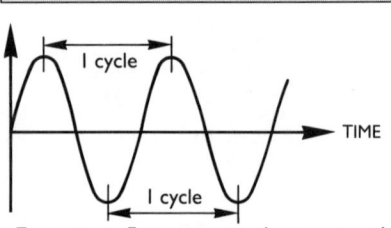

Frequency. Frequency = cycles per second. 1 cycle per second = 1 hertz (Hz).

WAVE ATTENUATION

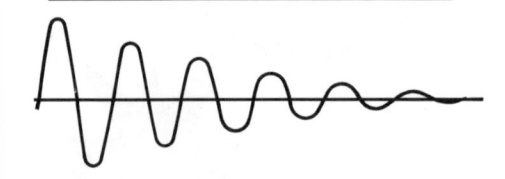

Wave attenuation. Energy is lost as a wave travels through a medium - the amplitude is reduced, and the wave is said to attenuated.

WAVELENGTH

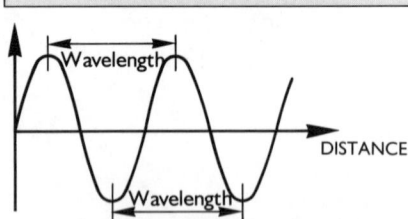

Wavelength. Wavelength is the distance between two successive points along a wave with similar amplitudes.

LONGITUDINAL WAVE

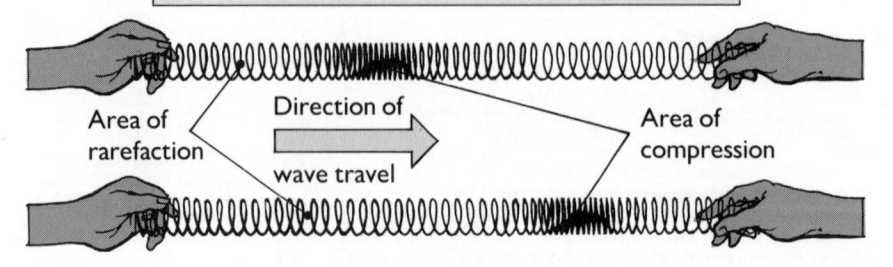

Area of rarefaction   Direction of wave travel   Area of compression

A longtidudinal wave in a 'slinky' spring.

hertz (symbol Hz) the SI unit of frequency. It is equivalent to one cycle per second. The hertz is named after the German physicist Heinrich Hertz (1857-94).

longitudinal waves periodic oscillations in which the vibrations are parallel to the direction of travel; see diagram.

mechanical waves travelling waves that propagate through a material - as, for example, happens when a metal rod is tapped at one end with a hammer. An initial disturbance at a particular place in a material will cause a force to be exerted on adjacent parts of the material.

An elastic force (see pp. 82-83) then acts to restore the material to its equilibrium position. In doing so, it compresses the adjacent particles and so the disturbance moves outward from the source. In attempting to return to their original positions, the particles overshoot, so that at a particular point a rarefaction follows a compression (or squeezing). The passage of the wave is observed as variations in the pressure about the equilibrium position or by the speed of oscillations. The change is described as oscillatory (like a pendulum) or periodic.

Wave motions transfer energy - for example, sound waves, seismic waves and water waves transfer mechanical energy. However, energy is lost as the wave passes through a medium. The amplitude (see diagram) diminishes and the wave is said to be attenuated. There are two distinct processes - spreading and absorption. In many cases there is little or no absorption - electromagnetic radiation from the Sun travels through space without any absorption at all, but planets that are more distant than the Earth receive less radiation because it is spreading over a larger area and so the intensity (the ratio of power to area) decreases according to an inverse-square law. The same applies to sound in the atmosphere. In some cases, however, energy is absorbed in a medium as, for example, when light enters and exposes a photographic film or when X-rays enter flesh.

oscillatory change see mechanical waves.

period (symbol T) the time taken for the wave to undergo one complete oscillation or cycle; also the time taken for the wave to travel one wavelength. See diagram.

periodic oscillation a travelling wave propagates by periodic oscillations either perpendicular or parallel to the direction of travel. There are two main types of periodic oscillation - longitudinal waves and transverse waves (q.q.v.). (See also mechanical waves.)

phase speed (of a wave) see speed of propagation.

rarefaction a stretching.

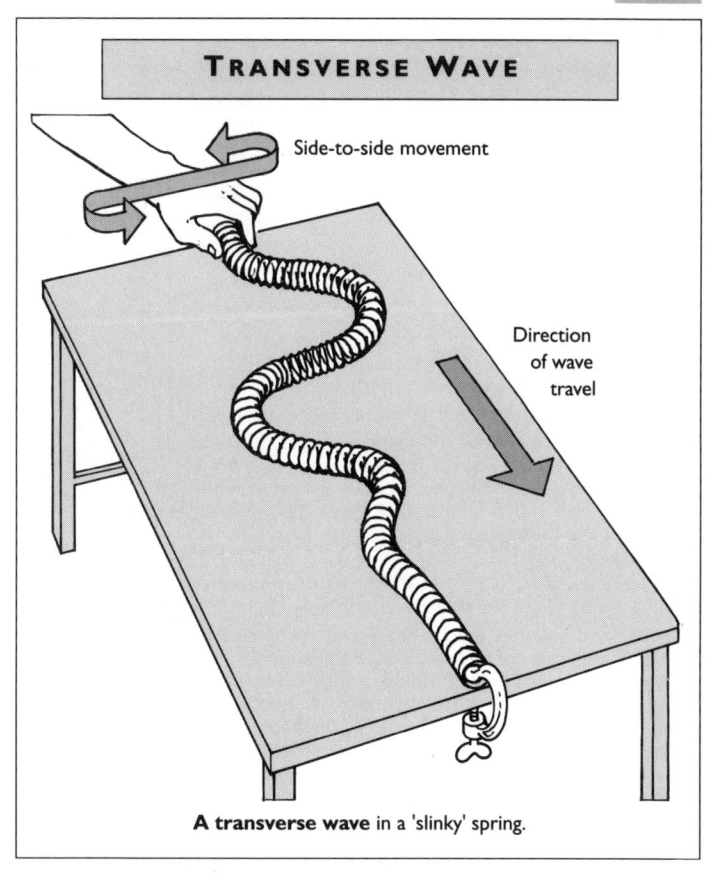

TRANSVERSE WAVE

A transverse wave in a 'slinky' spring.

speed of propagation (of the compressions of waves) or **phase speed** (of a wave) (symbol v) is equal to the product of the frequency and the wavelength.

sound waves see acoustics (pp. 102-03).

transverse waves periodic oscillations in which the vibrations are perpendicular to the direction of travel; see diagram.

travelling wave a disturbance that moves or propagates from one point to another.

wavelength (symbol λ) the distance between two successive peaks or troughs in the wave (or two successive compressions or rarefactions); see diagram.

water waves vibrations produced in water by the wind or some other disturbance. The particles move in vertical circles so there are both transverse and longitudinal displacements. The motion causes the familiar wave profile with narrow peaks and broad troughs. Water waves transfer mechanical energy.

wave attenuation occurs when energy is lost as a wave travels through a medium - the amplitude is reduced and the wave is said to be attenuated. See diagram.

REFLECTION AND REFRACTION OF WAVES

Reflection of waves is the return of part or all of the waves when they encounter the boundary between two materials or media. Refraction of waves is the change of direction of a wavefront as it passes obliquely from one medium to another in which its speed is altered. They are two different and distinct phenomena.

angle of incidence see reflection.
angle of reflection see reflection.
angle of refraction see refraction.
beat frequency see interference.
constructive interference see interference.
destructive interference see interference.
interference the phenomenon that occurs when two or more waves combine together in the manner described in the superposition principle (see below). If the resultant wave amplitude is greater than that of the individual waves then constructive interference is taking place; see diagram.

If the resultant wave is smaller, then destructive interference is taking place; see diagram.

If two sound waves of slightly different frequencies and similar amplitudes are played together (for example two tuning forks), then the resulting sound has varying amplitude. The varying amplitudes are called beats and their frequency is the beat frequency. This frequency is equal to the difference between the frequencies of the two original notes. (Listening for beats is an aid to tuning musical instruments; the fewer the beats, the more nearly in tune is the instrument.)

reflection (of plane waves at a plane surface) is as shown in diagram 1. The angle between the direction of the wavefront and the normal - i.e. a line perpendicular to the plane surface - is the angle of incidence (i). The

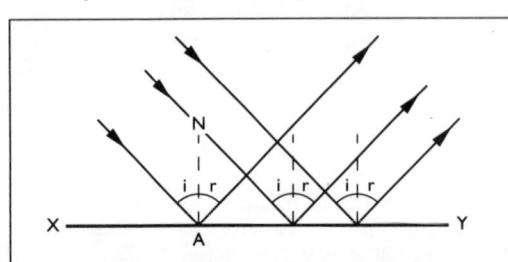

1. Reflection of plane waves at a plane surface. The waves are parallel as they approach XY and after they are reflected. AN is the normal to XY at A. i is the angle of incidence of the wave as it meets XY. The angle of reflection is r, and $i = r$.

angle between the reflected wave and the normal to the plane surface is the angle of reflection (r), and these angles i and r are equal.

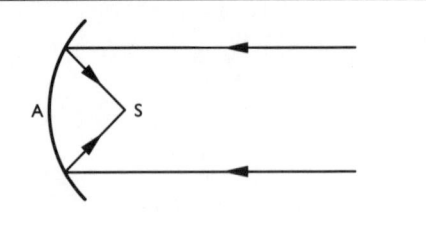

2. Waves reflected at a curved surface. Waves behave in the same way as light reflected in a concave mirror suggesting light is also a wave motion. S is the principal focus of the surface A.

The behaviour of waves reflected at curved surfaces is shown in diagram 2.

refraction (of plane waves at a plane surface) is as shown in diagram 3.

If a wave travels from one medium to another, the direction of propagation is changed or 'bent'; the wave is said to be refracted. The wave will travel in medium 1 with velocity v_1, and come upon the surface of medium 2 with angle of incidence i. Then the wave will be refracted, as in diagram 3, and r is the angle of refraction. The velocity will be v_2, which will be less than v_1 if medium 2 is more dense than medium 1, but greater than v_2 if medium 2 is less dense. The velocities are related by: $\frac{v_1}{v_2} = \frac{\sin i}{\sin r}$

and the ratio of $\frac{\sin i}{\sin r}$ is a constant. This constant is the refractive index of medium 2 with respect to medium 1. This relationship was formulated by the Dutch astronomer Willebrord Snell (1591-1626) and is known as Snell's Law.

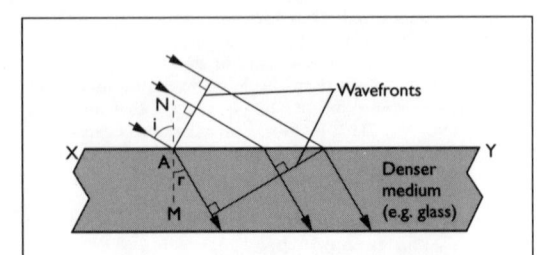

3. Refraction of a plane wavefront. MAN is the normal to XY; i is the angle of incidence; r is the angle of refraction. The waves are parallel after refraction.

refractive index see refraction.
Snell's Law see refraction.
superposition principle the state when several waves are travelling through a medium. The resultant at any point and time is the vector sum of the amplitudes of the individual waves. See also interference.
varying amplitude see interference.

PHYSICS

WAVEFRONTS

NORMALS AND WAVEFRONTS

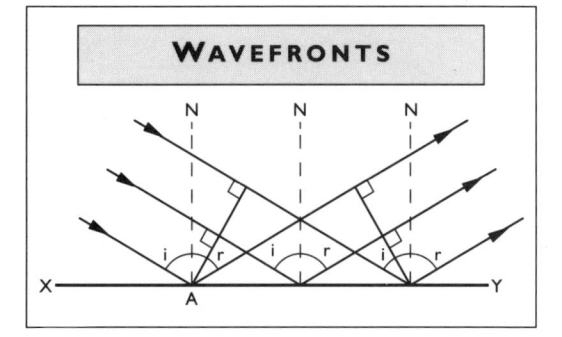

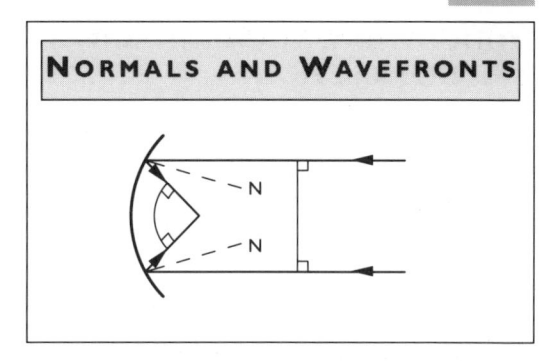

SPHERICAL AND PLANE WAVEFRONTS

Spherical and plane wavefronts. Wavefronts propagating outwards from point source O will be spherical in a three-dimensional context (such as light waves propagating from the Sun) or circular in a two-dimensional context (such as water waves propagating from a dropped pebble). Once far enough from the source, such wavefronts can for most purposes be considered as straight lines - plane wavefronts - much in the same way that the curvature of the earth is not noticeable to someone standing on it.

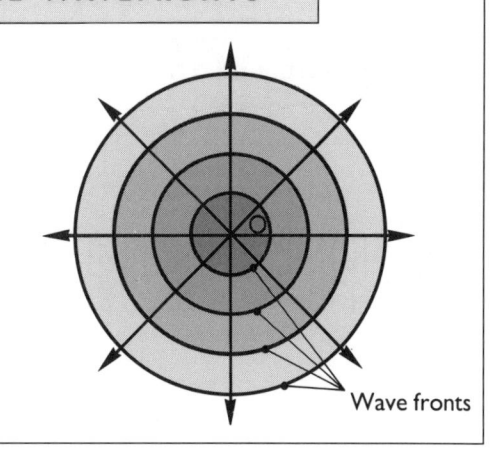

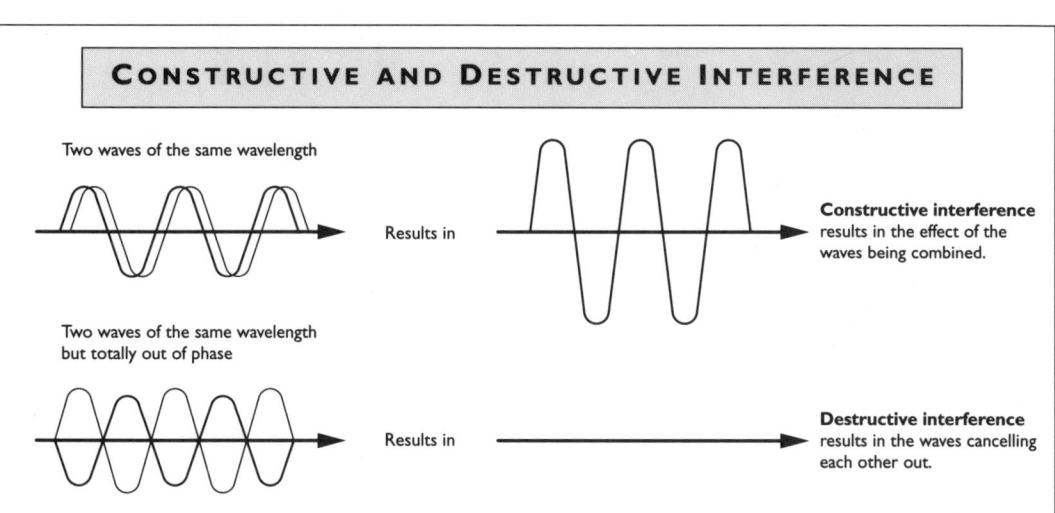

Wave fronts

CONSTRUCTIVE AND DESTRUCTIVE INTERFERENCE

Two waves of the same wavelength

Results in

Constructive interference results in the effect of the waves being combined.

Two waves of the same wavelength but totally out of phase

Results in

Destructive interference results in the waves cancelling each other out.

MODULATION, STANDING WAVES AND DIFFRACTION

Radio waves can be used to carry sound waves by superimposing the pattern of the sound wave onto the radio waves. This is called modulation and is one of the basic forms of radio transmission. There are two forms of modulating radio waves - amplitude modulation and frequency modulation.

AM see amplitude modulation.
amplitude the maximum displacement from the equilibrium position.
amplitude modulation (or **AM**) the most widely used form of modulating radio waves. The amplitude of the radio carrier wave is made to vary with the amplitude of the sound signal; see diagram.

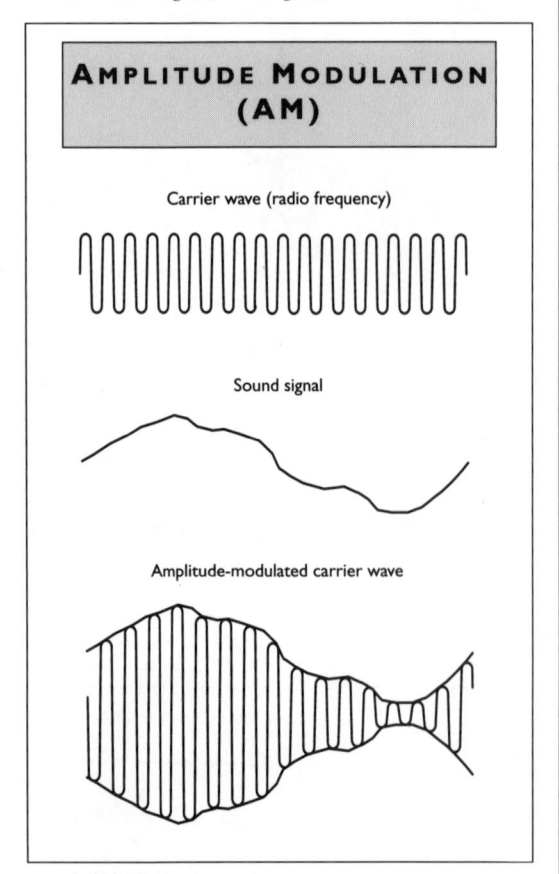

AMPLITUDE MODULATION (AM)

Carrier wave (radio frequency)

Sound signal

Amplitude-modulated carrier wave

antinodes see standing waves and pp. 102-03.
carrier wave an electromagnetic wave of specified frequency and amplitude that is emitted by a radio transmitter.

diffraction the bending or spreading of waves when they pass through a slit or aperture or round the edge of a structure. Waves will usually proceed in a straight line through a uniform medium. However, when they pass though a slit with width comparable to their wavelength, they spread out, i.e. they are diffracted (see diagram). Thus waves are able to bend round corners. For a sound wave of 256Hz the wavelength is about 1.3m (4ft 3in), comparable with the dimensions of open doors or windows.

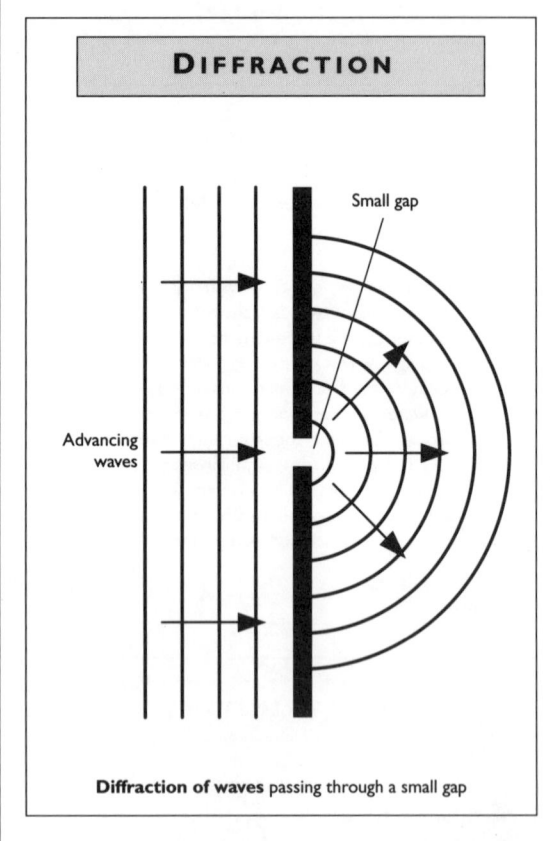

DIFFRACTION

Small gap

Advancing waves

Diffraction of waves passing through a small gap

If a beam of light is shone through a wide single slit onto a screen that is close to the slit, then a bright and clear image of the slit is seen. As the slit is narrowed there comes a point where the image does not continue getting thinner. Instead, a diffraction pattern of light and dark images is seen.

echo is produced by the reflection of sound. Sound is reflected off a hard surface such as a high wall of the sides of a ravine.

If you were to clap your hands loudly while standing at the foot of a high cliff, you would hear a reflection of that sound some time later. This would happen because sound obeys the same laws of reflection as light (see pp. 104-05).

In order for the human ear to detect the reflection of sound - as a noise that is distinct from the original sound - the reflected sound must arrive back at the hearer more than one-tenth of a second later. (As the speed of sound is 340m/s, a person must be at least 17m away from a wall or a similar object in order to detect reflected sound - an echo - from it.)

Echoes can measure the speed of sound in air with some accuracy.

FM see frequency modulation.

frequency modulation (or **FM**) the second and less widely used form of modulating radio waves. The frequency of the carrier wave is made to vary so that the variations are in step with the changes in amplitude of the sound signal; see diagram.

Huygens' principle (also known as **Huygens' construction**) a law proposed in 1676 by the Dutch physicist Christiaan Huygens (1629-95) to explain the laws of reflection (see pp. 98-99) and refraction (see pp. 98-99). He postulated that light was a wave motion. Each point on a wavefront becomes a new or secondary source. The new wavefront is the surface that touches all the wavefronts from the secondary sources. Diffraction describes the interference effects (see pp.98-99) observed between light derived from a continuous portion of a wavefront, such as that at a narrow slit; see diffraction (above). The work of the British physicist and physician Thomas Young (1773-1829) and others eventually supported Huygens' theory, which states that every point on a wavefront may itself be regarded as a source of secondary waves.

Hz the symbol for hertz; see pp. 96-97.

longitudinal waves periodic oscillations in which the vibrations are parallel to the direction of travel; see pp.96-97.

natural frequency the frequency of free oscillation in a system. If a periodic force is applied to a system with frequency at or near to the natural frequency of the system, then the resulting amplitude of vibration is much greater than for other frequencies. These natural frequencies are called resonant frequencies. When a driving frequency equals the resonant frequency then maximum amplitude is obtained.

The natural frequency of objects can be used destructively. High winds can cause suspension bridges to reach their natural frequency and vibrate, sometimes resulting in the destruction of the bridge. Soldiers marching in formation need to break step when crossing bridges in case they achieve the natural frequency of the structure and cause it to disintegrate.

nodes see standing waves and pp. 102-03.

reflection of sound see echo (above).

resonant frequency see natural frequency.

sound signal see amplitude modulation, frequency modulation and diagram.

standing waves (or **stationary waves**) are the result of confining waves in a specific region. When a travelling wave, such as a wave propagating along a guitar string towards the bridge, reaches the support, the string must be almost at rest. A force is exerted on the support that then reacts by setting up a reflected wave (see pp.98-99) travelling back along the string. This wave has the same frequency and wavelength as the source wave. At certain frequencies the two waves, travelling in opposite directions, interfere (see pp. 98-99) to produce a stationary - or standing - wave pattern. Each pattern or mode of vibration corresponds to a particular frequency.

The standing wave may be transverse (see pp. 96-97), as on a plucked violin string, or longitudinal (see pp. 96-97), as in the air in an organ pipe. The positions of maximum and minimum amplitude are called antinodes and nodes respectively; see pp. 102-03. At antinodes the interference is constructive; at nodes it is destructive.

stationary waves see standing waves.

transverse waves periodic oscillations in which the vibrations are perpendicular to the direction of travel; see pp. 96-97.

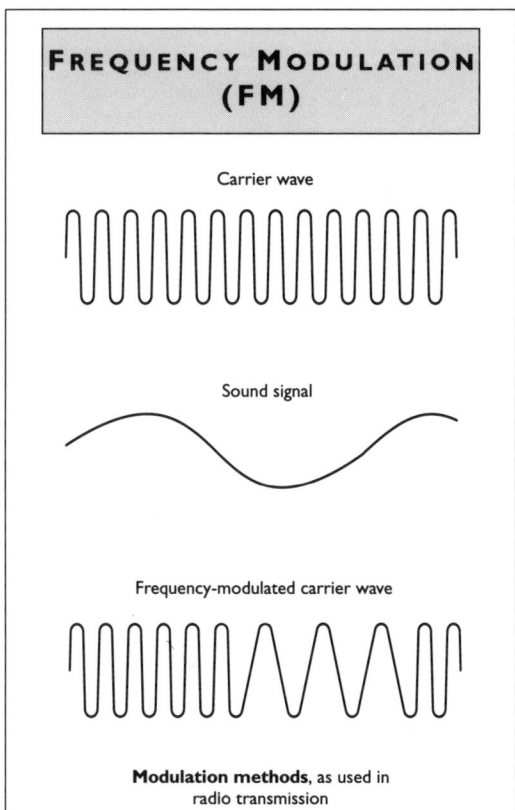

FREQUENCY MODULATION (FM)

Carrier wave

Sound signal

Frequency-modulated carrier wave

Modulation methods, as used in radio transmission

ACOUSTICS

The range of frequencies for which sound waves are audible to humans is from 20 to 20 000Hz (i.e. vibrations or cycles per second) - the higher the frequency, the higher the pitch. In music the A above middle C is internationally standardized at 440Hz. For orchestral instruments, the frequencies range between 6272Hz achieved on a handbell, and 16.4Hz on a sub-contrabass clarinet. On a standard piano (and a violin) the highest note is 4186Hz. For the organ, the highest frequency is 12 544Hz and the lowest is 8.12, using pipes of 1.9cm ($\frac{3}{4}$in) and 19.5m (64ft) respectively.

decibel (symbol db) the unit used to compare two power levels, applied to sound or electric signals. The decibel is one tenth of a bel, the original unit (named after Bell, the inventor of the telephone). The bel is graduated using a logarithmic scale but as the bel is a rather large unit, the decibel is more normally used.

Doppler effect (or **Doppler shift**) the apparent change in the observed frequency of a wave that results from motion between the source of the wave and the observer. The Doppler effect - first described by the Austrian physicist C.J. Doppler (1803-53) - is valid for all waves. It is more often noticed in acoustics and is particularly noticeable in the sirens used for emergency-service vehicles. The intensity and pitch of the siren seems to rise as the vehicle is approaching, then diminishes as it moves away.

This is explained by the fact that, as an observer moves towards a sound source, the pressure oscillations are encountered more frequently than if the observer were stationary. Thus the source seems to be emitting at a higher frequency. Conversely, if the observer is moving away from the source, the frequency seems to decrease. It also applies if the source is moving and the observer is stationary.

The Doppler effect is also observed in optics, and has proved to be of profound significance in our understanding of the universe. If a stellar spectrum is compared with an arc or spark spectrum of an element present in the star, then its spectral lines may be displaced. A shift in the red end of the spectrum - to the longer wavelengths - means that the star is moving away from the Earth. The US astronomer Edwin Hubble (1889-1953) studied the red shift in various galaxies, and determined that their velocity of recession was proportional to their distance from the Earth.

frequency (symbol f) the rate of repetition of a regular event. The number of waves or vibrations per second is expressed in hertz (Hz), cycles per second. Frequencies that are lower than the human audible range are referred to as infrasonic and those above as ultrasonic. Many mammals such as dolphins and bats have sensitive hearing in the ultrasonic range and they use high-pitched squeaks for echolocation. Large animals such as whales and elephants use frequencies in the infrasonic range to communicate over long distances. It is thought that migrating birds can use infrasonic sounds produced by various natural features.

harmonics see tones (below).

hertz (symbol Hz) the SI unit of frequency. It is equivalent to one cycle per second. The hertz is named after the German physicist Heinrich Hertz (1857-94).

human ear the organ of the body that perceives sound, in the human being an extremely sensitive detector. Its threshold of hearing corresponds to an intensity of sound of 10^{-12} watts per square metre (Wm^{-2}): this is a measure of the energy impinging on the ear, and is known as the threshold intensity. The range is enormous, so a logarithmic scale, to the base 10, is used.

The ear canal resonates slightly to sounds with frequencies of about 3200Hz. The human ear is most sensitive in the range 2500-4000Hz. Even then, only about 10 per cent of the population can hear a 0 db sound and then only in the 2500-4000Hz region. The response of the ear is not linear, i.e. there is no direct relationship with the intensity of the sound it detects. Sensitivity is related to frequency: it decreases strongly at the lowest audible frequencies, but less so at the highest.

The audible range of the normal human ear varies with age - the upper limit is about 20 000Hz in the mid-teens,

HARMONICS IN A VIBRATING STRING

1ST HARMONIC (FUNDAMENTAL)

2ND HARMONIC (1ST OVERTONE)

3RD HARMONIC (2ND OVERTONE)

but more likely to be 12 000-14 000Hz for someone 40 years old.

At the lower hearing threshold, the pressure fluctuations from sound waves are about 3×10^{-10} of atmospheric pressure. The eardrum - the tympanic membrane - vibrates at very low velocities. This may seem strange, given that it can vibrate at frequencies of up to 20 000Hz (cycles per second).

infrasonic frequency see frequency.

loudness see tones.

pitch the property of a sound that characterizes its highness or its lowness to a hearer. It is related to - but it is not identical with - frequency (see above). Pitch is closely related to frequency. In music (see tones), if frequency of vibration is doubled the pitch rises by one octave. In general, the higher the frequency the higher the pitch.

refraction of sound the change of direction by a wavefront as it passes obliquely from one medium to another in which its speed its altered.

At night the air near the ground is often colder than the air which is higher up, as the Earth cools after sunset. Thus a sound wave moving upwards will be slowly bent back towards the horizontal as it meets warmer layers of air. Eventually it will be reflected back downwards. Under these circumstances sound can be heard over long distances. This phenomenon is explained by Snell's law of refraction (see pp. 98-99); layers of air at different temperatures act as different media through which sound travels at different velocities. (During World War I the guns at the front line in northern France could sometimes be heard in southern England, although not in the intervening area This was a significant piece of evidence for the existence of the stratosphere, that part of the atmosphere above the troposphere (the lowest layer); in the middle and upper stratosphere the temperature increases with altitude.)

sonic booms a loud bang and large pressure variations caused by a supersonic plane. If the velocity of the source of the sound wave is greater than the velocity of sound - $300ms^{-1}$ (1080km/h or 670mph) in the upper troposphere (up to 10km/6mi above the Earth's surface) - then the wavefront produced is not spherical but conical. Different wave crests bunch together, forming a shock wave. A supersonic plane (travelling faster than the speed of sound) produces such a shockfront, causing the characteristic double loud bang.

sound waves are longitudinal compressions (squeezings) and rarefactions (stretchings) of the medium through which they are travelling, and are produced by a vibrating object. If a sound wave is travelling in any medium then the pressure variations formed along its path cause strains as a result of the applied stresses.(See also velocity of sound.)

tones (characteristics of notes played by musical instruments) include loudness, pitch and timbre or tone quality. Loudness would seem to be the most simple, but it is complicated by the non-linear response of the ear (see above). At 100Hz and 10 000Hz the hearing threshold is about 40db compared to the 0db at 2500-4000Hz. Thus the concept of loudness is not dependent just on the energy reaching the ear, but also on frequency.

Sounds created by musical instruments are not simple waveforms, but are the result of several waves combining. This complexity results in the tone quality or timbre of a note played by a particular stringed musical instrument. Even a 'pure' note may contain many waves of different frequencies. These frequencies are harmonics or multiples of the fundamental or lowest frequency, which has 2 nodes and 1 antinode, and is called the first harmonic (see diagram). The second harmonic has 3 nodes and 2 antinodes. The wavelength is halved, and the frequency is doubled. The third harmonic has 4 nodes and 3 antinodes. The wavelength is one third of the original wavelength, and the frequency has tripled. Different instruments emphasize different harmonics. (See also pitch.)

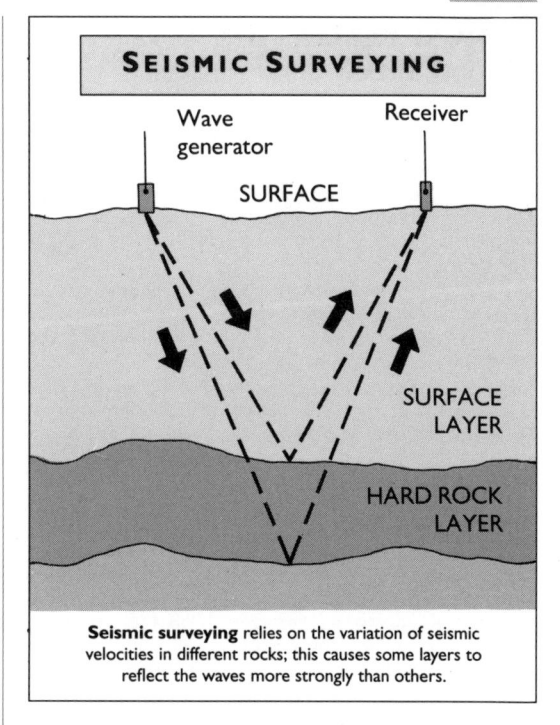

ultrasonic frequency see frequency.

velocity of sound is given by the square root of the appropriate elastic modulus (see pp. 83-84) divided by the density. The velocity of sound - as with other types of wave - differs in different media. In still air at 0°C, the velocity of sound is about $331ms^{-1}$ (1191.6km/h, or 740mph). If the air temperature rises by 1°C, then the velocity of sound increases by about $0.6ms^{-1}$. The velocity of sound in a metal such as steel is about $5060ms^{-1}$.

In ocean depths the combined effect of salinity, temperature and pressure results in a minimum velocity for sound. The channel that is centred around this minimum velocity at a depth of about 1000-1300m (3300-4250ft) allows sound waves, travelling at the minimum velocity, to propagate within it with relatively little loss over large horizontal distances. Signals have been transmitted in this way from Australia to Bermuda.

The fact that the velocity of sound varies in different media is one reason why seismic techniques can be used to probe layers of rock or minerals underground.

Similarly ultrasonic scanning can be used in medicine - for example, in the imaging of a baby in its mother's womb. In each case variations in materials are shown up through variations in the time it takes sound waves to travel to the detector.

SEISMIC SURVEYING

Seismic surveying relies on the variation of seismic velocities in different rocks; this causes some layers to reflect the waves more strongly than others.

OPTICAL REFLECTION AND REFRACTION

Optics is the branch of physics that deals with high-frequency electromagnetic waves that we call light. Optics is concerned with the way in which light propagates from sources to detectors via intermediate lenses, mirrors and other modifying elements. The electromagnetic spectrum (see pp. 112-13) includes a wide range of waves in addition to light (see below).

The region with wavelengths from 700 nanometres in the red region to 400nm in the violet is extended for practical optical systems into the ultraviolet and the mid-infrared regions (see pp. 112-13).

A beam of light may be considered to be made up of many rays, all travelling outwards from the source. This approach is used in ray diagrams. In geometrical simplifications, as in the diagrams used here, rays of light are shown as straight lines. The wavelength and amplitude of light are very short compared to the other dimensions of the system. The basic concept is a very simple one: light travels in straight lines unless it is refracted by a lens or a prism.

A point source of light emits rays in all directions. For an isolated point source in a vacuum the geometric wavefront will be a sphere.

Light is reflected and refracted (i.e. bent) in the same way as other waves.

amplitude the maximum difference of the disturbed quantity from its mean.

angle of incidence the angle between an incoming beam of light and a line at 90 degrees to the surface of the medium which that beam is coming into contact with.

beam (of light) see introduction (above).

deviation (of light) the turning away from the normal (q.v.), or perpendicular, of light when it travels from a medium to a less dense medium.

dispersion the effect of a prism (q.v.) on light whereby white light is separated into its component colours. This effect has been known for centuries. Newton used dispersion to produce and study the spectrum of sunlight.

enantioners see optical activity.

incident ray an incoming ray of light.

light that small part of the spectrum that can be detected by the human eye.

monochromatic beam (of light) a single-coloured beam of light.

normal the perpendicular from the surface at the point of incidence.

optical activity the property of some substances to rotate the plane of plane-polarized light as it passes through, for example, a crystal, a solution or a liquid. This happens because the molecules of the substance are asymmetric - they can occur in different structures that are exact mirror images of each other. There are two forms - one will rotate the light in one direction; the other will rotate the light by an equal amount in the opposite direction. The two forms are known as optical isomers and enantioners.

optical isomers see optical activity.

point of incidence the point at which an incident ray is reflected.

prism a block of glass - or any other transparent material - usually with a triangular base and always with a triangular cross-section.

The refractive index of optical glass is not constant for light of all frequencies. It is greater at the violet end and less at the red end of the spectrum. This means that a beam of light containing a mixture of different frequencies, for example, sunlight, will leave a prism with the different frequencies bent by different amounts.

A prism is used to deviate a beam of light by refraction. A beam of white light will be split into its component monochromatic coloured lights - from red to violet - which will form the familiar rainbow effect. Any light can be split up in this way; the display of separated wavelengths is called the spectrum of the original beam. (See dispersion, above.)

Under the right conditions dispersion occurring in spherical raindrops in the atmosphere produces a rainbow.

rainbow an arc in the sky, containing the colours of the spectrum in bands. Rainbows are formed by the refraction of the Sun's rays either in falling rain or in mist; the larger the drops, the stronger the colours.

rainbow effect see prism.

ray (of light) see introduction (above).

reflection (of light) one of the three results that can happen when one medium meets the surface of another medium; the other two possible results are refraction and absorption. In reflection of light the rays of light are 'thrown back'.

In the accompanying figure, a monochromatic beam of light falls or is incident upon a transparent material such as a block of glass. Angle i is the angle of incidence of the beam. Part of the beam is reflected at an angle t, the angle of reflection, and part is transmitted according to the law of refraction (q.v.).

The two laws of reflection allow a prediction to be made as to where an incident ray will go after it has been reflected from the point of incidence. (Angles are reflected from the normal (q.v.).) These laws state:

- that the reflected ray is in the same plane as the incident ray; and

- that the angle of reflection, r, equals the angle of incidence, i.

REFLECTION AND REFRACTION

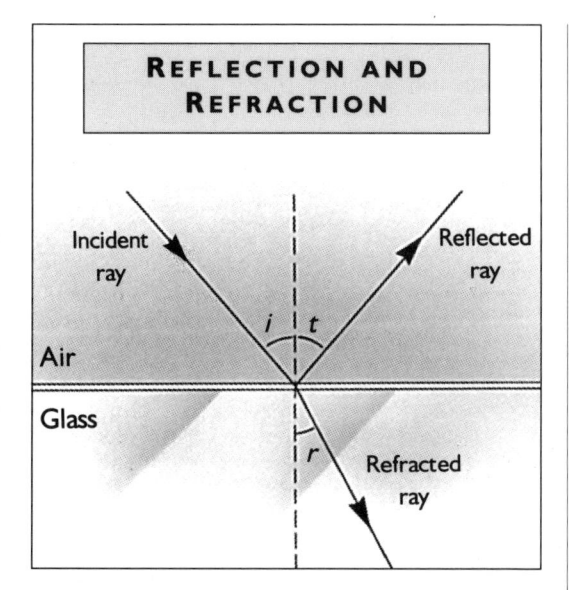

[Diagram: Incident ray and Reflected ray in Air at angles i and t from normal; Refracted ray in Glass at angle r]

refraction one of the three results that can happen when one medium comes into contact with the surface of another medium; the other two possible outcomes are reflection (q.v.) and absorption.

In refraction, the ray travels on through the second medium - in the accompanying diagram a block of glass - almost always taking a different direction. In this diagram r is the angle of refraction.

Snell's law of refraction can be stated as:

$$n_1 \sin i = n_2 \sin r$$

where n_1 and n_2 are the refractive indices of the materials; see also pp. 98-99.

refractive index is often expressed relative to another material; see also refraction (above). If no other material is quoted, the refractive index of a medium is assumed to be relative to air. The refractive index of a medium can also be derived as the ratio of the speed of light in a vacuum to the speed of light in the medium. The refractive index for a typical optical glass is 1.6, whereas the refractive index of diamond is about 2.4 in visible light.

Basically, the refractive index of a material determines how much it will refract light.

Snell's law (of refraction) see refraction.

spectrum a range of electromagnetic energies arranged in order of increasing or decreasing wavelength or frequency. This is often assumed to mean the visible spectrum which is, however, only a small part of the electromagnetic spectrum which ranges from X-rays to radio waves. See also prism and pp. 112-13.

speed of light varies in different materials - the speed of light (as of other electromagnetic waves) in a vacuum is $3 \times 10^8 ms^{-1}$ (300 000km or 186 000mi per second), but it travels more slowly through other media.

total internal reflection exceeds a certain angle, which depends on the refractive index.

When light travels from one medium to another less dense medium it is deviated or turned away from the normal - perpendicular to the interface at the point of incidence. This means the angle of refraction (r) is greater than the angle of incidence (i).

When the angle of refraction is less than 90°, some of the incident light will be refracted and some will be reflected. If the angle of incidence increases, the angle of refraction will increase more. It is possible to increase the angle of incidence to such a value that eventually the reflected ray disappears and all the light is refracted. This is total internal reflection; see diagram.

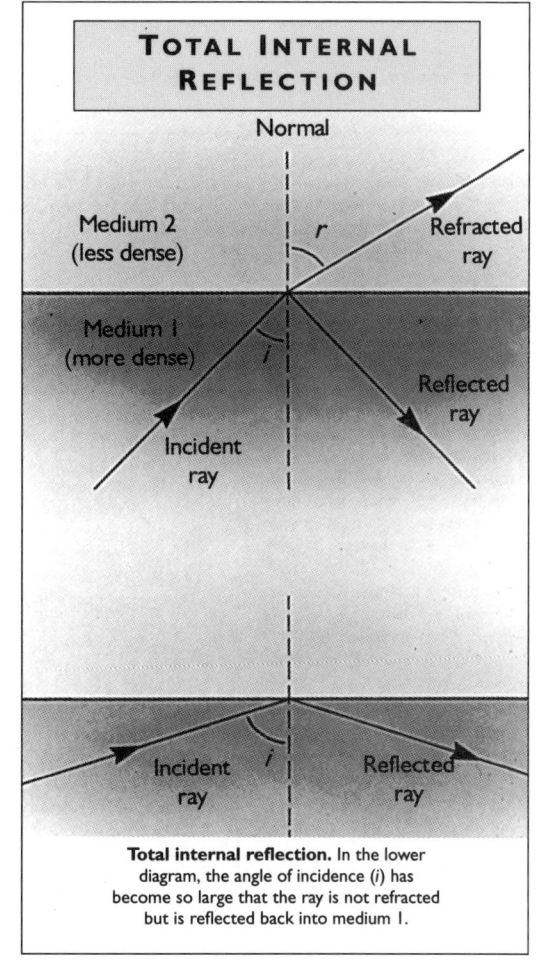

TOTAL INTERNAL REFLECTION

[Upper diagram: Incident ray in Medium 1 (more dense) at angle i; Refracted ray in Medium 2 (less dense) at angle r; Reflected ray]

[Lower diagram: Incident ray and Reflected ray at large angle i, no refracted ray]

Total internal reflection. In the lower diagram, the angle of incidence (i) has become so large that the ray is not refracted but is reflected back into medium 1.

wavelength the distance in metres between the successive points of equal phase in a wave.

LENS AND MIRRORS

A lens is a piece of transparent material made in a simple geometric shape. Usually at least one surface is spheric, and often both are.

Mirrors are reflecting optical elements. Plane mirrors are used to deviate light beams without dispersion or to reverse or invert images. Curved mirrors, which usually have spherical or parabolic surfaces, can form images, and are often used in illumination systems such as car headlamps.

aberration a fault on a lens.

achromatic lens see properties of lens.

chromatic of or having colour or colours.

chromaticity an objective description of colour quality. Chromaticity is expressed in terms of chromaticity coordinates.

convex curving outwards. A convex lens has at least one surface formed from the exterior surface of a sphere.

curved mirror see introduction (above).

dispersion the effects of a prism on light; see pp. 104-05.

focal length (of lens) the distance between the optical centre of the lens and the principal focus. The focal length is usually designated as f.

frequency the number of complete disturbances (cycles) in unit time.

microscope a device for making very small objects visible. It was probably invented by a Dutch spectacle-maker Zacharias Janssen (1580-1638), in 1609. Essentially it is an elaboration of the simple magnifying glass. The objective (q.v.) is used to form a highly magnified image of a small object placed close to its focal point (see diagram). This can be viewed directly, by means of another lens called the eyepiece. It can also be recorded directly on film or viewed via a video camera.

objective a lens with a short focus for a microscope but long for a telescope.

plane mirror see introduction (above).

principal focus see properties of lens.

properties of lenses see diagram. Under appropriate conditions a lens will produce an image of an object by refraction (see pp. 104-05) of light. It does by this bending rays of light from an object.

Some rays are refracted more than others, depending on how they arrive at the surface of the lens. The lens affects the velocity of the rays, since light travels more slowly in a dense medium such as the lens than in a less dense medium such as air. In this way, the expanding geometric wavefront that is generated by an object beyond the focal point is changed into a wavefront which for a convex or converging lens, converges to a point behind the lens. If the object is located a long distance from the lens - strictly an infinite distance but a star is an excellent approximation for practical purposes - this point is known as the rear focal point or principal focus (see diagram). (A lens has two principal foci - one on each side.)

For objects closer than the focal point the lens is unable to converge the waves - no 'real' image can be formed; instead a 'virtual' image is seen on the same side of the lens as the object. The lens is now being used as a 'magnifying glass'.

If a point source of light is placed at the principal focus of the convex lens, the rays of light will be refracted to form a parallel beam.

Because of the effects of dispersion (q.v.), the distance from the lens at which red light and blue light fro an object will be focused will be different. This can be demonstrated in the colour fringes that can be seen in simple hand magnifiers (small magnifying glasses). Such fringes are unacceptable, for instance, in camera lenses. A lens made from two different types of glass can be made to bring two colours to exactly the same focus. Such a lens is called achromatic. Single-element lenses are, therefore, used for only simple applications. Lenses for cameras, binoculars, telescopes and microscopes are made with many elements, with different curvatures. These lenses are made from glasses with different indices and dispersions (see pp. 104-05). The additional elements allow the lens designer greatly to reduce the aberrations of the lens.

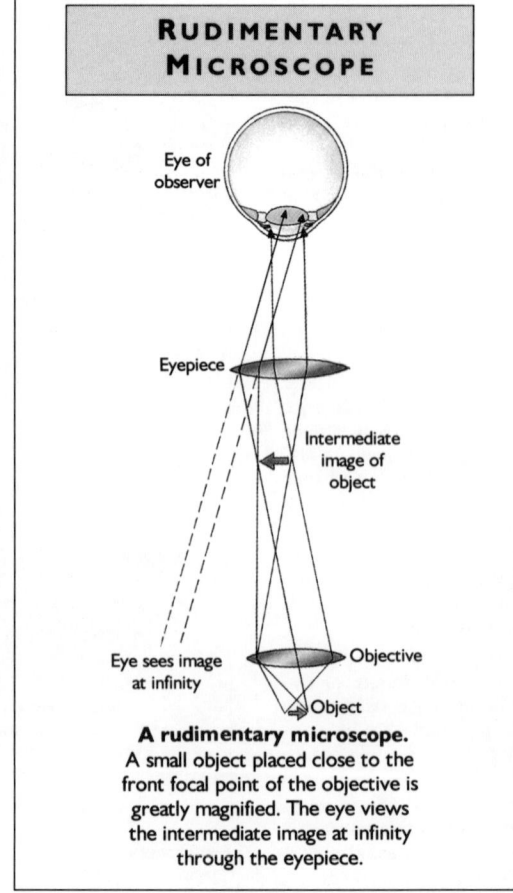

RUDIMENTARY MICROSCOPE

Eye of observer

Eyepiece

Intermediate image of object

Eye sees image at infinity

Objective

Object

A rudimentary microscope. A small object placed close to the front focal point of the objective is greatly magnified. The eye views the intermediate image at infinity through the eyepiece.

PROPERTIES OF LENSES

Parallel rays come to a focus on the far side of a convex lens.

Parallel rays leave a concave lens as if they had come from the focal point on the near side of it.

Properties of lenses. The lens at the top is *convex* (i.e. it is thicker in the middle), while the one below is *concave* (i.e. it is thinner in the middle).

properties of mirrors vary according to their form (see introduction) and manufacture. Mirrors can be coated with metals such as aluminium or silver, which have high resistance for visible light (or gold for the infrared). Alternatively, they may be coated with many thin layers of non-metallic materials for very high reflectances over a more restricted range of frequencies. A freshly coated aluminium mirror will reflect about 90% of visible light. Special mirrors, such as those used in lasers (see p. 108), can reflect over 99.7% of the light at one frequency.

Mirrors do not introduce any chromatic aberrations into optical systems.

real image is one through which rays of light actually pass, hence it can be formed on a screen. See also properties of lens (above).

rear focal point see properties of lens (above).

telescope a device used to form an enlarged image of an infinitely distant object. The enlarged object is viewed by the observer by means of an eyepiece (see diagram). The term 'infinite' is used relatively in this context: compared with the length of the telescope, the distance of the object can be considered as infinite. Telescopes are often made with reflecting mirrors instead of glass lenses, as large lenses sag under their own weight, thereby introducing distortions into the image and chromatic aberration is not caused by mirrors. The primary mirror is often a large concave paraboloid.

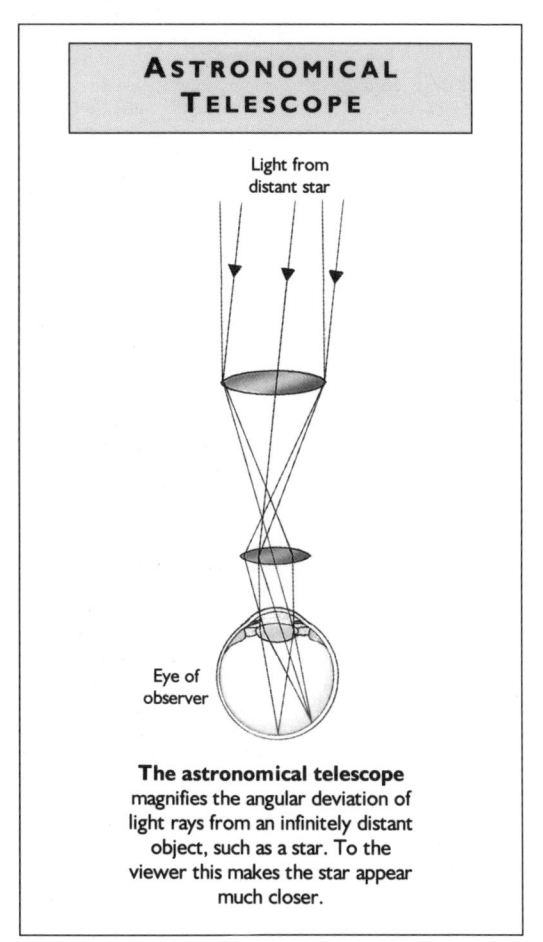

ASTRONOMICAL TELESCOPE

The astronomical telescope magnifies the angular deviation of light rays from an infinitely distant object, such as a star. To the viewer this makes the star appear much closer.

virtual image is one from which rays of light cannot be formed on a screen. Both convex and concave lenses can produce both types of image, depending on the position of the object. See also properties of lenses (above).

LASERS, HOLOGRAMS AND FIBRE OPTICS

The properties of prisms, lenses and mirrors have been known about for a long time. Lasers, holograms and fibre optics are, however, late 20th century developments in the realm of optics.

amplitude (of light) the maximum difference of the disturbed quantity (in this case light) from its mean value.

digital signals on-off signals.

external laser beam see laser.

fibre optics the use of flexible glass fibres with which light can be transmitted over great distances. These fibres are usually less than 1mm ($\frac{1}{25}$ in) in diameter, and can be used singly or in bunches.

Each fibre consists of a small core surrounded by a layer of 'cladding' glass with a slightly lower refractive index (see pp. 104-05). Certain rays experience total internal reflection (see p. 105), and this, coupled with the very low absorption of modern silica glasses, allows light to travel very long distances with little reduction in intensity.

Fibre optics provides the basis of endoscopy (a medical diagnostic technique) and are also used extensively in telecommunications, as light in a fibre optic cable can carry more digital signals (q.v.) with less loss of intensity than a copper wire carrying electrical digital signals.

hologram a 'three-dimensional' or stereoscopic image formed by beams of light.

Holography differs from conventional photography in that both the amplitude (see above) of the light and its phase (see below) are recorded on the film. It has many scientific uses as well as the more familiar display holograms that are, for instance, being used on credit cards.

A hologram is recorded as follows. When light beams from sources such as lasers overlap they produce interference fringes due to the wave nature of light. A hologram is produced by recording the fringes from the interference of two beams of laser light. The reference beam (RB) falls directly onto the film, while the object beam (OB) is scattered from the object. See diagram.

A hologram is replayed in the following manner. When the hologram is replayed by shining a beam of light onto it, the light is diffracted in such a way that it appears to come from the position of the original object. The image can be viewed from a range of angles and is a true three-dimensional reconstruction of the object. See diagram.

intensity (in light) the ratio of power to area.

interference the interaction of two or more wave motions.

laser a term derived from the technical name for the process Light Amplification by Stimulated Emission of Radiation.

When an amplifying material, such as a gas, crystal or liquid, is placed between appropriate mirrors, photons from a light beam repeatedly pass through it stimulating more photons and thus increasing their number with each pass. The additional photons all have the same frequency, phase and direction. One of the mirrors is made so that a small amount of light passes through it; this is the external laser beam, which can be continuous or pulsed. This beam can be focused onto very small areas and the intensity can be very great, enabling some lasers to burn through thick metal plates.

Lasers have a wide variety of applications, for example in surveying, communications and eye surgery.

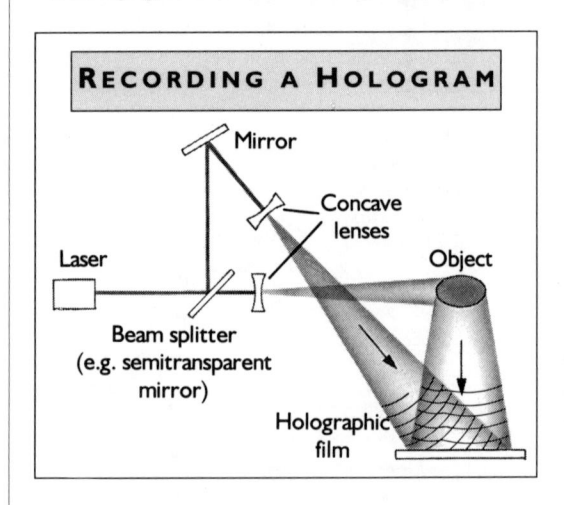

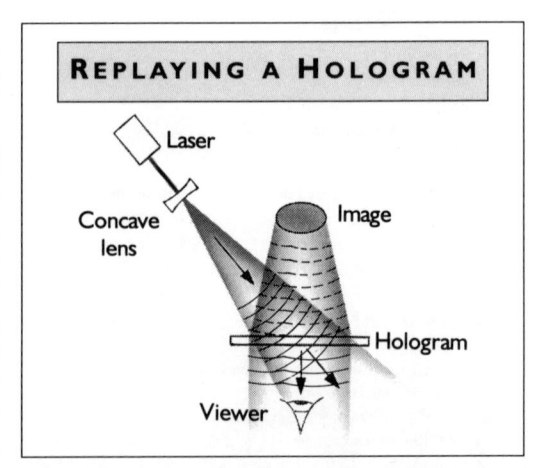

RECORDING A HOLOGRAM

REPLAYING A HOLOGRAM

phase (of light) a measure of the relative distance that light has travelled from an object.

photon a particle of light; a particle consisting of quantum electromagnetic radiation.

stimulated emission (of light) the emission of a photon (q.v.).

CATHODES

Cathodes - and cathode-ray tubes - lie behind one of the devices that we sometimes take for granted as one of the necessities of everyday modern life - the television set.

A black and white television set is basically a cathode-ray tube, in which a 'gun' fires a beam of electrons at a fluorescent screen. As they strike it, the screen lights up and the television signal is decoded. To make up the whole picture the beam is scanned to and fro in a series of lines (625 in modern sets), covering the entire screen in $\frac{1}{25}$ second.

In a colour television set there are three guns scanning a screen covered with phosphor dots that emit red, green or blue light when the cathode rays strike them.

anode a positive electrode.

cathode
1) a negative electrode in an electrolytic cell. Cations are attracted to the cathode in electrolysis.
2) in vacuum electronic devices, the electrons emitter (from which electrodes flow to the anode).
3) in batteries, the positive terminal.

cathode-ray oscilloscope a device based on the cathode-ray tube that provides visual images of electrical signals. An internal time base causes deflection of the beam which passes across a screen at a set rate.

cathode rays streams of electrons moving at high speed that are emitted at the hot cathode in an evacuated tube that contains a cathode and an anode. The first scientific descriptions of cathode rays were of those observed in low pressure gas discharge tubes. Electrons produced at the cathode are - under suitable conditions - accelerated down a cathode-ray tube towards the anode.

cathode-ray tube a device that provides a viewing screen for, for example, a radar viewer or a television set; see also cathode-ray oscilloscope.

The cathode-ray tube comprises an evacuated tube which contains a heated cathode and two or more anodes (which are ring-shaped). Cathode rays pass from the heated cathode through the ringed anodes to strike the enlarged end of the tube which is coated with fluorescent material to form a screen. When a cathode ray strikes this screen, the point at which it strikes becomes luminous.

The beam is varied by a control grid between the cathode and the anodes, thus controlling the brightness of the screen. Beams from the electron gun (q.v.) are focused and deflected by plates that form an electrical field (see pp. 110-11). The field concentrates the beam into a small point of light which produces an illuminated line as the beam sweeps the tube.

cation a positively charged ion. Cations are attracted to cathodes.

electron gun the assembly of cathode, control grid and anodes in a cathode-ray tube (q.v.).

screen see cathode-ray tube.

television tube a form of cathode-ray tube in which the beam is made to scan the screen 625 times to form a frame, with 25 new frames being produced per second. By means of the variations of intensity of the beams, each frame creates a picture. See cathode-ray tube.

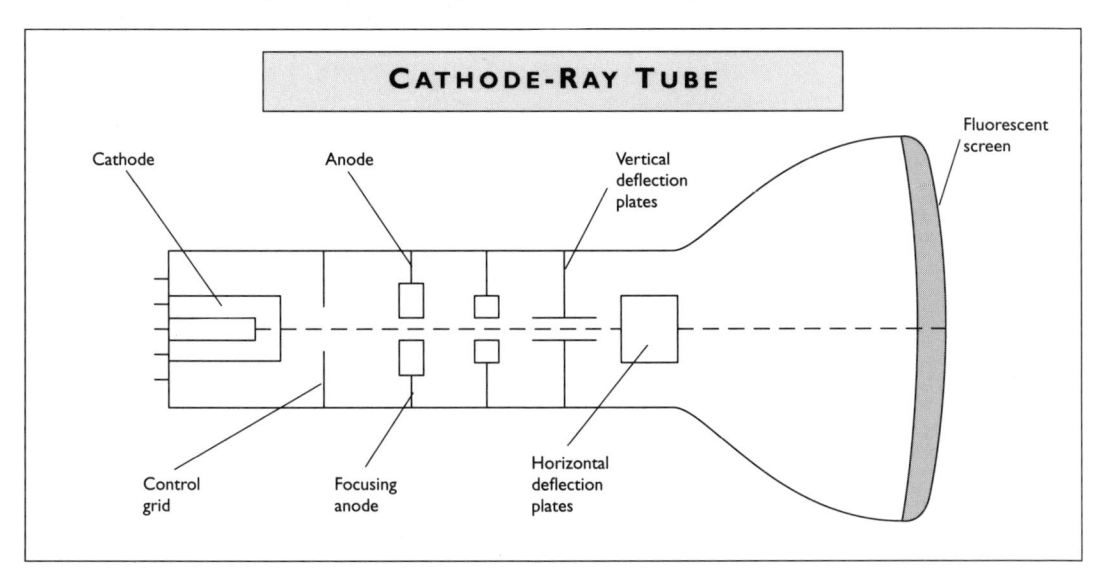

CATHODE-RAY TUBE

Cathode — Anode — Vertical deflection plates — Fluorescent screen

Control grid — Focusing anode — Horizontal deflection plates

MAGNETS AND ELECTRICITY

Magnetism and electricity are linked through their interaction.

Certain materials have the ability to attract iron. This property - known as magnetism - has been used since about 500 BC when metallic ores with magnetic properties were being used as compasses. These materials include lodestone (or magnetite), an iron ore as well as iron, steel, nickel, cobalt and a host of various alloys. Some alloys have recently been produced that make very strong magnets. They include alni, alcomax and ticonal.

If a bar magnet is placed on a cork which is floating on water, in such a way that the magnet is able to swing in a horizontal plane, it will come to rest with its axis in a north-south direction. It will lie in the magnetic meridian.

A temporary magnet can be made if a piece of iron is placed near a magnet. It will, however, lose its magnetism when it is removed. A lasting magnet can be created by electricity - see solenoid (below).

The interaction between magnetism and electricity - which were initially observed separately - was first noticed in the 19th century. It was discovered that electricity and magnetism were both manifestations of a single force, the electromagnetic force; see also pp. 112-13. Magnetic effects are now known to be caused by moving electric charges.

alcomax an alloy that makes a strong magnet; see introduction (above).

alni an alloy that makes a strong magnet; see introduction (above).

ampere the unit of electric current (q.v.).

attraction occurs between the N and S poles of two magnets; compare with repulsion (below). This may be summed up in the saying 'Like poles repel, unlike poles attract'.

coulomb the unit of charge; see Coulomb's law.

Coulomb's law the law that describes the electric force. This is an inverse-square law which is similar to the law for gravitational force.

Coulomb's law states that the attraction or repulsive force (F) between two point (or spherically symmetrical) charges is given by:

$$F = k \frac{Q_1 Q_2}{r^2}$$

where k is a constant, Q_1 and Q_2 are the magnitudes of the charges, and r is the distance between them. The force acts along the direction of r. The unit of charge is called the coulomb (C) and is the quantity of electric charge carried past a given point in one second by a current of one ampere.

electric charge a flow of charged particles, especially electrons, that constitutes an electric current. (Atomic electrons are in motion and thus all atoms exhibit magnetic fields.)

In dry weather, a nylon sweater being pulled off over the hair of the wearer may crackle; sparks may even be seen. This is caused by an electric charge, the result of electrons being pulled from one surface to the other. Objects can gain an electric charge by being rubbed against another material.

There are two types of charge. These are now associated with the negative and positive charges on electrons and protons respectively. Similar electric charges (i.e. two positives or two negatives) repel each other and unlike charges (i.e. a positive and a negative) attract. (Note that positive and negative are just conventional terms for opposite properties.)

No smaller charge than that on the electron has been detected - its value is 1.60×10^{-19} coulombs.

The force of repulsion or attraction is known as the electric force. It is described by Coulomb's law.

electric current consists of a flow of electrons, usually through a material but also through a vacuum, as in a cathode-ray tube in a TV set. Current flows where there is a potential difference or voltage between two ends of a conductor. Conventional current flows from the positive terminal to the negative terminal. However, electron flow is in fact from negative to positive; see pp. 114-15.

For measurement purposes, an electric current is defined as the rate of flow of charge. The unit of electric current is the ampere (A), often abbreviated to amp: one ampere = one coulomb per second.

electric field the region in which an electric charge experiences a force. Just as arrows can be plotted to show the magnitude and direction of the magnetic force that acts at points around a magnet (see below), so arrows can also plot the electric force that acts on a unit positive charge at each point. Such a chart would show the distribution of the electric field intensity. It is measured in terms of a force per unit charge, or newtons per coulomb.

In the same way that mass may have gravitational potential energy because of its position, so a charge can have electrical potential energy. This potential per unit charge is measured in volts (V) - named after the Italian physicist Alessandro Volta (1745-1827); see volt (below).

electric force the force of repulsion or attraction - see electric field.

electrical potential energy see electric field.

lines of force see magnetic field.

magnetic field the region around a magnet in which a force is exerted. Magnetic fields can be demonstrated in the following experiments.

If a bar magnet is placed on a piece of paper, a plotting compass may be used to mark out that magnet's magnetic field. Beginning at one end of the magnet the positions of the ends of the compass needle can be marked upon the paper by two dots. The compass can then be moved until the near end of the needle rests exactly over the dot which is furthest from the magnet and a third dot is made under the other end of the needle. This procedure is repeated many times until the other end of the magnet is reached and a line of dots is built up. Further lines can similarly be drawn. By convention these lines of force are labelled with an arrow which indicates the direction in which a N pole would move. The strength of the magnetic field is indicated by how far apart the lines of force are - the closer the lines, the greater the force.

In the accompanying diagrams of magnetic fields, figure 1 shows a magnetic field which is due to the earth alone. Figure 2 shows a magnetic field that is due to a magnet alone, while Figure 3 shows a magnetic field that is due to two attracting magnets being placed close to one another, end to end.

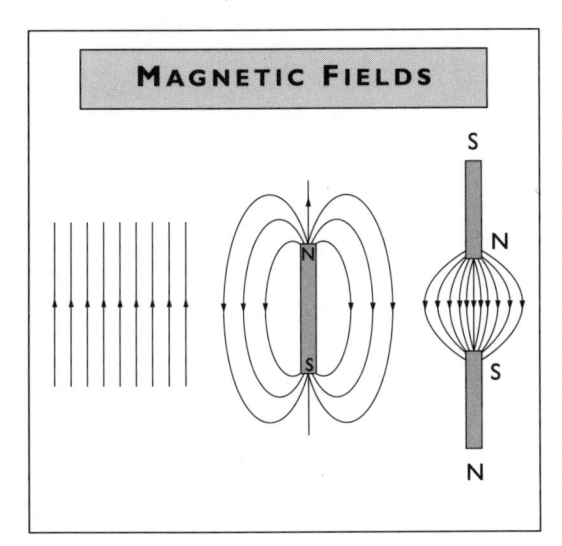

MAGNETIC FIELDS

magnetic meridian the north-south vertical plane in which a magnet lies.

mumetal an alloy that has been developed for the electromagnet.

negative charge see electric charge.

neutral point the place where two magnetic fields are equal and opposite such that the total field is zero.

north-seeking pole the end of a magnet which points towards north. It is called the N pole for short.

N pole see north-seeking pole.

positive charge see electric charge.

potential difference see electric current (above) and volt (below).

repulsion occurs when the N pole of a second magnet is brought near to the N pole of a magnet which is placed on a floating cork. The cork and magnet will swing round. Repulsion will also take place between two S poles in a similar fashion. See also attraction.

solenoid is a long cylindrical coil. Solenoids are used in magnetization by electrical methods - a form of magnetization which is quick and efficient.

A cylindrical coil is wound with a couple of hundred turns of insulated copper wire and is then connected to a direct current supply. A steel bar is then inserted into the coil and when the current is switched on it creates a strong magnetic field within the coil. The steel, thus, becomes a magnet and retains its magnetism even when the power is switched off.

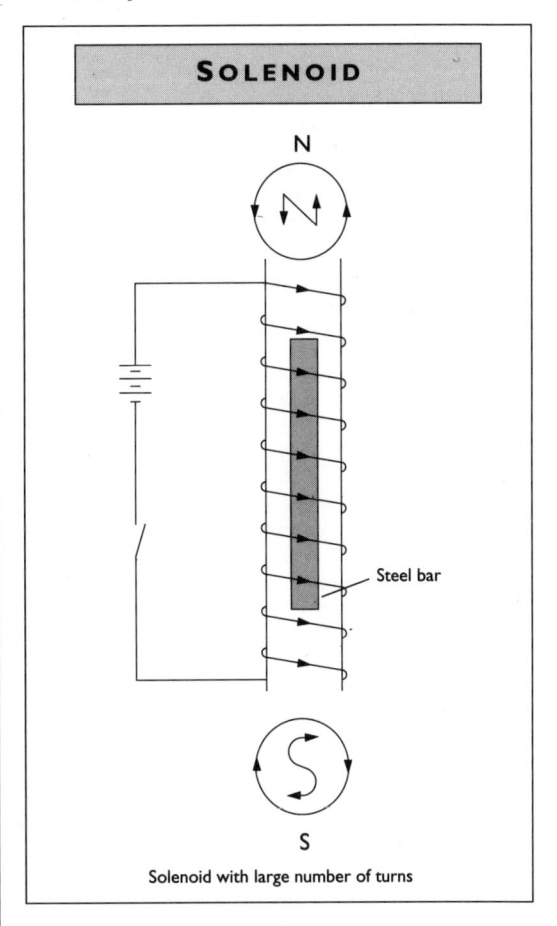

SOLENOID

Solenoid with large number of turns

south-seeking pole the end of a magnet which points towards south. It is called the S pole for short.

S pole see south-seeking pole.

ticonal an alloy that makes a strong magnet; see introduction (above).

volt (V) the unit of electric potential. The volt may be defined as follows: if one joule is required to move one coulomb of electric charge between two points, then the potential difference between the two points is one joule per coulomb = one volt. (See also electric field.)

ELECTROMAGNETISM

Electromagnetism is the study of the effects caused by stationary and moving electric charges. Electricity and magnetism were originally observed separately, but in the 19th century, scientists began to investigate their interaction. This work resulted in a theory that electricity and magnetism were both manifestations of a single force, the electromagnetic force.

The electromagnetic force is one of the fundamental forces of nature, the others being the gravitational force and the weak and strong nuclear forces. Recently the electromagnetic and weak forces have been shown to be manifestations of an electro-weak force.

In 1820 the Danish physicist Hans Christiaan Oersted (1777-1851) discovered that a copper wire bearing an electric current caused a pivoted magnetic needle to be deflected until it was tangential to a circle drawn around the wire, This was the first connection to be established between electrical and magnetic forces. Oersted's work was developed by the Frenchmen Jean-Baptiste Biot (1774-1862) and Felix Savart (1791-1841), who showed that the field strength of a current flowing in a straight wire varied with the distance from the wire. Biot and Savart were able to find a law relating the current in a small part of the conductor to the magnetic field. Ampère, at about the same time, found a more fundamental relationship between the current in a wire and the magnetic field about it.

cosmic rays see electromagnetic spectrum.

dynamo theory see magnetic fields.

electromagnetic force see Maxwell's theory.

electromagnetic induction began with the advance in 1831 when the English physicist Michael Faraday (1791-1867) found that an electric current could be induced in a wire by another, changing current in a second wire. Faraday published his findings before the US physicist Joseph Henry (1797-1878), who had first made the same discovery.

Faraday showed that the magnetic field at the wire had to be changing for an electric current to be produced. This may be done by changing the current in a second wire, by moving a magnet relative to the wire, or by moving the wire relative to a magnet. This last technique is that employed in a dynamo generator, which maintains an electric current when it is driven mechanically. An electric motor uses the reverse process, being driven by electricity to provide a mechanical result.

electromagnetic spectrum the range of wavelengths over which electromagnetic radiation extends.

Prior to Maxwell's discoveries it had been known that light was a wave motion, although the type of wave motion had not been identified. Maxwell was able to show that the oscillations were of electric and magnetic field. Hertz's waves had a wavelength of about 60cm; thus they were of much longer wavelength than light waves.

Nowadays we recognize a spectrum of electromagnetic radiation that extends from about 10^{-15}m to 10^9m. It is subdivided into smaller, sometimes overlapping ranges. The extension of astronomical observations from visible to other electromagnetic wavelengths has revolutionized our knowledge of the universe. See diagram.

Note that the cosmic rays continually bombarding the Earth from outer space are not electromagnetic waves, but high-speed protons and x-particles (i.e. nuclei of hydrogen and helium atoms), together with some heavier nuclei.

electromagnetic waves were demonstrated experimentally in 1887 by the German physicist Heinrich Hertz (1857-94) - who also gave his name to the unit of frequency. In his laboratory, Hertz transmitted and detected electromagnetic waves, and he was able to verify that their velocity was close to the speed of light. The electric and magnetic field components in electromagnetic waves are perpendicular to each other and to the direction of propagation.

(See also electromagnetic induction and Maxwell's theory.)

MAXWELL'S THEORY

Maxwell's theory was developed as a result of the work of the Scottish physicist James Clerk Maxwell (1831-79) in electromagnetism. It is of immense importance to physics.

It united the separate concepts of electricity and magnetism in terms of a new electromagnetic force. Maxwell extended the ideas of Ampère, then in 1864 he proposed that a magnetic field could also be caused by a changing electric field. Thus, when either an electric or magnetic field is changing, a field of the other type is induced. Maxwell predicted that electrical oscillations would generate electromagnetic waves, and he derived a formula giving the speed in terms of electric and magnetic quantities. When these quantities were measured he calculated the speed and found that is was equal to the speed of light in a vacuum. This suggested that light might be electromagnetic in nature - a theory that was later confirmed in various ways. Thus, when an electric current in a wire changes, electromagnetic waves are generated, which will be propagated with a velocity equal to that of light.

PHYSICS

Gamma rays have wavelengths less than 10^{-11}m. They are emitted by certain radioactive nuclei and in the course of some nuclear reactions.

infrared waves of different wavelengths are radiated by bodies at different temperatures. (Bodies at higher temperatures radiate either visible or ultraviolet waves.) The Earth and its atmosphere, at a mean temperature of 250K (−23°C or −9.4°F) radiates infrared waves with wavelengths centred at about 10 micrometres (μm) or 10^{-5}; see diagram.

magnetic fields lines showing the force one magnet exerts on another; see pp. 110-11. These lines of force can be demonstrated by means of small plotting compasses or iron filings. Magnetic fields are the spaces surrounding magnets where a magnetic force is experienced. Its direction at any point, shown by the lines of force, is that of the force that acts on the north pole. (See pp. 110-11.)

We now believe that the Earth's magnetic field is generated by the motion of charged particles in the liquid iron part of the Earth's core. This is known as the dynamo theory.

From Newton's third law (see pp. 75-77) and Oersted's observations it might be expected that a magnetic force can exert a force on a moving charge. This is observed if a magnet is brought up close to a cathode-ray tube in a TV set. The beam of electrons moving from the cathode to the screen is deflected. The force acts in a direction perpendicular to both the magnetic field and the direction of electron flow. If the magnetic field is perpendicular to the direction of the electrons, then the force has its maximum value. This is the second way in which electric and magnetic properties are linked.

magnetism a group of physical phenomena associated with magnetic fields. Magnetism has been known about since ancient times when magnetic ores with magnetic properties were used. It is now known that the Earth itself has magnetic properties. Investigation of the properties of magnetic materials gave birth to the concept of magnetic fields. (See pp. 110-11.)

Maxwell's theory see box.

microwaves are radio waves with shorter wavelengths, between 1mm and 30cm. They are used in radar and microwave ovens; see diagram.

radio waves have a large range of wavelengths - from a few millimetres to several kilometres; see diagram.

ultraviolet waves have wavelengths from about 380 nanometres down to 60nm. The radiation from hotter stars (above 25 000K/25 000°C/45 000°F) is shifted towards the violet and ultraviolet parts of the spectrum. See also diagram.

visible waves have wavelengths of 400-700 nanometres. The peak of the solar radiation (temperature about 6000K/6270°C/11323°F) is at a wavelength of about 550nm, where the human eye is not at its most sensitive. See also diagram.

X-rays have wavelengths from about 10 nanometres down to 10^{-4}nm; see diagram.

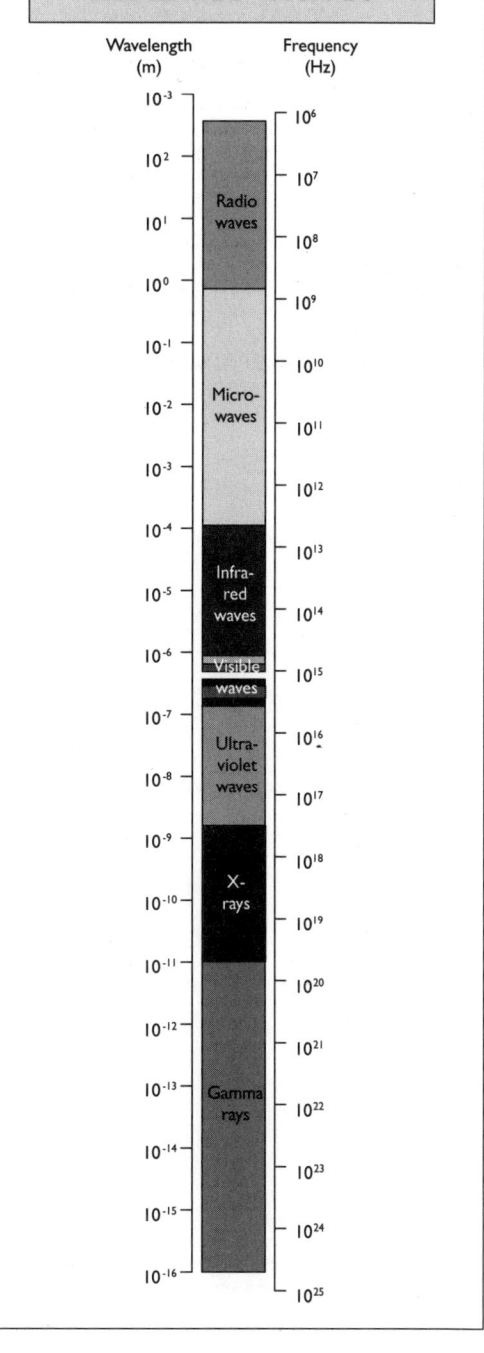

INFRARED WAVES

Wavelength (m)	Band	Frequency (Hz)
10^{-3}	Radio waves	10^{6}
10^{2}		10^{7}
10^{1}		10^{8}
10^{0}		10^{9}
10^{-1}		10^{10}
10^{-2}	Microwaves	10^{11}
10^{-3}		10^{12}
10^{-4}		10^{13}
10^{-5}	Infrared waves	10^{14}
10^{-6}	Visible waves	10^{15}
10^{-7}		10^{16}
10^{-8}	Ultraviolet waves	10^{17}
10^{-9}		10^{18}
10^{-10}	X-rays	10^{19}
10^{-11}		10^{20}
10^{-12}		10^{21}
10^{-13}	Gamma rays	10^{22}
10^{-14}		10^{23}
10^{-15}		10^{24}
10^{-16}		10^{25}

BATTERIES, CELLS AND CIRCUITS

There have been several key advances in the application of electricity towards developing our civilization. The first two were the dynamo and the electric motor. The dynamo provided a way of producing electricity in large quantities, and the electric motor provided a way of converting electric current into mechanical work. The evolution of electromagnetic theory (see pp. 112-13) provided the basis for the modern communications industry through radio and television, while the miniaturization of electronic components using semiconductor materials enabled powerful computers to be built for control purposes and to handle large amounts of information. Electric current is the flow of charges through a conductor. The first source of a steady electric current was demonstrated by the Italian physicist Alessandro Volta (1745-1827) in 1800 - the voltaic pile, the first battery.

accumulator a battery designed so that it can be 'recharged' by the passage of an electric charge back through it; a 12-volt car battery consists of six lead-acid cells in series.

anode see positive electrode.

battery a series of cells connected positive to negative; see cell (simple).

cathode see negative electrode.

cell (simple) a source of electric current; see diagram. The lamp in the diagram lights but soon goes out because bubbles of hydrogen cling to the copper electrode, thus decreasing the output of the cell. This is known as polarizing. The zinc electrode is eventually eaten away. A single cell can normally produce only a small voltage, but a number of them connected in a series (positive to negative) will give a higher voltage. A series of cells connected in this way is called a battery. (Similar principles to those used in cells are used in electrolysis and electroplating.)

circuit a complete conductive path between positive and negative terminals; conventionally power flows from positive to negative, although the direction of electron flow is actually from negative to positive.

electrodes the plates in a cell. They must be made of dissimilar materials, for example copper and zinc. Alternatively, one may be made of carbon.

electrolyte the solution used in a cell. In a dry cell it is a paste of ammonium chloride.

electromotive force (EMF) of a source (battery, generator, etc.) is the energy converted into electrical energy when unit charge passes through it. Its unit, like that of potential difference, is the volt; see volt.

filament a fine coiled tungsten wire of high resistance found in a light bulb (see diagram).

fuse a small resistance consisting of wire with a low melting point usually tinned copper. The *maximum* amount of current that can flow will be determined by its presence. If too much current flows, the fuse will overheat and melt, breaking the circuit.

gas discharge lamp a lamp that consists of a glass tube with electrodes sealed into each end. The tube contains a gas at low pressure such as neon, sodium or mercury vapour, which can be excited to emit light by the application of a high voltage to the electrodes.

heating (by electricity) is produced because when electrons pass through a wire they cause the atoms in it to vibrate and transfer kinetic energy thus generating heat - the greater the resistance, the greater the heat generated. This effect is used in electric heating devices. An electric radiant heater glows red hot. The temperature reached by using a special tough resistance wire is 900°C (1650°F). The connecting wires are of low resistance and stay cool. The power generated as heat in the resistance is given by

$$P = I^2 R$$

where I is the current and R is the resistance.

light bulb a device consisting of a glass envelope containing an inert (noble) gas, usually argon, at low pressure. The bulb has two electrodes connected by a filament (see above). The passage of a suitable electric current through the filament will raise its temperature sufficiently to make it glow white hot (2500°C/4500°F). The inert gas prevents the filament from evaporating. The efficiency of filament lamps is low. Gas discharge lamps (see above) are much more efficient.

negative electrode the electrode towards which power flows in a cell. It is called the cathode.

neon lamp see gas discharge lamp.

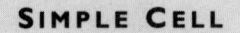

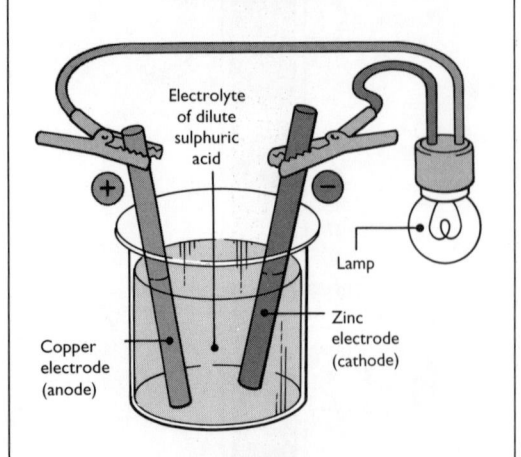

A simple cell. The lamp lights but soon goes out because bubbles of hydrogen cling to the copper electrode, thus decreasing the output of the cell. This is known as polarizing. The zinc electrode is eventually eaten away.

ohm (symbol Ω) the unit of resistance; the resistance of a conductor in which the current is 1 ampere when a potential difference of 1 volt is applied across it. It is named after the German physicist Georg Ohm (1787-1854). He discovered a relationship between the current (I) and the voltage (V) in a conductor known as Ohm's Law. This leads to the well known equation

$$V = IR.$$

Ohm's Law the current through a metallic conductor is directly proportional to the potential difference between its ends if the temperature and other physical conditions are constant. See ohm (above).

parallel connection the arrangement when electrical components such as bulbs and switches are connected side to side; see also series connection and diagram.

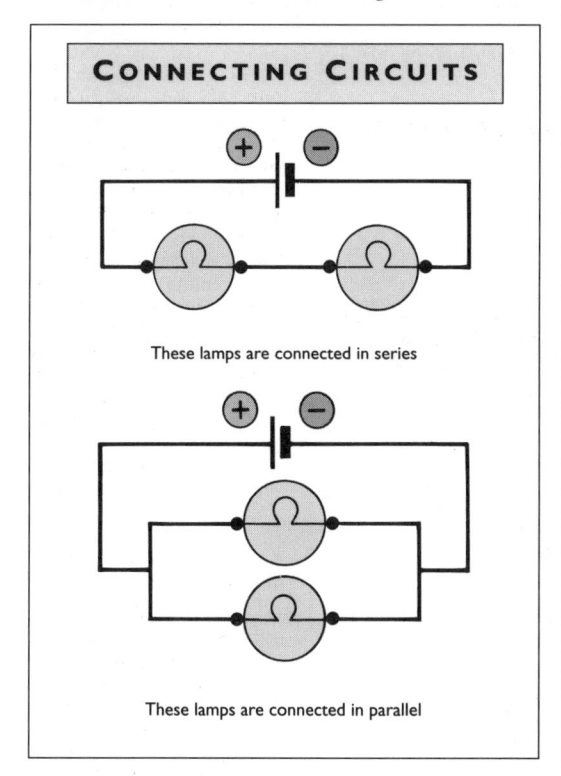

CONNECTING CIRCUITS

These lamps are connected in series

These lamps are connected in parallel

polarizing see cell (simple).

positive electrode the electrode from which power flows inside a cell. It is called the anode.

power the rate at which a body or system does work. The power in an electric conductor is measured in watts (see below); see also heating.

resistance the force that acts to reduce or resist the flow of an electric current passing through a conductor as well as the temperature (see temperature coefficient). Resistance is dependent upon the nature of the conductor and its dimensions. The unit of resistance is called the ohm (see above).

series connection the arrangement when electrical components such as bulbs and switches are joined end to end; see also parallel connection and diagram.

temperature coefficient of resistance is the fractional increase in the resistance of a material at 0°C per unit rise in temperature. However, graphite, semiconductors and most non-metals have a negative coefficient - their resistance decreases with temperature.

volt (symbol V) the unit of electric potential or potential difference; defined as the potential difference (p.d.) between two points in a circuit is the amount of electrical energy changed to other forms of energy when unit charge passes from one point to the other. That is

$$1 \text{ volt} = 1 \text{ joule per coulomb}$$

(that is equivalent to 1 watt per ampere.). The volt is named after Alessandro Volta; see above.

voltaic pile the first source of a steady electric current developed by Volta in 1800. The original voltaic pile used chemical energy to produce an electric current. The pile consisted of a series of pairs of metal plates (one of silver and one of zinc) piled on top of each other, each pair sandwiching a piece of cloth soaked in a dilute acid solution. The same principle is still used today (see diagram).

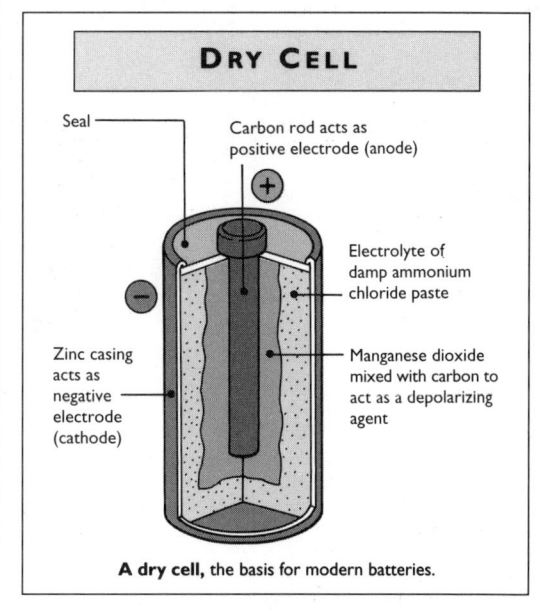

DRY CELL

Seal

Carbon rod acts as positive electrode (anode)

Electrolyte of damp ammonium chloride paste

Zinc casing acts as negative electrode (cathode)

Manganese dioxide mixed with carbon to act as a depolarizing agent

A dry cell, the basis for modern batteries.

watt (symbol W) the unit for measuring the power in an electric conductor. It is named after the British engineer James Watt (1736-1819). One watt is one joule per second, or the energy used per second by a current of one amp flowing between two points with a potential difference of one volt. In an electric conductor the power (W) is the product of the current (I) and the voltage (V).

ALTERNATING AND DIRECT CURRENT, MOTORS AND GENERATORS

There are two types of current electricity - direct current (DC) and alternating current (AC). Electrical power may be generated in either an AC generator (or alternator) or in a DC generator. An electric motor transfers electrical energy to mechanical energy.

alternating current (AC) the type of current used in most electrical appliances. In alternating current the direction of the drift velocity reverses, usually many times a second (see p. 118). The frequency of alternating current can vary over an enormous range. The electric mains operate at 50Hz (cycles per second) in the UK and Europe and at 60Hz in the USA. Today AC generators are more common than DC generators; they were developed after Faraday's discovery of the induction of a current in a circuit as a result of a changing magnetic field (see 112-13).

alternator a type of generator. Large alternators are used in power stations; small alternators, used to charge car batteries, produce AC that is then rectified to DC using semiconductor diodes.

anode a positive electrode; the electrode from which power normally flows in a cell.

armature part of an electric motor, or generator comprising the whole assembly of coils and the soft iron core on which they are wound.

cathode a negative electrode; the electrode to which power normally flows in a cell.

diode (thermionic) a device in which electrons are emitted by a heated cathode and flow through a vacuum to the anode (see diagram). A diode allows the passage of electricity in one direction only.

direct current (DC) the type of electric current in which the drift velocity of the charge carriers is in one direction only. A battery produces a steady DC whereas rectified AC changes continually.

dynamo (see diagram) an electrical current generator, consisting of a coil that is rotated in a magnetic field by some external means. The source of the rotation may be a turbine in which blades are moved by the passage through them of water, as in a hydroelectric plant, or steam, produced by a boiler heated by nuclear fission or by burning fossil fuels. Wind turbines spin as a result of the passage of air through the large rotors. Different types of generators produce either AC or DC, while alternators produce AC. In a large generator the armature is fixed and the magnetic field is rotated.

electric motor a device similar to a generator. However, it works in reverse. An electric current is applied to the coil windings, causing rotation of the armature.

electron emission the escape of electrons when, for example, the filament of a light bulb is heated and the energy of some of the electrons in the filament is greatly increased by thermal motion, although the average increase for all the electrons is very small. If their energy reaches an adequate level, many are able to escape. This process of electron emission is called thermionic emission. If another electrode is put in the evacuated bulb and placed at a higher potential than the filament, this will act as an anode and will attract electrons towards it. A current will then flow in an external circuit; the device so formed is called a thermionic diode.

generator see dynamo and diagram in which as the commutator rotates, each carbon brush makes contact first with the positive end of the rotating wire, then with the negative end, so continually reversing the direction of current flow in the coil, while maintaining the direction in the external circuit.

photoelectric effect occurs when electromagnetic radiation of a sufficiently high frequency shines onto a metal, causing electrons to be emitted from its surface (see photons, p. 94).

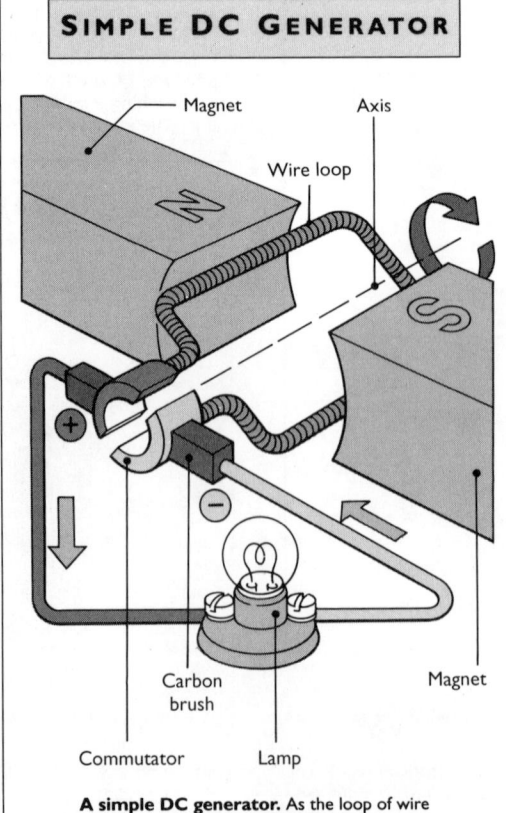

SIMPLE DC GENERATOR

A simple DC generator. As the loop of wire rotates within the magnetic field, an electrical current is induced within the circuit, illuminating the lamp. This simple device shows the basic principle by which all electricity is generated.

PHYSICS

THERMIONIC TRIODE

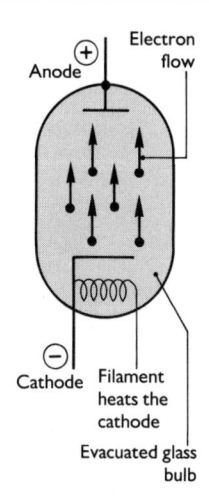

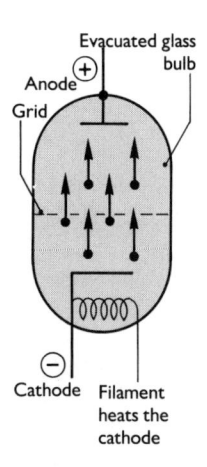

Diagram 1: Diode tube
- Anode (+)
- Electron flow
- Cathode (−)
- Filament heats the cathode
- Evacuated glass bulb

Diagram 2: Triode tube
- Anode (+)
- Grid
- Evacuated glass bulb
- Cathode (−)
- Filament heats the cathode

RMS root mean square values. The value of an alternating current and EMF varies from one instant to the next. The effective or RMS value is the steady direct current which converts electrical energy to other forms of energy in a given resistance at the same rate as the AC. For the usual AC it can be shown to be equal to the peak value.

thermionic emission see electron emission.

transformer a device used to transfer electrical energy from AC circuit to another with a change of voltage or a change of current or phase. If, for example, two insulated coils of wire are wound on the same soft iron core and an alternating current is passed through one of the coils, a current will be induced in the other coil. The ratio of the number of turns in the input coil (N_1) and the output coil (N_2) will determine the ratio of the output voltage (V_2) to the input voltage (V_1). The relationship is:

$$\frac{V_2}{V_1} = \frac{N_2}{N_1}$$

In this way transformers can either step voltage up or step it down. Note that they have the reverse effect on current. This principle is used for efficient long-distance power transmission at very high tension.

triode a device in which a third electrode in the shape of a grid is placed in a tube between a filament and an anode. In this case the anode current is so sensitive to changes in the grid voltage that the whole device can act as an amplifier; see diagram.

CONDUCTORS

A conductor is a substance that has high electrical conductivity. The conduction within a conductor is the result of the movement of free electrons. A metal - acting as a conductor - consists of an array of positive ions in a 'sea' of free electrons. The electrons move randomly with mean speeds of around $10^6 ms^{-1}$. When a potential difference is applied across a metal a small drift velocity is added. The metal atoms are thought to give up one or more electrons, which can then migrate freely through the material. These electrons move in a zigzag manner along a conductor. As a result their typical velocity, called the drift velocity, is small, in the order of $10^{-4} ms^{-1}$. Thus it would take more than an hour to move one metre. Note that the electric signals that drive the electrons travel with a speed in the order of $10^8 ms^{-1}$ - i.e. the speed of light - in some circuits.

This classical picture of electron conduction explains some but not all conduction phenomena. For these a quantum mechanical model is required (see p. 94). This model explains the basis of semiconductors, which now play such an important part in electronics.

Metals are good conductors of electricity because there are always many unoccupied quantum states into which electrons can move. Non-metallic solids and liquids have nearly all their quantum states occupied by electrons, so it is difficult to produce large currents. If the numbers of unoccupied states and of electrons free to move into them is small the material is an insulator. If there are more free electrons and unoccupied states the substance is called a semiconductor. See also pp. 114-15 and 116-17.

charge carrier the entity that transports electric charge in an electric current. The type of carrier depends upon the type of conductor: in metals the carriers are electrons; in semiconductors the carriers are either electrons (n-type) or positive holes (p-type).

conductor any material that is able to pass an electric charge with ease. (The only materials that are able to do this are carbon and metals. Metals are able to do so because they contain free electrons that are able to move about; see pp. 104-05.)

conduction (electrical) the process by which a charge is transferred through a medium. Conduction is usually accompanied by a transfer of energy.

conductivity (electrical) the reciprocal of the resistivity of a material. Conductivity is measured in siemens per metre.

drift velocity the typical velocity of electrons moving along a conductor; see the introduction to this section (above). The drift velocity (v) can be calculated using the expression

$$v = \frac{I}{n A e}$$

where I is the current, n is the number of free charge-carriers per unit volume, each carrying charge e and A the cross-section area.

hole a vacant electron position.

insulator any material in which the numbers of unoccupied states and of electrons free to move into them are small; see also semiconductor.

light-emitting diode a device made from material such as gallium arsenide in which the p-n junction (see below) will emit light whenever an electric current passes through it. Light-emitting diodes are used in digital displays in clocks and radios. The light is emitted when an electron and a hole meet at the junction and annihilate each other - they cancel each other out.

n-type doping negative-type doping; see semiconductor.

n-type semiconductor a semiconductor (see below) in which the doping results in the charge carriers being negative electrons.

p-n junction the boundary between p-type and n-type materials in a semiconductor (see below). In some materials, such as gallium arsenide, a p-n junction will emit light whenever an electric current passes through it. Such a device is called a light-emitting diode.

p-type doping positive-type doping; see semiconductor.

p-type semiconductor a semiconductor (see below) in which the doping results in the charge carriers being electron deficiencies or holes.

photodiode a semiconducting device whose resistance decreases as the incident light increases owing to the production of more electron-hole pairs.

phototransistor can be thought of as a photodiode giving current amplification due to transistor action.

photovoltaic effect an effect which occurs when light is absorbed by a p-n or n-p junction. Electrons are liberated at the junction by an incident photon and diffuse through the n-type region. The hole drifts through the p-type layer until it recombines with an electron flowing round the external circuit. See solar cells (below).

rectifier a device used to convert alternating current to direct current.

resistivity (symbol p) the measure of the ability of a material in opposing the flow of electric current. Resistivity is defined by

$$R = \frac{L}{A}$$

where R is the resistance, L the length and A the area of cross-section of the conductor. Resistivity is measured in ohm-metres.

semiconductor any material in which there are more free electrons and unoccupied states than in an insulator. Semiconductors have a charge-carrier density that lies between those of conductors (see above) and insulators. Two metal-like elements, silicon and germanium, are the two semiconductors used most frequently. These may be 'doped' with an impurity to modify their conduction behaviour - n-type doping increases the number of free electrons, p-type doping increases the number of unoccupied states. Most semiconductor devices are made from materials that are partly n-type (see above) and partly p-type (see above). The boundary between them is known as a p-n junction. Such a device is known as a semiconductor diode and acts as a rectifier (see above). See also temperature coefficient, pp.114-15.

semiconductor diode a device made from materials that are partly n-type and partly p-type and which acts as a rectifier.

solar cell the first practical photovoltaic device (made in 1954). In essence a solar cell is a light-emitting diode acting in reverse. It converts light into electric current, which is the basis of solar power.

Most solar cells comprise a single-crystal silicon p-n junction (q.v.). Photons of light energy from the Sun falling at or near the semiconductor junction create electron-hole pairs which are forced to separate by the electric field at the junction. After separation the holes pass to the p-region and the electrons pass to the n-region. When a load is connected across the terminals of the device, the displacement of free charge (described above) creates an electric current.

solar constant the rate at which solar energy (q.v.) is received per unit area at the outer edge of Earth's atmosphere (at the mean distance between Earth and the Sun).

The value is 1.353kWm^{-2}.

solar energy electromagnetic energy radiated from the Sun. The very small percentage that falls upon the Earth - about 5×10^{-10} - is indicated by the solar constant (q.v.).

The total amount of solar energy falling upon our planet in one year is in the region of:

$$4 \times 10^{18} \text{J}.$$

The total amount of energy consumed by all of the Earth's inhabitants in a year is in the region of:

$$3 \times 10^{14} \text{J}.$$

Solar energy could therefore, in theory, satisfy all of the planet Earth's energy requirements.

solar heating domestic or industrial heating that relies upon solar energy.

superconductivity the absence of measurable electrical current in materials at critical low temperatures. Superconductivity was discovered by the Dutch physicist Kamerlingh Onnes (1853-1926) in 1911. Below a certain critical temperature, various metals show zero resistance to current flow. Once a current is started in a closed circuit, it keeps flowing as long as the circuit is kept cold. The critical temperature for aluminium is 1.19K (−272°C/−457°F), and similar values hold for other metals. Some alloys have higher critical temperatures. Up to 1986 the highest transition temperature known was about 25K (−248°C/−414°F). More recently a new class of copper oxide and other materials have shown superconductivity at up to at least 125K (−148°C/−234°F). These developments promise enormous savings in energy.

thermistor a semiconducting device whose resistance varies markedly with temperature change - usually their resistance decreases with increasing temperature. They can be used as thermometers.

PHYSICS

TRANSISTORS

Since their invention in 1948 transistors have revolutionized the application of electricity. A transistor - a small solid-state electronic device - consists of semiconductor material (see p. 118). It is used instead of a thermionic valve (see pp. 116-17). Its best known everyday use is in transistor radios but it is also the basic unit in television sets and computers.

base the third contact in a point-control transistor (see below) or the central part of the p-n-p structure in a bipolar transistor (see below).

bipolar transistor any example of a type of transistor that has majority and minority carriers. Bipolar transistors include junction transistors, for example the p-n-p transistor (see below and diagram). Bipolar transistors were first developed in 1949-50. In bipolar transistors, two p-type semiconductor regions occur on either side of a thinner n-type region - this gives the typical p-n-p structure. In some cases an n-p-n structure is used instead. The central part in either type is called the base. On one side of the base is the emitter; on the other side is the collector - the former is forward-biased; the latter is reverse biased. In the p-n-p transistor the forward bias means that holes (see below) in the emitter flow across the junction into the base. Because the base is thin, most holes pass across it into the collector. The holes that do not pass from the base combine with electrons in the n-type base - an action which is balanced by a small electron flow in the base circuit.

charge carrier the entity that transports electric charge in an electric current. The type of carrier depends upon the type of conductor (see p. 118). In metals the carriers are electrons; in semiconductors (see p. 118) the carriers are either electrons (n-type) or positive holes (p-type).

collector that part of the p-n-p structure which is on the opposite side of the base to the emitter.

drain part of a field-effect transistor to which current flows.

emitter that part of the p-n-p structure which is on the opposite side of the base to the collector.

field-effect transistor a unipolar transistor. There are two types: the junction FET (JFET or JUGFET) and the insulated-gate FET or IGFET. In the JFET and IGFET (see diagram) the current flows through a narrow channel which is between two electrodes (the gate). The current flows from the source to the drain and a modulating signal is applied to the gate. In the JFET the channel comprises semiconductor material of quite low conductivity; this is situated between two areas of high conductivity of opposite polarity. In the IGFET the central wafer of semiconductor material has two regions of opposite polarity forming the source and the drain.

gate part of a unipolar transistor formed by two electrodes.

hole a vacant electron position. It is found in the lattice structure of a solid that behaves like a mobile positive charge carrier with a negative rest energy.

junction transistor a bipolar transistor (see above); the type of transistor that replaced the point-contact transistor. The p-n-p junction transistor (see diagram) is an example.

p-n-p junction transistor a type of junction transistor; see diagram and bipolar transistor (above).

point-contact transistor the first type of transistor that was invented. It comprises a small crystal of germanium with two rectifying point contacts attached to it. The base - a third contact - makes a low-resistance non rectifying contact with the germanium crystal. Current flows through the device between the point contacts and is modulated by the signal fed to the base. This type of transistor is now obsolete.

rest energy the rest mass of a body in energy terms.

source the part of a field-effect transistor from which current flows.

unipolar transistor a type of transistor in which the current is carried by majority carriers only; for example, the FET (see above).

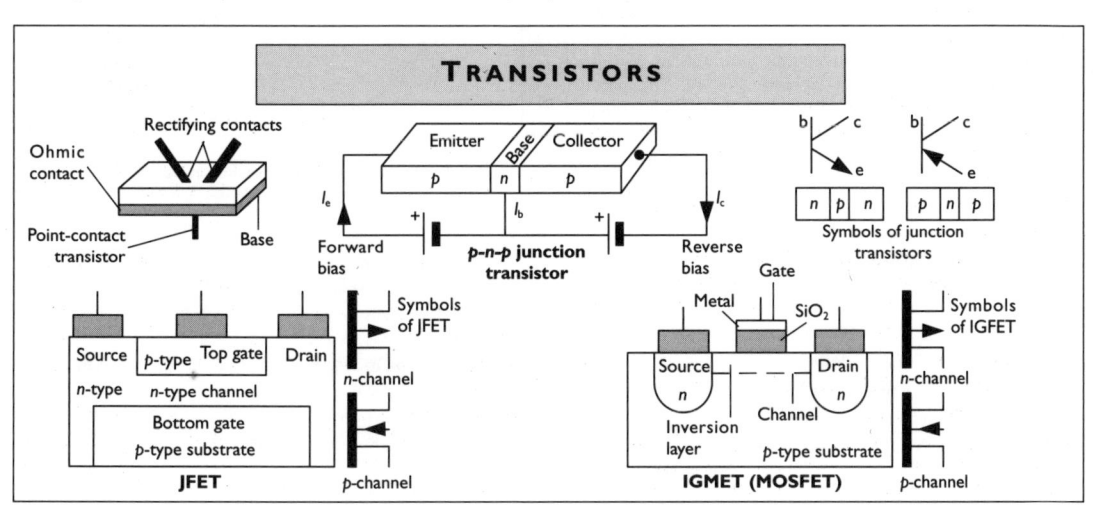

TRANSISTORS

Point-contact transistor — Ohmic contact, Rectifying contacts, Base

p-n-p junction transistor — Emitter (p), Base (n), Collector (p); Forward bias, Reverse bias; I_e, I_b, I_c

Symbols of junction transistors — n p n, p n p

JFET — Source (n-type), p-type Top gate, Drain, n-type channel, Bottom gate, p-type substrate; Symbols of JFET, n-channel, p-channel

IGMET (MOSFET) — Metal, SiO_2, Gate, Source (n), Drain (n), Channel, Inversion layer, p-type substrate; Symbols of IGFET, n-channel, p-channel

ATOMIC AND SUBATOMIC PARTICLES

Of the fundamental forces that are the most important in the natural world, the gravitational force (see p. 78) is the dominant long-range force when the motion of planets and other celestial bodies is considered. When the smallest entities are considered, the other fundamental forces - the electromagnetic force (see pp. 112-13), the strong force (which holds together the atomic nuclei) and the weak force (which is involved in nuclear decay) - become important.

The word atom is derived from an ancient Greek word for a particle of matter so small that it cannot be split up. In his atomic theory of 1803, the British chemist John Dalton (1766-1844) defined the atom as the smallest particle of an element (see pp. 128-35) that retained its chemical properties. Various phenomena could be explained using this hypothesis - which still holds true today. However, no physical description of the atom was available until after the discovery of the electron.

allowed orbit see shell.

alpha decay produces nuclei of helium that each contain two neutrons (q.v.) and two protons (q.v.). They are called alpha-particles and are formed in spontaneous decay of the parent nucleus. Thus uranium-238 decays to thorium-234 with emission of an alpha particle. See also radiation and radioactivity (below).

alpha-particles see alpha decay (above).

antineutrino see beta decay (below).

antiparticle a particle with the same mass but opposite in some other characteristic such as charge. Thus the positron (q.v.) with positive charge is the antiparticle of the negatively charged electron (q.v.). Some particles such as the photon (q.v.) may be their own antiparticle. See nuclear particle (below).

antiquark see quark (below).

baryon a hadron resulting from the combination of three quarks.

beta decay is characterized by the emission of particles that are either electrons or positrons (q.v.). The parent nucleus retains the same number of nucleons (q.v.) but its charge varies by plus or minus 1. In these processes another kind of particle is produced - a neutrino or an antineutrino.

binding energy see strong nuclear force.

Bose-Einstein particle see boson.

boson (or **Bose-Einstein particle**) a particle that can be produced and destroyed freely, provided the laws of conservation of charge and of mechanics are obeyed. See fermion and nuclear particle (below).

bottom quark see quark (below).

carbon dating see half-life (below).

charmed quark see quark (below).

colour charge a type of charge of quarks (q.v.). The force associated with the colour charge binds the quarks together and is thought to be the source of the strong force binding the hadrons (q.v.) together. Thus the colour force is the most fundamental force.

critical mass the level above which a chain reaction will be set in nuclear fission (q.v.).

cyclotron see nuclear accelerator (below).

down quark see quark (below).

eight-fold way see quark (below).

electron a small subatomic particle. They carry a negative charge and move around the nuclei of an atom. When an atom gains or loses an electron, ions are formed (see p. 128).

Electrons have an electric charge of -1, and a mass (MeV) of 0.511.

The discovery of the electron was in 1897 by J.J. Thompson. The nuclear model was proposed by the physicist Ernest Rutherford (1871-1937) in 1911 (see diagram). His model consists of a small but dense central nucleus (see below), which is positively charged, orbited by negatively charged electrons.

The electron was recognized by its behaviour as a particle; see wave-particle duality.

electronic configuration the way in which electrons are located in an atom.

electronvolt (symbol eV) the increase in energy of an electron when it undergoes a rise in potential of 1 volt. This may be given as:

$$1\text{eV} = 1.6 \times 10^{-19} \text{ joules.}$$

THE RUTHERFORD MODEL

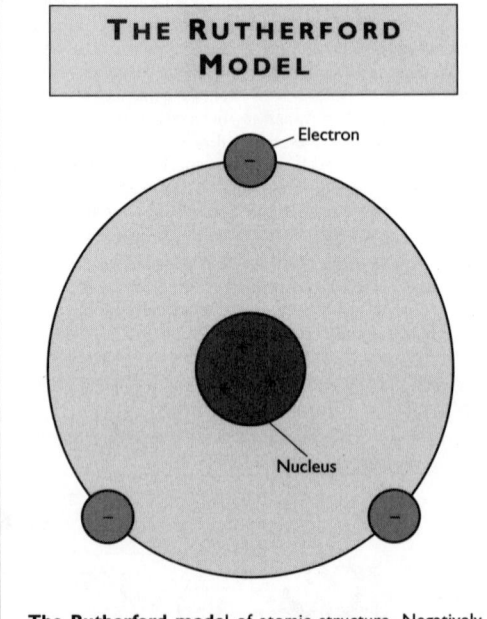

The Rutherford model of atomic structure. Negatively charged electrons orbit a positively charged nucleus.

electroweak force see weak force (below).

Fermi-Dirac particle see fermion.

fermion (or **Fermi-Dirac particle**) a particle that has a permanent existence. Leptons (q.v.) are fermions. See also boson (above) and nuclear particle (below).

fission a nuclear reaction from which nuclear power comes. See also fusion.

In the fission process (see diagram), a large nucleus, such as uranium-235 (^{235}U), splits to form two smaller nuclei that have greater binding energies than the original uranium. Thus energy is given out in the process. Fission is used in nuclear reactors and in atomic weapons. There are other isotopes in addition to uranium-235, such as plutonium-239, that give rise to fission.

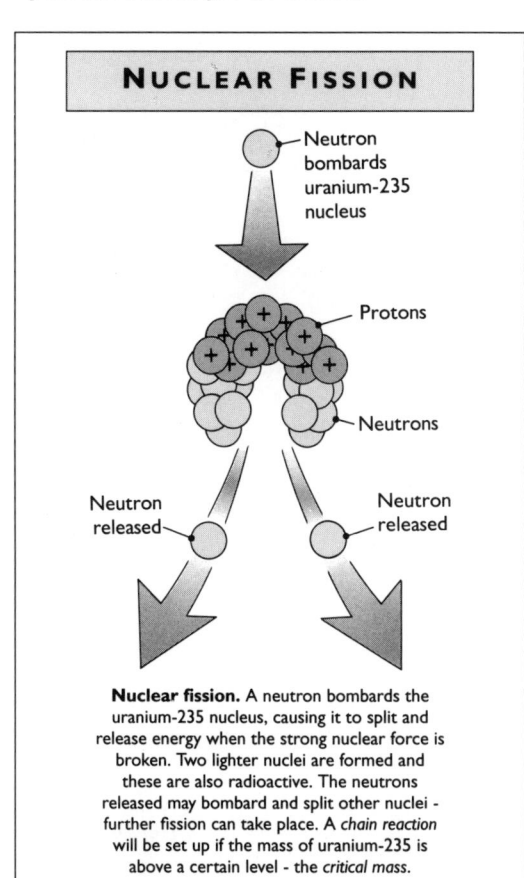

NUCLEAR FISSION

Nuclear fission. A neutron bombards the uranium-235 nucleus, causing it to split and release energy when the strong nuclear force is broken. Two lighter nuclei are formed and these are also radioactive. The neutrons released may bombard and split other nuclei - further fission can take place. A *chain reaction* will be set up if the mass of uranium-235 is above a certain level - the *critical mass*.

flavour (of quarks) see quark (below).

fusion nuclear reaction from which nuclear power comes. See also fission.

In the fusion process (see diagram), two light nuclei fuse together to form two particles, one smaller and one larger

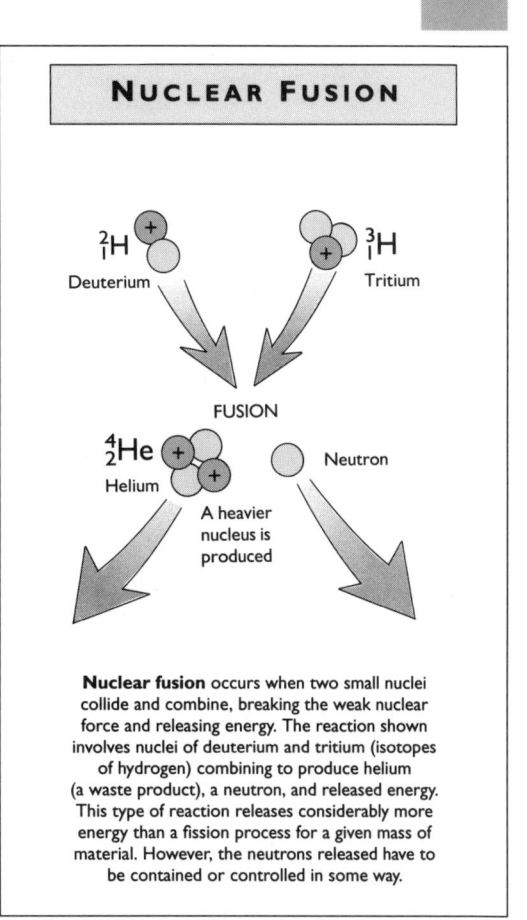

NUCLEAR FUSION

Nuclear fusion occurs when two small nuclei collide and combine, breaking the weak nuclear force and releasing energy. The reaction shown involves nuclei of deuterium and tritium (isotopes of hydrogen) combining to produce helium (a waste product), a neutron, and released energy. This type of reaction releases considerably more energy than a fission process for a given mass of material. However, the neutrons released have to be contained or controlled in some way.

than the original nuclei. Usually one of them is sufficiently strongly bound to give a great release of energy. The fusion of hydrogen to form helium is a power source in stars such as the Sun, although solar fusion processes differ in detail from the simple process described here. Nuclear fusion is the basis of the hydrogen bomb.

gamma decay a type of radiation. In gamma decay high-energy photons (q.v.) may be produced in a process of radioactive decay if the resultant nucleus jumps from an excited energy state to a low-energy state. The rate at which radioactive decay takes place depends only on the number of radioactive nuclei that are present. Thus the half-life is characteristic for that type of nucleus. Decay can result in a series of new elements being produced, each of which may in its turn decay until a stable state is achieved.

hadron (from the Greek for 'bulky') a heavy particle (for example, a proton or a neutron) that is affected by the strong force. Hadrons - unlike leptons which are thought to be fundamental particles - are thought to be made up of quarks (q.v.). See nuclear particle and lepton.

half-life the time taken for half a given number of radioactive nuclei to decay. The isotope carbon-14 has a half-life of 5730 years, and measurement of its decay is used in carbon-dating of organic material.

isotopes different atoms of the same element containing equal numbers of protons (q.v.) but different numbers of neutrons (q.v.) in their nuclei hence having different atomic mass. Isotopes of an element contain the same nuclear charge, and their chemical properties are identical, but they display different physical properties. An isotope may be represented in a number of ways, such as uranium-235 or U-235 or ^{235}U, where the 235 refers to the total number of the protons and neutrons in a nucleus.

lepton (from the Greek for 'small') a (generally) light particle - such as an electron or a neutrino (q.q.v.) - that is not subject to the strong force. Leptons are fermions (q.v.). See also nuclear particle and hadron.

Leptons are thought to be fundamental particles, while the hadrons are thought to be made up of quarks (q.v.).

linear accelerator see nuclear accelerator.

mass defect the difference between the masses of the nucleus and its component parts.

meson a subatomic particle resulting from the combination of a quark (q.v.) and an antiquark. A meson is a boson (q.v.); it is a short-lived particle that jumps between protons and neutrons, thus holding them together.

neutrino a particle produced in beta decay (q.v.). The neutrino has no charge (the word means 'little neutral one') and a mass that - if it could be measured at rest - would probably be zero.

neutron a subatomic particle of similar size to a proton but electrically neutral. Protons and neutrons in the atomic nucleus are held tightly together by the strong nuclear force (q.v.). See also proton and electron.

nuclear accelerator a large machine that can accelerate particle beams to a very high speed, so enabling research into particle physics. Electric fields are used to accelerate the particles, either in a straight line (linear accelerator) or in a circle (cyclotron, synchrotron or synchro-cyclotron). Powerful magnetic fields are used to guide the beams. Energy levels of the particles may be as high as several hundred giga electronvolts.

Nuclear accelerators have provided experimental evidence for the existence of numerous subatomic particles predicted in theory.

nuclear particle any of over 200 elementary particles that are now known. They may be divided into two types: hadrons (q.v.) and leptons (q.v.).

A further very important distinction is that between fermions and bosons (q.q.v.).

Every type of particle is thought to have a companion antiparticle (q.v.).

nuclear reaction see fission and fusion.

nuclear structure see nucleus (below).

nucleons collectively, protons and neutrons (q.q.v.).

nucleus the central part of an atom where mass is concentrated - the nucleus contains over 99.9% of the mass but its diameter is in the order of 10^{-15}m (compared with the much larger size - about 10^{-10}m - of the atom.

The nucleus has a positive charge - in the neutral atom the positive nucleus is balanced by the negatively charged electrons which orbit around it.

The nucleus comprises protons and neutrons. (NB. The isotope hydrogen–1 comprises atoms that contain no neutrons.) Proteins and neutrons are known collectively as nucleons. See also electron (above).

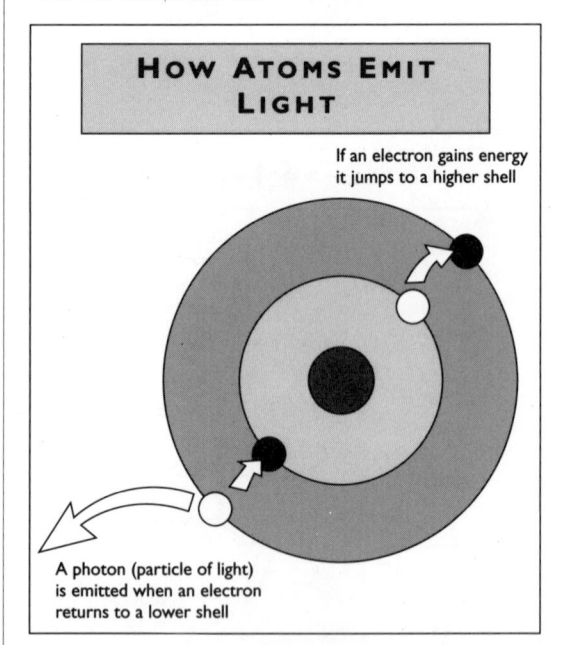

HOW ATOMS EMIT LIGHT

If an electron gains energy it jumps to a higher shell

A photon (particle of light) is emitted when an electron returns to a lower shell

orbit (of electrons) see shell.

positron a particle that is identical to an electron but with a positive charge.

proton a subatomic particle carrying a positive charge, equal in magnitude to that of the negatively charged electron. See electron, strong nuclear force and neutron.

quantum (plur. **quanta**) an elemental unit; a discrete amount.

quark a subatomic particle that is thought to be the 'building blocks' for hadrons (q.v.). (The work quark is borrowed from James Joyce's novel *Finnegan's Wake*.) Quarks may have fractional electrical charge. It is probable that free quarks do not exist. If three quarks combine the resulting hadron is called a baryon.

In the same way that Mendeleyev's table of chemical elements (see pp. 128-30) predicted new elements such as germanium and gallium that were subsequently discovered, so a pattern of hadrons may be drawn up based on combinations of the different types of quark. This pattern is called the eight-fold way - a term borrowed from Buddhism.

It predicted the existence of the omega-particle, the discovery of which in 1963 helped to validate the theory.

There are believed to be six types of flavours of quark - up, down, charmed, strange, top and bottom. Evidence for the existence of all except the top quark is now available.

Quarks carry electrical charge and another type of charge that is called colour (q.v.).

radiation radiant energy. Radiation is either the result of spontaneous emission of particles or is an electromagnetic wave. There are three types of radiation - from alpha decay, from beta decay and from gamma decay (q.q.v.). See also radioactivity.

radioactivity giving off radiant energy in particles or rays as a result of the disintegration of atomic nuclei.

shell the outer layer or layers of an atom.

The Danish physicist Niels Bohr (1885-1962) had suggested that electrons were allowed to move in circular orbits or shells around the nucleus, but that only certain orbits were allowable (see diagram).

THE RUTHERFORD-BOHR MODEL

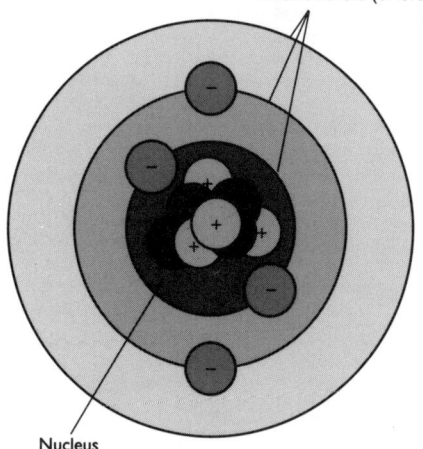

The Rutherford-Bohr model of atomic structure. The number of electrons orbiting the nucleus is equal to the number of positively charged protons within the nucleus. The number of electrons within each shell is also limited - no more than 2 in the first shell, 8 in the second, 18 in the third, etc.

This theory was able to explain many of the features of the spectrum of light emitted by excited hydrogen atoms. The wavelengths of the spectral lines are related to the energy levels of the allowed orbits. These would be those whose circumference was a multiple of the electron's wavelength.

When Rutherford showed experimentally that an atom must consist of a small nucleus surrounded by electrons, there was a fundamental problem. To avoid collapsing into the nucleus, the electrons would have to move in orbits - as Bohr had proposed. This means that they must have continuous acceleration towards the nucleus. But according to the electromagnetic theory, an accelerated charge must radiate energy, so no permanent orbit could exist. Bohr, therefore, argued that energy could not be lost continuously but only in quanta (discrete amounts) equivalent to the difference in energies between allowed orbits. Thus light would be emitted when an electron jumps from one allowed level to another level of energy; see diagram.

strange quark see quark (above).

strong nuclear force the force that holds together protons and neutrons in the atomic nucleus. The strong nuclear force overcomes the much weaker electromagnetic force of repulsion between positively charged protons.

The mass of a nucleus is always less than the sum of the masses of its constituent nucleons (q.v.). This is explained using the relationship derived by Einstein. If the nucleus is to be separated into protons and neutrons then the strong nuclear force needs to be overcome and energy has to be supplied to the nucleus from an external source to break it up. This energy is equal in value but opposite in sign to the binding energy and is related to the mass defect. Those nuclei with large binding energies per nucleon are most stable; these have about 50-75 nucleons in the nucleus.

In fission, unstable U-235 splits into two more stable nuclei; in fusion H-2 or H-3 join to make more stable nuclei. In both cases energy is released equal to the charge in binding energies.

synchrocyclotron see nuclear accelerator (above).

synchrotron see nuclear accelerator (above).

top quark see quark (above).

up quark see quark (above).

wave-particle duality was proposed by de Broglie in 1923 to describe the movement of particles. The wavelength of a particle would be equal to the Planck constant (see pp. 93-95) divided by its momentum. As the wavelength is dependent upon momentum it can take any value. For an electron the wavelength can be of the order of the atomic diameter (this principle is used in electron microscopes). At suitable energy levels the wavelength of electrons and neutrons can be equivalent to the atomic spacing in solids.

weak force is associated with the radioactive beta-decay of some nuclei. It has been shown - in the theory of the electroweak force - that the electromagnetic and weak forces are linked. This theory predicted the existence of the W and Z° particles, which were discovered at the CERN nuclear accelerator at Geneva during 1982-83.

W particle see weak force.

Z° particle see weak force.

Chemistry

INTRODUCTION TO THE PRINCIPLES AND METHODS OF CHEMISTRY

Modern chemistry derives its name from alchemy which originated in ancient Egypt. One of the aims of alchemy was the transmutation of metals: alchemists strove for a 'philosopher's stone' that could be used to convert easily corrupted 'base' metals such as iron, copper and lead into the 'noble' metal gold. They thought that the philosopher's stone would also be the 'elixir' of immortality. Much experimentation followed, which - although not leading to the desired ends - led to the development of techniques that form the basis of modern chemistry.

In modern chemistry the philosopher's stone has been replaced by a fundamental belief in the importance of understanding the physical laws that govern the behaviour of atoms and molecules. Such an understanding has resulted in the development of methods of converting cheaply available and naturally occurring minerals, gases and oils into substances that have high commercial or social value.

Elements are a convenient starting point for discussing chemical phenomena and methods. Elements (see pp. 130-135) have been called the building blocks of matter. They may occur singly or in compounds and mixtures (see below).

compound a substance formed by combination of two or more elements. In a compound of oxygen and hydrogen the two elements are made to react or combine with one another. The reaction yields the liquid water, hydrogen oxide, which is a compound of hydrogen and oxygen. If equal volumes of hydrogen and oxygen had been present in the reaction vessel not all of the oxygen would be combined into the compound, water, but some oxygen would still be left unreacted. Some oxygen would be left because water is formed by the reaction (see below) of exactly twice as much hydrogen by volume as oxygen. The 'recipe' - or formula - for hydrogen oxide (water) is unique: two 'parts' hydrogen to one 'part' oxygen. Molecules of water have three atoms (see p. 130) - two hydrogen atoms and one oxygen atom. The properties of compounds differ from the properties of the elements of which they are made. Compare compounds with mixtures (see below).

concentration the strength or density of a solution. The concentration of a solution is a measure of the quantity of solute dissolved in the solution. This is expressed in terms of mass - for example in grams - or the number of particles - for example moles - per unit volume of the solution.

element a substance that cannot be broken down into simpler substances by chemical methods. See pp. 128-35.

fraction a part separated - for example by distillation (see below) - from a mixture.

mixture is composed of two or more different substances not chemically combined. Air, for instance, is a mixture of elements (nitrogen, oxygen, argon, etc.) and compounds (carbon dioxide, water vapour, etc.). The substances in the mixture do not affect one another. Each substance behaves in exactly the same way that it would if the other substances in the mixture were absent. The properties of a mixture are the sum of the properties of all the components of that mixture. The proportions of the substances that occur in a mixture can vary and these substances can be separated by physical means. However, no energy is either absorbed or released when a mixture is made. Compare mixtures with compounds (see above).

molecule the smallest particle of either an element or a compound that can exist independently - apart from the noble gases whose molecules consist of single atoms, a molecule contains two or more atoms that are bonded together in small whole numbers, for example O_2, a molecule of oxygen.

rate of reaction see reaction.

reaction (chemical) the action of one substance with another to produce a chemical change. The speed or rate of reaction depends upon several factors: temperature, particle size of a solid, concentration of a solution and catalysts, for instance.

Temperature: the higher the temperature the faster the reaction will be because at a high temperature the particles will move at greater speed, with the result that they collide more frequently and more energetically.

Particle size: the smaller the particles of a solid reactant the greater will be the total surface area that is available for chemical reactions to take place and therefore the reaction will be faster.

Concentration: the more concentrated the closer together the particles of solute and, hence, the more frequently they collide with particles of the other reactant(s).

Catalysts: the presence of catalysts will change the rate as they provide an alternative means by which a reaction may take place.

solute the substance that is dissolved in a solvent to produce a solution.

solvent the substance that dissolves a solute to produce a solution.

solution the dispersion of one or more solutes in a solvent. A solution is a homogenous mixture not a compound. The substances in a solution can be separated by physical processes (see below).

Methods of Separating Mixtures

To separate the elements of a compound a chemical reaction is required because the elements are bonded together. However, to separate components in a mixture various physical processes may be used. Mixtures may be made of two or more elements (for example alloys), or of two or more compounds (for example crude oil) or of a mixture of elements and compounds (for example air). The following methods will sometimes result in the separation of more than one component of a mixture.

centrifugation a physical method of separating a solid from a liquid in which it does not dissolve. Centrifugation achieves the same result as decantation and filtration but - unlike those two methods - requires the use of a piece of specialist machinery, a centrifuge, a machine that uses centrifugal force to separate particles of varying density by spinning. Centrifugation would be used in the case of a very fine solid mixed with a liquid where filtration would be impeded by the fine solid blocking the pores of the filter paper. The rapid spinning motion in a centrifuge would cause the solid to settle quickly at the bottom of a tube and the liquid could then be poured off by decantation.

chromatography a physical method of separating several components from a mixture in solution. This is done by running the solution through an adsorbent material on which the different substances are separated out as spots or bands. depending on the type of chromatography. Paper chromatography may be used for example to separate dyes in ink. A spot of the ink is placed on filter paper which is then placed upright in a beaker that contains a small amount of water. The level of the water should be below the ink mark. The experiment is then left until the water has risen up to the top of the filter paper. The different dyes in the ink will be carried up the filter paper at different rates and will therefore be separated on the paper. Chromatography works with very small quantities and is used to separate proteins, enzymes and forensic samples. See also gas chromatography (below).

crystallization a physical method of separating a solid from its solution in a solvent. Crystallization initially

CHROMATOGRAPHY

Chromatogram showing six different components in a mixture

involves evaporation of solvent up to the point at which a saturated solution is formed. This solution is then cooled to room temperature and crystals are naturally produced. The apparatus that is employed in crystallization is exactly the same as that used in evaporation (see below). Crystallization is used in cases where complete evaporation of solvent from a solution would not result in the desired crystals - this is because many crystals decompose into a powder when heated.

decantation a physical method of separating a solid from a liquid in which it does not dissolve. The liquid (usually a solution) is carefully poured from one beaker into another. The solid is usually heavy and will remain at the bottom of the first beaker. Such a method could be used to pour a saturated solution away from crystals which would remain behind as the residue.

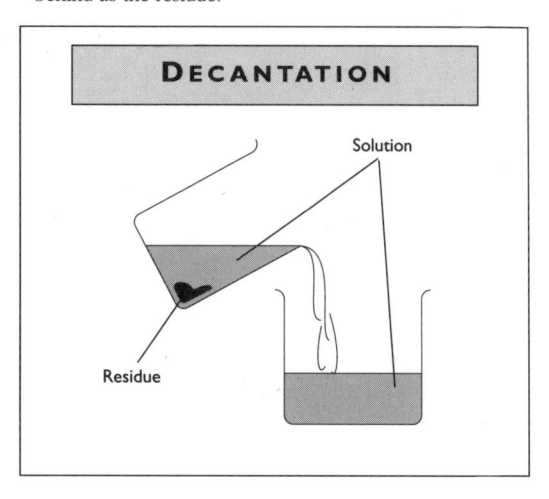

DECANTATION

distillation a physical method of separating a liquid from a solution. There are two basic forms of distillation - simple distillation and fractional distillation.

Simple distillation is used to separate a single liquid from one or more solids that are dissolved in it; for example it could be used to obtain pure water from sea water. In the accompanying diagram it can be seen that heat is applied to the solution, the liquid rises as water vapour and is condensed and falls as a liquid - the distillate - in the jar on the far side. The solids are left behind.

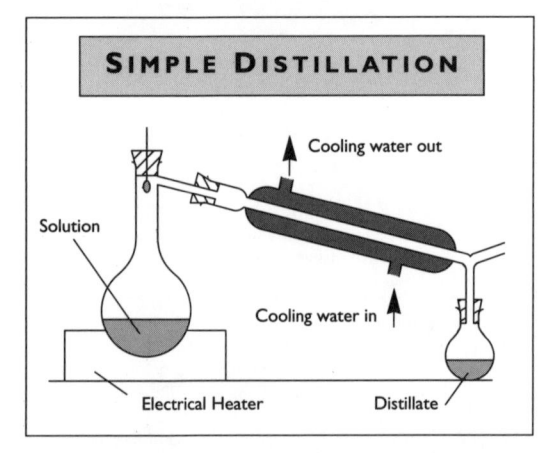

SIMPLE DISTILLATION

Fractional distillation is used to separate several liquids from a mixture of liquids. Fractions (see above) are obtained as the vapours of different liquids rise at different rates up the fractionating column. The rest of the method is as above.

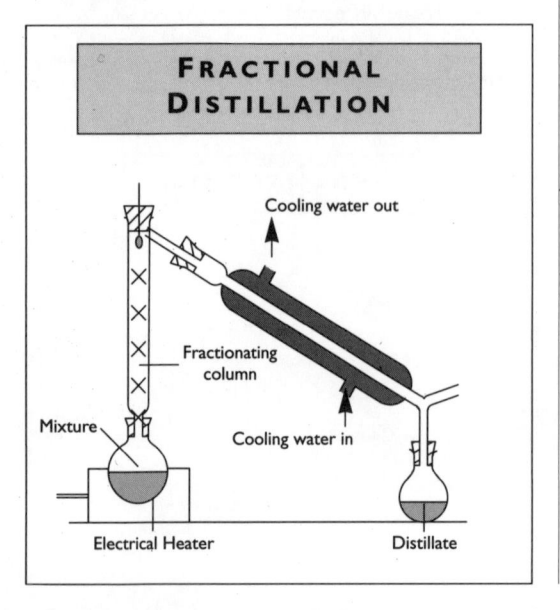

FRACTIONAL DISTILLATION

evaporation a physical method of separation a solid from its solution in a solvent. Evaporation is used when a solid is to be extracted from a solution - for example, a solution evaporates and is lost into the air as water vapour while the solid part of the solution remains behind. Evaporation would be used when a chemist is interested in the solid part of the solution but when it is the liquid part of the solution that is required the vapour would be condensed using simple forms of distillation described above.

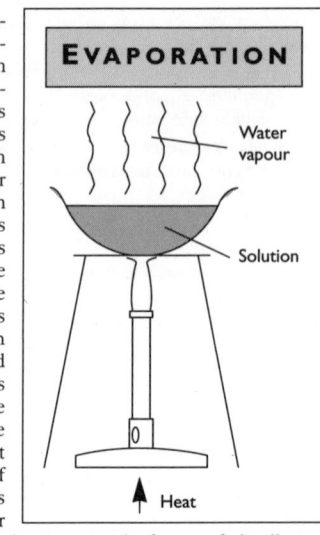

EVAPORATION

filtration a physical method of separating a solid from a liquid in which it does not dissolve. This is achieved by straining out solid particles from a liquid, usually a solution in which the solid is insoluble, for example sand from a salt solution. The solution is strained through a filter paper into a container placed below. The sand that remains behind on the filter paper is termed the residue.

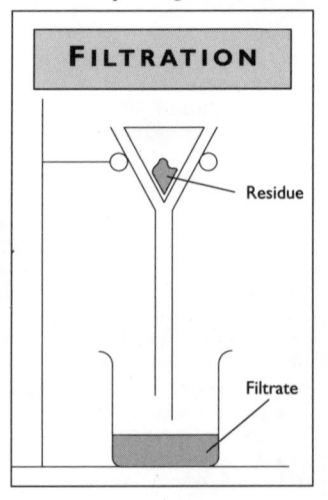

Filtration may also take place in a vacuum and vacuum filtration is a common technique.

FILTRATION

fractional distillation see distillation.

gas chromatography a physical method of separating the components of a mixture of substances in the gaseous state. Carrier gas from a cylinder is passed at a controlled rate into the gas chromatograph. The mixture is introduced into the carrier gas via the inlet system. The column is usually a tube which is packed with stationary phase - an adsorbent solid or an oil or wax coating a solid - and the components of the sample are separated as they elute through the column by adsorption or partition between the moving and

CHEMISTRY

stationary phases. The separated materials pass into the detector which monitors the eluting carrier gas continuously. Electrical signals from the detector are amplified and recorded to give a chromatogram on which the separation can be read.

HPLC high performance liquid chromatography, which is partition column chromatography under high pressure in which there is distribution between two liquids in a column.

reaction techniques the different techniques used to carry out different chemical reactions. They may be summarized to include:

heating solids or liquids;

reacting solids with a solution in order to collect a gas;

precipitation;

passing a gas or a vapour over a heated solid;

combustion - burning a solid or a liquid in a gas.

reflux a process in which liquid is obtained from partial condensation of vapour which is returned to the top of a fractionating column and then allowed to flow down the column in the opposite direction to the ascending vapour.

simple distillation see distillation.

sublimation a physical method of separating one solid from another solid. This is an unusual method and has relatively few uses as few substances will sublime. One of the few substances that will do so is iodine. Sublimation is similar to distillation but is used for purifying solids rather than liquids. In this process the heated solid becomes a vapour without first melting. Upon cooling it again becomes a solid.

vacuum filtration see filtration.

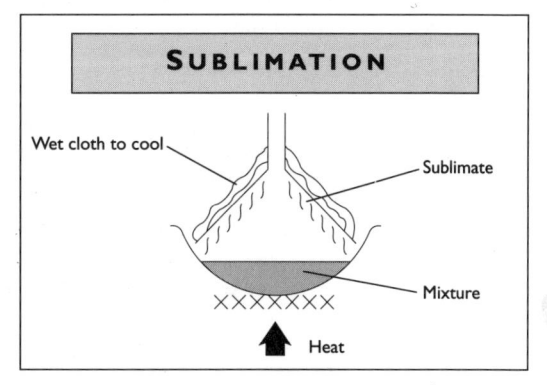

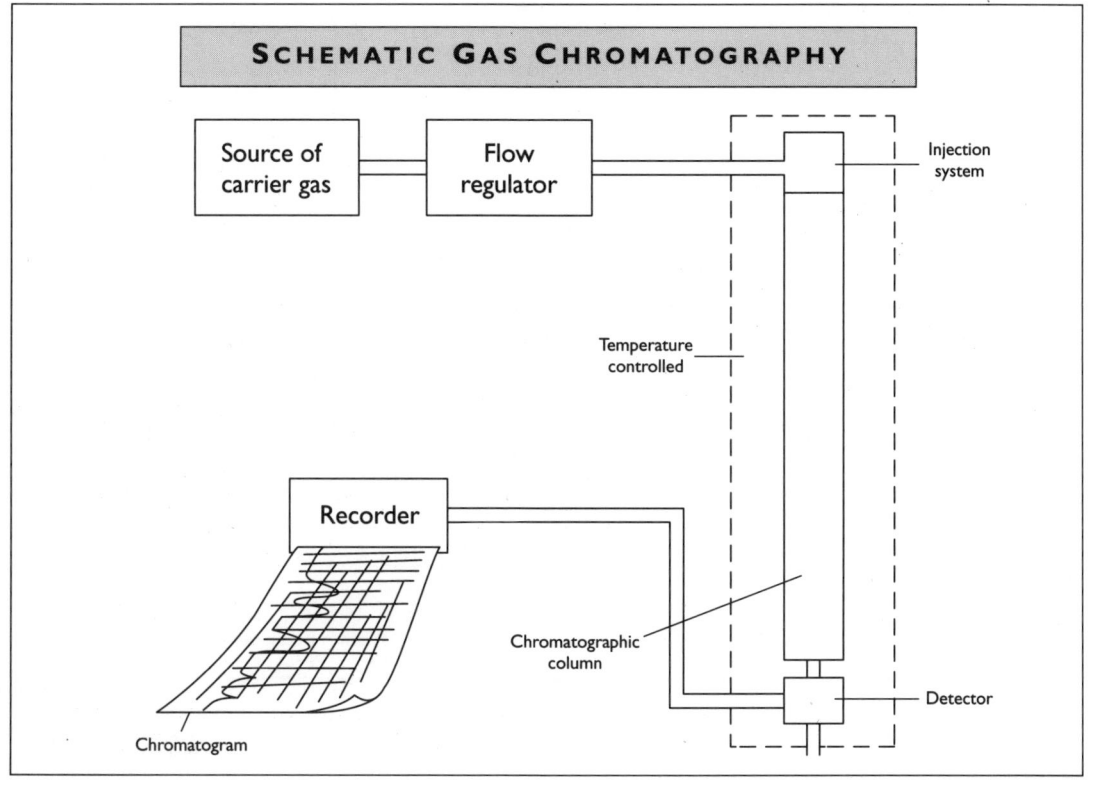

SUBLIMATION

Wet cloth to cool — Sublimate — Mixture — Heat

SCHEMATIC GAS CHROMATOGRAPHY

Source of carrier gas — Flow regulator — Injection system — Temperature controlled — Chromatographic column — Detector — Recorder — Chromatogram

ELEMENTS AND THE PERIODIC TABLE

The world is made up of a limited number of chemical elements. In the Earth's crust there are 82 stable elements and a few unstable (radioactive) ones. Among the stable elements, there are some, such as oxygen and silicon, that are very abundant, while others - the metals ruthenium and rhodium, for example - are extremely rare. Indeed 98 per cent of the Earth's crust is made up of just eight elements - oxygen, silicon, aluminium, iron, calcium, sodium, magnesium, and potassium. Each element is associated with a unique number, called its atomic number.

atom the smallest particle of an element that can exist. Atoms are the 'building blocks' of everything. They are, in turn, made up of subatomic particles - protons, neutrons and electrons.

atomic number the figure that represents the number of protons (positively charged particles) in the nucleus of each atom of an element. Each hydrogen atom has one proton, so hydrogen is the first and lightest of the elements and is placed first in the Periodic Table; each helium atom has two protons and helium is thus the second lightest element and is placed second in the Periodic Table; and so one can continue through each of the elements, establishing their place on the Periodic Table according to their atomic numbers. The atomic number of bismuth is 83, and this number of protons represents the upper limit for a stable nucleus. Beyond 83 all elements are unstable. The largest atomic number observed so far is 109, but only a few atoms of this element - named unnilennium - have been made artificially, so little is known about it.

d sub-level one of the sub-levels (see electron). The d sub-level identifies one of the main blocks or groups

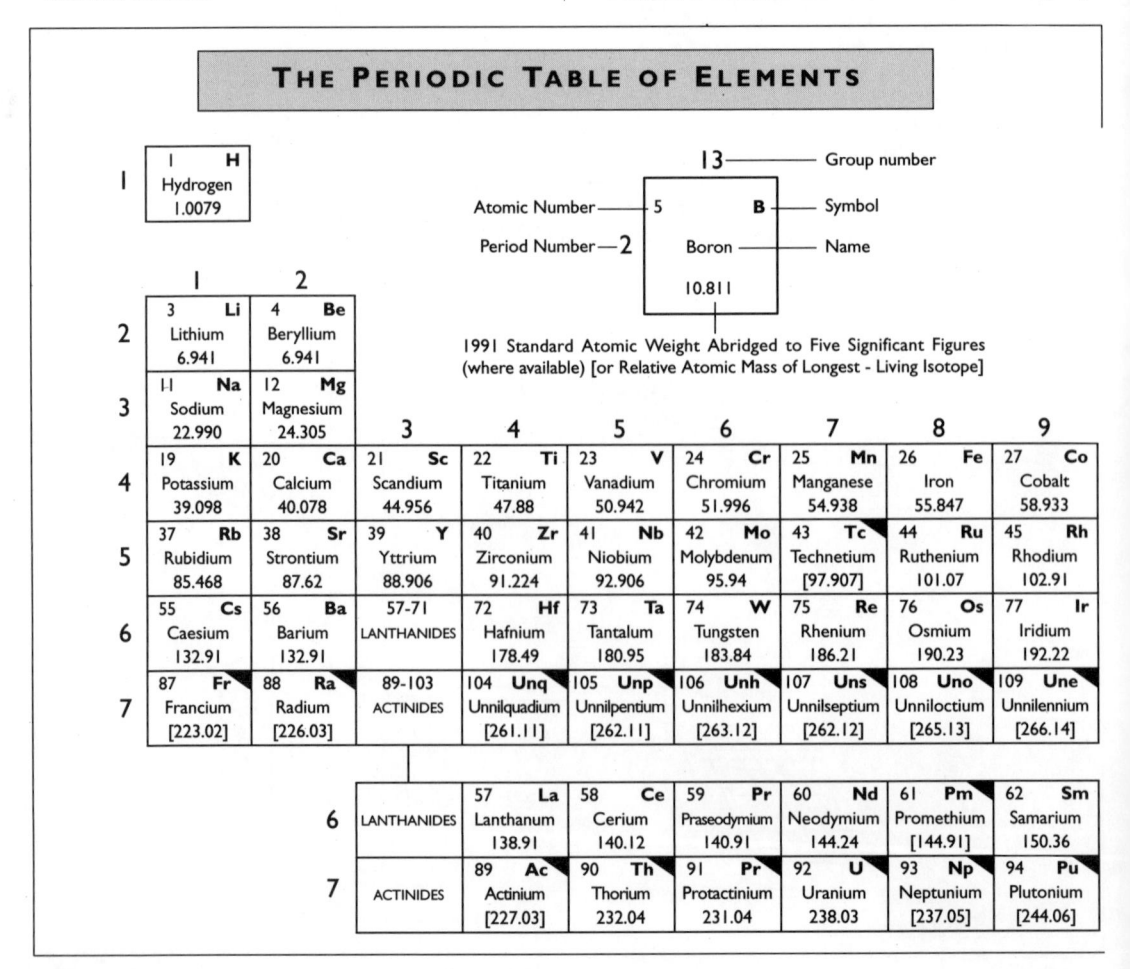

THE PERIODIC TABLE OF ELEMENTS

Atomic Number — 5
Period Number — 2
Group number — 13
Symbol — B
Name — Boron
10.811

1991 Standard Atomic Weight Abridged to Five Significant Figures
(where available) [or Relative Atomic Mass of Longest - Living Isotope]

	1	2	3	4	5	6	7	8	9
1	1 H Hydrogen 1.0079								
2	3 Li Lithium 6.941	4 Be Beryllium 6.941							
3	11 Na Sodium 22.990	12 Mg Magnesium 24.305							
4	19 K Potassium 39.098	20 Ca Calcium 40.078	21 Sc Scandium 44.956	22 Ti Titanium 47.88	23 V Vanadium 50.942	24 Cr Chromium 51.996	25 Mn Manganese 54.938	26 Fe Iron 55.847	27 Co Cobalt 58.933
5	37 Rb Rubidium 85.468	38 Sr Strontium 87.62	39 Y Yttrium 88.906	40 Zr Zirconium 91.224	41 Nb Niobium 92.906	42 Mo Molybdenum 95.94	43 Tc Technetium [97.907]	44 Ru Ruthenium 101.07	45 Rh Rhodium 102.91
6	55 Cs Caesium 132.91	56 Ba Barium 132.91	57-71 LANTHANIDES	72 Hf Hafnium 178.49	73 Ta Tantalum 180.95	74 W Tungsten 183.84	75 Re Rhenium 186.21	76 Os Osmium 190.23	77 Ir Iridium 192.22
7	87 Fr Francium [223.02]	88 Ra Radium [226.03]	89-103 ACTINIDES	104 Unq Unnilquadium [261.11]	105 Unp Unnilpentium [262.11]	106 Unh Unnilhexium [263.12]	107 Uns Unnilseptium [262.12]	108 Uno Unniloctium [265.13]	109 Une Unnilennium [266.14]

			57 La Lanthanum 138.91	58 Ce Cerium 140.12	59 Pr Praseodymium 140.91	60 Nd Neodymium 144.24	61 Pm Promethium [144.91]	62 Sm Samarium 150.36
6	LANTHANIDES							
7	ACTINIDES		89 Ac Actinium [227.03]	90 Th Thorium 232.04	91 Pa Protactinium 231.04	92 U Uranium 238.03	93 Np Neptunium [237.05]	94 Pu Plutonium [244.06]

of the Periodic Table containing the elements of atomic number 21 to 30, 39 to 48, 71 to 80 and 103 to 109: scandium, titanium, vanadium, chromium, manganese, iron, cobalt, nickel, copper, zinc, yttrium, zirconium, niobium, molybdenum, technetium, ruthenium, rhodium, palladium, silver, cadmium, lutetium, hafnium, tantalum, tungsten, rhenium, osmium, iridium, platinum, gold, mercury, lawrencium, unnilquadium, unnilpentium, unnilhexium, unnilseptium, unniloctium, and unnilennium.

electron a subatomic negatively charged particle that moves around the nucleus of an atom. The number of electrons always equals the number of protons in an atom when that atom is electrically neutral. Thus, for example, an electrically neutral atom of calcium contains 20 protons and 20 electrons. Electrons can be thought of as moving around the nucleus in certain fixed orbits or 'shells', the electrons in a particular shell being associated with a particular energy level. With regard to an atom's chemical behaviour, it is the electrons in the outer shell that are most important and it is these that fix the group position (see below) of the atom in the Periodic Table. The shells and the major energy levels of electrons are numbered 1, 2, 3, etc., counting outwards from the nucleus. This number is called the principal quantum number and is given the symbol **n**. Each shell/energy level can hold only a certain number of electrons; the further out it is, the more it can accommodate. This capacity is related to the value of **n**. The nearest shell to the nucleus can hold only 2 electrons, the next 8, then 18, then 32, and so on. Each energy level is divided into sub-levels, called **s**, **p**, **d** and **f**, which hold a maximum of 2, 6, 10 and 14 electrons respectively.

f sub-level one of the sub-levels (see electron). The f sub-level identifies one of the main blocks of the Periodic Table containing the elements of atomic numbers 57 to 70, and 89 to 102: cerium, praseodymium, neodymium, promethium, samarium, europium, gadolinium, terbium, dysprosium, holmium, erbium, thulium, ytterbium, lutetium, thorium, protactinium, uranium, neptunium, plutonium, americium, curium, berkelium, californium, einsteinium, fermium, mendelevium, nobelium and lawrencium.

group position the position of an element in a group of similar elements on the Periodic Table.

The groups of the Periodic Table are numbered 1 to 18

with the **f**-block not included. Members of the same group have the same number of electrons in the outer shell of the atom and consequently behave in a similar manner chemically. This fact is reflected in the composition of their chemical compounds (which in turn can be explained in terms of their oxidation states).

As we go from left to right across the table we can see particular properties change in a regular fashion. It was this periodic rise and fall in such properties as density and melting point that led to the term 'Periodic Table'.

noble gases (or **inert gases**) monatomic gases whose configurations are very stable because of their electron arrangement. A hydrogen atom has one electron in the first principal energy level, while a helium atom has two - the maximum capacity for this level. The possession of one extra electron may seem a trivial difference, but a world of difference separates hydrogen and helium; hydrogen is very reactive and forms compounds with many other elements; helium combines with nothing. These two elements are rather exceptional in all their chemical behaviour and are usually either given a small section of their own on the table or are placed apart at either end of the table with hydrogen above group 1 and helium in group 18. Helium is placed in the same group as the other chemically unreactive gases - the noble gases - even though it does not have the eight outer electrons that they do.

To the left of neon is fluorine (configuration 2.7), a reactive element. Fluorine (like the other elements in group 17 - the halogens) is one electron short of a noble gas electron arrangement. Fluorine's tendency to combine with other elements in order to achieve a noble gas electron arrangement makes it one of the most reactive of all elements - so reactive that it will even combine with the noble gases krypton and xenon.

The noble gases with their stable electron arrangement make a natural break in the arrangement of the Periodic Table. After the **p** sub-shell has been filled, the next electron starts another shell further out from the nucleus. This lone electron makes the elements of group 1 - the alkali metals - highly reactive, because they tend to lose the extra electron in order to achieve a noble gas electron arrangement. They are indeed so reactive that some of them, such as caesium, explode when dropped into water.

neutron a subatomic particle found at the centre of the nucleus of most atoms. It has the same mass as a proton but carries no electric charge.

oxidation to become oxidized by losing an electron (or by gaining oxygen or losing hydrogen).

Periodic Table a table that represents all of the elements in such a manner as to show similarities and differences in their chemical properties. The elements are arranged in increasing number of atomic number as one reads from left to right across the table.

The discovery of the Periodic Table was made possible by an Italian chemist, Stanislao Cannizzaro (1826-1910), who in 1858 published a list of fixed atomic weights (now known as relative atomic masses) for the 60 elements that were then known. By arranging the elements in order of increasing atomic weight, a curious repetition of chemical properties at regular intervals was revealed. This was noticed in 1864 by the English chemist John Newlands (1838-98), but his 'law of octaves' brought him nothing but ridicule. It was left to the Russian chemist Dmitri Ivanovich Mendeleyev (1834-1907) to make essentially the same discovery five years later. What Mendeleyev did, however, was so much more impressive that he is rightly credited as the true discoverer of the Periodic Table.

While working on his *Principles of Chemistry* (1869), Mendeleyev wrote the names and some of the main features of the elements on individual cards, to help establish a suitable order in which to discuss their chemistry. It was while he was arranging this pack of cards that he stumbled upon the pattern we now recognize as the Periodic Table. Mendeleyev laid out his cards in order of atomic weights of the elements, placing together elements that formed similar oxides. By arranging similar elements in columns, he established the arrangement of the table that has been followed ever since.

Mendeleyev's genius lay in that he recognized that there was an underlying order to the elements - he did not design the Periodic Table, he discovered it. If he was right, he knew that there were places in his table for new elements. He was so confident in his discovery that he predicted the properties of these missing elements - and his predictions were subsequently shown to be accurate. In some cases, Mendeleyev also swapped the order of the atomic weights, so that similar elements appeared in the same group.

Since 1869 when Mendeleyev published his table, a further 40 elements have been found or produced by nuclear reactions, and the Periodic Table has been redesigned to accommodate them. Mendeleyev lived long enough to learn of the discovery of the electron, but not long enough to know how the arrangement of electrons about the nucleus of the atom explains the structure of the Table.

p sub-level one of the sub-levels (see electron). The p sub-level identifies one of the main blocks of the Periodic Table containing the elements of atomic number 5 to 10, 13 to 18, 31 to 36, 49 to 54 and 81 to 86 (or the elements of groups 13 to 18): boron, carbon, nitrogen, oxygen, fluorine, neon, aluminium, silicon, phosphorus, sulphur, chlorine, argon, gallium, germanium, arsenic, selenium, bromine, krypton, indium, tin, antimony, tellurium, iodine, xenon, thallium, lead, bismuth, polonium, astatine, and radon.

proton a positively charged subatomic particle found in the nucleus of an atom. The number of protons in an atom is its atomic number and the atomic number of each element is unique.

s sub-level one of the sub-levels (see electron). The s sub-level identifies one of the main blocks of the Periodic Table containing the elements of groups 1 and 2: lithium, beryllium, sodium, magnesium, potassium, calcium, rubidium, strontium, caesium, barium, francium, and radium.

CHEMISTRY

TABLE OF THE 109 ELEMENTS

Atomic Number	Symbol	Element Name	Discoverers	Year	Atomic Weight	Density At 20°C (Unless Otherwise Stated) (g/cm³)	Melting Point (°C)	Boiling Point (°C)	Number Of Nuclides
1	H	Hydrogen	H. Cavendish (UK)	1766	1·007 94	0·0871 (solid at mp) 0·000 089 89 (gas at 0°C)	−259·198	−252·762	3
2	He	Helium	J. N. Lockyer (UK) and P. J. C. Janssen (France)	1868	4·002 602	0·190 8 (solid at mp) 0·000 178 5 (gas at 0°C)	−272·375 at 24·985 atm	−268·928	8
3	Li	Lithium	J. A. Arfwedson (Sweden)	1817	6·941	0·5334	180·54	1339	8
4	Be	Beryllium	N. L. Vauquelin (France)	1798	9·012 182	1·846	1287	2471	9
5	B	Boron	L. J. Gay Lussac and L. J. Thenard (France) and H. Davy (UK)	1808	10·811	2·333 (*b* Rhombahedral)	2130	3910	13
6	C	Carbon	Prehistoric		12·011	2·266 (Graphite)	3530 3·515 (Diamond)	3870	15
7	N	Nitrogen	D. Rutherford (UK)	1772	14·006 74	0·9426 (solid at mp) 0·001 250 (gas at 0°C)	−210·000	−195·798	13
8	O	Oxygen	C. W. Scheele (Sweden) and J. Priestley (UK)	1772–1774	15·9994	1·359 (solid at mp) 0·001 429 (gas at 0°C)	−218·792	−182·954	15
9	F	Fluorine	H. Moissan (France)	1886	18·998 403	1·780 (solid at mp)	−219·673	−188·191	14
10	Ne	Neon	W. Ramsay and M. W. Travers (UK)	1898	20·179 7	1·434 (solid at mp) 0·000 899 9 (gas at 0°C)	−248·594	−246·053	17
11	Na	Sodium	H. Davy (UK)	1807	22·989 768	0·9688	97·82	882	17
12	Mg	Magnesium	H. Davy (UK)	1808	24·3050	1·737	650	1095	17
13	Al	Aluminium	H. C. Oerstedt (Denmark) and F. Wöhler (Germany)	1825–1827	26·981 539	2·699	660·323	2516	18
14	Si	Silicon	J. J. Berzelius (Sweden)	1824	28·0855	2·329	1414	3190	21
15	P	Phosphorus	H. Brand (Germany)	1669	30·973 762	1·825 (White) 2·361 (Violet) 2·708 (Black)	44·13 597 at 45 atm 606 at 48 atm	277 431 sublimes 453 sublimes	21
16	S	Sulphur	Prehistoric 'sulphurum'		32·066	2·070 (Rhombic)	115·18	444·614	22
17	Cl	Chlorine	C. W. Scheele (Sweden)	1774	35·4527	2·038 (solid at mp) 0·003 214 (gas at 0°C)	−100·97	−33·97	20
18	Ar	Argon	W. Ramsay and Lord Rayleigh (UK)	1894	39·948	1·622 (solid at mp) 0·001 784 (gas at 0°C)	−189·344	−185·848	20
19	K	Potassium (Kalium)	H. Davy (UK)	1807	39·0983	0·8591	63·58	758	20
20	Ca	Calcium	H. Davy (UK)	1808	40·078	1·526	842	1495	19

TABLE OF THE 109 ELEMENTS CONT.

Atomic Number	Symbol	Element Name	Discoverers	Year	Atomic Weight	Density At 20°C (Unless Otherwise Stated) (g/cm³)	Melting Point (°C)	Boiling Point (°C)	Number Of Nuclides
21	Sc	Scandium	L. F. Nilson (Sweden)	1879	44·955 910	2·989	1541	2830	16
22	Ti	Titanium	M. H. Klaproth (Germany)	1795	47·88	4·504	1672	3360	20
23	V	Vanadium	N. G. Sefström (Sweden)	1830	50·9415	6·099	1928	3410	19
24	Cr	Chromium	N. L. Vauquelin (France)	1798	51·9961	7·193	1860	2680	21
25	Mn	Manganese	J. G. Gahn (Sweden)	1774	54·938 05	7·472	1246	2051	21
26	Fe	Iron (Ferrum)	Earliest smelting	c. 4000 BC	55·847	7·874	1538	2840	22
27	Co	Cobalt	G. Brandt (Sweden)	1737	58·933 20	8·834	1495	2940	22
28	Ni	Nickel	A. F. Cronstedt (Sweden)	1751	58·6934	8·905	1455	2890	24
29	Cu	Copper (Cuprum)	Prehistoric (earliest known use)	c. 8000 BC	63·546	8·934	1084+62	2570	25
30	Zn	Zinc	A. S. Marggraf (Germany)	1746	65·39	7·140	419·527	908	25
31	Ga	Gallium	L. de Boisbaudran (France)	1875	69·723	5·912	29·765	2203	24
32	Ge	Germanium	C. A. Winkler (Germany)	1886	72·61	5·327	938·2	2770	25
33	As	Arsenic	Albertus Magnus (Germany)	c. 1220	74·921 59	5·781	817 at 38 atm	603 sublimes	23
34	Se	Selenium	J. J. Berzelius (Sweden)	1818	78·96	4·810 (Trigonal)	221·14	685	23
35	Br	Bromine	A. J. Balard (France)	1826	79·904	3·937 (solid at mp) 3·119 (liquid at 20°C)	−7·25	59·74	26
36	Kr	Krypton	W. Ramsay and M. W. Travers (GB)	1898	83·80	2·801 (solid at mp) 0·003 749 (gas at 0°C)	−157·374	−153·340	25
37	Rb	Rubidium	R. W. Bunsen and G. R. Kirchhoff (Germany)	1861	85·4678	1·534	39·29	687	28
38	Sr	Strontium	W. Cruikshank (UK)	1787	87·62	2·582	769	1388	28
39	Y	Yttrium	J. Gadolin (Finland)	1794	88·905 85	4·468	1522	3300	24
40	Zr	Zirconium	M. H. Klaproth (Germany)	1789	91·224	6·506	1854	4360	24
41	Nb	Niobium	C. Hatchett (UK)	1801	92·906 38	8·595	2472	4860	25
42	Mo	Molybdenum	P. J. Hjelm (Sweden)	1781	95·94	10·22	2623	4710	24
43	Tc	Technetium	C. Perrier (France) and E. Segrè (Italy/USA)	1937	(97·9072)	11·40	2180	4860	23
44	Ru	Ruthenium	K. K. Klaus (Estonia)	1844	101·07	12·37	2333	4310	25

CHEMISTRY

TABLE OF THE 109 ELEMENTS CONT.

Atomic Number	Symbol	Element Name	Discoverers	Year	Atomic Weight	Density At 20°C (Unless Otherwise Stated) (g/cm³)	Melting Point (°C)	Boiling Point (°C)	Number Of Nuclides
45	Rh	Rhodium	W. H. Wollaston (UK)	1804	102·905 50	12·42	1962	3700	24
46	Pd	Palladium	W. H. Wollaston (UK)	1803	106·42	12·01	1554·7	2970	27
47	Ag	Silver (Argentum)	Prehistoric (earliest silversmithery)	c. 4000 BC	107·8682	10·50	961·78	2167	29
48	Cd	Cadmium	F. Stromeyer (Germany)	1817	112·411	8·648	321·068	768	33
49	In	Indium	F. Reich and H. T. Richter (Germany)	1863	114·818	7·289	156·599	2019	32
50	Sn	Tin (Stannum)	Prehistoric (intentionally alloyed with copper to make bronze	c. 3500 BC	118·710	7·288	231·928	2590	33
51	Sb	Antimony (Stibium)	Near historic	c. 1000 BC	121·757	6·693	630-636	1635	29
52	Te	Tellurium	F. J. Muller (Baron von Reichenstein) (Austria)	1783	127·60	6·237	449·81	989	33
53	I	Iodine	B. Courtois (France)	1811	126·904 47	4·947	113·6	185·1	33
54	Xe	Xenon	W. Ramsay and M. W. Travers (UK)	1898	131·29	3·410 (solid at mp) 0·005897 (gas at 0°C)	−111·774	−108·083	36
55	Cs	Caesium	R. W. von Bunsen and G. R. Kirchoff (Germany)	1860	132·905 43	1·896	28-46	668	36
56	Ba	Barium	H. Davy (UK)	1808	137·327	3·595	729	1740	31
57	La	Lanthanum	C. G. Mosander (Sweden)	1839	138·9055	6·145	921	3410	30
58	Ce	Cerium	J. J. Berzelius and W. Hisinger (Sweden) and M. H. Klaproth (Germany)	1803	140-115	6·688 (beta) 6·770 (gamma)	799	3470	30
59	Pr	Praseodymium	C. Auer von Welsbach (Austria)	1885	140·907 65	6·772	934	3480	29
60	Nd	Neodymium	C. Auer von Welsbach (Austria)	1885	144·24	7·006	1021	3020	30
61	Pm	Promethium	J. Marinsky, L. E. Glendenin, and C. D. Coryell (USA)	1945	(144·9127)	7·141	1042	3000	28
62	Sm	Samarium	L. de Boisbaudran (France)	1879	150-36	7·517	1077	1794	30
63	Eu	Europium	E. A. Demarçay (France)	1901	151·965	5·243	822	1556	29
64	Gd	Gadolinium	J. C. G. de Marignac (Switzerland)	1880	157·25	7·899	1313	3270	27
65	Tb	Terbium	C. G. Mosander (Sweden)	1843	158·925 34	8·228	1356	3230	26
66	Dy	Dysprosium	L. de Boisbaudran (France)	1886	162·50	8·549	1412	2570	29
67	Ho	Holmium	J. L. Soret (France) and P. T. Cleve (Sweden)	1878–1879	164·930 32	8·794	1474	2700	29
68	Er	Erbium	C. G. Mosander (Sweden)	1843	167·26	9·064	1529	2810	29

TABLE OF THE 109 ELEMENTS CONT.

Atomic Number	Symbol	Element Name	Discoverers	Year	Atomic Weight	Density At 20°C (Unless Otherwise Stated) (g/cm³)	Melting Point (°C)	Boiling Point (°C)	Number Of Nuclides
69	Tm	Thulium	P. T. Cleve (Sweden)	1879	168·934 21	9·319	1545	1950	31
70	Yb	Ytterbium	J. C. G. de Marignac (France)	1878	173·04	6·967	817	1227	30
71	Lu	Lutetium	G. Urbain (France)	1907	174·967	9·839	1665	3400	35
72	Hf	Hafnium	D. Coster (Netherlands) and G. C. de Hevesy (Hungary/Sweden)	1923	178·49	13·28	2250	4700	31
73	Ta	Tantalum	A. G. Ekeberg (Sweden)	1802	180·9479	16·67	3020	5490	31
74	W	Tungsten (Wolfram)	J. J. de Elhuyar and F. de Elhuyar (Spain)	1783	183·84	19·26	3420	5860	33
75	Re	Rhenium	W. Noddack, Fr. I. Tacke and O. Berg (Germany)	1925	186·207	21·01	3185	5610	33
76	Os	Osmium	S. Tennant (UK)	1804	190·23	22·59	3127	5020	35
77	Ir	Iridium	S. Tennant (UK)	1804	192·22	22·56	2446	4730	33
78	Pt	Platinum	A. de Ulloa (Spain)	1748	195·08	21·45	1768·1	3870	35
79	Au	Gold (Aurum)	Prehistoric	c. 1600 BC	196·966 54	19·29	1064·18	2870	32
80	Hg	Mercury (Hydrargyrum)	Near historic		200·59	14·17 (solid at mp) 13·55 (liquid at 20°C)	−38·829	356·661	33
81	Tl	Thallium	W. Crookes (UK)	1861	204·3833	11·87	303	1468	29
82	Pb	Lead (Plumbum)	Prehistoric		207·2	11·35	327·462	1748	34
83	Bi	Bismuth	C. F. Geoffroy (France)	1753	208·980 37	9·807	271·402	1566	30
84	Po	Polonium	Mme. M. S. Curie (Poland/France)	1898	(208·9824)	9·155	254	948	27
85	At	Astatine	D. R. Corson and K. R. Mackenzie (USA) and E. Segrè (Italy/USA)	1940	(209·9871)	7·0	302	377	28
86	Rn	Radon	F. E. Dorn (Germany)	1900	(222·0176)	4·7 (solid at mp) 0·010 04 (gas at 0°C)	−64·9	−61·2	31
87	Fr	Francium	Mlle. M. Perey (France)	1939	(223·0197)	2·8	23	650	31
88	Ra	Radium	P. Curie (France), Mme. M. S. Curie (Poland/France), and M. G. Bemont (France)	1898	(226·0254)	5·50	707	1530	28
89	Ac	Actinium	A. Debierne (France)	1899	(227·0278)	10·04	1230	3600	26
90	Th	Thorium	J. J. Berzelius (Sweden)	1829	232·0381	11·72	1760	4660	25
91	Pa	Protactinium	O. Hahn (Germany) and Fr. L Meitner (Austria); F. Soddy and J. A. Cranston (UK)	1917	231·03588	15·41	1570	4490	24
92	U	Uranium	M. H. Klaproth (Germany)	1789	238·0289	19·05	1134	4160	20

TABLE OF THE 109 ELEMENTS CONT.

Atomic Number	Symbol	Element Name	Discoverers	Year	Atomic Weight	Density At 20°C (Unless Otherwise Stated) (g/cm³)	Melting Point (°C)	Boiling Point (°C)	Number Of Nuclides
93	Np	Neptunium	E. M. McMillan and P. H. Abelson (USA)	1940	(237·0482)	20·47	637	4090	18
94	Pu	Plutonium	G. T. Seaborg, E. M. McMillan, J. W. Kennedy and A. C. Wahl (USA)	1940–1941	(244·0642)	20·26	640	3270	17
95	Am	Americium	G. T. Seaborg, R. A. James, L. O. Morgan and A. Ghiorso (USA)	1944–1945	(243·0614)	13·76	1176	2023	13
96	Cm	Curium	G. T. Seaborg, R. A. James and A. Ghiorso (USA)	1944	(247·0703)	13·68	1340	3180	14
97	Bk	Berkelium	S. G. Thompson, A. Ghiorso and G. T. Seaborg (USA)	1949	(247·0703)	14·65	1050	2710	11
98	Cf	Californium	S. G. Thompson, K. Street Jr., A. Ghiorso and G. T. Seaborg (USA)	1950	(251·0796)	15·20	900	1612	18
99	Es	Einsteinium	A. Ghiorso et al (USA)	1952	(252·0829)	9·05	860	996	14
100	Fm	Fermium	A. Ghiorso et al (USA)	1953	(257·0951)	9.42	852	1077	18
101	Md	Mendelevium	A. Ghiorso, B. G. Harvey, G. R Choppin, S. G. Thompson and G. T. Seaborg (USA)	1955	(258·0986)	–	–	–	13
102	No	Nobelium		1958	(259·1009)	–	–	–	11
103	Lr	Lawrencium		1961	(262·11)	–	–	–	10
104	Unq	Unnilquadium*	A. Ghiorso, M. Nurmia, J. Harris, K. Eskola and P. Eskola (USA/Finland)	1969	(261·1087)	–	–	–	10
105	Unp	Unnilpentium*	A. Ghiorso, M. Nurmia, K. Eskola, J. Harris and P. Eskola (USA/Finland)	1970	(262·1138)	–	–	–	8
106	Unh	Unnilhexium*	A. Ghiorso et al (USA)	1974	(263·1182)	–	–	–	4
107	Uns	Unnilseptium*	G. Münzenberg (Germany)	1981	(262·1229)	–	–	–	2
108	Uno	Unniloctium*	G. Münzenberg (Germany)	1984	(265·1302)	–	–	–	2
109	Une	Unnilennium*	G. Münzenberg (Germany)	1982	(266·1376)	–	–	–	1

*The names for elements 104 to 109 are those currently used. The I.U.P.A.C. (International Union of Pure and Applied Chemistry) will eventually name these elements.

DIFFUSION

Diffusion may be defined as the movement of particles in order to achieve an even distribution of concentration. Particles collide with one another thus slowing down their progress.

Diffusion only takes place very rapidly when a substance is present in a vacuum.

Diffusion plays an important role in spreading pollutants in the atmosphere. Pollutants are usually formed in relatively confined areas - for example a factory chimney or the exhaust pipe of a car. Diffusion spreads pollution into the surrounding atmosphere; it does not, however, remove the problem - it merely spreads it to larger area.

concentration of particles see rate of diffusion.

diffusion in gases may be demonstrated as described in the following experiment.

A piece of cotton wool is placed at either end of a tube - one piece of cotton wool is soaked in concentrated ammonia solution; the other piece is soaked in concentrated hydrochloric acid. The tube is corked to prevent draughts from moving the gases involved in the experiment and is maintained in a level position so that the effect of gravity is avoided.

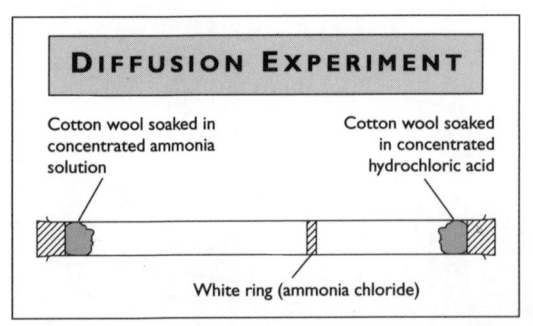

DIFFUSION EXPERIMENT

Cotton wool soaked in concentrated ammonia solution

Cotton wool soaked in concentrated hydrochloric acid

White ring (ammonia chloride)

Gases are given off from each piece of cotton wool and a white compound - which is ammonium chloride - is formed in the tube at the point at which the ammonia and hydrogen chloride meet. The compound does not, however, form halfway along the tube. It forms nearer the hydrogen chloride concentration. It can thus be deduced that the ammonia molecules have moved further, and therefore faster, than the hydrogen chloride molecules over the same period of time.

diffusion in liquids may be demonstrated by the following experiment.

If a crystal of potassium manganate (VII) is placed in a beaker of water diffusion will slowly occur. After about a quarter of an hour a purple colour will begin to spread from the crystal as it starts to dissolve. After 24 hours, a solution with a uniform purple colour will have resulted.

Because the particles in a liquid are closer together than the particles in a gas, the rate of diffusion in the liquid is, therefore, much slower than in a gas.

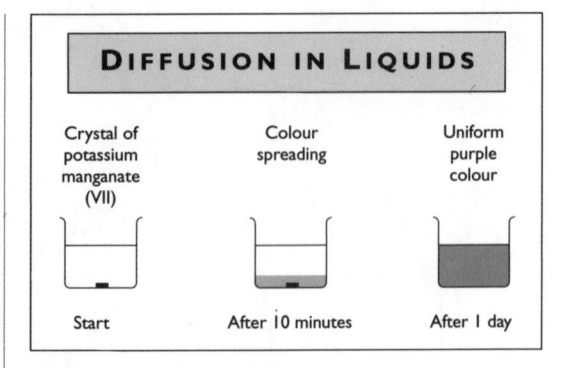

DIFFUSION IN LIQUIDS

Crystal of potassium manganate (VII)

Colour spreading

Uniform purple colour

Start — After 10 minutes — After 1 day

diffusion in solids occurs but is a very slow process indeed although it can be speeded up by the introduction of heat.

energy of particles see rate of diffusion.

kinetic energy energy that a moving object processes. The amount of energy depends upon the object's mass and its speed and can be given in the following equation:

$$\text{kinetic energy} = \tfrac{1}{2} \text{mass} \times \text{speed}^2$$

When considering a single substance, the kinetic energy decreases gas → liquid → solid because these states exist at different temperatures.

When comparing different substances, the kinetic energy of their particles depends only on their temperature. Thus a solid, a liquid and a gas, *all at the same temperature*, will contain particles with the *same* average kinetic energy.

Kinetic energy can be converted into heat energy - for example, when a vapour condenses into a liquid, heat energy is given off.

mass of particles see rate of diffusion.

pollutant a substance or effect that adversely alters the environment because it changes the growth rate of a species, interferes with a food chain, is toxic or interferes with health, comfort, amenities or property values.

rate of diffusion depends upon the speed at which particles move and the closeness of the particles. The quicker particles move, the faster diffusion occurs. Thus, the collision of particles slows down the rate of diffusion.

The rate of diffusion depends upon three factors: the concentration of the particles, the speed of the particles and the mass of the particles.

The concentration of particles has an important effect upon the rate of diffusion - the fewer particles present, the greater the rate of diffusion. This is because the particles do not collide with one another so often.

The speed of particles depends upon the temperature of the substance. The higher the temperature, the greater the rate at which diffusion occurs.

The mass of particles also has an effect upon the rate of diffusion, particularly in a gas. Thus, the lighter the particles, will be and the faster the rate of diffusion. At a particular temperature hydrogen molecules, being the lightest of all particles, will diffuse most quickly.

CHEMISTRY

STATES OF MATTER

Solid, liquid and gas (or vapour) are the three states in which substances can exist. The state in which any particular substance exists depends upon the temperature and upon the pressure being exerted upon it.

Virtually all substances are able to exist in more than one of these states. At normal temperature or pressure water is a liquid but can exist as a vapour (steam) or as a solid (ice) at different temperatures and pressures.

In chemical equations, initials are sometimes used to indicate the state of a particular substance:

- (g) = gas;
- (l) = liquid;
- (s) = solid;
- (aq) = aqueous solution is also used but this is not a state.

The particles in a gas move very quickly, are not close together and consequently have relatively little attraction for one another. If the temperature of that gas is reduced, the particles in it will move more slowly. The slower particles take up less space and, because they are closer, there will be a stronger attraction between them. If the temperature is reduced enough, the particles adhere to one another and the gas becomes a liquid. In the liquid, particles move randomly and more slowly than in the gas. If the temperature is decreased further, the particles may cease to move randomly and may then vibrate in fixed positions. The liquid becomes a solid.

These states of matter - plotted against temperature and also against time - are shown in the accompanying diagram.

The reverse situation occurs when the temperature of a solid is raised. The solid will be turned to a liquid and, with a further increase in temperature, the liquid will be turned into a gas.

bond the chemical link which holds atoms together.

gas (or **vapour**) the normal state of a substance whose particles have very high energy.

When heat is applied to a liquid, the kinetic energy of the atoms, ions and molecules within that liquid is increased (see p. 136). The increased energy may be enough to weaken the forces of attraction that bond these atoms, ions and molecules together as a liquid. If this happens, the liquid will boil and it turns into a gas.

In a container, collisions of gas particles with the container's walls exert a pressure. Gas particles move very quickly - usually at a rate of hundreds of metres a second; they also move in random directions and, in so doing, fill all the available spaces.

The density of gases is much less - averaging about $\frac{1}{1000\text{th}}$ - of that of a solid or of a liquid.

liquid the state of matter in which the particles are held together, but not in a rigid lattice, which is the case in a solid.

During evaporation and boiling the particles acquire enough energy to overcome the forces that hold them together and become a gas.

DIFFERENT STATES OF MATTER WHEN TEMPERATURE DECREASES

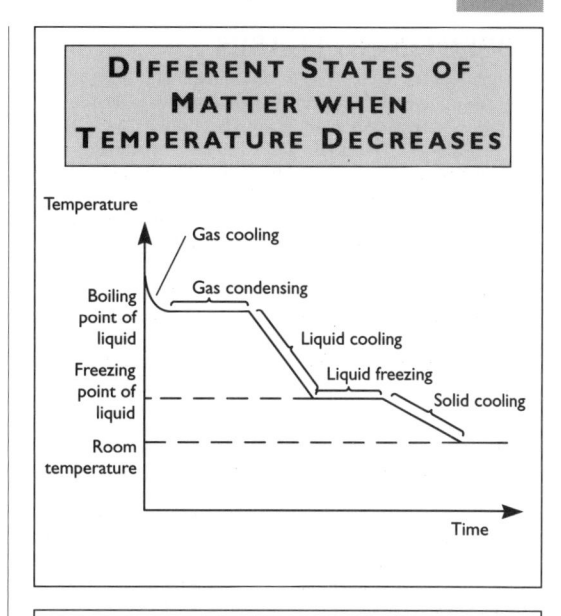

Temperature

- Gas cooling
- Boiling point of liquid
- Gas condensing
- Liquid cooling
- Freezing point of liquid
- Liquid freezing
- Solid cooling
- Room temperature

Time

DIFFERENT STATES OF MATTER WHEN TEMPERATURE INCREASES

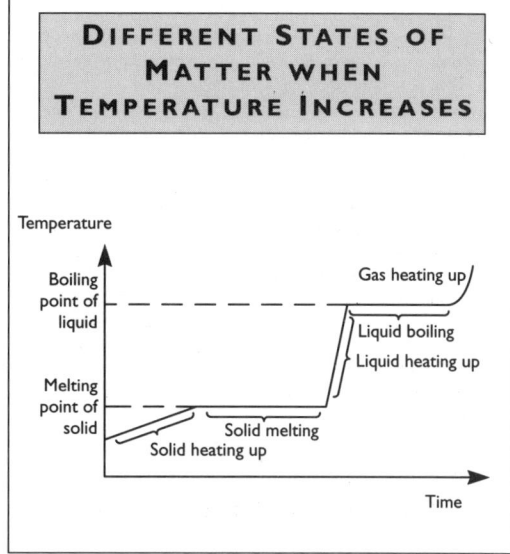

Temperature

- Boiling point of liquid
- Gas heating up
- Liquid boiling
- Liquid heating up
- Melting point of solid
- Solid melting
- Solid heating up

Time

solid the state of matter in which ions or molecules are fixed in position and do not have the freedom to move that atoms, ions and molecules do in a liquid or a gas. The particles vibrate about fixed positions. The atoms, ions and molecules are held together in a lattice of bonds. Movement of atoms, ions and molecules in a solid is only possible when these bonds are partly broken, for example by the application of heat to such an extent that the solid melts.

vapour see gas.

NOBLE (INERT) GASES

The noble gases form a group of almost totally inert gases. They are the elements helium, neon, argon, krypton, xenon and radon - the elements of group 18 of the Periodic Table (see pp. 128-30). The noble gases are all monatomic (q.v.). They do not need to combine with any other atom in order to obtain stable electronic configuration because they already have that configuration. (From shell 3 onwards, eight is the maximum number of electrons that an electron can accommodate.)

The noble gases all occur in the air, but, with the exception of argon, they are all relatively rare. Helium is sometimes found in natural gas and so may be obtained from it; the other noble gases are obtained by fractional distillation of liquid air.

argon a noble or inert gas. Argon - which is odourless - forms nearly one per cent of the atmosphere. Atomic number: 18. Symbol: Ar. (See also p. 131.) Argon is the principal gas to be used to fill electric light bulbs where it prevents the tungsten filament from burning. As argon is readily available in the atmosphere it is a relatively cheap gas to produce commercially. It is, therefore, commonly used where a non-reactive atmosphere is needed - it is used, for example, in welding where welded metal might fail if it were to react with oxygen in the atmosphere. By directing argon onto the weld, oxygen is kept away and the weld is secure.

helium a noble or inert gas. Helium is light and colourless. Because it is so light, helium is used as a non-flammable substitute for inflating balloons. (It is not, however, used in so-called 'hot air' balloons.) Helium mixed with oxygen is given to deep-sea divers in order to prevent the 'bends'. (Nitrogen in air dissolves in the blood as a diver descends and reverses that process when the diver ascends again, forming painful gas bubbles in the capillaries. Helium is used, therefore, in place of nitrogen because it does not act in the same way.) Helium has the smallest molecule of the noble gases - this means that its molecules attract one another least, consequently, helium has the lowest melting and boiling points of any group 18 element. Atomic number: 2. Symbol: He. (See also p. 131.)

inert having few or no active properties.

krypton a rare noble or inert gas. Atomic number: 36. Symbol: Kr. (See also p. 132.) Krypton is used as a filling for light bulbs to prevent the filament of the bulb from burning. The use of krypton - for example in the bulbs found in miners' helmets - allows the filament to be run at a much higher temperature and thus to produce a much brighter light.

lack of reactivity (of noble gases) is due to the inability of the atoms of these elements to transfer or to share electrons. This is because each of the noble gas elements has atoms with a stable electronic configuration. Noble gases, thus, do not need to react with other atoms - not even their own - in order to obtain a stable configuration. They are, therefore, inert with other substances.

No compound of a noble gas was obtained until the 1960s when xenon was found to be capable of reacting with fluorine (which is the most reactive non-metal that is known). The increased reactivity of xenon illustrates that elements at or near the bottom of a group are more capable of losing an electron (or electrons) from the outer shell. The most reactive noble gases are, therefore, krypton, xenon and radon.

monatomic molecule molecules that contain a single atom.

neon a rare noble or inert gas. Neon is colourless. Atomic number: 10. Symbol: Ne. (See also p. 131.) Neon is used in the familiar red illuminated signs and is also employed as a 'starter gas' in sodium street lamps - the characteristic red glow of neon is seen when these lamps are first turned on. Electric current is conducted by the neon and sodium is vaporized as the lamp warms up. When the sodium is fully vaporized it conducts the current. The yellow glow of sodium, when mixed with the red of neon, gives the familiar orange glow of street lamps.

physical properties (of noble gases) include an increase in melting point, boiling point and density down the group, from helium to radon.

radon a noble or inert gas. This radioactive gas is formed in the atomic disintegration of radium. Radon has the largest molecules of any of the noble gases. It has the highest boiling and melting points in the group. Atomic number: 86. Symbol: Rn. (See also p. 134.)

xenon a noble or inert gas. Xenon - which is heavy and colourless - is present in the atmosphere in minute quantities. Atomic number: 54. Symbol: Xe. (See also p. 133.) See also lack of reactivity (of noble gases) - above. Xenon is used to fill light bulbs to prevent the filament from burning. It is particularly used in the bulbs found in lighthouses. Xenon allows the filament to be run at a very high temperature and, therefore, to produce a much more intense light.

PHYSICAL PROPERTIES OF NOBLE GASES

Noble gas	Density/g per litre	Melting point °C	Boiling point °C
helium	0.17	−270	−269
neon	0.84	−248	−246
argon	1.66	−189	−186
krypton	3.46	−157	−152
xenon	5.45	−112	−107
radon	8.90	−71	−62

CHEMISTRY

CHEMICAL BONDS

Although there are only 109 known elements, there are millions of chemical substances found in nature or made artificially. These substances are not simply mixtures of two or more elements: they are chemical compounds, formed or combining two or more elements together in a chemical reaction. The chemical 'glue' that holds these compounds together is chemical bonding.

The properties of compounds vary very widely. Some are highly reactive, others inert; some are solids with high melting points, others are gases. Furthermore, the properties of a compound are very different from those of its constituent elements. To understand how and why these differences arise, we need to understand the different types of chemical bond.

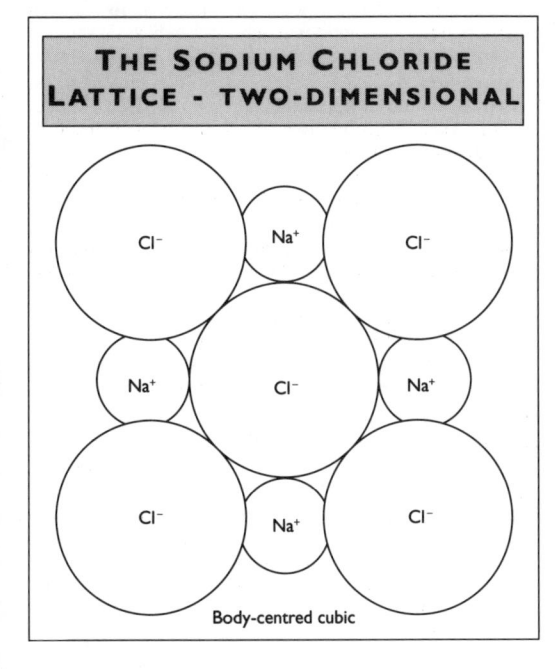

THE SODIUM CHLORIDE LATTICE - TWO-DIMENSIONAL

Body-centred cubic

allotrope an allotropic form; allotropy is the property that certain chemical elements have of existing in two or more different forms. (See covalent bonding, below.)

anion a negatively charged ion, e.g. F^-.

buckminsterfullerene an allotropic form of carbon (q.v.). Fullerenes are hollow clusters of carbon atoms that are joined into geometrical shapes. Before the discovery of the fullerenes in the 1980s and 1990s, elemental carbon was only known to exist in tetrahedral form (diamond) and in a hexagonal form (graphite). The most publicized fullerene is buckminsterfullerene (C_{60}). Nicknamed 'buckyballs', buckminsterfullerene molecules consist of 60 carbon atoms that are bonded together to resemble a soccer ball.

'buckyball' see buckminsterfullerene.

carbon a non-metallic element, which is found in group 4 of the Periodic Table. Atomic number: 6. Symbol: C. (See also covalent bond (below), giant molecules (below) and p. 131.)

Carbon is found in three different forms or allotropes (q.v.) - diamond (q.v.), graphite (q.v.) and fullerene (see buckminsterfullerene; above). All living tissue contains carbon compounds - organic compounds - such as carbohydrates, proteins and fats, without which life would not be possible.

cation a positively charged ion, e.g. Na^+.

colloid a substance composed of insoluble, nondiffusable particles that remain in suspension in a medium of different matter. When a colloid does diffuse it does so slowly.

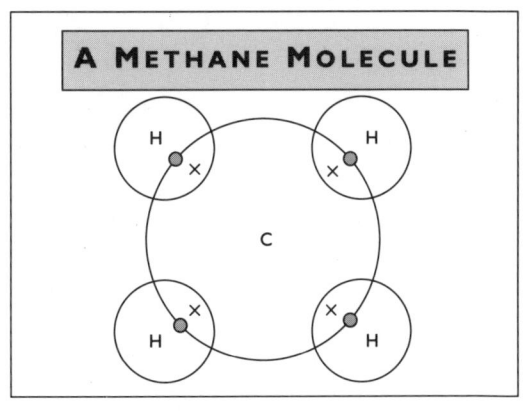

A METHANE MOLECULE

compound a substance that is made of atoms of two or more elements bonded together.

The properties of compounds are quite different from the properties of the elements from which they are made. For example, sodium chloride is a solid and is better known as common salt, which is used in cooking. Sodium, by contrast, is a corrosive metal which reacts violently with water, while chlorine is a poisonous gas which has a characteristic choking smell.

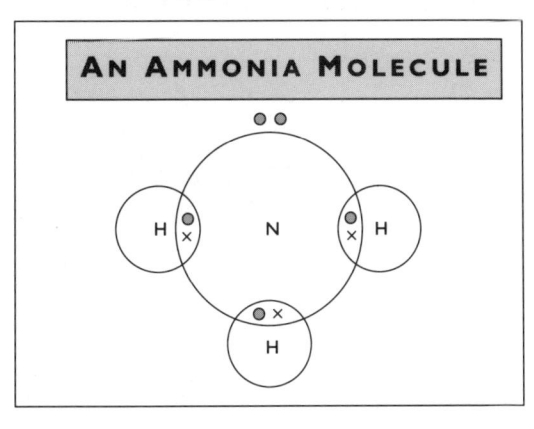

AN AMMONIA MOLECULE

The atoms in a compound may be held together by either ionic bonds (q.v.) or covalent bonds (q.v.).

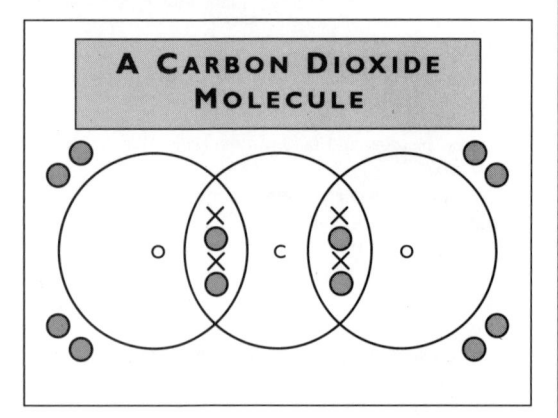

A CARBON DIOXIDE MOLECULE

covalent bonding the type of bond that is formed when two atoms share electrons.

If we bring together two fluorine atoms, each having seven outer electrons (one less than neon), the formation of two ions with the noble-gas configuration is not possible. If, however, they share a pair of electrons - one from each atom - then both achieve the noble-gas configuration and a stable molecule results:

There is a force of attraction between the shared pair of electrons and both positive nuclei, and this is what is known as a covalent bond. The stronger the attraction of the nuclei for the shared pair, the stronger the bond.

An atom of oxygen, having two electrons less than neon, must form two covalent bonds to attain a share in eight outer electrons.

Example: A molecule of water (H_2O), consisting of two hydrogen atoms (H) and one oxygen atom (O), has two covalent O-H bonds.

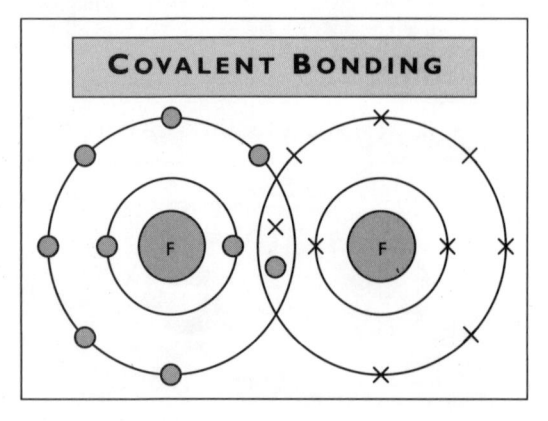

COVALENT BONDING

Another way for oxygen for to achieve the stable noble-gas configuration is to form two bonds with the same atom. Thus two oxygen atoms bond covalently to one another by sharing two pairs of electrons. This is known as double bonding.

Like oxygen, sulphur (now usually known as sulfur) has six outer electrons and again needs to form two bonds to attain a share in eight electrons. There are two ways in which sulphur atoms join together - either in rings of eight atoms (S_8) or in long chains of many atoms bonded together. The different forms in which elemental sulphur exits are known as allotropes. Other elements found in allotropic forms include carbon (graphite, diamond and fullerene) and oxygen (oxygen and ozone).

Atoms of nitrogen (N), containing five outer electrons, need to form three covalent bonds to attain a share in eight outer electrons. This may be done, for example, by forming one bond to each of three hydrogen atoms to give ammonia (NH_3). Another possibility is to form all three bonds to a second nitrogen atom, which produces a nitrogen molecule (N_2), containing a triple covalent bond.

The carbon atom (C), which has four outer electrons, needs to form four bonds to attain the noble-gas configuration. Thus a carbon atom forms one bond to each of four hydrogen atoms to give methane (CH_4). Although carbon is not known to form a quadruple bond to another carbon atom, some other elements, such as the heavy metal rhenium, do form quadruple bonds.

covalently bonded simple molecular elements and compounds (properties of) include low melting points <250°C and low boiling points <500°C. They are non-conductors of electricity in any state, are usually insoluble in water and soluble in organic solvents. (Note, however, that giant molecules include diamond and silica which have very high melting points and boiling points.)

crystal lattice see giant structure.

crystalloid a substance - usually crystallizable - which, when in solution, readily passes through membranes. Crystalloids diffuse quickly.

diamond an allotropic mineral that consists of nearly pure carbon (q.v.) in a tetrahedral crystalline form. (Note that diamond is pure carbon, although the mineral may not be absolutely so.) Diamond is the hardest naturally-occurring mineral known and is used for cutting and abrasives as well as a gem.

double bond see covalent bonding.

electrovalent bonding see ionic bonding.

fullerene an allotropic form of carbon; see buckminsterfullerene.

giant molecule a structure in which a large and indefinite number of atoms are present. These particles may form a crystal lattice in which each particle has a strong attraction for the particles around it, but other structures are also possible, see polymers pp. 156-60.

Although two carbon atoms do not form a quadruple bond (see covalent bond, above) to one another, carbon atoms can combine to form a giant crystal lattice if each atom is bonded to four others by single covalent bonds. This is the structure of diamond (q.v.), one of the allotropes of elemental carbon. Many other elements and compounds exist as giant covalent crystal lattices, including quartz, which is a form of silicon dioxide (SiO_2).

Crystals of these substances contain many millions of atoms held together by strong covalent bonds, so that a large amount of energy is needed to break them. Thus these substances all have high melting points and are hard solids.

giant structure a structure in which a very large number of atoms or ions are present in a regular arrangement, known as a lattice. Ionic compounds, metals and giant molecular elements and compounds have giant structures.

graphite an allotropic soft black hexagonal form of carbon (q.v.), which is used in lubricants and in pencils.

hydrogen bond a force of attraction between hydrogen and oxygen, nitrogen or fluorine atoms in neighbouring molecules.

Some small molecules have much higher melting points and boiling points than would be expected on the basis of their size. One such example is water (H_2O), which has about the same mass as a neon atom but has a much higher melting point. There must therefore be unusually strong intermolecular forces between the water molecules. Although the oxygen and hydrogen atoms share a pair of electrons in a covalent bond (q.v.), the oxygen atom exerts a stronger 'pull' on these electrons and so becomes electron-rich, leaving the hydrogen atom electron-poor. As a result, there is a force of attraction between hydrogen and oxygen atoms on neighbouring molecules. This is known as hydrogen bonding.

As well as accounting for the surprisingly high melting point of water, hydrogen bonding is responsible for the rigid open structure of ice crystals, and is very important in influencing the structures and properties of biological molecules.

Although hydrogen bonds are stronger than van der Waals forces, they are still much weaker than covalent bonds.

intermolecular forces forces of attraction between molecules; see hydrogen bond and van der Waals force.

ion an atom or group of atoms that has become electrically charged by either gaining or losing one or more electrons. Atoms tend to lose or to gain electrons to produce an ion with the same stable configuration as a noble gas (see p. 138). See also ionic bonding, cation and anion.

ionic bonding (or **electrovalent bonding**) a chemical bond that occurs because of electrostatic attractive forces between positively and negatively charged ions.

The atoms of the element neon have eight electrons in their outer shell, with the electron arrangement 2.8. This arrangement is very stable and neon is not known to form a chemical bond with any other element. An atom of the element sodium (Na) has one more electron than neon (configuration 2.8.1), while an atom of the element fluorine (F) has one electron less than neon (configuration 2.7). If an electron is transferred from a sodium atom to a fluorine atom, two species are produced with the same stable electron configuration as neon. Unlike neon, however, the species are charged and are known as ions. The sodium atom, having lost a (negative) electron, has a net positive charge and is known as a cation (q.v.; written Na^+), while

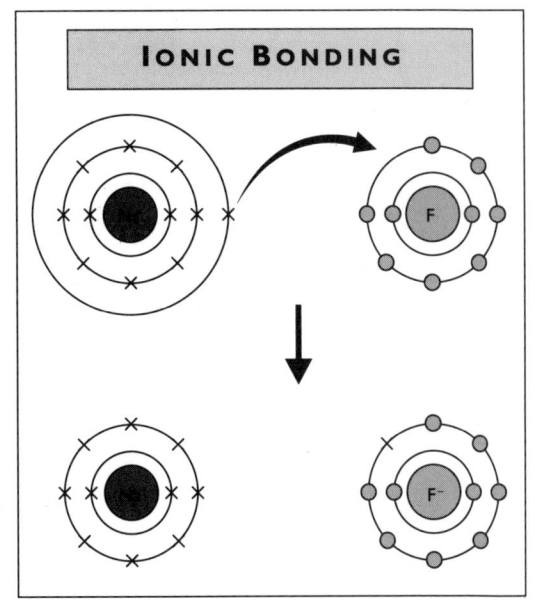

IONIC BONDING

the fluorine atom, having gained an electron, has a net negative charge and is called a fluoride anion (q.v.; written F^-).

When oppositely charged ions such as Na^+ and F^- approach one another, there is a strong attraction between them; a large amount of energy is released - the same amount of energy as would have to be supplied in order to separate the ions again. This force of attraction is called an ionic (or electrovalent) bond. The energy released more than compensates for the energy input required to transfer the electron from the sodium atom to the fluorine atom. Overall there is a net release of energy and a solid crystalline compound - sodium fluoride (NaF) - is formed.

Atoms that have two more electrons than the nearest noble gas (such as magnesium, configuration 2.8.2) or two less (such as oxygen, configuration 2.6) also form ions having the noble-gas configuration by transfer of electrons - in this case Mg^{2+} and O^{2-}. The ionic compound magnesium oxide (MgO) has the same arrangement of ions as NaF, but since the ions in MgO have a greater charge, there is a stronger force between them. Thus more energy must be supplied to overcome this force of attraction, and the melting point of MgO is higher than that of NaF.

Although the ions are fixed in position in the crystal, they become free to move when the solid is melted. As a liquid, therefore, the compound becomes electrolytic and is able to conduct electricity.

Many other complex ionic structures are known. The formula of any ionic compound can be worked out by balancing the charges of its ions.

Examples:

Mg^{2+} and F^- form MgF_2

Na^+ and O^{2-} form Na_2O.

ionic compounds (properties of) include high melting points >250°C and high boiling points >500°C. As solids ionic compounds are non-conductors of electricity. In a molten state ionic compounds are good conductors of electricity. They are usually soluble in water, but are normally insoluble in organic solvents.

large molecule a type of macromolecule which is characterized by the very great number of atoms contained. Large molecules have thousands rather than billions of atoms per molecule. Polymers (proteins, starches and plastics) are large molecules.

Large molecules have lower melting points than giant molecular structures, but they have higher melting points than simple molecules. Large molecules do not conduct electricity (because they lack ions). They are not particularly soluble in water.

macromolecule a molecule containing up to billions of atoms.

quadruple bond see covalent bond.

radical a group of atoms that acts as a single atom. A radical goes through a chemical reaction unchanged.

triple covalent bond see covalent bond.

valency the number of bonds which an atom forms with another atom.

The valency of an element is the number of electrons that it needs to form a compound or a radical. These electrons may be lost, or gained, or shared with another atom.

Some elements always have the same valency; for example, the valency of sodium is always 1. This means that - in compounds such as sodium chloride - it gives one electron away when it forms the sodium ion (q.v.) Na^+. Oxygen, on the other hand, always has a valency of 2. Thus oxygen accepts two electrons when it forms the oxide ion (q.v.) O^{2-} or shares two of carbon's electrons in a covalent compound such as CO_2 (carbon dioxide).

Some elements (for instance, transition elements) have more then one valency. Copper, for example, has a valency of 1 or 2. Cobalt has a valency of 2 or 3, while iron has a valency of either 2 or 3.

The lists below give the valencies of some common atoms and radicals and the ions that they form.

van der Waals forces a weak intermolecular force whose strength depends upon the number of electrons in the molecule involved.

Two neon atoms do not form covalent bonds with one another because of their stable configuration of electrons. There are, however, weak forces of attraction between two neon atoms. This can be shown because, when neon gas is compressed or cooled, it eventually turns into a liquid in which the atoms are

VALENCY

1	2	3
Ammonium (NH_4^+)	Carbonate (CO_3^{2+})	Aluminium (Al^{3+})
Bromine (Br^-)	Calcium (Ca^{2+})	Iron (Fe^{3+})
Chlorine (Cl^-)	Copper (Cu^{2+})	Phosphate (PO_4^{3-})
Copper (I) (Cu^+)	Iron (Fe^{2+})	
Fluorine (F^-)	Lead (Pb^{2+})	
Hydrogen (H^+)	Magnesium (Mg^{2+})	
Hydroxide (OH)	Oxygen (O^{2+})	
Iodine (I^-)	Sulphate (SO_4^{2-})	
Lithium (Li^+)	Sulphur (S^{2-})	
Nitrate (NO_3)	Zinc (Zn^{2+})	
Potassium (K^+)		
Silver (Ag^+)		
Sodium (Na^+)		

weakly attracted to one another. These weak forces are called van der Waals forces.

Bromine (Br) is made up of large covalently bonded molecules that - because they have more electrons per molecule than neon has - have much stronger van der Waals forces between them than exist in atoms of neon. Thus at room temperature bromine exists as a mixture of liquid and vapour. However, the forces *between* the bromine molecules are much weaker than covalent bonds, so that - while it is easy to separate the bromine molecules from one another and vaporize the liquid - it requires much more energy to separate the bromine atoms by breaking the covalent bond between them. The forces of attraction between polar molecules are also known as van der Waals' forces.

weak forces see van der Waals forces.

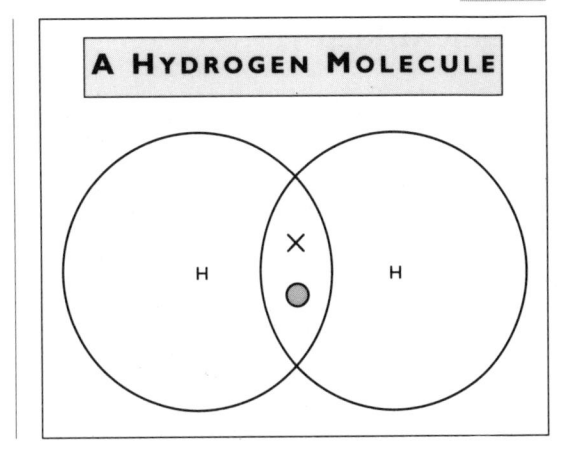

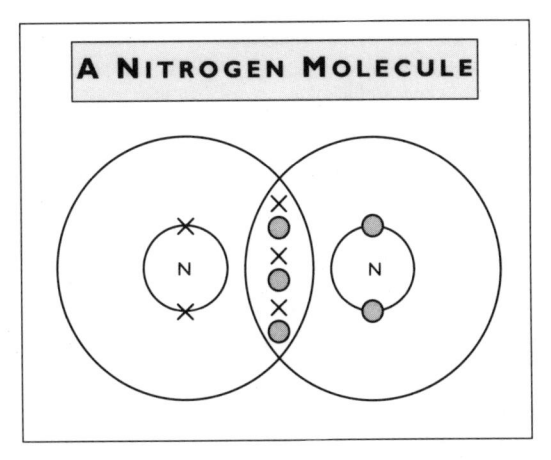

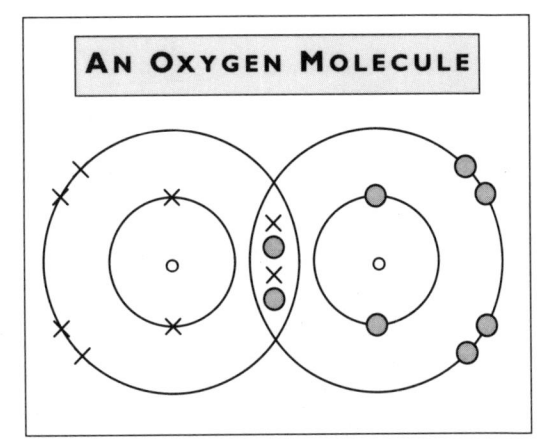

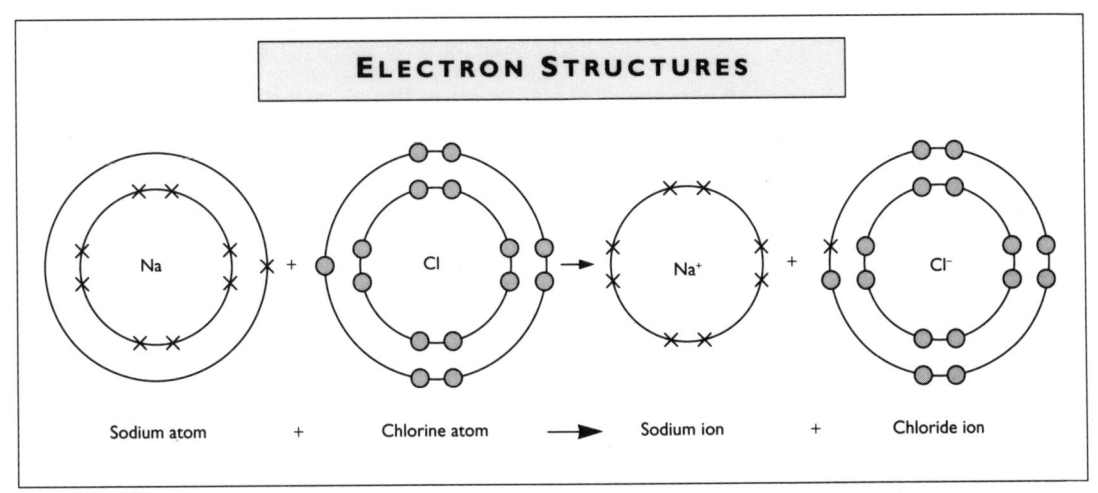

A Hydrogen Molecule

A Nitrogen Molecule

An Oxygen Molecule

Electron Structures

Sodium atom + Chlorine atom → Sodium ion + Chloride ion

CHEMICAL REACTIONS

Chemical reactions are the means by which new substances are formed from old ones. Among the chemical reactions occurring everywhere around us are the changes that take place when fuels are burnt, the industrial means by which metals are extracted from their ores, and the processes controlling life itself. Since chemistry is centrally concerned with the means by which substances change, the study of reactions lies at the heart of the subject.

An essential characteristic of all chemical reactions is that there is an exchange of energy between the reacting system and the surroundings. For instance, a burning fuel liberates heat to the surroundings, whereas the surroundings must supply heat to bring about the decomposition of calcium carbonate:

$$CaCO_3(s) \rightarrow CaO(s) + CO_2(g)$$

Similarly, the chemical reaction that occurs in a battery liberates electrical energy. When the battery needs recharging, however, electrical energy has to be supplied to bring about the desired chemical change.

acid a substance that tends to donate protons (H⁺s) (the Brønsted-Lowry definition).

Example: Hydrogen chloride (HCl) readily dissolves in water. The HCl donates its proton to water in a reversible reaction:

$$HCL(aq) + H_2O(l) \rightleftharpoons H_3O^+(aq) + Cl^-(aq)$$

The products of the forward reaction are ions:
H_3O (the oxonium ion) and Cl^- (the chloride ion).
Many non-metal oxides form acids when added to water.

Example: Sulphur trioxide gas (SO_3) reacts with water to form sulphuric acid (H_2SO_4) - the reaction that occurs in the formation of acid rain.

alkali a soluble base.

balanced chemical equation an equation that gives information concerning a reaction. By convention the reactants appear on the left-hand side and the products appear on the right-hand side. The states of matter are indicated by the abbreviations found on p. 137; the elements are represented by the symbols found on pp. 131-35. See also law of constant composition.

A balanced equation is a quantitative statement about the chemical reaction concerned. Such an equation - combined with the mole concept (q.v.) - enables us to predict how much of a product will be formed from a given mass of reactants. This provides valuable information that can be put to use, for example, in industrial production processes and in the analysis of chemical samples of unknown composition.

base a proton acceptor. Bases are capable of accepting protons off oxonium ions present in a solution. A good example of a Brønsted-Lowry base is the hydroxide ion OH⁻, which reacts reversibly with the oxonium ion to produce two molecules of water:

$$H_3O^+(aq) + OH^-(aq) \rightleftharpoons 2H_2O(l)$$

Examples of bases include sodium and potassium hydroxides (NaOH and KOH), which liberate aqueous hydroxide ions in solution. Many metal oxides, such as calcium oxide (CaO; 'lime'), are also basic. CaO reacts violently with water to form calcium hydroxide - $Ca(OH)_2$; 'slaked lime'. Soluble bases are known as alkalis.

basic pertaining to a base.

Brønsted-Lowry definition (of acids) see acid.

catalyst a substance that is chemically unchanged at the end of a reaction or at equilibrium but whose presence serves to alter its rate. (NB. Catalysts *do* change during the course of a chemical reaction, but go back to their original composition at the end of the reaction or at equilibrium.)

cation a positive ion.

chemical equation see balanced chemical equation.

disproportionation reaction a reaction in which a single chemical species is simultaneously oxidized and reduced.

Example: Copper (I) sulphate (Cu_2SO_4) dissolves in water to produce copper metal and copper (II) sulphate:

$$Cu_2SO_4(aq) \rightarrow Cu(s) + CuSO_4(aq)$$

Here one Cu^+ ion is oxidized to Cu^{2+}, while the second is reduced to Cu metal.

equilibrium see reaction equilibrium.

indicator a dye upon which the effects of acids and alkalis can be observed. Indicators are used as detectors of acids and alkalis. The best-known is litmus.

ion electrically charged species.

law of constant composition states that matter cannot be created or destroyed during a chemical reaction. Thus in the reaction of potassium with water (see reactants; below), the number of atoms of potassium, hydrogen and oxygen (calculated by multiplying each symbol in the equation by the number placed before the chemical formula and the number immediately after the symbol) is the same before and after the reaction, and the equation is said to be balanced.

litmus an indicator (q.v.); a dye derived from lichen, which is turned red by acids and blue by alkalis.

molar quantity the number of moles (q.v.).

mole a measure of the amount of substance, based on the atomic theory of matter (see pp. 120-21). A mole is defined as the amount of substance that contains as many entities as there are carbon atoms in 12 grams of the isotope carbon-12. The mass of 1 mole of an entity (atom, molecule, ion, etc.) is the relative mass of that entity, expressed in grams. Every chemical compound has a fixed relative molecular mass, which is determined by the relative atomic masses of its constituent elements (see pp. 131-35), so that molar quantities of any substance can be found using simple arithmetic. The mole concept has important industrial applications.

Example: Titanium (Ti) is one of the most useful metals - it is light, strong and durable, with a very high melting point, making it a vital material in aircraft manufacture. Titanium is usually produced by converting its oxide into the chloride ($TiCl_4$), which is then reduced by sodium (Na), in an inert atmosphere:

$$TiCl_4 + 4Na \rightarrow Ti + 4NaCl$$

Sodium is an expensive metal, so costs have to be taken into account in the manufacturing process. Suppose a manufacturer needs to make 10kg (10 000g) of titanium. To find the number of moles, we divide this mass by the relative atomic mass (RAM) of titanium, which is approximately 48. Thus the number of moles of titanium = 10 000/48 = 208.3.

From the equation we see that four moles of sodium produce one mole of titanium. In this case this means 208.3 × 4, or 833.2 moles of sodium are needed. To find out the mass of sodium needed we multiply the number of moles by the RAM of sodium (23). This gives us 833.2 × 23g or 19.163kg. Thus from the equation we can reckon that approximately 20kg of sodium are needed to manufacture 10kg of titanium.

neutralization see salt (below).

oxidation a class of reactions that includes all combustion processes such as those occurring when fuels burn in air, as well as the reactions that cause metals to corrode in air.

Example: Magnesium metal burns with an incandescent white flame in air because of a vigorous reaction with oxygen, forming magnesium oxide:

$$2Mg(s) + O_2(g) \rightarrow 2MgO(s)$$

The product in this case is an ionic solid, which we could write more specifically as $Mg^{2+}O^{2-}$. During the reaction, magnesium loses two electrons to form the cation Mg^{2+}; the electrons are accepted by oxygen. The transfer of electrons between chemical entities is a common process in many chemical reactions, so the term oxidation has come to possess a wider meaning than that implying solely the addition of oxygen atoms to an element or compound. As in the case of the formation of magnesium oxide, oxidation means the loss of electrons by an entity; the opposite process is reduction.

precipitation reactions involve the formation of an insoluble material from the reaction of two soluble substances on mixing their two solutions.

Example: Silver nitrate ($AgNO_3$) dissolves readily in water, producing a colourless solution of aqueous silver (Ag^+) and nitrate (NO_3^-) ions. If solutions of silver nitrate and sodium chloride are mixed, a white turbidity (cloudiness) instantly forms; this is due to the precipitation of fine particles of highly insoluble silver chloride. The precipitate gradually accumulates at the bottom of the vessel, leaving colourless sodium nitrate in solution:

$$NaCl(aq) + AgNO_3(aq) \rightarrow AgCl(s) + NaNO_3(aq)$$

The overall reaction is one in which ions are exchanged between partners. It is possible to predict the outcome of such precipitation reactions from our knowledge of the solubilities of the various species involved.

products the substances that result from the rearrangement of atomic constituents during a chemical reaction.

proton a positively-ionized hydrogen atom (H^+).

reactants substances that react together during a chemical reaction. The atomic constituents of reactants are rearranged to produce other substances (products). Thus, for example, during the reaction of potassium (K) with water (H_2O), potassium hydroxide (KOH) and hydrogen gas (H_2) are formed. This information can be represented by a chemical equation:

$$2K(s) + 2H_2O(l + aq) \rightarrow 2KOH(aq) + H_2(g)$$

So much heat is liberated during the reaction of water and potassium that the highly flammable hydrogen gas frequently ignites.

reaction equilibrium the end of a reaction. The point at which there is no further change in the amount of products formed or reactants destroyed. At equilibrium, there may still be appreciable amounts of reactants present.

Example: When ethanoic acid (CH_3CH_2OH) and ethanol (CH_3COOH) react, the products of that reaction can themselves react to give back ethanoic acid and ethanol so that a mixture of reactants and products is obtained.

$$CH_3COOH(l) + CH_3CH_2OH(l) \rightleftharpoons CH_3COOCH_2CH_3(l) + H_2O$$

The two half-headed arrows indicate a reversible reaction.

At equilibrium, the rate of the forward reaction equals the rate of the backward reaction, and the quantities of all the substances become constant.

redox reaction a reaction in which oxidation (q.v.) and reduction (q.v.) take place. Thus in the equation:

$$2Na(s) + Cl_2(g) \rightarrow 2NaCl(s)$$

the sodiums are oxidized once they each lose an electron to the chlorines, which are thereby reduced. If an electric current is passed through an electrolyte such as an aqueous solution of copper (II) chloride ($CuCl_2$), a redox process known as electrolysis occurs. (Electrolysis is the basis of electroplating.) Positively charged Cu^{2+} ions are attracted to the negative electrode (cathode), where they take up two electrons each and are thereby reduced to copper metal, which is deposited at the cathode. The negative chloride ions (Cl^-) are attracted to the positive electrode (anode) where they each lose an electron and are thereby oxidized to atoms before pairing off to give chlorine gas (Cl_2).

reduction the process of losing oxygen or gaining electrons.

relative molecular mass the mass of a molecule on a scale where the mass of an atom of carbon-12 is exactly 12 units.

RMM the abbreviation for relative molecular mass.

salt a compound formed by the replacement of some or all of the hydrogens in an acid by a metal or ammonium ion.

Thus the reactions between the base (q.v.) sodium hydroxide and sulphuric acid yield two salts, sodium hydrogen sulphate and sodium sulphate.

$$NaOH(aq) + H_2SO_4(aq) \rightarrow NaHSO_4(aq) + H_2O(l)$$
$$2NaOH(aq) + H_2SO_4(aq) \rightarrow Na_2HSO_4(aq) + 2H_2O(l)$$

stoichiometry the numerical proportions in which substances combine to form the products of a chemical reaction.

strong acid/base an acid/base that is ionized to a large extent in aqueous solutions. Thus hydrochloric acid and sodium hydroxide are both strong.

$$HCl(aq) \rightarrow H^+(aq) + Cl^-(aq)$$
$$Na^+OH^-(aq) \rightarrow Na^+(aq) + OH^-(aq)$$

transfer (of electrons) see oxidation and reduction.

weak acid/base an acid/base that is ionized to an small extent in aqueous solution. Thus ethanic acid and ammonia are weak.

$$CH_3COOH(aq) \rightleftharpoons CH_3COO^-(aq) + H^+(aq)$$
$$NH_3(aq) + H_2O(l) \rightleftharpoons NH_4 + (aq) + OH^-(aq)$$

The larger backward pointing half-headed arrow indicates that, at equilibrium, most of the material present is in ionized form.

METALS

Metals are usually defined by their physical properties, such as strength, hardness, lustre, conduction of heat and electricity, malleability and high melting point. They can also be characterized chemically as elements that react with (or whose oxides react with) acids, usually to free positively charged metal ions. By either definition, more than three quarters of the known elements can be classified as metals. They occupy all but the right-hand top corner of the Periodic Table. A few elements - such as germanium, arsenic and antimony - are borderline, and are often classed as metalloids.

Given such a large number of metals, it is not surprising that some of them have rather untypical properties. For instance, mercury is a liquid at room temperature, and - with the exception of lithium - all the alkali metals (see pp. 147-48) melt below 100°C (212°F). The alkali metals are also quite soft - they can easily be cut with a knife - and extremely reactive: rubidium and caesium cannot be handled in air and may react explosively with water.

alkali metals see pp. 147-48.

alkaline earth metals see pp. 147-48.

conduction of heat and electricity characterizes metals. This property is due to their unique type of bonding. The solid metals behave as though they were composed of arrays of positively charged ions, with electrons free to move throughout the crystalline structure of the metal. This results in high electrical conductivity. The conduction of heat can also be seen in terms of the motion of electrons, which becomes faster as temperature rises. Since the electrons are mobile, the heat can be conducted readily through the solid.

The majority of metals are good conductors of electricity, but germanium and tin (in the form stable below 19°C/64°F) are semiconductors.

corrosion see tarnishing.

extraction of metals is thought to have begun with the accidental extraction of copper. The discovery of copper is thought to have occurred when pieces of the metal ore used in early fireplaces came into contact with the hot charcoal, so releasing the metal. Essentially the same process (smelting) is used in modern furnaces. Any of the metals from manganese (Mn) to zinc (Zn) in the Periodic Table can be obtained by roasting their oxides with coke at temperatures of up to about 1600°C (2912°F).

The ores of the lighter, more reactive metals cannot be reduced by carbon at practical temperatures, because their ions are more strongly bonded in the ore. These metals are usually obtained by electrolysis or by the reaction of their compounds with an even more reactive metal. For instance, aluminium is obtained by the electrical reduction of aluminium oxide in a mixture of cryolite and calcium fluoride at about 950°C (1742°F). On the other hand, titanium is obtained by converting its oxide into the chloride, which is then reduced with elemental sodium or magnesium.

mechanical strength is the reason why many metals are used for purposes such as construction. However, most pure metals are actually quite soft. In order to obtain a tough hard metal, something else has to be added. For instance, the earliest useful metal was not copper but bronze, which is copper alloyed to tin. Similarly, iron is never used in the pure form but as some form of steel.

The softness of a metal results from a lack of perfection in the crystal framework formed by its ions. Even when the most rigorous conditions are employed, it is impossible to grow any material in perfect crystalline form. There will always be some ions in the wrong place or missing from their proper place. When solidification occurs fairly rapidly, as when a molten metal is cooled in a mould, even more defects occur. Under bending or shearing stresses, such defects can move and allow the metal to change shape easily. When the foreign atoms of an alloying element are present, they usually have a different size from those of the host and cannot easily fit into the crystal lattice. They therefore tend to site themselves where the lattice is irregular, i.e. where the defects are. The effect of this is to prevent the defects from moving, and so to increase the rigidity of the metal.

metalloids the borderline elements - for instance, germanium, arsenic and antimony - that are not true metals.

natural occurrence (of metals) is mostly as oxides. Some metals - mostly the heavier ones such as mercury and lead - occur as sulphides, and the more reactive ones as chlorides and carbonates. Only a few - the noble and coinage metals - are found in the metallic state. Their chemical unreactiveness makes them useful in coinage and jewellery, since they do not corrode.

A few metals do not occur naturally at all, because they are radioactive and have decayed away. Technetium and all the elements with higher atomic numbers than plutonium (Pu, 94) are made by the 'modern alchemy' of nuclear reactors or accelerators, while promethium is found only in minute amounts as a product of the spontaneous fission of uranium. The very heaviest elements have been obtained only a few atoms at a time, and are immensely radioactive.

smelting see extraction of metals.

tarnishing (and **corrosion**) occur in nearly all metals, which are prone to surface oxidation, i.e. the surface reacts with oxygen or other components of the atmosphere. The major exceptions are the coinage metals and those in group 10 in the Periodic Table, and even these react with sulphur compounds and turn black. All other metals should, in principle, react with moisture and the oxygen in the air, yet some corrode badly and others appear to be inert. In fact, they all oxidize, but in many cases a thin layer of oxide adheres firmly to the metal surface and prevents further reaction. This is the case with aluminium and titanium. On the other hand, iron forms porous oxides that readily break away, allowing corrosion to continue.

transition metals are not a group of the Periodic Table - they do not all have the same number of electrons in the outer shell of their atoms. They are rather a 'family' of metals - a block of metals in the middle of the Periodic Table which share certain properties in common. They have special properties (coloured compounds, variable oxidation state, etc.).

Modern texts no longer consider the terms 'transition metals' and 'd-block elements' to be synonymous.

ALKALINITY, THE ALKALI METALS & THE ALKALINE EARTH METALS

The alkali metals are the most reactive metals in the Periodic Table (see pp. 130-35). They are the elements of group 1 of the Periodic Table. These elements - lithium, sodium, potassium, rubidium, caesium and francium - are soft metals. Their softness and low melting point are both the result of the weakness of the metallic bonding in these elements. Their low melting point is a characteristic which - along with their low density - makes them quite distinct from other metals. The alkali metals can be cut easily with a knife to reveal a shiny surface which quickly tarnishes. The elements of group 1 are called alkali metals because of the alkalis that are formed when they are in reaction with either air or water (see below). The pure metals themselves are not alkalis.

The alkaline earth metals are the elements of group 2 - beryllium, magnesium, calcium, strontium, barium and radium. Of these elements, calcium and magnesium are the most common. The alkaline earth metals are not so reactive as the alkali metals. They are harder than the alkali metals - they are not so easy to cut with a knife - and also have higher melting points than the elements in group 1. As with all groups of metals, the melting points of the metals in group 2 decrease going down on the Periodic Table. By contrast, their density increases down the group.

acidity see pH (below).

alkali a base that is soluble in water. They are usually metal hydroxides, such as sodium hydroxide.

Properties of alkalis:

Alkalis are corrosive.

Alkalis turn red litmus blue.

In solution, they have a pH of over 7; see pH (below).

Alkalis neutralize acids.

Alkalis react with acids and, in so doing, produce only water and a salt.

barium an alkaline earth metal. Atomic number: 56. Symbol: Ba. (See p. 133.)

beryllium an alkaline earth metal. Atomic number: 4. Symbol: Be. (See p. 131.)

caesium an alkali metal, Atomic number: 55. Symbol: Cs. (See p. 133.)

calcium an alkaline earth metal. Atomic number: 20. Symbol: Ca. (See p. 131.)

compounds of alkali metals are formed when alkali metals react strongly with halogens (see p. 149) or with oxygen. As a result, the compounds alkali metal halides or alkali metal oxides are formed. In these reactions the alkali metals, atoms lose their single outer shell electron.

The chloride compounds - for example, LiCl (lithium chloride), NaCl (sodium chloride), KCl (potassium chloride) - are all ionic salts. These can be electrolysed when in the molten state, the alkali metal being given at the cathode and the chlorine being given at the anode. (Electrolysis initiates the decomposition.)

The hydroxide compounds - for example, LiOH (lithium hydroxide), NaOH (sodium hydroxide), KOH (potassium hydroxide) - are all strong alkalis.

compounds of alkaline earth metals are formed when alkaline earth metals react with halogens or oxygen. However, they react less vigorously than the alkali metals do with halogens or oxygen.

francium the most reactive of the alkali metals. Atomic number: 87. Symbol: Fr. (See p. 134.) Francium is a synthetic element and has only been made in very small quantities. Its reactions, however, have been just as those predicted from an observation of the properties of the other elements within the group.

ionic salt a salt containing ionic bonds (see compounds of alkali metals (above) and pp. 139-43).

kalium an alternative name for potassium.

lithium the least reactive of the alkali metals. Atomic number: 3. Symbol: Li. (See p. 131.)

magnesium a light silvery-coloured alkaline earth metal. Atomic number: 12. Symbol: Mg. (See p. 131.)

pH scale is a measure of acidity or alkalinity. The lower the value on the pH scale, the more acidic the solution i.e. the greater the concentration of hydrogen ions in it. By contrast, the higher the value on the pH scale, the more alkaline the solution. A neutral solution has equal concentrations of hydrogen and hydroxide ions and has a pH of 7.

potassium (or **kalium**) an alkali metal. Atomic number: 19. Symbol: K. (See p. 131.)

radium an alkaline earth metal. Atomic number: 88. Symbol: Ra. (See p. 134.)

reactions of the alkali metals include the following examples.

The alkali metals are so reactive that - in laboratory conditions - they must not be allowed to contact other elements with which they react. They are, therefore, normally stored under a liquid hydrocarbon - for example, paraffin oil - to prevent them from making contact with water, carbon dioxide and oxygen. Alkali metals react with all non-metals (except for the noble gases) and with most compounds that contain these non-metals.

The reactivity of the alkali metals increases down the group - thus lithium is the least reactive of the alkali metals and francium is the most reactive of the group.

The reaction of alkali metals with water is distinctive. Sodium, lithium and potassium float on water because they are less dense than water.

Sodium will react with water in the following manner:

If a small piece of sodium is placed in water it will immediately melt as the reaction takes the sodium above its melting point (98°C) - in so doing it forms a small ball of metal which will move about on the surface of the water producing a hissing noise. The reaction may finish with a small bang. What has happened is that the sodium has reacted with the water to produce sodium hydroxide and the sodium hydroxide itself dissolves in the water. At the same time, hydrogen gas is given off.

The reaction is:

$2Na(s) + 2H_2O(l) + aq \rightarrow 2NaOH(aq) + H_2(g)$
sodium + water $\rightarrow$ sodium hydroxide + hydrogen

Lithium will react with water in the following manner:
Lithium is harder than sodium and has a higher melting point. For this reason, lithium does not melt in a similar experiment. It does, nevertheless, float. Lithium is less reactive than sodium but it will produce an alkaline solution of lithium hydroxide and also hydrogen gas, as in the experiment using sodium.

Potassium will react with water in the following manner:
Potassium is softer than sodium and has a lower melting point. For this reason, potassium will melt, forming a floating globule on the water - much in the same manner as sodium. Potassium is more reactive than sodium and will, in fact, appear to catch fire. What is happening is that the hydrogen which is given off burns with a lilac flame. Potassium will produce an alkaline solution of potassium hydroxide (KOH) and also hydrogen gas.

Sodium will react with air in the following manner:
If heated, sodium will burn in air with a yellow flame. In so doing, sodium oxide is formed.
The reaction is: $4Na(s) + O_2(g) \rightarrow 2Na_2O(s)$
sodium + oxygen $\rightarrow$ sodium oxide

If water is present, the residue may be tested with litmus paper or another indicator - this will reveal the alkaline nature of the sodium hydroxide (formed when the oxide reacts with water).

Lithium will react with air in the following manner:
Lithium will burn in air to produce the alkaline oxide.

Potassium will react with air in the following manner:
Potassium will burn in air to produce the alkaline oxide.

Lithium will react with chlorine in the following manner:
Lithium will burn in chlorine. The result will be a white solid, lithium chloride.

Potassium will react with chlorine in the following manner:
Potassium will burn in chlorine. The result will be a white solid, potassium chloride.

Sodium will react with chlorine in the following manner:
Sodium will burn vigorously in chlorine with a bright yellow flame.
The reaction is: $2Na(s) + Cl_2(g) \rightarrow 2NaCl(s)$
sodium + chlorine $\rightarrow$ sodium chloride

reactions of the alkaline earth metals
In reaction, the group 2 metals lose two electrons from their outer shell and, in so doing, gain a double positive charge.
Compared with the elements of group 1 (the alkali metals), the alkaline earth metals all tarnish more slowly.
Their reactions with water are also much slower than those of the elements in group 1. In reaction with water, group 2 elements produce hydrogen and a metal oxide or hydroxide.
Alkaline earth metals are, nevertheless, reactive and - except for magnesium and beryllium - have to be kept under, for example, paraffin, to avoid reactions with water vapour, oxygen or carbon dioxide.

Calcium will react with air in the following manner:
Calcium, a very reactive metal, is kept under oil to reduce its reaction with air or water vapour. Calcium reacts quickly in air, burning with a red flame to produce a white oxide coating.
The reaction is: $2Ca(s) + O_2(g) \rightarrow 2CaO(s)$
calcium + oxygen $\rightarrow$ calcium oxide

Calcium will react with water in the following manner:
If a small piece of calcium is placed in water it will sink. As it does so, a steady stream of gas bubbles (hydrogen) will be produced and a white residue (calcium hydroxide) will appear in the water.
The reaction is: $CaO(s) + H_2O(l) \rightarrow Ca(OH)_2(s)$
calcium oxide + water $\rightarrow$ calcium hydroxide

If the resulting solution is then filtered and the filtrate is shaken with some carbon dioxide, a milky precipitate (calcium carbonate) will be formed. The reaction is:

$Ca(OH)_2(aq) + CO_2(g) \rightarrow CaCO_3(s) + H_2O(l)$
calcium + carbon $\rightarrow$ calcium + water
hydroxide dioxide carbonate

Calcium oxide will react with water in the following manner:
Calcium oxide - a base - reacts with water to produce its hydroxide, which is an alkali (commonly known as slaked lime).

Magnesium will react with water in the following manner:
Magnesium - a light silvery metal - will not react with unheated water unless it is in a finely powdered form when the greater surface area of the powder will aid a reaction. The reaction will, however, be very slow - taking days rather than minutes.

A more rapid reaction can be obtained by combining magnesium with steam. The reaction is:
$Mg(s) + H_2O(g) \rightarrow MgO(s) + H_2(g)$
magnesium + steam $\rightarrow$ magnesium oxide + hydrogen

Magnesium will react with air in the following manner:
Magnesium will burn in either air or oxygen to form magnesium oxide.
The reaction is: $2Mg(s) + O_2(g) \rightarrow 2MgO(s)$
magnesium + oxygen $\rightarrow$ magnesium oxide

Strontium will react with water in the following manner:
Strontium - which is below calcium in group 2 of the Periodic Table - would be expected to be more reactive than calcium. This is the case in its reaction with water. Strontium reacts quickly with water, producing a fast stream of hydrogen bubbles. The result is a cloudy alkaline solution.

Strontium will react with air in the following manner:
Strontium burns in air more vigorously than either magnesium or calcium.

reactive metals see compounds of alkali metals.

rubidium an alkali metal. Atomic number: 37. Symbol: Rb. (See p. 132.)

salt a compound formed when the hydrogen of an acid is either totally or partially replaced by a metal.

slaked lime see reactions of the alkaline earth metals.

sodium an alkali metal. Atomic number: 11. Symbol: Na. (See reactions with alkali metals (above) and p. 131.)

strontium an alkaline earth metal. Atomic number: 38. Symbol: Sr. (See p. 132.)

HALOGENS

Halogens are reactive non-metals of group 17 - fluorine, chlorine, bromine and iodine (plus astatine, see below). As a group they share certain characteristics (see below).

To achieve a noble gas electron structure of eight electrons in its outer shell, each halogen atom must accept an electron. This is accomplished by either accepting an electron by transfer from a metal atom or by accepting a share of an electron from another non-metal atom. An electron that is being accepted by a halogen will be most strongly held if it becomes part of an electron shell that is close to the nucleus, which strongly attracts electrons. The halogen with an outer electron shell nearest to the nucleus is fluorine, which is the most reactive halogen and will readily form compounds. Reactivity decreases down the group as the outer electron shells get further away from the nucleus.

All of the halogens are oxidizing agents; their oxidizing power - like their chemical reactivity - decreases as you go down the group on the Periodic Table.

They react vigorously with metals and hydrogen to form halides (see below).

astatine a synthetic halogen which is placed below iodine in the Periodic Table. It is the least reactive of the halogens. Atomic number: 85. Symbol: At. (See also p. 134.)

bleaching properties see reactions of the halogens.

bromine a halogen that is a liquid at room temperature. Atomic number: 35. Symbol: Br. (See also p. 132.)

characteristics of halogens include:
- they are all diatomic; that is, they have two atoms for each molecule;
- they are coloured (chlorine and fluorine are green gases, bromine is a brown liquid, and iodine is grey as a solid and purple as a vapour);
- reactivity decreases down the group (this a reversal of the situation with the metals).

chlorine a halogen that is a gas at room temperature. Atomic number: 17. Symbol: Cl. (See also p. 131.)

The usual chemical test for chlorine relies upon the formation of acid and bleach upon the reaction of chlorine and water. With chlorine, a piece of damp litmus paper will first turn red and then will be become colourless. A similar reaction will occur with bromine and iodine but the reaction will be very much slower.

The reaction is:

$Cl_2(g) + H_2O(l) + aq \rightarrow HCl(aq) + HOCl(aq)$
chlorine + water $\rightarrow$ hydrochloric + chloric (I) acid acid (bleach)

fluorine a halogen that is a gas at room temperature. Atomic number: 9. Symbol: F. Fluorine is a dangerous highly reactive substance, the most reactive member of the group. It displaces chlorine, bromine and iodine from their ionic compounds. (See also p. 131.)

halides compounds of a halogen and one other element. Thus, hydrogen bromide, phosphorus pentachloride and chromium (III) fluoride are all halides.

iodine a halogen that is a solid when at room temperature. Atomic number: 53. Symbol: I. (See also p. 133.)

melting point (of halogens) increases down the group. Thus fluorine has the lowest melting point and iodine the highest. (The artificial halogen astatine has a higher melting point than iodine.)

reactions of the halogens are described as follows: The halogen that is most commonly used for study in experimentation is chlorine. Fluorine, the most reactive, would not be studied in a school or college laboratory.

Chlorine reacts with water in the following manner:

There is no vigorous reaction when chlorine comes into contact with water. A similar result is experienced in the cases of iodine and bromine. Chlorine, bromine and iodine react to decreasing extents to form an acidic solution. These solutions have bleaching powers that increase as the reactivity of the halogen decreases.

Chlorine reacts with iron in the following manner:

Iron is chosen for this experiment because it is of average reactivity.

A stream of dry chlorine passed over heated iron wool produces a strongly exothermic reaction. The wool becomes red hot and a brown cloud of iron (III) chloride smoke forms. If collected on the sides of a cool container a black solid is formed.

The reaction is: $2Fe(s) + 3Cl_2(g) \rightarrow 2FeCl_3(s)$
iron + chlorine $\rightarrow$ iron (III) chloride

This reaction is an oxidation of iron by chlorine.

Chlorine reacts with sodium in the following manner:

A much more vigorous result occurs when chlorine reacts with sodium. The product is sodium chloride, common salt. Each chlorine atom accepts one electron from a sodium atom. The two elements no longer exist as atoms but as ions (see pp. 140-41) of sodium and chlorine.

The reaction is: $2Na(s) + Cl_2(g) \rightarrow 2NaCl(s)$
sodium + chlorine $\rightarrow$ sodium chloride

A similar reaction occurs when the other halogens become halides.

reaction of the halogens with halides are described as follows. A more reactive halogen may displace a less reactive halogen from a salt of the latter. Thus chlorine can displace bromine from ionic bromides (or iodine from ionic iodides).

The reaction is:

$Cl_2(g) + 2Br^-(aq) \rightarrow 2Cl^-(aq) + Br_2(aq)$
chlorine + bromide ions $\rightarrow$ chloride ions + bromine

If sodium bromide is used the reaction is:

$Cl_2(g) + 2NaBr(aq) \rightarrow 2NaCl(aq) + Br_2(aq)$
chlorine + sodium bromide $\rightarrow$ sodium chloride + bromine

The reaction of chlorine with iodides is similar.

The reaction is:

$Cl_2(g) + 2KI(aq) \rightarrow 2KCl(aq) + I_2(aq)$
chlorine + potassium iodide $\rightarrow$ potassium chloride + iodine

SMALL MOLECULES

Although the Earth's atmosphere consists almost entirely of two gases - nitrogen and oxygen - a number of other gases are also present at low concentration, together with varying amounts of water vapour. With the exception of the noble gases (see p. 138), most other components of air form part of natural cycles, each remaining in the atmosphere only for a limited period. Not only are these gases of major importance in relation to industrial processes that dominate economies throughout the world, but cyclical processes involving water, oxygen, carbon dioxide and nitrogen - together with solar radiation - are essential to plant and animal life.

Although these small molecules are simple in the sense that they are composed of few atoms, their structures and - for those with three or more atoms - their shapes vary. In most cases, their atoms are held together in the molecule by two, four or six electrons, resulting in single, double or triple covalent bonds (see pp. 139-43). Three of these molecules - nitrogen monoxide, nitrogen dioxide and oxygen - are paramagnetic because of the arrangement of their electrons.

acid rain see sulphur trioxide.

ammonia a colourless gas with a penetrating odour. It is less dense than air. Ammonia is highly soluble in water, giving an alkaline solution. World production is in the order of 100 million tonnes - 80% of which is turned into fertilizers.

Formula: NH_3.

Concentration in unpolluted air (parts per million): variable.

Industrial production: in a catalytic reaction of nitrogen and hydrogen.

Major industrial uses: fertilizers, plastics, explosives.

carbon dioxide a colourless gas with slight odour and an acid taste. It is available as a gas, as liquid and as the white solid known as 'dry ice'. Its cycle in nature is tied to that of oxygen (q.v.), the relative levels of the two gases in the atmosphere (apart from human activity) being regulated by the photosynthetic activity of plants. It is produced on a vast scale, mostly as a by-product of other processes.

With the ever-increasing input of carbon dioxide to the atmosphere, due largely to the burning of fossil fuels and forests, and the manufacture of cement, the natural 'sinks' for carbon dioxide - chiefly photosynthesis and transfer to the oceans - can no longer keep pace with the total input. If this imbalance continues, it is thought that levels will be reached where the infrared-absorbing properties of carbon dioxide will result in a progressive warming of the Earth's atmosphere, accompanied by melting of the polar ice and flooding of what is now dry land - the so-called greenhouse effect.

Formula: CO_2.

Concentration in unpolluted air (parts per million): c. 315.

Industrial production: as a by-product of other processes, e.g. the burning of fossil fuels, fermentation and calcining limestone.

Major industrial uses: as a coolant and in fire extinguishers.

carbon monoxide is a colourless, odourless, toxic gas. The input to the atmosphere due to human activity is about 360 million tonnes per year, mostly from the incomplete combustion of fossil fuels. The natural input is about 10 times this figure and results from the partial oxidation of biologically produced methane. The background level of 0.1 parts per million can rise to 20ppm at a busy road junction, and a five-minute cigarette gives an intake of 400ppm. Since the atmospheric level of carbon monoxide is not rising significantly, there must be effective sink processes, one being its oxidation in air to carbon dioxide. In addition, there are soil micro-organisms that utilize carbon monoxide in photosynthesis.

Formula: CO.

Concentration in unpolluted air (parts per million): 0.1.

Industrial production: action of steam on carbonaceous material.

Major industrial uses: as a fuel; as a reducing agent in metallurgy; in chemical production, e.g. of methanol.

dinitrogen monoxide a colourless, odourless gas.

Formula: N_2O.

Concentration in unpolluted air (parts per million): 0.5.

Industrial production: in the heating of ammonium nitrate.

Major industrial uses: as an anaesthetic; as a propellant in the food industry.

'fix' (nitrogen) to convert nitrogen into inorganic nitrogen compounds (i.e. ammonium and nitrate salts) that can be easily assimilated by plants. This is achieved by some kinds of bacteria, including what were formerly known as the blue-green algae (see pp. 18-19 and nitrogen - below), and, industrially, by the catalysed reaction between nitrogen and hydrogen to produce ammonium.

greenhouse effect see carbon dioxide.

hydrogen the simplest of all stable molecules, consisting of two protons and two electrons. It is a colourless, odourless gas and is lighter than air. The last of these properties led to its use in lifting airships, but this use was discontinued - because of its explosiveness when ignited - following the *Hindenberg* disaster in 1937. Most hydrogen is used on site where it is produced, but it is also transported as compressed gas in steel cylinders and in liquid form at very low temperatures.

Formula: H_2.

Atomic number (of the element): 2; see also p. 131.

Concentration in unpolluted air (parts per million): 0.5.

Industrial production: in the reaction of coal or petroleum with steam in the presence of a catalyst, and also in the electrolysis of water.

Major industrial uses: in synthesis of ammonia and methanol (the simplest alcohol); in the removal of impurities such as sulphur-containing compounds from natural gas, oil and coal.

methane a colourless, odourless, flammable gas.

Formula: CH_4.

Concentration in unpolluted air (parts per million): 1-1.6.

Major industrial uses: a main constituent of natural gas which has many chemical and other uses, e.g. as a source of power.

nitrogen monoxide occurs at high levels in the atmosphere. This is closely connected with the internal combustion engine. At the high temperature reached when petroleum and air ignite, nitrogen and oxygen combine to form nitrogen monoxide, which slowly reacts with more oxygen to form nitrogen dioxide (q.v.).

Formula: NO.

Concentration in unpolluted air (parts per million): variable.

Industrial production: in the catalytic oxidation of ammonia.

Major industrial uses: vital in the chemical industry, e.g. as an intermediate in the production of nitric acid.

nitrogen is a colourless, odourless gas. Although it is very stable and chemically unreactive, it cycles both naturally and as a result of its use in the chemical industry. The natural cycle results from the ability of some types of bacteria (see pp. 18-19), including what were formerly known as the blue-green algae, (in the presence of sunlight) to 'fix' nitrogen. Since World War I human activity has increasingly contributed to the cycling of nitrogen, because of the catalytic production of ammonia (used mainly in nitrate fertilizers), which ultimately reverts to nitrogen gas.

Formula: N_2.

Atomic number (of the element): 7; see also p. 131.

Concentration in unpolluted air (parts per million): 780 900.

Industrial production: liquefying and distilling air.

Major industrial uses: in the synthesis of ammonia; in processes that make use of its inertness (e.g. in metallurgy and in the food and chemical industries).

nitrogen dioxide is present at a high level in the atmosphere. This is closely connected to the internal combustion engine, which produces nitric acid (q.v.). This, in turn, reacts with oxygen to produce nitrogen dioxide. Most internal combustion engines produce some unburnt or partially burnt fuel. In the presence of sunlight, this reacts with nitrogen dioxide by a sequence of fast reactions, forming organic peroxides, which are harmful constituents of photochemical smog.

Formula: NO_2.

Concentration in unpolluted air (parts per million): 0.02

Industrial production: in the oxidation of nitric oxide.

Major industrial uses: as nitric oxide (q.v.).

oxygen a highly reactive colourless, odourless and tasteless gas. At low temperature, it condenses to a pale blue liquid, slightly denser than water. Oxygen supports burning, causes rusting and is vital to both plant and animal respiration.

Formula: O_2.

Atomic number (of the element): 8; see also p. 131.

Concentration in unpolluted air (parts per million): 209 400.

Industrial production: in liquefying and distilling air.

Major industrial uses: in steel manufacture; in all combustion processes; many uses in the chemical industry.

ozone is a highly toxic, unstable colourless gas. Its primary importance stems from its formation in the stratosphere. In this layer of the atmosphere, temperature increases with height, principally because of the reaction of high-energy ultraviolet solar radiation with oxygen. This can be expressed in two equations:

$$O_2 + \text{sunlight} \rightarrow 2O$$
$$O_2 + O \rightarrow O_3$$

Ozone in the stratosphere functions as a very effective filter for high-energy ultraviolet solar radiation. Radiation in this energy range is sufficiently high to break bonds between carbon and other atoms, making it lethal to all forms of life. It is currently thought that the introduction of CFCs (used in sprays and refrigerants) and the related 'halons' (used in fire extinguishers) may contribute to the partial destruction of the ozone layer. These classes of compounds are highly volatile, chemically very stable and essentially insoluble in water, so that they are not washed out of the atmosphere by rain. When, by normal convection, they reach the stratosphere, they react destructively with ozone.

Formula: O_3.

Concentration in unpolluted air (parts per million): c. 0.01

Industrial production: ultraviolet irradiation of air; electric discharge through air.

Major industrial uses: in the treatment of drinking water; bleach for clay materials.

paramagnetic attracted to a magnet.

sulphur dioxide (modern chemical usage **sulfur dioxide**) a pungent-smelling acidic gas, which is produced through volcanic eruptions and - to a small extent - through the burning of fossil fuels. The level of sulphur dioxide in unpolluted air is 0.0002ppm but in the London smog of 1952 it rose to 1.54ppm.

Formula: SO_2.

Concentration in unpolluted air (parts per million): 0.0002.

Industrial production: in the oxidation of sulphur.

Major industrial uses: an intermediate in the production of sulphuric acid.

sulphur trioxide (modern chemical usage **sulfur trioxide**) a pungent-smelling acidic gas, which is produced by volcanic action. In the atmosphere sulphur dioxide is slowly oxidized to sulphur trioxide, reactions that are catalysed by sunlight, water droplets and particulate matter in the air. Ultimately, sulphur trioxide is deposited as dilute sulphuric acid - acid rain.

Formula: SO_3.

Concentration in unpolluted air (parts per million): variable.

Industrial production: oxidation of sulphur dioxide.

Major industrial uses: as with sulphur dioxide (q.v.).

water a covalent molecule consisting of two atoms of hydrogen and one atom of oxygen.

The total amount of water on Earth is fixed, and most is recycled and re-used. The largest reservoirs are the oceans and open seas, followed by glaciers, ice caps and ground water. Very little is contained within living organisms although water is a major constituent of most life forms.

Water is one of the most remarkable of all small molecules. On the basis of its relative molecular mass (18), it should be a gas; its high boiling point - 100°C (212°F) - is due to the interaction of water molecules with each other (hydrogen bonding; see pp. 139-43), which effectively increases its relative molecular mass. Water is also unusual in that - as ice - it is less dense than the liquid at the same temperature.

Formula: H_2O.

Concentration in unpolluted air (parts per million): variable.

Major industrial uses: as a solvent.

ORGANIC CHEMISTRY - NATURAL COMPOUNDS

The molecular basis for life processes, which have evolved with such remarkable elegance around carbon as the key element, is beginning to be understood, thanks to the combined triumphs of biological, chemical and physical scientists during the last hundred years.

Although the chemist can now make synthetically almost any chemical compound that nature produces, the challenge remains to achieve this objective routinely with the efficiency and precision that characterizes the chemistry of living systems.

acetone see propanone.

acyclic skeleton an open-chain skeleton of atoms when, for example, carbon combines with other atoms. See also cyclic skeleton and functional group.

alanine an amino acid (q.v.). Many organic compounds, such as alanine (2-aminopropanoic acid) and limonene (q.v.) are built up asymmetrically around a central carbon atom and can exist in two mirror-image forms, known as enantiomers (q.v.). In the case of naturally-occurring amino acids such as alanine, one enantiomer predominates greatly over the other, the latter having a small role in nature. This apparently superficial difference in form can have a startling effect on the properties of the compound concerned - see also limonene.

See also structural formulae.

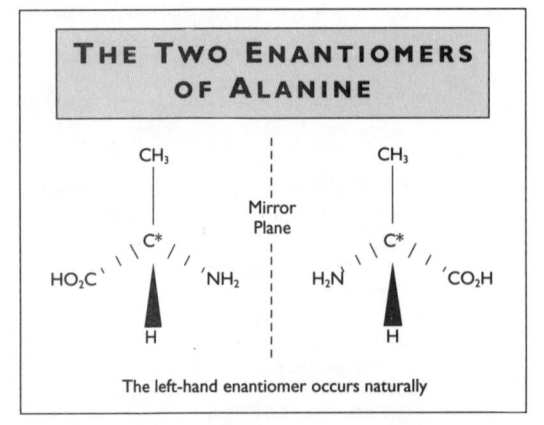

THE TWO ENANTIOMERS OF ALANINE

The left-hand enantiomer occurs naturally

alcohols an homologous series (q.v.). Alcohols are colourless volatile liquids. Examples include methanol and ethanol (q.q.v.).

aldehydes an homologous series (q.v.); colourless volatile fluids that are obtained from alcohol by distillation. An important example is methanal.

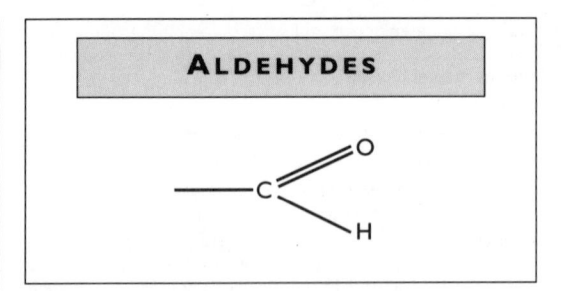

ALDEHYDES

aliphatic compound a compound with an open chain of carbon atoms. The chain may be branched or unbranched and may contain single, double or triple bonds or combinations of these.

alkanes an homologous series (q.v.). Alkanes, which are hydrocarbons, include methane and ethane. They are referred to as saturated hydrocarbons - they contain no double bonds. Alkanes are used as fuels and to make alkenes by a process known as cracking.

Alkanes react with chlorine and bromine in the presence of sunlight. This may be given as:

$CH_4 + Cl_2 \rightarrow CH_3Cl + HCl$
methane + chlorine → chloromethane + hydrogenchloride

This is a substitution reaction (q.v.).

Alkanes may be compared with alkenes. The differences between these two homologous series include:

- alkanes are characterized by a C–C bond; alkenes are characterized by a C=C bond;
- alkanes do not react with bromine solution, although they make bromine lose colour very slowly; alkenes make bromine lose colour rapidly.

See also alkenes.

alkenes an homologous series (q.v.). Alkenes are hydrocarbons (q.v.) that contain one or more carbon-carbon double bonds. Alkenes with just one double bond form a series including ethene (ethylene; C_2H_4), propene and butene.

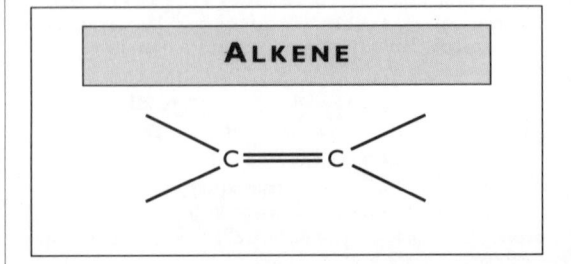

ALKENE

amides an homologous series (q.v.). The most important type of amide group is that formed in protein synthesis, when the carboxyl group of one amino acid condenses with the amine group of another to give what, in this context, is known as a peptide link (see polymers, pp. 156-60).

AMIDES

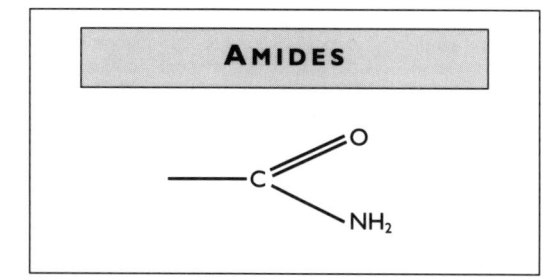

[Structure: C with =O and -NH₂]

amines an homologous series (q.v.). Amines are derivatives of ammonia in which hydrogen atoms are replaced by groups that contain hydrogen and carbon atoms. Amines occur in all amino acids.

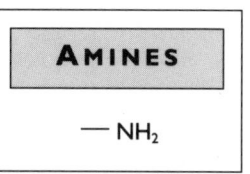

AMINES

—NH₂

amino acid a building block of protein, consisting of one or more carboxyl groups (–COOH) and one or more amino groups (–NH₂) attached to a carbon atom. There are over 80 naturally occurring amino acids, 20 of which occur in proteins. Each amino acid is distinguished by a different side chain (an 'R' group). Organisms can sythesize many amino acids, but there are some that have to be obtained from the diet.

aromatic compounds compounds containing at least one benzene ring.

benzene ring a ring of six carbons in which the fourth electron in one outer shell of each carbon shell is delocalized.

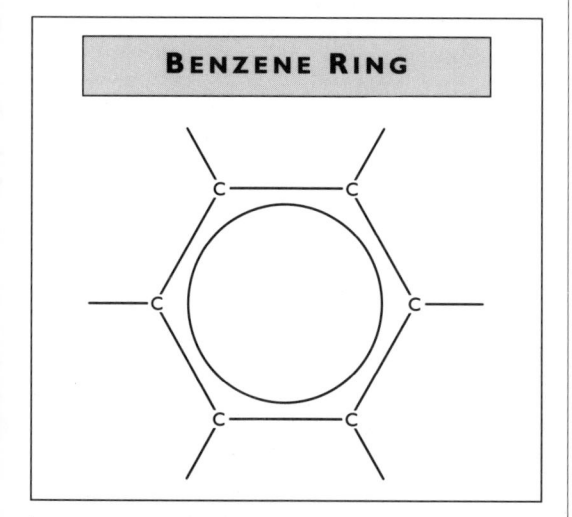

BENZENE RING

[Hexagonal ring structure with circle inside]

bivalent having two valences; see valency pp. 139-43.

butane an alkane (q.v.); CH₃CH₂CH₃. A hydrocarbon in the methane series. It is used as a fuel, for example as camping gas and as lighter fuel.

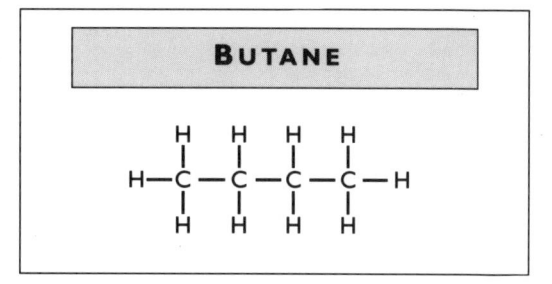

BUTANE

[Structure of butane: H-C-C-C-C-H with H atoms]

butanol 4-carbon, straight-chain alcohols (q.v.); CH₃CH₂CH₂CH₂OH is butan-1-ol, CH₃CH₂CH(OH)CH₃ is butan-2-ol. It is used as a fuel, for example, as camping gas and as lighter fuel.

butene 4-carbon straight-chain alkenes (q.v.); CH₃CH₂CH=CH₂ is but-1-ene, CH₃CH=CHCH₃ is but-2-ene.

carbon (Atomic number: 6; see also pp. 131 and 139-43) is unique in the readiness with which it forms bonds both with other carbon atoms and with the atoms of other elements.

Having four electrons in its outer shell, a carbon atom requires four more electrons to achieve a stable noble-gas configuration (see p. 138). It therefore forms four covalent bonds with other atoms, each of which donates a single electron to each bond. In this way electronic requirements are satisfied and a 'tetracovalent' environment is built up around the carbon atom.

Carbon bonds are found both in pure forms (diamond, graphite and buckminsterfullerene; see pp. 139-43) and in association with other atoms in a vast array of compounds. Carbon bonds readily with hydrogen, oxygen, nitrogen, sulphur (sulfur), phosphorus and the halogens (such as chlorine and bromine). Often the covalent bonds between carbon and other atoms are stable enough for us to handle the resulting compounds at room temperature; yet these compounds are not so strongly bonded that they cannot be manipulated by means of well-known chemical reactions.

Carbon combines with itself and other atoms to produce acyclic and cyclic skeletons, into which are built functional groups (q.v.).

carboxylic acids an homologous series (q.v.). As well as occurring in organic acids, such as ethanoic acid (vinegar), the functional group (–COOH) of this homologous series is also found in all the amino acids, including alanine.

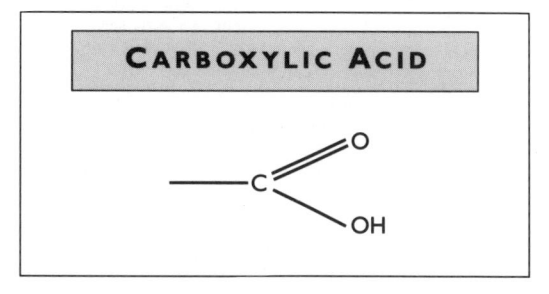

CARBOXYLIC ACID

[Structure: C with =O and -OH]

chiral see enantiomer.

compound a substance that is made of atoms of two or more elements bonded together. See pp. 139-43.

In organic chemistry, the first part of the name of compounds indicates the number of carbon atoms that are in that compound:

meth- indicates 1 carbon atom (e.g. methane);

eth- indicates 2 carbon atoms (e.g. ethene);

prop- indicates 3 carbon atoms (e.g. propane);

but- indicates 4 carbon atoms (e.g. but-2-ene);

pent- indicates 5 carbon atoms.

configuration (chemical) the spatial arrangement of groups bonded to a central carbon atom in such a way as to define a tetrahedron in three dimensions. If four different atoms or groups are attached to the central carbon, this spatial arrangement can exist in two forms - enantiomers - which are non-superimposable mirror images.

covalent bond the type of bond that is formed when two atoms share electrons; see also pp. 139-43.

cracking the process of breaking long-chain molecules into small molecules. The reaction needs a high temperature and a catalyst.

cyclic skeleton a ring-shaped skeleton formed by atoms when, for example, carbon combines with other atoms. See also acyclic skeleton and functional group.

electrophilic electron-pair seeking.

empirical formula the simplest possible formula showing stoichiometric proportions only.

Example: 2-hydroxypropanoic acid ($CH_3CH(OH)COOH$) has the empirical formula CH_2O.

enantiomer either of the two non-superimposable mirror-images that a configuration can take. The spatial arrangement, or configuration, of four different groups bonded to a central carbon atom can take two forms (enantiomers), one the non-superimposable image of the other. They differ as our right hand does from our left, so the central carbon atom is said to be chiral (from the Greek for hand) or asymmetric.

ethanol an alcohol (q.v.); C_2H_5OH. Ethanol is a colourless liquid which is completely miscible with water. In an excess of air, it burns to give carbon dioxide and water.

Ethanol, which can be manufactured by reacting ethene with steam in the presence of a catalyst, is used as a solvent. It is present in beers and spirits (alcohol). It is used as fuel (as the principal component of methylated spirits), in tincture of iodine (dissolved with iodine) and as a liquid polish (dissolved with shellac).

ethane an alkane (q.v.); C_2H_6. It contains a pair of electrons between the carbon atoms. The molecule contains a single bond: carbon-carbon.

ethanethiol a thiol (q.v.); CH_3CH_2SH. As North Sea gas contains methane which - unlike coal gas - is odourless, tiny amounts of ethanethiol - an evil-smelling chemical - are added to aid its detection from open taps and gas leaks.

ethene an alkene (q.v.); C_2H_4. An ethene molecule consists of two pairs of shared electrons between carbon atoms. The molecule contains a carbon-carbon double bond. Ethene is a flammable gas which is used as a fuel and an anaesthetic. Ethene is used to manufacture ethanol (q.v.) and to make a number of polymers including poly(ethene) (also known as polythene), poly(phenylethene) (also known as polystyrene) and poly(chloroethene) (also known as P.V.C.) - see pp. 156-59.

formaldehyde see methanal.

functional group an atom or group of atoms that give a compound its particular chemical properties. Thus ethanol (CH_3CH_2OH), ethanal (CH_3CHO), ethanoic acid (CH_3COOH), and ethene ($CH_2=CH_2$) behave differently because they have different functional groups (–OH, –CHO, –COOH and C=C respectively).

The diverse but predictable chemical behaviour of the different functional groups is a consequence of their ability either to attract or to repel electrons compared with the rest of the carbon skeleton. The overall effect of the resulting charge distribution is to create a molecule in which some regions are nucleophilic (q.v.) and others are electrophilic (q.v.).

Most organic reactions involve the electrophilic and nucleophilic centres of different molecules coming together as a prelude to the formation of new covalent bonds. An appreciation of how particular compounds behave towards others and of the various mechanisms by which such reactions occur forms the basis of classical organic synthesis. This allows chemists to build up large molecules, containing many different functional groups and with a great diversity of chemical properties, in a controlled and predictable manner.

homologous series a family of compounds containing the same functional group.

Homologous series include alkenes, alcohols, ketones, aldehydes, carboxylic acids, amines, amides, and thiols.

hydrocarbon any compound consisting of hydrogen and carbon only. Hydrocarbons are extremely important, notably as the principal components of fossil fuels and the starting materials in the synthesis of most man-made organic compounds.

isomers compounds with the same molecular formula. Thus ethanol (CH_3CH_2OH) and methoxymethane (CH_3OCH_3) are isomers since they have the molecular formula (C_2H_6O).

ketones an homologous series (q.v.). Ketones are organic chemical compounds containing the functional group CO in combination with two hydrocarbon radicals. Examples include propanone (CH_3COCH_3) and MVK.

limonene an organic compound which, like alanine (q.v.), is built up asymmetrically around a central carbon atom and can exist in two non-superimposable mirror-image forms, known as enantiomers (q.v.). The superficial difference between the mirror-images has a startling effect upon the enantiomers of limonene, one of which smells strongly of lemons, the other of oranges.

See also structural formulae.

KETONES

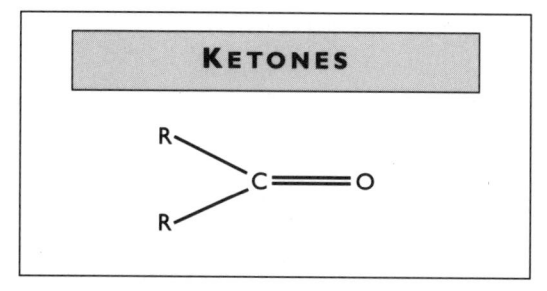

$$R_2C=O$$

methanal an aldehyde (q.v.) also known as formaldehyde; HCHO. It is used in the production of formalin (a disinfectant) and of synthetic resins.

methane an alkane (q.v.); CH_4. Methane is a colourless, odourless gas, which is less dense than air. It is neutral and is insoluble in water. See also pp. 150-51.

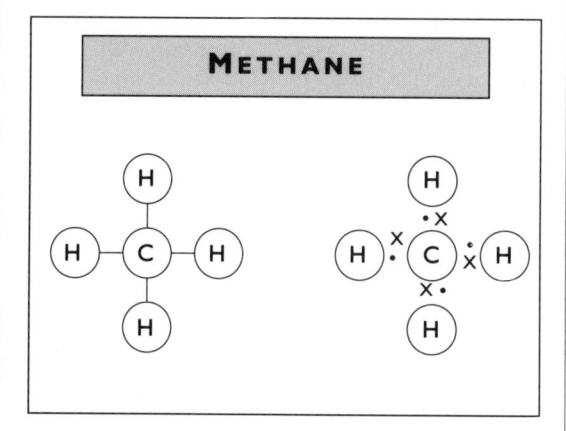

METHANE

methanol an alcohol (q.v.); CH_3OH. A poisonous liquid, which is obtained by the destructive distillation of wood. It is used as a fuel.

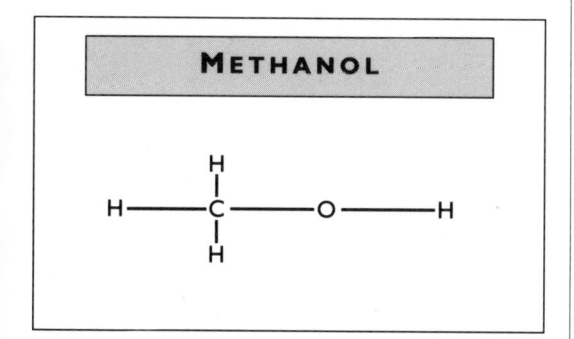

METHANOL

methylated spirit see ethanol.
methylvinylketone see MVK.

miscible mixable; usually applied to liquids, e.g. water and ethanol are completely miscible.

molecular formula the formula showing the molecular composition, but not the sequence of atoms.
Example: 2-hydroxypropanoic acid (CH_3OH) has the molecular formula HCHO.

MVK the abbreviation for methylvinylketone (more correctly known as **but-3-en-2-one** ($CH_3COCH=CH_2$); a ketone.

nucleophilic electron-pair donating.

octane an alkane (q.v.). A hydrocarbon which is present in petrol.

organic chemistry the branch of chemistry dealing with carbon compounds, excluding the oxides of carbon and the carbonates.
There is something very special about the chemistry of carbon that has singled it out as the atomic building block from which all naturally occurring compounds in living systems are constructed. The subject that deals with this important area of science, nestling between biology and physics, has become so vast and significant that it has earned recognition as a separate field of scientific investigation. As it was originally thought that such carbon-based compounds could be obtained only from natural sources (e.g. organisms), this field of study became known as organic chemistry.

propane an alkane (q.v.); C_3H_8.

propan-1-ol and propan-2-ol alcohols (q.v.) with the molecular formula C_3H_8O and with the structure $CH_3CH_2CH_2OH$ and $CH_3CH(OH)CH_3$ respectively.

propanone (acetone) a ketone (q.v.); CH_3COCH_3.

propene an alkene (q.v.); C_3H_6. Propene polymerizes to form poly(propene).

radical a group of atoms that operates as a single atom and either goes through a reaction unchanged or is replaced by another atom.

retrosynthetic analysis a design method in which the desired product is broken down theoretically, or 'disconnected', into smaller and smaller fragments until a convenient starting material is reached.

structural formulae a formula which shows the sequence and arrangement of the atoms in a molecule, for example, CH_3COCH_3 for propanone.

substitution reaction a reaction in which an atom or group of atoms is replaced by another atom or group of atoms.

thiols an homologous series (q.v.). The functional group is characterized by a strong disagreeable odour. An example is ethanethiol.

THIOLS

— SH

POLYMERS

Polymers are chemicals composed of large molecules in which a group of atoms is repeated. They include naturally occurring substances such as starch and cellulose and synthetic substances such as nylon and poly(ethene).

One of the most far-reaching influences chemistry has had on our lives in the 20th century has been the introduction of increasingly advanced materials based upon synthetic polymers.

The structure of many polymers is complicated. They can be made up of very many molecules - sometimes the total runs into thousands of molecules - arranged in a chain of monomers (q.v.). The way in which these chains interact with one another depends upon the monomers from which the polymers are composed, and can differ greatly. This accounts for the wide range of properties that polymers can possess, including flexibility, strength and heat resistance.

Polymerization is the process in which many small molecules of monomers combine together to form large molecules of polymers. See addition polymerization and condensation polymerization (below).

Demand for improved materials, both for technological and household application, is still increasing, and - although already widespread - there is still a lot more polymer-based technology to come into our lives.

Along with the emergence of excellent new dyes and paints, the fashion and display industries have been revolutionized by the advent of polyester fibres such as Dacron and polyamide fibres such as Nylon 6 and Nylon 66. These early synthetic polymers were all composed of moderately flexible chains. More recently, stiff-chain polymers - often incorporating aromatic rings - have found many applications. These include Kevlar and Victrex PEEK.

Note that the term polymer refers to the chemical and not the polymer molecule.

addition polymerization polymerization that involves one type of molecule containing a carbon-carbon double bond.

The carbon-carbon double bond in, for example, ethene is very reactive. Electrons are transferred from these double bonds, allowing ethene molecules to link together, a process that allows the formation of a long chain of single carbon-carbon bonds in the polymer poly(ethene), which is also known as polythene. The requirements for this reaction to occur are an initiator (q.v.), a temperature of 200°C and high pressure.

In any such reaction the number of monomer molecules is variable and for this reason the exact molecular formula may not always be known. By convention the number of monomer units in a polymer is shown by the number n, which represents the number of times that the unit is repeated. This is usually a large number. As an example, poly(ethene) may be represented as:

$$\text{\textendash}(CH_2 - CH_2)_n\text{\textendash}$$

Poly(ethene) molecules have very large relative molecular masses which are in excess of 100 000. The polymer that is formed in this manner has none of the properties of ethene and the double carbon bond is, in fact, absent from the product. The monomer is an unsaturated hydrocarbon while the resulting polymer is a saturated hydrocarbon.

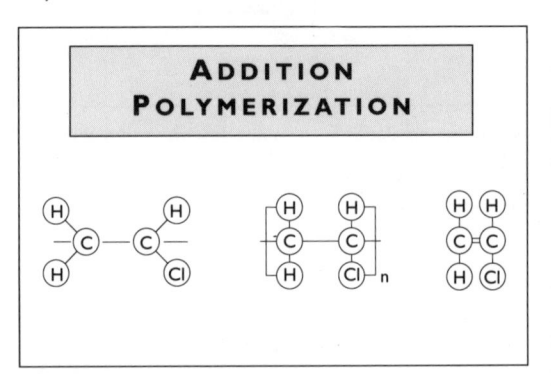

ADDITION POLYMERIZATION

Any other molecule that has the carbon-carbon double bond will also polymerize in exactly the same manner. The polymers that are derived from such monomers are simply chains of monomer units that are linked together, but in that process they lose the carbon-carbon double bond.

The monomer from which a polymer is made can be discovered by identifying the repeating unit, by inserting a double bond between the carbon atoms and eliminating the unused bonds at the end of the unit in the polymer (1, in the diagram), repeated unit (2) and (3) repeat unit.

Some of the polymers that are formed by addition polymerization can be converted back into their monomers by heating. This is referred to as depolymerization.

amino acid a building block of protein (q.v.), consisting of one or more carboxyl groups (–COOH) and one or more amino groups (–NH$_2$) attached to a carbon

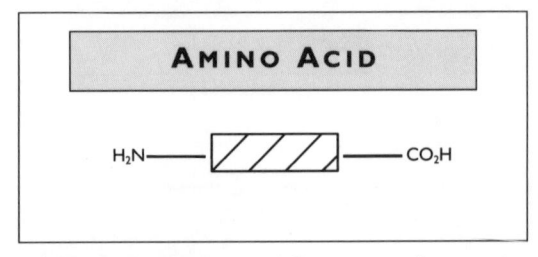

AMINO ACID

H_2N——▨——CO_2H

atom. There are over 80 different naturally occurring amino acids, 20 of which occur in proteins. Each amino acid is distinguished by a different side chain (an 'R' group). Organisms can synthesize many amino acids, but there are some that have to be obtained from the diet.

FORMATION OF AMIDE LINKAGE

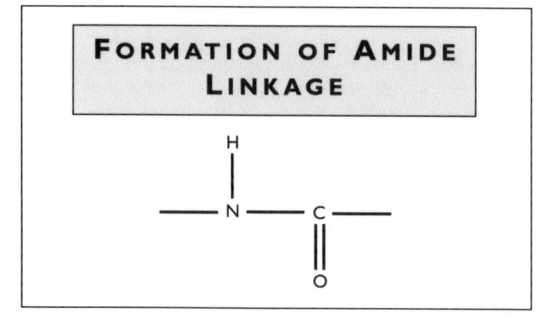

```
         H
         |
 ——— N ——— C ———
             ‖
             O
```

Araldite resin a thermosetting synthetic polymer. Araldite, like all thermosets, cannot be softened once set; it is used as a strong adhesive.

Bakelite a thermosetting synthetic polymer. Bakelite, like all thermosets, cannot be softened once set; it is used in electrical fittings.

cellulose a polysaccharide (q.v.) carbohydrate made up of unbranched chains of many glucose molecules. Its fibrils form the framework of plant cell walls and it is the principal constituent in the shells of certain protoctists (see pp. 18-19). Cellulose is generally important in the diets of herbivorous animals, many of which cannot produce the enzymes necessary for its digestion and therefore enlist the help of cellulose-digesting bacteria and protoctists.

chloroethene monomer molecules combine to give the polymer poly(chloroethene) (which is also known as P.V.C.).

condensation polymerization the type of polymerization that occurs with monomers with reactive groups at each end of the molecule.

This is a quite different type of polymerization to that involving the carbon-carbon double bond. It involves two kinds of molecule that condense together to form long chains in a process that involves the loss of a small molecule by the monomers during linking. The molecule that is lost as each link is formed is usually either a water molecule or a hydrogen chloride molecule.

Monomers must be capable of reacting together in order for condensation polymerization to occur - there must be a reactive group at each end. In the reaction, these groups link together to form polymer molecules and the polymer molecules can continue to grow at either end at reactive points.

The compounds thus formed tend to be more complex than those formed by addition polymerization. Nylon (q.v.) and terylene (q.v.) are examples of condensation polymers.

depolymerization see addition polymerization.

enzyme any protein that acts as a catalyst in a biochemical reaction.

ethene a monomer whose molecules combine to give the polymer poly(ethene) (which is also known as polythene).

fat (or **lipid**) any polymer formed from the reaction of fatty acids and glycerol.

Fats are organic compounds formed from carbon, hydrogen and oxygen. They normally comprise one glycerol molecule linked to three different fatty acid molecules.

Fat molecules are large. The linkages between the glycerol and fatty acid molecules of which fat molecules are composed are ester linkages.

glycogen the form in which carbohydrates are stored in the liver and muscles of animals. It is a polysaccharide (q.v.) carbohydrate consisting of glucose units and is readily converted to glucose.

glycerin see glycerol.

glycerol (or **glycerin**, or more correctly **propane-1,2,3-triol**) is the alcohol to which three carboxylic acid molecules are attached by ester linkages in fats.

high-density polythene or **HD polythene** see poly(ethene).

hormone a chemical secretion (a natural polymer) produced by endocrine cells or glands. Hormones regulate many metabolic functions. Insulin, for example, controls the body's sugar metabolism.

hydrolysis the breaking down of a substance into other substances by reaction with water.

initiator a catalyst used in the reaction of polymerization.

insulin a natural polymer; a hormone secreted by the pancreas in vertebrates. Insulin - which regulates blood-sugar levels - is regarded as a relatively simple protein although it is composed of no fewer than 51 molecules and 14 different amino acids.

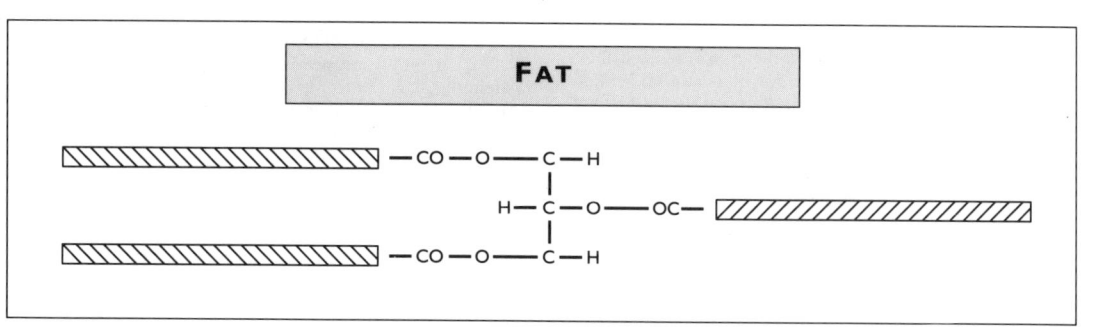

FAT

```
▨▨▨▨▨▨▨▨▨▨ —CO—O——C—H
                    |
              H—C—O——OC— ▨▨▨▨▨▨▨▨▨▨
                    |
▨▨▨▨▨▨▨▨▨▨ —CO—O——C—H
```

Kevlar a synthetic polymer.

Kevlar has a tensile strength higher than that of steel but it also has a strength-to-weight ratio that is six times better. Because of their low flammability, high thermal stability and great tensile strength, Kevlar and similar compounds are becoming very widespread; they may be found in an enormous range of products, from electrically heated hair styling brushes to bullet-proof vests. Composite materials containing Kevlar have been used to build the tail sections of jumbo jets - it is transparent to radar, it eliminates corrosion, and its lightness lowers fuel costs.

KEVLAR

lipid an alternative name for a fat.

low-density polythene or **LD polythene** see poly(ethene).

melamine a thermosetting plastic polymer. Because thermosets like melamine cannot be softened by heat once set, it is used for heat-proof containers and heat-resisting surfaces.

methyl methacrylate a monomer whose molecules combine to give poly(methyl methacrylate) (which is also known as perspex).

monomer a chemical whose molecules react to make up the chain of a polymer molecule.

natural polymers natural compounds that are made up of a large number of repeated units.

Natural polymers differ from synthetic polymers in their constituents. Whereas synthetic polymers can be formed from a variety of elements, natural polymers are formed principally from just four elements: carbon, hydrogen, nitrogen and oxygen. Chlorine, which is an element commonly used in the formation of synthetic polymers, is not found in any known natural polymer.

Natural polymers may be divided into three main types: starches, proteins, and fats (q.q.v.).

nylon a synthetic condensation polymer.

The advent of synthetic fibre nylons in the 1930s caused a major revolution in the fashion industry. Nylon can be synthesized by polymerizing hexane-1,6-diamine (hexamethylenediamine) with hexanedioyl dichloride (adipyl chloride). In the reaction an $-NH_2$ group and a $-COCl$ group react together to form an amide linkage. A molecule of hydrogen chloride is formed in the reaction. Nylon is a polyamide.

Nylon has many useful properties. It has a similar structure to natural protein and has considerable strength even when occurring in thin threads. It repels water and resists bacterial and fungal attack. It is a thermosoftening plastic (q.v.) and is made by forcing fibres of molten nylon through tiny holes and cooling them.

Uses include nylon fabric, carpet fibre, 'nylons' and tights, anoraks, string and rope, and various lightweight gears.

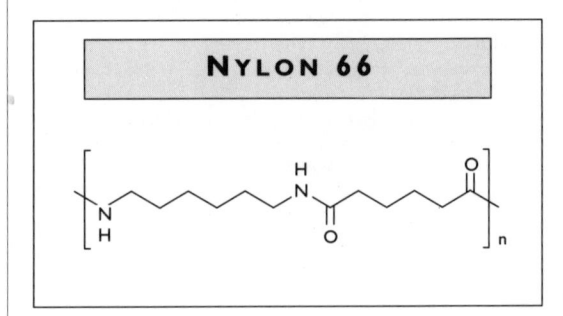

NYLON 66

nylon 66 a synthetic condensation polymer.

perspex see poly(methyl methacrylate).

phenylethene a monomer whose molecules combine to give the polymer poly(phenylethene) (which is also known as polystyrene).

plastic any synthetic polymer that can be formed into a desired form or shape. Plastics are either thermosetting plastics or thermosoftening plastics (q.q.v.).

Plastics are easy to make, cheap, lightweight, resistant to chemicals, waterproof and tough. Many plastics are, however, flammable and the disposal of such resistant substances is a major pollution problem.

polyamide a polymer containing many amide linkages. See nylon.

polychloroethene (better known as **P.V.C.**) a synthetic addition polymer. It is formed from the combination of chloroethene monomer molecules.

P.V.C. is tough, rubbery and strong, and yet remarkably flexible. It is flame resistant and resists solvents. P.V.C. is also an electrical insulator. It is, therefore, used for electrical wire insulation. It is also found in raincoats and wellington boots, and is used to make artificial leatherware.

polyester a polymer with many ester linkages. See terylene.

poly(ethene) (also known as **polythene**) a synthetic addition polymer which is formed from the combination of ethene monomer molecules in addition polymerization.

CHEMISTRY

The polymer poly(ethene) is waterproof and is a cheap, strong substance that resists acids found in most foods and also resists solvents. It is, therefore, used in food wrapping, in kitchenware, containers, bags, pipes, etc.

There are two types of polythene - high-density polythene (HD polythene) and low-density polythene (LD polythene). In HD polythene the molecules lie closely parallel to each other which increases the force of attraction between them. HD polythene is, therefore, stronger and more rigid than LD polythene and also has a higher softening temperature than LD polythene. HD polythene is, therefore, used in situations where strength is important, e.g. buckets.

polymerization see introduction (above), addition polymerization and condensation polymerization.

poly(methylmethacrylate) (also known as perspex) a synthetic addition polymer which is formed from the combination of methyl methacrylate monomer molecules.

Perspex is a tough, glass-like substance that is shatterproof. One of its most useful properties is that perspex is easily moulded. It is used for camera lenses, for the windows of aircraft and for the windshields of motor cycles. It is also used for baths and similar objects.

polypeptide any natural polymer made up of amino acids (q.v.).

poly(phenylethene) (also known as **polystyrene**) a synthetic addition polymer which is formed from the combination of phenylethene monomer molecules.

The polymer polystyrene is flexible and strong and retains its strength at low temperatures. It is, however, somewhat brittle. One of the useful properties of polystyrene is that it can be worked easily with hand tools. It is used for making insulation for homes, for heat insulation in general, for packing materials, and for some parts in refrigerators. It is also quite widely used for making toys and food containers.

poly(propene) a synthetic addition polymer which is formed by the combination of propene monomer molecules.

The polymer poly(propene) is both harder and more rigid than poly(ethene). One of its most useful properties is a high softening point which is >100°C (212°F). It is used in carpet fibre and in moulded furniture. Poly(propene) is also found in sterilizable bottles used in hospitals.

polysaccharide any of a class of large carbohydrate molecules, including cellulose, glycogen and starch.

polystyrene see poly(phenylethene).

poly(tetrafluoroethene) (also known as **PTFE**) a synthetic addition polymer which forms from the combination of tetrafluoroethene monomer molecules.

PTFE is heat-resistant and solvent-resistant. It has low friction and is generally 'non-stick'. One of its most useful properties is that it is unaffected by most chemicals that are used in normal industrial and domestic situations. For this reason it is used in acid-resisting containers in chemical and industrial plants. It also used in the coating of 'non-stick' frying pans.

MONOMERS - POLYMERS

Monomer	Polymer
H H │ │ C=C │ │ H H ethene	[H H │ │ C—C │ │ H H] poly(ethene)
H H │ │ C=C │ │ H Cl chloroethene	[H H │ │ C—C │ │ H Cl]$_n$ poly(chloroethene)
H H │ │ C=C │ │ H C$_6$H$_5$ phenylethene	[H H │ │ C—C │ │ H C$_6$H$_5$]$_n$ poly(phenylethene)
H H │ │ C=C │ │ H CH$_3$ propene	[H H │ │ C—C │ │ H CH$_3$]$_n$ poly(propene)
H CO$_2$CH$_3$ │ │ C=C │ │ H CH$_3$ methyl methacrylate	[H CO$_2$CH$_3$ │ │ C—C │ │ H CH$_3$]$_n$ poly(methylmethacrylate)
F F │ │ C=C │ │ F F tetrafluorethene	[F F │ │ C—C │ │ F F]$_n$ poly(tetra fluorethene)

polythene see poly(ethene).

propene a monomer whose molecules combine to give molecules of the polymer poly(propene).

protein any of a number of natural polymers - found in plants and animals.

Proteins contain nitrogen, hydrogen, carbon, and oxygen (and sometimes sulphur). The structure of proteins is complex and protein molecules are large. These polymers are made up of long chains of different amino acids (q.v.). Chains may be folded and are joined to other chains in several different ways. No part of a protein polymer repeats itself. The linkage in a protein polymer is known as either an amide or a peptide linkage.

Proteins 'build' cells and tissues of plants and animals and are constituents of animal skin, bone and muscle. Enzymes and hormones, for example insulin, are proteins.

Protein may be broken down by hydrolysis into the monomers that form it - amino acids.

PTFE see poly(tetrafluoroethene).

P.V.C. see poly(chloroethene).

radical a group of atoms that acts as a single atom and goes through a chemical reaction unchanged.

starch a natural polymer which is a compound of carbon, hydrogen and oxygen - the hydrogen and oxygen elements are present in the same ratio as in water. The reaction to form the linkage involves the −OH group at each end of the monomer. Its structure may be given as:

STARCH

Starch is a polysaccharide carbohydrate into which glucose is converted for storage in plants. (Starch is formed in all plants by photosynthesis; see pp. 20-25.)

Glucose is the repeating monomer in chains of starch. Under hydrolysis starch may be broken down with glucose as the only product.

sugar any of a group of carbohydrates of low molecular weight, including mono-, di- and some oligosaccharides. Sugars include glucose, maltose and lactose, which act as monomers.

synthesize to form a complex compound by the combination of two or more simpler compounds, elements or radicals.

terylene a synthetic condensation polymer. Terylene is made by the polymerization of benzene-1,4-dicarboxylic acid (terephthalic acid) and ethane-1,2-diol (ethylene glycol). The reaction is between the $-CO_2H$ group of the terephthalic acid with the −OH group of ethylene glycol.

In the reaction a water molecule is lost and an ester linkage is formed. Terylene is a polyester.

Terylene is a thermosoftening polymer (q.v.). It is made into fibres by forcing molten terylene through tiny holes and cooling these continuous threads.

Terylene has many useful properties including its resistance to wear and to chemical attack. Uses include fibre in cloth - terylene will take a 'permanent' crease. Terylene is also used in plastic bottles and in video and audio tapes.

tetrafluorethene monomer molecules combine to form poly(tetrafluoroethene) (which is also known as PTFE).

thermoplastics see thermosoftening plastics.

thermosets see thermosetting plastics.

thermosetting plastics thermosets; plastics that cannot be melted by heat once they have set. This useful property is possessed by Araldite resin, Bakelite and melamine (q.q.v.). See also plastic and thermosoftening plastics.

In the case of thermosets the polymer is more rigid as molecules are built up from polymer molecules by a series of 'cross-links' rather than the 'end-to-end' links that characterize thermoplastics.

thermosoftening plastics thermoplastics; plastics that have the property of thermoplasticity, that is they soften when heated. When softened by heat, these polymers can be shaped. See also plastic and thermosetting plastics.

Thermoplastics are made up of long polymer molecules which are in constant motion and are attracted to one another by the weak force (see pp. 120-23), which decreases the distance between individual molecules. Heat increases the motion of these molecules and increases the spaces and reduces the attraction between them. The plastic can then change shape more easily when force is applied.

Victrex PEEK a synthetic polymer which is characterized by low flammability, high thermal stability and great tensile strength.

The outstanding mechanical and electrical properties of Victrex PEEK, particularly when employed in fibre-reinforced composites with glass, carbon or Kevlar, are such that it can be used to replace metal alloys to great advantage in many engineering situations.

VICTREX PEEK

CHEMISTRY

SYNTHETIC PRODUCTS

Our knowledge of natural compounds will ultimately be dwarfed by that relating to synthetic products. The reason for this is that chemistry is not only the science that deals with molecules, but also an exercise in design and engineering at molecular level. In so far as chemistry creates its own world, its practice in an abstract form has infinite possibilities. This section gives some examples of synthetic products.

high-temperature superconductors include the material that is often referred to as '1-2-3', on account of the ratios of the metals involved.

The amazing ability of a material to transmit an electrical current without showing any electrical resistance is known as superconductivity. In 1911 the Dutch physicist Onnes discovered that the resistance of mercury falls to zero in liquid helium, which boils at 4.2K (−268.8°C/−452°F). Progress was subsequently made in increasing the transition temperature to the superconducting state, notably by using certain alloys of niobium and titanium, but the major breakthrough came in 1986, when it was announced that a ceramic metal oxide of lanthanum, barium and copper loses its resistance at 30K (−243°C/−405°F). Later research resulted in which yttrium replaces lanthanum. This oxide - sometimes popularly known as 1-2-3 (see above) - was demonstrated to be superconducting at just below 100K (−173°C/−279°F).

liquid crystals have been heralded as the fourth state of matter, occurring at the interface between the solid and the liquid phase. They were discovered in 1888 by the Austrian botanist Friedrich Reinitzer.

About 5% of crystalline compounds do not simply melt when heated: they form turbid (cloudy) liquids, which may also exhibit marked colour changes as the temperature is raised, before becoming normal liquids. The process is reversible upon cooling, the liquid passes back through the so-called liquid-crystalline state before solidifying again. Generally, it is rod-shaped molecules, such as 4-n-hexyl-4-cyanobiphenyl, that exhibit liquid-crystal behaviour.

Because of the relatively weak forces between the molecules of a liquid crystal, the interactions between the molecules, and hence their relative orientations can be changed not only by temperature and pressure but also by electric and magnetic fields. The effect of temperature on the colour of liquid crystals has led to their use in the detection of tumours that are 'warmer' than surrounding healthy tissue. Their most familiar application is in the liquid-crystal displays used in watches and calculators, where their optical properties are controlled by applying electric fields to change the orientation of molecules in the liquid crystal.

superconductivity see high-temperature superconductors.

synthetic zeolites are fashioned by chemists to convert methylbenzene (toluene) into benzene and 1,4-dimethylbenzene (paraxylene), an intermediate in the production of polyester fibres. Natural zeolites are highly porous crystalline minerals, consisting mainly of silicon, aluminium and oxygen. They are built up of three-dimensional networks of silicate and aluminate tetrahedrons, arranged in such a way that they form tiny submicroscopic channels. In the natural state these channels hold water molecules; however, if the water is driven off by heat, other molecules of appropriate size can enter. Zeolites can, therefore, be used as 'molecular sieves', which separate mixtures of compounds purely on the basis of their different sizes. By chemical means, a new generation of zeolites is being fashioned to accomplish different tasks.

THE LIQUID-CRYSTAL PHASE

Crystalline (solid) ⇌ Smetic ⇌ Nematic ⇌ Isotropic (liquid)

The liquid crystal phase occurs between the solid (ordered) and liquid (disordered) phases. The intermediate phase may take a number of slightly different forms, of which 'smetic' and 'nematic' are examples. The characteristic orientation of the molecules of a liquid-crystalline substance are due to the long, rod like shape of the molecules, which allow a weak, long-range order to develop between each molecule and its neighbours.

A typical example is 4-n-hexyl-4-cyanobiphenyl:

H$_3$C—(CH$_2$)—4—(3,2)—1,1'—(2',3')—4'—CN

Mathematics

NUMBERS

Natural numbers or whole numbers are those that we use in counting. Natural numbers - one, two, three, four and so on - are learned at an early age. Important features of our number system, these numbers can be used to count sets of objects, and form a naturally ordered progression that has a first member, the number 1, but no last number. No matter how big a number you come up with, it can always be capped with a bigger one - simply by adding 1.

addition a mathematical operation by which the sum (see below) of two numbers is calculated. The operation is normally indicated by the symbol +.

Example: 3 + 4 = 7.

cubic number any number which has been formed by multiplying a whole number by itself and then multiplying the result by that whole number.

Example: 4 × 4 × 4 = 64, thus 64 is a cubic number.

4 × 4 × 4 may be written 4^3, that is to say 'four cubed'. (See also square number.)

difference (of numbers) the result when the smaller of two numbers is taken away from the larger.

division a mathematical operation in which the quotient (see below) of two numbers is calculated. The operation is normally indicated by the symbol ÷.

Example: 20 ÷ 5 = 4. Note that this is the inverse operation of multiplication.

even number any number which will divide by 2 exactly.

Examples: 2, 4, 6, 8, 10, 12, 14, 16, 18, 20, ...

exponent another name for an index (see below).

factor any number that divides exactly into another number.

Examples: 6 divides into 48 exactly (that is, without leaving a remainder) 8 times. Thus both 6 and 8 are factors of 48.

Similarly 2 divides into 6 exactly 3 times - thus both 2 and 3 are factors of 6 - and 2 divides into 8 exactly 4 times - thus 2 and 4 are factors of 8.

HCF the abbreviation for highest common factor, see below.

highest common factor (or **HCF**) the largest number that will divide into two or more numbers exactly (that is, without leaving a remainder).

Examples: 78 = 3 × 26.
182 = 7 × 26.

Thus 26 is the highest common factor of 78 and 182. (See also the lowest common multiple.)

index (or **exponent**) a number placed above the line after another number to show how many times the number is to be multiplied by itself. The number of times that the number occurs in the operation is called the power.

Example: 7 × 7 × 7 × 7 × 7 is the fifth power of 7. In index form, this is shown as 7^5.

integer any negative or positive whole number.

Examples: −3, −2, −1, 0, +1, +2, +3, +4, ...

Note that 0 is an integer, but it is neither a positive nor a negative integer.

irrational number any number which cannot be expressed as a fraction or ratio.

Examples: $\sqrt{3}$.

π (*pi*), which to seven decimal places is 3.141 592 6. π has been calculated to 1 073 740 000 decimal places, but the decimal (probably) has no end.

LCM the abbreviation for lowest common multiple, see below.

lowest common multiple (or **LCM**) the lowest common multiple of two or more numbers is the smallest number which is exactly divisible by them without leaving a remainder.

Example: to find the lowest common multiple of 4, 8 and 20.

Factors of 4 = 2 × 2, that is (2^2).

Factors of 8 = 2 × 2 × 2, that is (2^3).

Factors of 20 = 2 × 2 × 5, that is (2^2 × 5).

Thus the lowest common multiple of 4, 8 and 20 = 2^3 × 5 = 40. (See also the highest common factor.)

multiple any number that is the product of a given number and any other whole number.

Examples: 14 is a multiple of 2, 14 is also a multiple of 7. 15 is a multiple of 3. 15 is also a multiple of 5.

MATHEMATICS

multiplication a mathematical operation by which the product of two or more numbers is calculated. This operation is the inverse operation to division (see above). The operation is normally indicated by the symbol ×.
Example: 5 × 4 = 20.

natural number (or **whole number**) any number which is a positive integer (q.v.).
Examples: 1, 2, 3, 4, 5, 6, 7, 8, 9, 10, 11, 12, 13, ...
Note that 0 is not a natural number.

odd number any integer which will not divide by 2 exactly.
Examples: 1, 3, 5, 7, 9, 11, 13, ...

perfect number any number that is equal to the sum of its factors, excluding the number itself.

The first perfect number is 6 whose factors (excluding 6 itself) are 1, 2 and 3. As 1 + 2 + 3 = 6, 6 is a perfect number.

The next perfect number is 28. The factors of 28 are 1, 2, 4, 7 and 14. 1 + 2 + 4 + 7 + 14 = 28.

Pythagoras knew these first two perfect numbers in the sixth century BC. In the third century BC Nicomachus of Alexandria discovered the next two perfect numbers - 496 and 8128. The fifth perfect number - 33 550 336 - was not discovered until over 1000 years later. Until the 1950s only seven perfect numbers were known. Today, even with the help of computers, only 30 perfect numbers have been discovered.

power see index (above).

prime number any natural number that can only be divided by itself and 1. The number 1 is not considered to be a prime number.

Prime numbers can be found by taking a sequence of numbers such as

1, 2, 3, 4, 5, 6, 7, 8, 9, 10, 11, 12, 13, 14, 15, 16, 17, 18, ...

and first deleting all the numbers divisible by 2 (excluding 2 itself, which is only divisible by itself and 1), then all those divisible by 3, then (since anything divisible by 4 has already been deleted) all those numbers divisible by 5, and so on.

All non-prime numbers must by definition be divisible by other numbers apart from themselves and 1; these other numbers can be repeatedly divided until you are left with a series of prime numbers. All non-prime numbers can be expressed as the product of a series of primes - in fact, for each number the series is unique.

product (of numbers) the result of multiplying numbers together.

quotient (of numbers) the result of dividing one number by another number.

rational number any number that can be expressed as a fraction or a ratio.
Examples: $0.5 = \frac{1}{2}$
$8:5 = 1.6 = \frac{8}{5}$

remainder the amount left over when one number cannot be exactly divided by another.

square number any number which has been formed by multiplying a whole number by itself.
Examples: 4 × 4 = 16 thus 16 is a square number.
4 × 4 may be written as 4^2, that is to say 'four squared'.

square root any number which, when multiplied by itself, gives a specified number. The symbol for square root is √.
Examples: $\sqrt{36} = 6$,
$\sqrt{81} = 9$.

subtraction a mathematical operation in which one number is taken from another. The operation is normally indicated by the symbol −.
Example: 7 − 3 = 4.

sum (of numbers) the added total of given numbers.

whole number see natural number.

PRIME NUMBERS FROM 1 TO 1000

2	3	5	7	11	13	17	439	443	449	457	461	463	467
9	23	29	31	37	41	43	479	487	491	499	503	509	521
47	53	59	61	67	71	73	523	541	547	557	563	569	571
79	83	89	97	101	103	107	577	587	593	599	601	607	613
109	113	127	131	137	139	149	617	619	631	641	643	647	653
151	157	163	167	173	179	181	659	661	673	677	683	691	701
191	193	197	199	211	223	227	709	719	727	733	739	743	751
229	233	239	241	251	257	263	757	761	769	773	787	797	809
269	271	277	281	283	293	307	811	821	823	827	829	839	853
311	313	317	331	337	347	349	857	859	863	877	881	883	887
353	359	367	373	379	383	389	907	911	919	929	937	941	947
397	401	409	419	421	431	433	953	967	971	977	983	991	997

FRACTIONS, DECIMALS AND PERCENTAGES

Fractions, decimals and percentages are three ways of showing the same information: $\frac{1}{2}$, 0.5 and 50% are, for example, recording the same part of a larger amount.

A fraction is any quantity that is defined in terms of a numerator (the term above the line in a fraction which indicates how many parts of a specified number of parts of a unit are being taken) and a denominator (the term below the line in a fraction which indicates the number of equal parts into which the whole is divided).

A decimal is a fraction with an unwritten denominator of some power of ten. This is shown by the decimal point which is placed before the numerator (see above).

A percentage (shown by the symbol %) can also be regarded as a fraction. It shows a quantity as a part of 100.

decimal see introduction (above).

decimal point separates the whole numbers from the fractions in a decimal.

Examples: 17.3 represents 17 whole units and 3 tenths.
(0.3 is another way of saying $\frac{3}{10}$).
5.25 represents 5 whole units, 2 tenths and 5 hundredths.
(0.25 is another way of saying $\frac{25}{100}$, or $\frac{1}{4}$.)
9.837 represents 9 whole units, 8 tenths, 3 hundredths and 7 thousandths; and so on.

By convention a space is left after every three units following a decimal point, for example 3.677 245 21.

denominator see introduction (above).

equivalent fractions two fractions that express the same relationship of quantities.
Example: $\frac{3}{4}$, $\frac{6}{8}$, $\frac{9}{12}$ and $\frac{12}{16}$ are equivalent fractions.

Of these equivalent fractions $\frac{3}{4}$ is any of $\frac{6}{8}$, $\frac{9}{12}$ and $\frac{12}{16}$ in its lowest terms - the numerator and the denominator cannot be further divided by the whole numbers.

fraction see introduction (above).

improper fraction a fraction in which the numerator is larger than the denominator, for example $\frac{3}{2}$, $\frac{7}{5}$ and $\frac{18}{11}$. See also vulgar fraction (below).

irrational number a number which cannot be expressed as a fraction; such numbers have decimals which neither stop nor recur.
Examples: π and $\sqrt{2}$.

LCM lowest common multiple. For two or more integers, their LCM is the smallest integer into which all can be divided exactly.

lowest common denominator the LCM of the integers forming the denominators of the separate fractions.
Example: To compare $\frac{1}{2}$, $\frac{3}{5}$ and $\frac{7}{10}$ we need to express all three fractions in terms of the same denominator.

$$\frac{1}{2} = \frac{5}{10}$$
$$\frac{3}{5} = \frac{6}{10}$$
$$\frac{7}{10} = \frac{7}{10}$$

Thus 10 is the lowest common denominator and all three fractions can be converted into tenths.

lowest terms see equivalent fraction (see above).

mixed number a number that comprises an integer and a fraction, for example $3\frac{3}{4}$ (in other words $3 + \frac{3}{4}$).

numerator see introduction (above).

percentage see introduction (above).

proper fraction an alternative name for a vulgar fraction.

rational number any number which can be expressed as a fraction (or 'ratio').

recurring (of fractions, decimals and percentages) a pattern which repeats indefinitely.

vulgar fraction (or **proper fraction**) a fraction in which the numerator is smaller than the denominator, for example in $\frac{3}{7}$, $\frac{6}{12}$ and $\frac{5}{14}$. See also improper fraction (above).

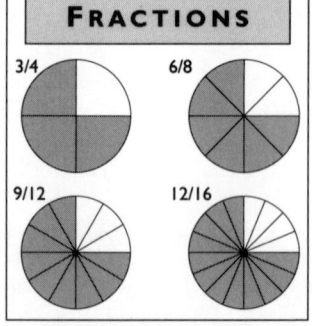

FRACTIONS

3/4, 6/8, 9/12, 12/16

CORRESPONDING FRACTIONS, DECIMALS AND PERCENTAGES

Fraction	Decimal	Percentage	Fraction	Decimal	Percentage
$\frac{1}{20}$	0.05	5%	$\frac{1}{2}$	0.50	50%
$\frac{1}{10}$	0.10	10%	$\frac{3}{5}$	0.60	60%
$\frac{1}{9}$	0.111 11*	$11\frac{1}{9}$%	$\frac{5}{8}$	0.625	62.5%
$\frac{1}{8}$	0.125	12.5%	$\frac{2}{3}$	0.666 67*	$66\frac{2}{3}$%
$\frac{1}{7}$	0.142 86	$14\frac{1}{7}$%	$\frac{3}{4}$	0.75	75%
$\frac{1}{6}$	0.166 67*	$16\frac{2}{3}$%	$\frac{4}{5}$	0.80	80%
$\frac{1}{5}$	0.20	20%	$\frac{5}{6}$	0.833 33*	$83\frac{1}{3}$%
$\frac{1}{4}$	0.25	25%	$\frac{7}{8}$	0.875	87.5%
$\frac{1}{3}$	0.333 33*	$33\frac{1}{3}$%	$\frac{8}{9}$	0.888 89*	$88\frac{8}{9}$%
$\frac{3}{8}$	0.375	37.5%			

* = recurring; by convention a recurring digit over 5 is rounded up.

ALGEBRA

In simple algebra, we generalize arithmetic by using letters to stand for unknown numbers whose value is to be discovered, or to stand for numbers in general. Certain conventions are followed.

Usually letters from the beginning of the alphabet are used to stand for numbers in general - for example, to stand for a general truth about numbers, such as

$$a + b = b + a.$$

The letters at the end of the alphabet are generally used to represent unknown numbers.

Example: A farmer has 42 cows and goes to market and purchases an unknown number of cows. When these additional cows are added to the herd the farmer has a total of 53 cows. This may be expressed as:

$$42 + x = 53,$$

where x is the unknown number of additional cows.

The operations of addition, subtraction, multiplication and division - that are used in normal arithmetic - can also be used in algebra, with certain basic restrictions. In addition and subtraction in algebra it is only possible to collect like terms (that is, terms that are expressed as parts or multiples of the same unknown quantity).

Examples: $3b + 2b + b + 7b = 13b.$
$3x + 2y + 5x + y + 2y + 6x = 14x + 5y.$
$2a + 7b - 3c + 3a + 2b + 6c - 8a + 2b - 2c =$
$= -3a + 11b + c.$

Multiplication and division in algebra also follow certain simple rules - the multiplication and division signs may be omitted and the numbers placed in juxtaposition to indicate multiplication.

Examples: $7 \times a = 7a.$
$-b \times c = -bc.$
$12y \div 4 = 3y.$

brackets (in algebra) are used to enclose a classified grouping within specified limits.

Examples: $(-2) + (-3),$
$(4x + 5x)^2.$

In the removal of brackets in an expression each term inside the bracket has to be multiplied by the number or term outside the bracket.

Example: $7(z + 2) = 7$ multiplied by $z + 7$ multiplied by 2
$= 7z + 14.$

change of subject see transformation of formulae (below).

equation a statement of equality between two quantities (expressed as numbers and/or letters). The equality of the two parts of the equation is shown by the equal sign (=).

Working out the answer to an equation is said to be 'solving' it. In solving a simple equation, the aim is to group the letters on one side of the equation and the numbers on the side.

Example: $9y - 3 = 7y + 3$
becomes $2y = 6$
giving $y = 3.$

Note that when terms are moved from one side of an equation to the other the addition and subtraction signs change. Thus in the above example:

$$9y - 7y = 3 + 3$$
giving $2y = 6.$

expression a symbol or symbols that express some mathematical fact - for example, $3a + 3b$. (Note that when an equal sign is introduced it becomes an equation - see above.)

factorization of an expression is basically the reverse operation of removing brackets from an expression - see brackets (above).

Examples: $4x + 4y$ becomes $4(x + y),$
$6a - 3b$ becomes $3(2a - b)$ - note that in this example 3 is the common factor (see pp. 162-63).

formula a set of symbols that express a mathematical rule.

Example: Simple interest is calculated using the formula

$$I = \frac{PRT}{100}$$

where
  I = simple interest,
  P = principal (the amount borrowed),
  R = the rate of interest, and
  T = the time for which the amount is borrowed (in years).

index (plur. **indices**) another word for power (see below).

inequality the use of one or other of the symbols < and >. The statement $x < y$ means that x is less than y; the statement $x > y$ means that x is greater than y.

inequation the use of one or other of the symbols ≤ and ≥. The symbol ≤ means less than or equal to; the symbol ≥ means greater than or equal to.

negative index (plur. **negative indices**) are common features in many equations.

$$b^2 = b + b \text{ but}$$
$$b^{-2} = \frac{1}{b^2}.$$

like terms see introduction (above).

power (multiplication and division) in equations and expressions the powers of the same number are added together in multiplication and subtracted in division. See also pp. 162-63.

Examples: $6n \times 6q = 6^{(n+q)}$
$p^9 \div p^3$ can be given as $p^{(9-3)} = p^6.$

transformation of formulae (or **change of subject**) is a simple rearrangement of a formula.

Example: The simple interest formula - see formula (above) - can be transformed in the following manner. As $I = \frac{PRT}{100}$, I is the subject, but this can be transformed to make T the subject.

If both sides of the formula are multiplied by 100 it gives

$$100I = PRT.$$

T is then isolated to become the subject of the formula by dividing both sides of the formula by PR. This gives

$$\frac{100I}{PR} = T.$$

RATIO AND PROPORTION

A ratio is the relation between quantities - two or more - of the same kind. Ratios are another way of expressing the same relationship that can be seen in a fraction. Proportion is the comparative relation between things - their price, size or amount, etc.

direct proportion the relationship between two quantities whereby an increase in one quantity is matched by an increase in a second quantity. Two quantities are also in direct proportion if a decrease in one quantity is matched by a decrease in the second quantity.

> *Example:* 5 plastic card index boxes cost £14. How much will 12 boxes cost?
>
> If 5 boxes cost £14, one box will cost £14 divided by 5 = £2.80.
>
> Therefore 12 boxes will cost £2.80 times 12 = £33.60.

inverse proportion the relationship whereby an increase in one quantity produces an increase in a second quantity in the same ratio. Two quantities are also in indirect proportion where a decrease in one quantity results in a decrease in a second quantity in the same ratio.

> *Example:* If it takes 9 days for 3 council workers to prune all the trees in a park, how long will it take 6 workers? As it is reasonable to assume that a greater number of council workers will complete the task in a shorter period of time, the two quantities are related in inverse proportion. In the example above it is reasonable to assume that 6 workers will complete the task in half the time taken by 3 workers, i.e. 4.5 days.

ratio the relation between two (or more) quantities of the same kind. A ratio is a comparison of sizes (of, for example, masses, prices of items, lengths or heights, etc.) and therefore no units are needed in this comparison.

> *Example:* The ratio of a journey of 5km to a journey of 25km = 5km:25km, or - as in this comparison of units of the same kind no units are required 5:25. This ratio cancels down to 1:5 in its most simple form as both 5 and 25 can be divided by 5.

Ratios and fractions are linked.

> *Examples:* the ratio 2:10 may also be expressed as the fraction $\frac{2}{10}$;
>
> the ratio 3:15 may also be expressed as the fraction $\frac{3}{15}$;
>
> likewise, the ratio 7:35 may also be expressed as the fraction $\frac{7}{35}$.
>
> All of these fractions cancel down to $\frac{1}{5}$; therefore the ratios 2:10, 3:15 and 7:35 may be expressed in the simplest form as 1:5.

NB: Only units of the same kind can be compared in a ratio.

> *Example:* The ratio of 1 sq metre to 1 hectare is not 1:1 as the two quantities are not of the same kind. It is 1 sq metre:1000 sq metres (1 hectare = 1000 sq metres) or 1:1000.

Ratios may be used to solve problems.

> *Example:* 120 tonnes may be divided in the ratio 1:5.
>
> The ratio 1:5 means that altogether 6 parts (1 plus 5) are involved in the division of the 120 tonnes. Each share is therefore worth 20 tonnes (that is, 120 divided by six); one share is 20 tonnes and five shares is 100 tonnes. Therefore 120 tonnes divided in the ratio of 1:5 is 20 tonnes:100 tonnes.

> *Example:* The ratio of the length of a piece of cloth to its width is 8:3. The width of the cloth is 1.5m. What is its length?
>
> As the ratio of length to width is 8:3, the width represents three shares.
>
> 1.5 divided by 3 = 0.5; one share is, therefore, 0.5.
>
> The length is 8 shares; 8 times 0.5 = 4.
>
> The length is, therefore, 4m.

scale the proportion that a map or model bears to the thing that it represents.

> *Example:* If a model train is made at a scale of $\frac{1}{150th}$ of the full size, what length on the real train would be represented by 2cm on the model?
>
> The proportion of model:real train is 1:150, therefore 150cm on the real train is represented by 1cm on the model.
>
> Hence 2cm on the model would represent 300cm on the real train.

In the case of map scales (or map ratios), measurements on the land are scaled down to be represented on a sheet of paper, a map. In atlases, sheet maps and town plans, a variety of scales are used. The terms small scale and large scale are often misunderstood. The larger the map scale, the larger the area of actual land that is being represented; the smaller the map scale the smaller the area of land that is being represented.

> *Example:* What is the actual distance on the ground represented by a line that is 4cm long on a map:
> (a) in which the scale is 1:600 000, and
> (b) a map in which the scale is 1:2 400 000?
>
> In map (a) 1cm on the map represents 600 000cm on the ground, therefore 4cm on the map represents 2 400 000cm on the ground, that is 24 000m or 24km.
>
> In map (b) 1cm on the map represents 2 400 000cm on the ground, therefore 4cm on the map represents 9 600 000cm on the ground, that is 96 000m or 96km.

NUMBER PATTERNS AND BINOMIALS

Some numbers are said to be rectangular, square or triangular. A number pattern may be either a representation of a particular number in the form of dots or some other symbol arranged in a pattern or a diagram in which different numbers are related (see Pascal's triangle, below).

binomial a mathematical expression consisting of two terms connected by a plus or minus sign.
Examples: $a + b$
$7 - 5$
Binomials are used in the laws of arithmetic - the commutative law, the associative law and the distributive law (see box).

binomial theorem a general mathematical formula expressing any power (see pp. 162-63) of a binomial as a sum.
Example: $(x + y)^2 = x^2 + 2xy + y^2$.

coefficient a number used as a multiplier.

laws of arithmetic see binomial (above) and box.

rectangular numbers are composite numbers, that is any number that is not prime (see pp. 162-63), Any composite number may be represented in the form of a rectangle of dots, Thus $6 =$

$2 \times 3 = 6$

square numbers are numbers with a pair of equal factors (see p. 162), and may therefore be represented as a square. Thus $4 =$

$2 \times 2 = 2^2$
$= 4$

and $9 =$

$3 \times 3 = 3^2$
$= 9$

1, 4, 9, 16, 25, 36, 49, 64, 81, 100, 121, 144, 169 are respectively the squares of 1, 2, 3, 4, 5, 6, 7, 8, 9, 10, 11 and 12. They - and all square numbers - are positive.

triangular numbers are numbers that can be formed into a series of equilateral triangles. Triangular numbers - such as 3, 6, 10 and 15 - can be represented by a triangular pattern of dots. Thus $6 =$

6

The differences between successive triangular numbers are the natural numbers (see pp. 162-63).

PASCAL'S TRIANGLE

Pascal's triangle is one of the most famous and most important of all number patterns.
Although it was long known before Pascal - who died in 1626 - he was the first to make ingenious and wide use of its properties. The numbers in Pascal's triangle appear in the binomial theorem, in problems about the selection of combinations of objects, and therefore in the theory of probability and statistics (see pp. 175-77).

Each number is formed by adding the two numbers diagonally above it, e.g. $10 = 6 + 4$, as ringed in the table. The numbers in the rows so formed are then the coefficients of the terms in the binomial theorem. Thus the numbers in the fourth row (1 3 3 1) are the coefficients of the terms in the expansion of $(a + x)^3$, while those in the sixth row would be the expansion of $(a + x)^5$, i.e. 1 5 10 10 5 1.

```
            1
           1 1
          1 2 1
         1 3 3 1
        1 4 6 4 1
       1 5 10 10 5 1
```

Totals
First line	1	$= 2^0$
Second line	2	$= 2^1$
Third line	4	$= 2^2$
Fourth line	8	$= 2^3$
Fifth line	16	$= 2^4$
Sixth line	32	$= 2^5$

PROBABILITY

At its most basic probability may be described as the study of chance and choice.

Not all actions and happenings have completely predictable results. Often we know that there is only a limited range of possible outcomes, but we do not know with certainty which of these to expect.

Probability theory enables us to describe with mathematical rigour the chance of an action or happening having a particular outcome. We may not, even then, make the right choice, but at least we shall have made a justifiable choice.

addition rule of probability see box.

***a posteriori* probability** an alternative name for empirical probability (q.v.).

***a priori* probability** the theoretical probability. See also frequency (below).

empirical probability is based on observation and experiment.

Here the probability of a particular outcome is calculated from the proportion of times it has been observed to have happened before under the same conditions - that is its relative frequency. Thus, if you tossed a coin ten times and the coin came up heads three times, the empirical probability that one of these throws comes up heads is 3/10.

equiprobable (of outcomes) see frequency.

frequency (mathematical) the number of times that any event, value, etc., is repeated in a given period or group.

When we toss a coin, we cannot predict which side will land facing upwards - this, after all, is the point of tossing coins, assuming we accept the fairness of the coin and the way it is tossed, we know it is just as likely to come up heads or tails, and there is no other possible outcome. The outcomes are said to be equiprobable.

The *a priori* probability of a coin coming up heads is 1 in 2 - or in the case of throwing a 6 on a single die the *a priori* probability is 1 in 6 or 1/6. See also empirical probability.

law of averages see law of large numbers.

law of large numbers states that as the number of trials increases, the observed empirical probability comes closer and closer to the theoretical value. There is no law of averages.

Suppose we toss a coin ten times and the outcome is only three heads. The probability of a head is 1/2, so why do we not get five heads? We try a total of 100 tosses of the coin, and the outcome is now, say, 40 heads, the last six being all heads. A gambler might back the chance of the 101st toss of the coin being tails, because, previously, there had been more tails than heads. Another gambler might back heads because there seemed to be 'a run' of heads, which would conform with the so-called 'law of averages'. However, we know that the probability of a head or a tail at any one toss of the coin is 1/2 and a coin cannot remember - it cannot be influenced by what has gone before.

PROBABILITY SCALE

When an outcome is certain, it occurs every time: 1 in 1, 2 in 2, etc. Expressing this as a fraction, we say the probability is 1/1, that is, one. When an outcome is impossible, it occurs no times in any number of tests, so we say the probability is zero.

For example, when throwing a die, the probability of throwing a number greater than six is zero, and the probability of throwing a number between 1 and 6 is one. Probabilities between certainties and impossibilities are expressed as fractions. So if, for example, we know that the six sides of a die are equiprobable, and the probability of throwing one of them is 1, the probability of each must be 1/6. Furthermore, if we consider only two possible outcomes, an odd number or an even number, the probability of each must be 1/2. The fact that there are three odd outcomes each with a probability of 1/6 - and 1/6 + 1/6 + 1/6 = 1/2 - demonstrates, very simply, the addition rule of probability: we can add up the individual probabilities of the different possible outcomes in a particular trial to get the combined probability.

In this example it means that only in a very long run will the relative frequency settle down towards 1/2.

likelihood ratio (or '**odds**') means the proportion of favourable to unfavourable possibilities or outcomes, and is a different way of expressing probability.

As we can see in the example quoted in the box, the probability of throwing, say, a 4 with a die is 1/6. Therefore, the probability of not throwing a 4 is 5/6. The 'odds' are thus expressed as 1 to 5 on throwing a 4 (or 5 to 1 against throwing a 4).

odds (a scale of measuring chance particularly used in gaming and gambling). Odds are more formally known as likelihood ratio (q.v.).

probability scale see box.

relative frequency see empirical probability.

Probability scale see box.

SETS

Sets can be considered simply as collections of objects. However, in the early 20th century, when attempts were made to formalize the properties of sets, contradictions were discovered that have affected mathematical thinking ever since. A set can be specified by stipulating some property for an object as a condition of membership of a set, or by listing the members of a set in any order. Conventionally this is written within braces i.e. curly sets of brackets { } see examples below.

complement see relative complement.

disjoint sets sets that have no members in common.

empty set see nullset.

intersection the set of members that is common to one or more sets, denoted by $\cap$.

member an object belonging to a set. $x \in A$ denotes that x is a member of set A.

null set a set containing no members, denoted by $\emptyset$.

relative complement those items in a universal set that are not part of a particular set. (See also the example, below.) The complement of set A is denoted by A'.

subset a set in which all of the members are also members of another set. Thus a set may be said to be a subset of another set where one set is contained within another. If the larger set is X and the smaller set - the subset - is Y, the equation $Y \subset X$ means that Y is a subset of X. The equation $Y \supset X$ means that set X contains subset Y. (See also the example below.)

union the total of all of the members in one or more sets, denoted by $\cup$.

universal set a set containing all of the items under consideration in a particular situation. (See also the example, below.) It is denoted by the symbol ξ.

Example: Consider the Smith family which has a bicycle, a motorcycle, a van, a family car and a sports car. We could represent the vehicles ridden or driven by Mrs Smith as

{motorcycle, van, family car}.

The number of elements in a set are indicated by the notation n(s) where s is the set. In this case n(s) = 3.

Sets are often shown by drawing a circle around representations of their members.

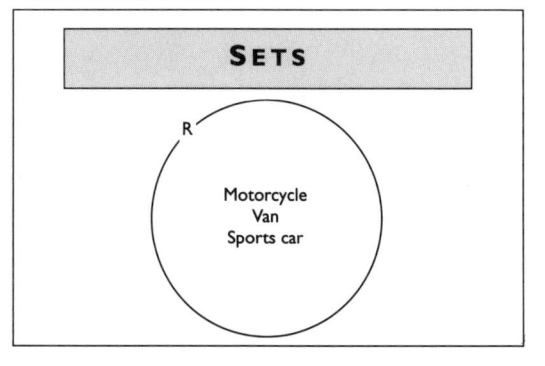

SETS

R
Motorcycle
Van
Sports car

We can use circles to represent the relationship between two or more sets.

If Mr Smith drives the sports car and the motorcycle but also shares the use of the van with Mrs Smith the set of vehicles used by him is

{motorcycle, van, sports car}

This can be represented as follows.

If R is the set of vehicles used by Mr Smith and S is the set of vehicles used by Mrs Smith R and S can be shown as two intersecting circles.

The set of all the vehicles used by the Smiths is

$R \cup S$ = {bicycle, motorcycle, van, family car, sports car}.

The two sets have one member in common, the van

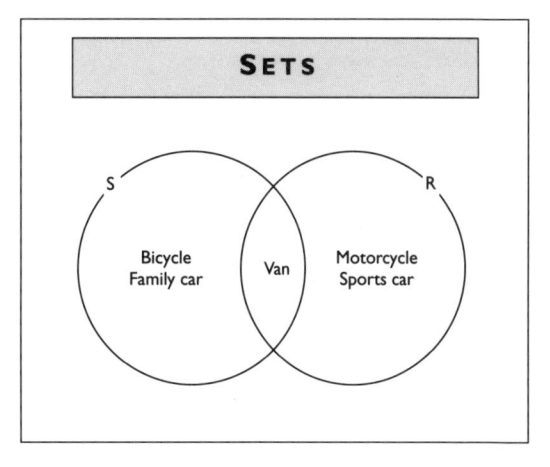

SETS

S
Bicycle
Family car
Van
Motorcycle
Sports car
R

The set of members that belong to both sets is known as their intersection. In the example given here it is the set whose only member is the van. This is written

$R \cap S$ = {van}.

In this example, among the Smiths' vehicles {van} is a subset of {bicycle, motorcycle, van, family car, sports car}. All of the vehicles used by the Smith family constitute a universal set.

In the universal set - the vehicles used by the Smiths - the vehicles not driven by Mrs Smith form the relative complement of the set of the vehicles she uses. Thus

S' = {bicycle, sports car}.

The set of vehicles driven by the Smiths' young child would be a null set. It would be written

{child's vehicle's} = $\emptyset$.

Suppose the young child has a bicycle - obviously a small one that only she can ride. The set of vehicles used by the child - C - is {child's bicycle}, a set of vehicles that does not intersect with the set of vehicles used by her mother. Thus C and S have no members in common and their intersection is an empty set. C and S are, therefore, disjoint sets, and we can write $C \cap S = \emptyset$.

COORDINATES AND GRAPHS

A graph is a diagram that shows the relationship between sets of numbers. It may be thought of as a picture that shows the values taken by a function. A function can be represented by a curve, or by a straight line, so allowing us to picture how a process changes and develops.

See also Calculus p. 172.

Graphs can be used for many purposes: for showing rates of change, for example in temperature in a scientific experiment, or for rates of exchange of currencies to convert one currency to another. A common use of graphs is the time/distance graph which shows how far a traveller, or a particular form of transport, can go in a particular period of time.

Example: The following graph shows the distance travelled by a car over a certain period of time. From the graph the speed of the vehicle can be worked out.

The speed can be calculated by dividing the distance travelled by the time taken. (This gives the same equation as the gradient (q.v.) of the graph.)

In this case, the speed of the vehicle
= 20km ÷ 45 minutes (that is, 0.75hr)
= 26.67km per hr.

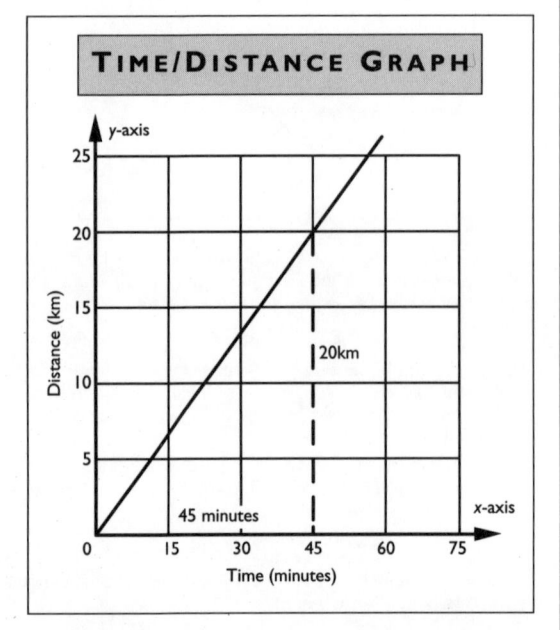

TIME/DISTANCE GRAPH

numbers are the coordinates (see below) of the point. Thus the coordinates of the point P in the diagram are (1,2).

Here the independent and the dependent variables of a function are represented by two lines at right angles (the x-axis and the y-axis) that cross at the origin (q.v.).

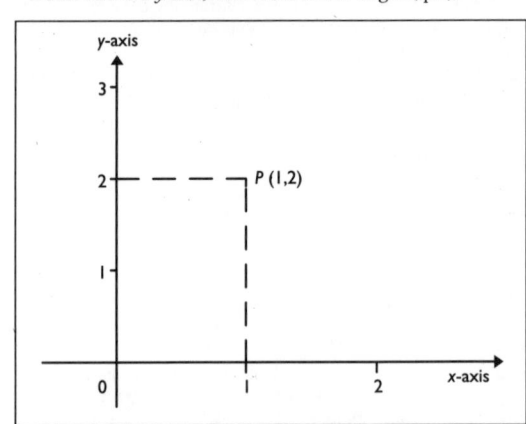

Cartesian coordinates see coordinate.

coordinate any of two or more magnitudes used to define the position of a point on a graph. Coordinates are known as an ordered pair of numbers as they are always given in the same form (x, y). The x value is always given first, followed by the y value.

Example: The curve representing the function $y = x^2$ is the set (see p. 169) of the pairs (x, y), of real numbers for which y is the square of x; thus, for example, (2,4), (−1,1), (−2,4), ($\sqrt{2}$,2), etc., are all in the graph of the function. The curve responding to this function is shown below.

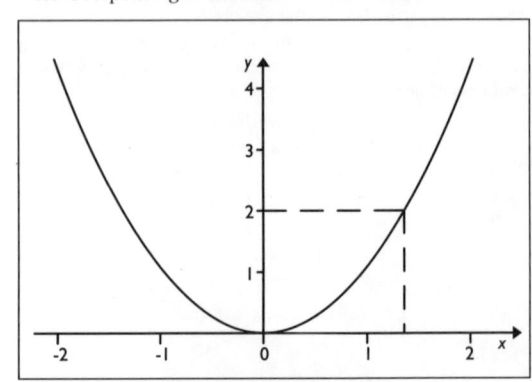

This system of coordinates is named after the French philosopher and mathematician René Descartes (or des Cartes, whence the adjective Cartesian).

function a relationship between two quantities or 'variables', whereby the values of one (the dependent variable) are uniquely determined by values of the other (the independent variable); see also the introduction (above).

axis (plur. **axes**) a straight line of reference, as in the horizontal and vertical axes of a graph. Real numbers can be represented geometrically by a line (axis) marked off from the origin (0) using some numerical scale.

Any point in a plane can be similarly represented by the pair of numbers that correspond to its respective distances from two such axes, as shown in the diagram; these

MATHEMATICS

gradient the rate of ascent or descent of a line on a graph. Gradients may be either positive or negative depending upon which direction the line slopes. See also positive gradient and negative gradient.

The gradient of a straight line can be worked out by dividing the change in the upward distance by the change in distance to the right horizontally.

In this example the gradient of the line
= 3 (upwards) ÷ 2 (downwards).
Thus the gradient = $\frac{3}{2}$ or $1\frac{1}{2}$.
This is a positive gradient.
Negative gradients are shown in the following example:
The gradient of line 1
= –2 (downwards) ÷ 2 (to the right).
Thus the gradient of line 1 = $\frac{-2}{2}$.
The gradient of line 2
= –3 (downwards) ÷ 4 (to the right).
Thus the gradient of line 2 = $\frac{-3}{4}$.

intercept the x-intecept is the coordinate at which a line or curve crosses the x-axis; the tan y-intercept is defined similarly for the y-axis.

negative gradient a gradient on a graph in which the line slopes downwards from left to right. See also gradient.

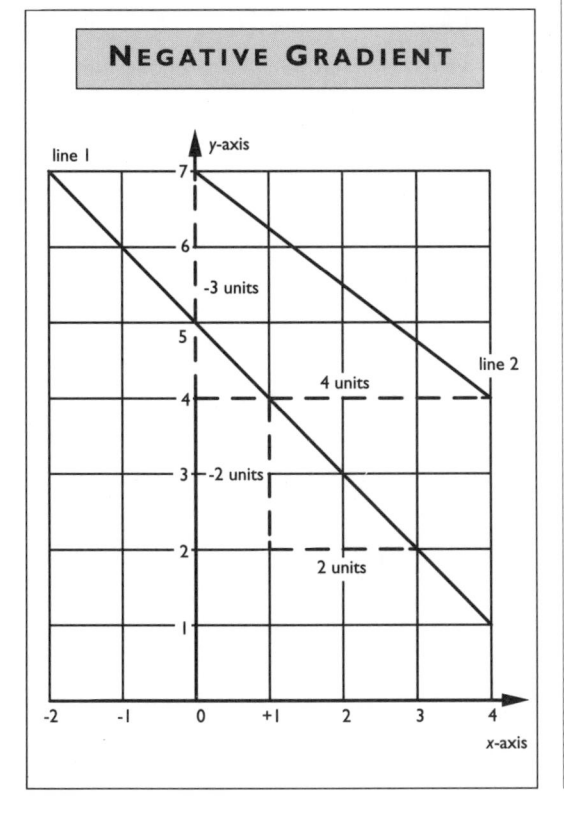

NEGATIVE GRADIENT

ordered pair (of numbers) see coordinates.

origin the point at which the axes of a graph cross. It has often, but has by no means always, the coordinates 0,0.

plane a two-dimensional area.

positive gradient a gradient on a graph in which the line slopes upwards from left to right. See also gradient.

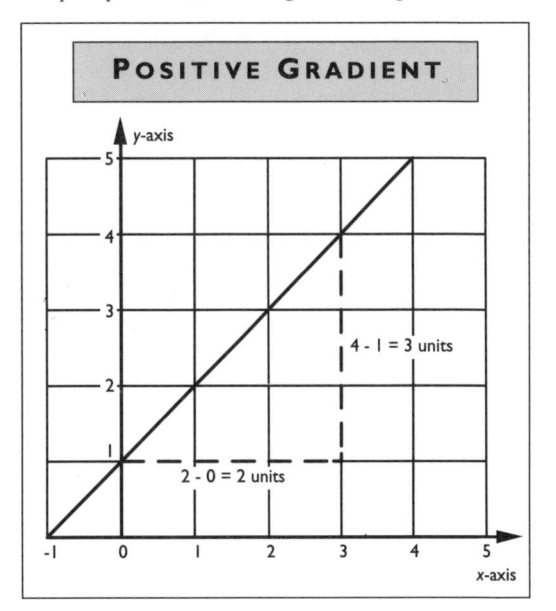

POSITIVE GRADIENT

straight line equation the equation $y = mx + c$ represents a straight line on a graph, for which m is the value of the gradient and c is the intercept.

Example: $y = 4x + 5$ represents a straight line of gradient 4 through the point (0,5).

table of values a chart or table of values from the equation of a line, or of a curve, from which the graph of a line, or of a curve, may be plotted.

Example: Plotting a table of values for the equation
$$y = 4x + 5.$$
For x values ranging from –5 to +5.
Where $x = -5$, $2x = -10$. Thus, $2x + 5 = -5 = y$.
Where $x = -4$, $2x = -8$. Thus, $2x + 5 = -3 = y$, and so on.
Thus, the following table of values can be drawn up:

x	–5	–4	–3	–2	–1	0	+1	+2	+3	+4	+5
y	–5	–3	–1	+1	+3	+5	+7	+9	+11	+13	+15

From these values a graph may be drawn.

x-axis the horizontal axis of a graph. The x value is always given first.

y-axis the vertical axis of a graph. The y value is always given secondly, that is following the x value.

CALCULUS

Calculus is the branch of mathematics that studies continuous change in terms of the mathematical properties of the functions that represent it, and these results can be interpreted in geometric terms relating to the graph (see pp. 170-71) of the function.

Calculus was developed independently by Newton and Leibniz in the late 17th century. Because their presentation involved paradoxical references to infinitesimals, many scientists rejected their 'infidel mathematics', but at the same time there was considerable dispute about who should have the credit for its discovery.

coefficient a number or symbol which is used as a multiplier.

coordinate any of two or more magnitudes used to define the position of a point; see also pp. 170-71.

The real numbers can be represented geometrically by a line (an axis) on a graph; see pp. 170-71.

curve see graph (below).

dependent variable see function.

derivative the instantaneous rate of change of one variable in relation to another variable.

differential calculus the branch of mathematics concerned with differentiation (q.v.).

differential coefficient of a product if

$$y = uv$$

where

u and v are functions of x, then

$$\frac{dy}{dx} = u\frac{dv}{dx} + v\frac{du}{dx}$$

differential coefficient of a quotient if

$$y = \frac{u}{v}$$

where

u and v are functions of x, then

$$\frac{dy}{dx} = \frac{v\frac{du}{dx} - u\frac{dv}{dx}}{v^2}$$

differentiation see box.

function a relationship between two quantities or 'variables' whereby the values of one (the dependent variable) are uniquely determined by the values of the other (the independent variable).

Suppose we go for a cycle ride and keep up a speed of 15km/h. Then our distance from home is determined by how long we have been travelling. For example, after half an hour we will have travelled 7.5km; after an hour 15km; after 2 hours 30km, and so on. We can express this relationship by saying that the distance is a function of the time we have been travelling. Here the two quantities, time and distance, might be represented by the variables t and d, and the mathematical relationship between them means that for any number of units of time, t, we can work out the number of units of distance travelled, d, by multiplying t by 15.

In general the notation for a function is $y = f(x)$, which indicates that the value of y depends upon the values of x; in that case, y is called the dependent variable and x is called the independent variable. The variables are thought of as running through a range of values - for example, if our journey takes a total of 3 hours, the domain of t is the interval (0,3), and the range of d is the interval (0,45).

Because a function associates elements of one set (see p. 169) with those of another, it defines the set of all pairs of elements, (x,y), in which x is the value of the independent variable and y is the value of the function for the argument x. Another way of expressing this is that any point that satisfies the function $y = f(x)$ can be represented by the point $(x, f(x))$.

Since a function must be a many-one relation, every such pair has a different first element, so the pairs can be listed in a unique order. The function can be thought of as moving through the values of the independent variable as the value of the independent variable increases. This is what is represented by a graph in the Cartesian coordinate system (see pp. 170-71):

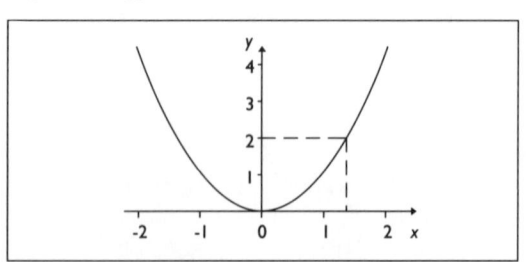

If we now draw a line joining the points $(x, f(x))$ as x increases, this line passes through all and only the points whose coordinates satisfy the function. Such a line is usually called a graph (see below and pp. 170-71).

graph (short for **graphic formula**) a diagram that shows the relationship between certain sets of numbers. Mathematicians tend to use the term graph for the set of values of variables and call the diagram a curve. Since this way of representing change and dependency is equivalent to the function itself, curves provide us with a way of visualizing processes of change.

See also pp. 170-71.

increment the amount of increase.

independent variable see function.

infinitesimal an infinitely small quantity.

interval see function.

DIFFERENTIATION

differentiation the process of finding the derivative of a function. This process can also be interpreted as geometry. However, it is not always necessary to work out a derivative by means of a graph. Instead, certain general principles apply.

The derivative of a function can itself be differentiated: for example, acceleration is the rate of change of velocity, and the derivative of the velocity function with respect to time can be worked out. This is the second derivative of the displacement function.

If y is any function of x, and Δy, Δx are corresponding increments of y and x, then the differential coefficient of y with respect to x (written $\frac{dy}{dx}$) is defined as

$$\operatorname*{Lt}_{\Delta x \to 0} \frac{[f(x - \Delta x) - f(x)]}{\Delta x}$$

$\frac{dy}{dx}$ gives the gradient of the curve, i.e. it measures the rate of change of one variable with respect to another.

Thus since velocity is the rate of change of displacement with respect to time, it may be expressed in calculus terms as

$$\frac{ds}{dt}$$

where

s in the displacement of a body from a fixed point and the equation of motion of the body is of the form $s = f(t)$.

Similarly, since acceleration is the rate of change of velocity with time, it may be expressed as

$$\frac{dv}{dt}$$

or as

$$\frac{d^2s}{dt^2}$$

i.e. as the second differential of s with respect to t. Acceleration may also be expressed as

$$v \frac{dv}{ds}$$

i.e. as the velocity multiplied by the rate of change of velocity with distance. In general, if

$$y = ax^n$$

then

$$\frac{dy}{dx} = nax^{n-1}$$

Since $\frac{dy}{dx}$ gives the gradient of a curve it may be used to find the maximum and minimum values of a function. thus if

$$y = f(x)$$

then when

$$\frac{dy}{dx} = 0$$

the tangents to the curve will be parallel to the x-axis, and will indicate the positions of the critical values (the maximum or minimum) but without distinguishing them. However, if

$$\frac{d^2y}{dx^2} \text{ is +ve}$$

the critical value of x gives a minimum value of a function, while if

$$\frac{d^2y}{dx^2} \text{ is -ve}$$

the critical value gives a maximum value of a function, and if

$$\frac{d^2y}{dx^2} = 0$$

and changes sign as x increases through the point, the curve is passing through a point of inflection.

MECHANICS

Mechanics is that branch of physics which is concerned with the motion of bodies and the action of forces on bodies; see pp. 74-83 (further equations will be found in those pages). This section deals with the mathematics of mechanics.

circular motion (where a body is moving in a circle with uniform speed) may be represented by the following equation:

$$vr\omega$$

where
v = linear velocity,
r = the radius of the circle, and
ω = the angular velocity.

The body will nevertheless have an acceleration (since a force is acting on it to make it move in a circle) but this will be directed towards the centre.

The acceleration will be:

$$r\omega^2$$

The force will be:

$$mr\omega^2$$

where
m is the mass of the body.

component of a force see force (below).

conservation of momentum principle see momentum.

force changes an object's motion, thus making it move more or less quickly and/or causing it to change direction. A force F causes a mass m to accelerate at a rate a according to the equation

$$F = ma$$

The effect of a force F in a direction at an angle θ to the direction in which the force is applied is called its component in that direction and is given by:

$$F\cos\theta$$

friction equations see pp. 82-83.

impact of elastic bodies if the bodies are smooth (for example, snooker balls) and only the forces between the bodies are considered, the equations determining the velocities and directions of the bodies after impact will be found under momentum (q.v.) and the velocity of separation (q.v.).

kinetic energy (symbol KE) of a particle of mass m moving with velocity v is: $\frac{1}{2}v^2$.

momentum the product of a mass and its velocity. The conservation of momentum principle states that the total momentum of bodies in any given direction is the same before and after impact.

The following equation can be given:

$$m_1v_1 + m_2v_2 = m_1u_1\cos\alpha + m_2v_2\cos\beta$$

See velocity of separation for an explanation of the terms used in the above equation.

Newton's laws of motion (equations) see pp. 74-78.

potential energy (symbol PE) gained by a mass m as it is raised through a height h is mgh.

projectiles formulae for calculations include the following.

For simple cases in which air resistance is neglected and the vertical velocity is subject only to the acceleration of

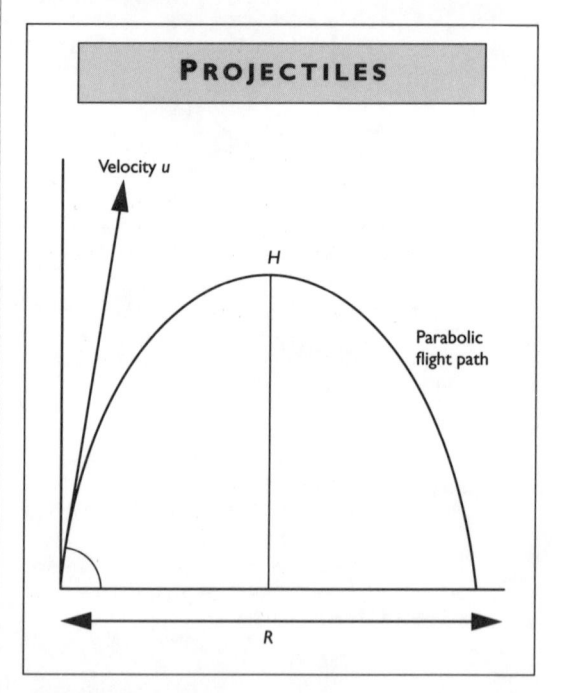

PROJECTILES

Parabolic flight path

gravity, the results below may be derived from the fundamental equations of motion:

The time of flight

$$T = \frac{2u\sin\theta}{g}$$

The time to the greatest height

$$= \frac{T}{2} = \frac{u\sin\theta}{g}$$

The greatest height obtained

$$H = \frac{u^2\sin^2\theta}{2g}$$

The range on a horizontal plane

$$R = \frac{u^2\sin 2\theta}{g}$$

For a given velocity of projection u there are, in general, two possible angles of projection to obtain a given horizontal range. These directions will make equal angles with the vertical and horizontal respectively. For maximum range the angle makes 45° with the horizontal.

relative velocity may be calculated in the following manner.

To find the velocity of a body A relative to a body B, combine with the velocity of A a velocity equal and opposite to that of B. The sides of the triangle represent the velocities of magnitude and direction. Thus to a person on a ship at B, the ship A would appear to be

RELATIVE VELOCITY

(Diagram: triangle with sides labelled "Velocity of A", "Velocity equal and opposite to that of B", and "Velocity of A relative to B")

moving in the direction (and at the speed) represented by the double-arrowed line.

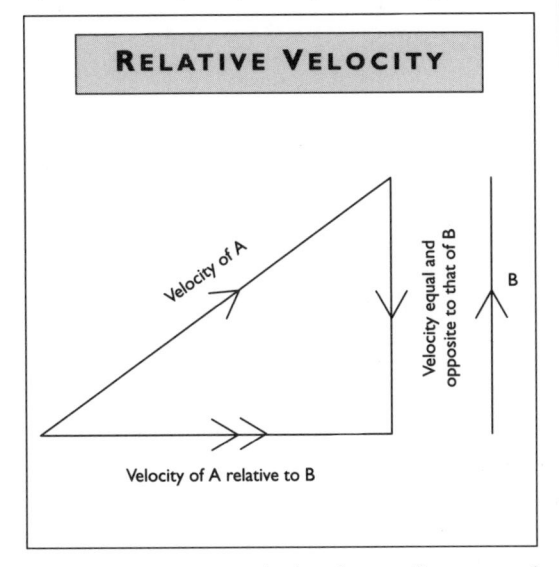

The triangle ABC shows how the track (i.e. the actual direction) and velocity relative to the ground (the ground speed) of an aircraft or a boat may be found from the course set and the wind or current.

TRIANGLE OF VELOCITIES

(Diagram: triangle ABC with sides labelled "Course set", "Wind or current", and "Track and ground speed")

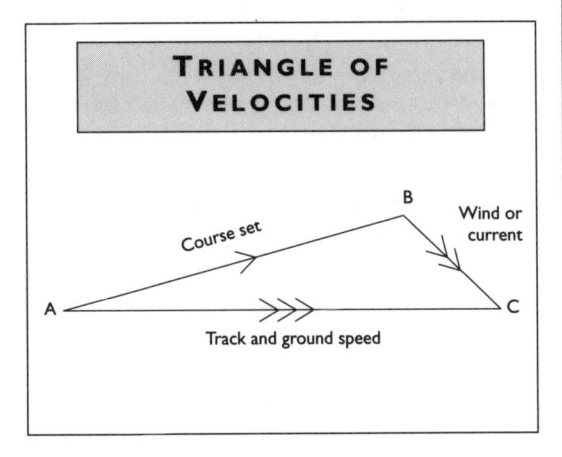

In vector terms,
$$\vec{AB} + \vec{BC} = \vec{AC}$$

simple harmonic motion occurs when a particle moves so that its acceleration is directed towards a fixed point in its path, and is proportional to its distance from that point.

The fundamental equation is:
$$\frac{d^2x}{dt^2} = -\omega^2 x$$

By integrating the corresponding equation
$$v\frac{dv}{dx} + \omega^2 x = 0$$

the velocity at any displacement of x is given by:
$$v = \omega\sqrt{a^2 - x^2}$$

where

a is the maximum value of x.

The displacement at time t is given by

$x = a \sin\omega t$ if timed from the centre, or

$x = a \cos\omega t$ if timed from the maximum displacement.

triangle of velocities see relative velocity.

uniform acceleration equations include the following:

If a particle accelerates at a constant rate (a) from a velocity u to a velocity v over a displacement s and during time t, then its motion is governed by the equations:

$$v = u + at$$
$$v^2 = u^2 + 2as$$
$$s = ut + \tfrac{1}{2}at^2$$
$$s = \tfrac{1}{2}(u + v)t$$

vector either a physical quantity with magnitude and direction (such as velocity or force) or a line that represents such a quantity.

velocity of separation is equal to the velocity of approach (also measured along the line of centres) multiplied by the coefficient of elasticity between the two bodies. If the impact is oblique (and the bodies are smooth) the velocities at right angles to the line of centres is unchanged.

If u_1 and u_2, m_1 and m_2 are the initial velocities and masses of the two spheres, and α and β are the angles these velocities make with the line of centres, and v_1 and v_2 are the components of velocities along the line of centres after impact, the above statements may be represented by the following equation:

$$v_2 \pm v_1 = e(u_1 \cos\alpha - u_2 \cos\beta)$$

See also momentum.

work done by a force F acting on a body as it covers a displacement s is

$$Fs.$$

MATRICES

A matrix is an array of numbers, of rectangular shape, which presents information in a concise form. Matrices serve many purposes, and according to the circumstances they may be added or subtracted or multiplied.

Two matrices can be multiplied if there are the same number of *rows* in the second matrix as there are *columns* in the first, but they may only be added or subtracted if they have the same number of columns and rows, i.e. they are exactly the same shape.

A 2 × 3 matrix is one with two rows (such as C below) and three columns. Thus a 2 × 3 matrix can be multiplied by a 3 × 4 matrix or a 3 × 2 matrix or a 3 × n, where *n* is any number.'

Examples:

If $A = \begin{pmatrix} a & b \\ c & d \end{pmatrix}$, $B = \begin{pmatrix} p & q \\ r & s \end{pmatrix}$ and $C = \begin{pmatrix} u & v & w \\ x & y & z \end{pmatrix}$

then, $AB = \begin{pmatrix} a & b \\ c & d \end{pmatrix}\begin{pmatrix} p & q \\ r & s \end{pmatrix} = \begin{pmatrix} ap+br & aq+bs \\ cp+dr & cq+ds \end{pmatrix}$

$A + B = \begin{pmatrix} a & b \\ c & d \end{pmatrix} + \begin{pmatrix} p & q \\ r & s \end{pmatrix} = \begin{pmatrix} a+p & b+q \\ c+r & d+s \end{pmatrix}$

$AC = \begin{pmatrix} a & b \\ c & d \end{pmatrix}\begin{pmatrix} p & q \\ r & s \end{pmatrix} = \begin{pmatrix} au+bx & av+by & aw+bz \\ cu+dx & cv+dy & cw+dx \end{pmatrix}$

A and C cannot be added because they are not the same shape.

determinant of a 2 × 2 matrix for the matrix $\begin{pmatrix} a & b \\ c & d \end{pmatrix}$, the determinant is ad-bc.

The value of the determinant of a transformation matrix represents the ratio by which the area of the original figure has been changed. If the determinant is zero, all the points will be moved to lie on a line, and the matrix is said to be singular.

identity matrix is the result when a matrix is multiplied by its inverse. It has the effect of leaving unchanged any matrix by which it is multiplied. See also box.

inverse of a 2 × 2 matrix is the matrix which exactly reverses the transformation performed by the original matrix. If A denotes a matrix, A^{-1} denotes its inverse.

isometric transformation a transformation which does not change either the shape or the size of a figure. See also box.

singular matrix see determinant of a matrix.

transformation matrix see box.

TRANSFORMATION MATRICES

Any 2 × 2 matrix represents some transformation of points in a plane. To perform the transformation, write the points as columns of a matrix and pre-multiply by the transformation matrix, e.g.

$\begin{pmatrix} 1 & 0 \\ 0 & -1 \end{pmatrix}\begin{pmatrix} 4 & 4 & 1 \\ 1 & 2 & 2 \end{pmatrix} = \begin{pmatrix} 4 & 4 & 1 \\ -1 & -2 & -2 \end{pmatrix}$

shows the points (4,1), (4,2) and (1,2) being transformed to the points (4,–1), (4,–2) and (1,–2) by the matrix

$\begin{pmatrix} 1 & 0 \\ 0 & -1 \end{pmatrix}$

which has reflected them in the x-axis.

The following are the principal transformation matrices:

Reflection in the *x*-axis
$\begin{pmatrix} 1 & 0 \\ 0 & -1 \end{pmatrix}$

Reflection in the *y*-axis
$\begin{pmatrix} -1 & 0 \\ 0 & 1 \end{pmatrix}$

Reflection in the line $y = x$
$\begin{pmatrix} 0 & 1 \\ 1 & 0 \end{pmatrix}$

Reflection in the line $y = -x$
$\begin{pmatrix} 0 & -1 \\ -1 & 0 \end{pmatrix}$

Rotation through 90° about the origin in a +ve (anticlockwise) direction
$\begin{pmatrix} 0 & -1 \\ 1 & 0 \end{pmatrix}$

Rotation through 180° (+ve or –ve)
$\begin{pmatrix} -1 & 0 \\ 0 & -1 \end{pmatrix}$

+ve rotation about the origin through an angle θ
$\begin{pmatrix} \cos\theta & -\sin\theta \\ \sin\theta & \cos\theta \end{pmatrix}$

The identity matrix
$\begin{pmatrix} 1 & 0 \\ 0 & 1 \end{pmatrix}$

This leaves the elements of the multiplied matrix unchanged. The following matrices change the shape of the figure.

An enlargement, factor E
$\begin{pmatrix} E & 0 \\ 0 & E \end{pmatrix}$

if E = 3 the figure will have its linear dimensions trebled.

A stretch parallel to the *x*-axis, factor S
$\begin{pmatrix} S & 0 \\ 0 & 1 \end{pmatrix}$

A stretch parallel to the *y*-axis, factor S
$\begin{pmatrix} 1 & 0 \\ 0 & S \end{pmatrix}$

A shear parallel to the *x*-axis
$\begin{pmatrix} 1 & S \\ 0 & 1 \end{pmatrix}$

A shear parallel to the *y*-axis
$\begin{pmatrix} 1 & 0 \\ S & 1 \end{pmatrix}$

STATISTICS

Statistics is sometimes treated as a branch of mathematics, but it can also be regarded as a separate science - the science that is concerned with the collection, study and analysis of numerical data.

The data studied can be the result of surveys, forms, questionnaires or censuses, or may derive from systematic experiments and the recording of results or from observation and the recording of information. Governments have statistical departments recording and predicting finance, trade, population figures, industrial and agricultural production and so on.

Data has to be sorted into a form that is useful. Different categories have to be identified so that the data can be presented in a form that is meaningful. At its simplest, this is achieved by means of frequency distribution tables and tally charts (q.q.v.).

When data has been processed - into a frequency distribution table, for example, it can then be presented visually in a diagram. The forms most commonly used are the pictogram, the bar chart, the pie chart and the histogram (see below).

bar chart a popular method of displaying statistical data. It is also, probably, the easiest of the methods considered here.

A bar chart consists of columns that are arranged either vertically or horizontally. (When the columns are arranged vertically a bar chart is sometimes referred to as a column graph.)

Each bar is the same width and, where there are spaces between them, those spaces are uniform. The length (in a horizontal bar chart) or the height (in a vertical bar chart) depends upon the size of the section of the data that it represents.

Example: The following bar chart shows the gross national products (GNPs) of the world's leading economic powers - the figures are 1992 estimates. (The GNP of a country is the money value of the total amount of goods and services produced by a country over a one-year period, plus profits, interest and dividends from abroad.)

This chart is compiled from the following table:

Country	GNP (billion US$ - 1992 est.)
USA	5 670
Japan	4 000
Germany	1 950
France	1 360
Italy	1 170
UK	1 000
Canada	588
Spain	540
China	476
Brazil	393
Mexico	382
South Korea	354
Netherlands	330
Australia	309
Argentina	273
Sweden	260
Taiwan	241
Switzerland	240
India	236
Belgium	224

The bar chart shows far more dramatically than a table of figures just how dominant the US economy is worldwide. It graphically reveals the challenge of Japan in second place, the large gap between the size of the Japanese and German economies, and the greater size of the German economy compared with the economies of the other leading European states - France, Italy, the UK and Spain.

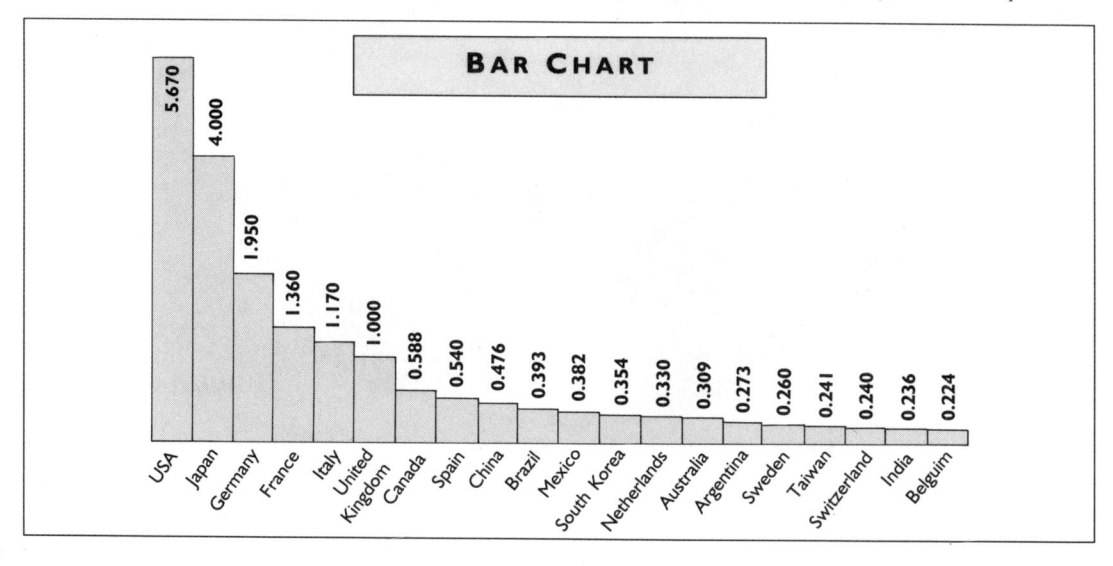

BAR CHART

PICTOGRAM

🍗 = 100 kilos of meat sold

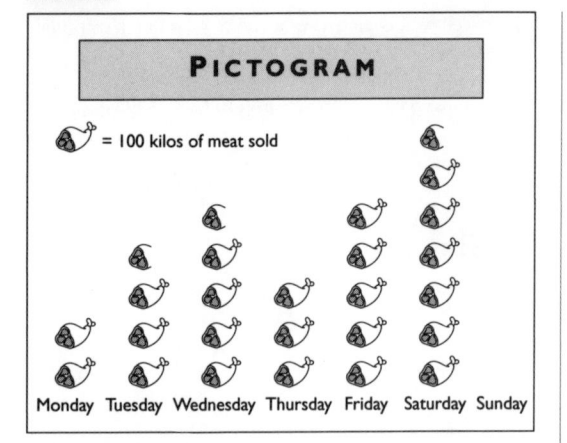

Monday Tuesday Wednesday Thursday Friday Saturday Sunday

column graph see bar chart.

frequency distribution table a table derived from information shown on the final column on a tally chart (q.v.).

grouped data see tally chart.

histogram a visual method of showing data which resembles a vertical bar chart. It differs, however, in that it is the area rather than the height of each bar that represents the data. In a histogram - which is often but not always drawn on squared graph paper - there are no gaps between the bars.

ideograph another name for a pictogram.

pictograph another name for a pictogram.

pictogram (or **pictograph** or **ideograph**) a simple method of showing frequencies by means of symbols or small pictures. In a pictogram all the symbols are drawn the same size and are placed regularly in appropriate columns or blocks. Data of less than the scale size is shown by a part of the symbol, for example one half or one quarter of the symbol.

Example: The following example records the amount of meat sold by a butcher during one week.

Monday	(when weekend joints are still being used)	200 kilos
Tuesday		350 kilos
Wednesday		450 kilos
Thursday		300 kilos
Friday	(when weekend shopping begins)	500 kilos
Saturday	(when weekend joints are bought)	650 kilos
Sunday	(when the shop is closed)	0

pie chart (or **pie graph**) a visual method of showing data in which a circle is drawn and divided into slices of a 'pie' to show the different shares of the total. The size of each slice depends upon the size of each item of data. A pie chart can only be used to show the sizes of parts of a known total.

Example: The following pie chart shows the world's major

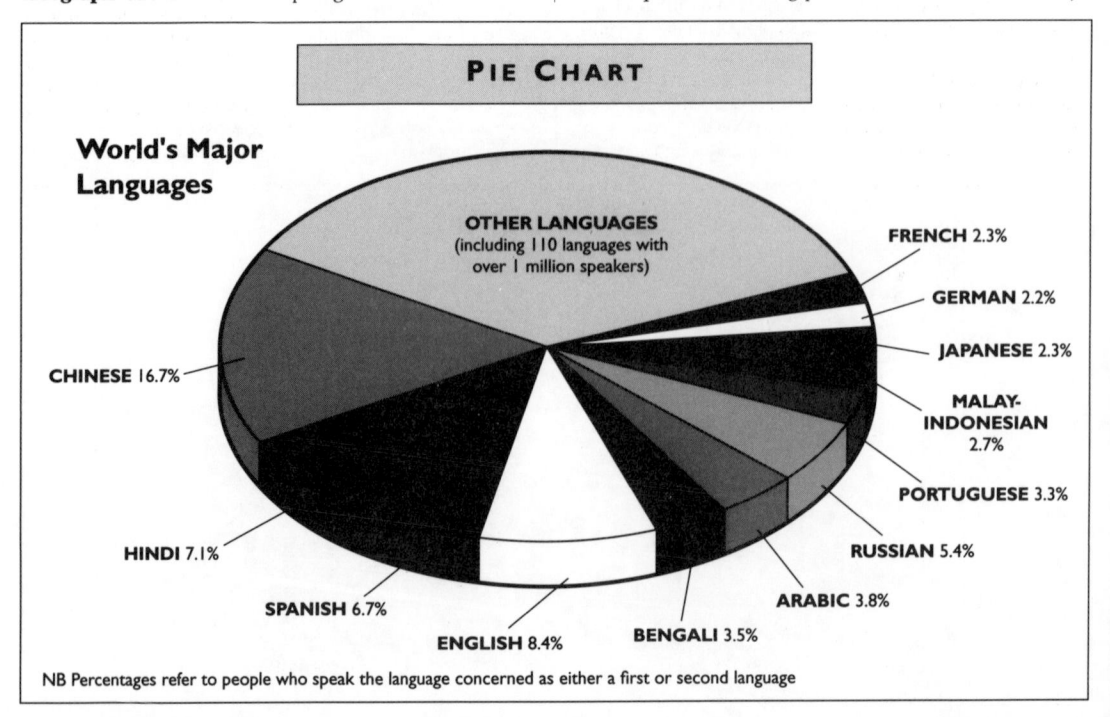

PIE CHART

World's Major Languages

- OTHER LANGUAGES (including 110 languages with over 1 million speakers)
- FRENCH 2.3%
- GERMAN 2.2%
- JAPANESE 2.3%
- MALAY-INDONESIAN 2.7%
- PORTUGUESE 3.3%
- RUSSIAN 5.4%
- ARABIC 3.8%
- BENGALI 3.5%
- ENGLISH 8.4%
- SPANISH 6.7%
- HINDI 7.1%
- CHINESE 16.7%

NB Percentages refer to people who speak the language concerned as either a first or second language languages as a percentage of the total world population. In the chart 16.7% of the total is shown as speaking Chinese as either a first or second language. The entire pie chart has an angle of 360° at the centre. Therefore the slice represented by 16.7% of the total will be 360 ÷ 100 = 3.6 × 16.7 = 60.1°. Similarly the percentage of the world's population speaking English as either a first or a second language is 8.4%; this is represented by as slice whose angle at the centre of the circle is 30.2° (that is, 8.4 × 3.6°).

tally chart a table in which data is sorted into meaningful categories.

Example: In the cricket Test match between New Zealand and Pakistan held at Hamilton on 2, 3, 4, 5 January 1993 all of the batsmen in both sides batted in both innings. Pakistan scored 216 (including 8 extras) in the first innings and 174 (including 4 extras) in the second innings; New Zealand scored 264 runs in the first innings (including 33 extras) and only 93 in the second innings (including 22 extras).

The number of runs scored by each batsman was as follows:

4 0 0 92 14 23 27 32 13 2 1 133 43 2 2 14 6 12 16 0 3 0 8 0 11 12 0 75 15 33 4 10 2 4 8 9 19 0 9 9 4 13 0 0 0.

In this form it is not easy to make very much use of these figures, but if a tally chart such as the one shown here is drawn up then the information becomes more meaningful.

From the list of scores one stroke is marked on the tally chart each time a score occurs within one of the chosen categories. Every fifth score is drawn across the previous four, giving handy blocks of five.

In this case the cricket scores are shown on the tally chart in categories, because it would be totally unwieldy to have a separate line for every line between 0 (the lowest score) and 133 (the highest score). Because it has been necessary to use these groups, the chart shows what is called grouped data. (In the case of a tally chart showing, for example, the marks out of ten for a French test taken by a class there would be no need to group the scores achieved - a separate line could be used for each mark between 0 and 10. Such a tally chart would show ungrouped data.)

A final column can be added to the above example to record the frequency with which scores fall into each category. It would appear as follows:

TALLY CHART

Number of runs scored by individual batsmen in the New Zealand v. Pakistan Test Match (January 1993)

Number of runs	Tallies
under 10	JHT JHT JHT JHT JHT II
11-20	JHT JHT
21-30	II
31-40	II
41-50	I
51-60	
61-70	
71-80	I
81-90	I
91-100	
101-110	
111-120	
121-130	
131-140	I

TALLY CHART WITH FREQUENCY CHART

Number of runs scored by individual batsmen in the New Zealand v. Pakistan Test Match (January 1993)

Number of runs	Tallies	Frequency
under 10	JHT JHT JHT JHT JHT II	27
11-20	IIII IIII	10
21-30	II	2
31-40	II	2
41-50	I	1
51-60		0
61-70		0
71-80	I	1
81-90	I	1
91-100		0
101-110		0
111-120		0
121-130		0
131-140	I	1

By omitting the central column - tallies - a frequency distribution table can be drawn up.

ungrouped data see tally chart.

GEOMETRY AND TRIGONOMETRY

Geometry is that branch of mathematics which deals with the properties of lines, points, surfaces and solids. The rules of geometry - which have been discovered not invented - are used to derive angles, areas, distances, etc., which may not otherwise be measured directly.

A simple example of a mathematical model is the representation of a portion of the Earth's surface by a set of interlocking triangles, from the measurement of which maps may be constructed. Geometry establishes that two triangles each have angles of the same sizes if, and only if, corresponding pairs of sides are in the same proportions.

TRIANGLE ABC

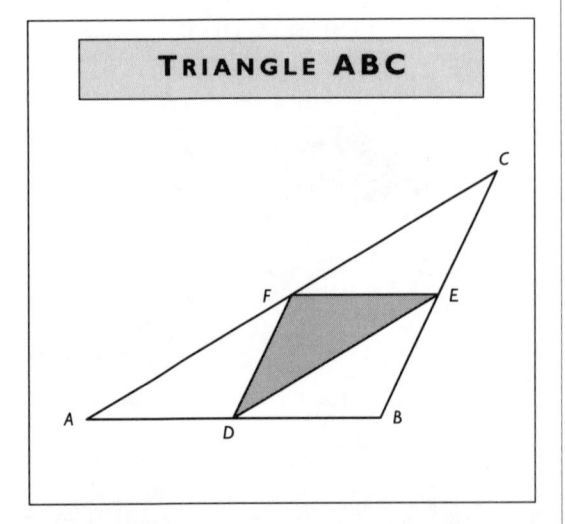

Here D, E and F are the centre-points of sides AB, BC and CA respectively. So, DE is half the length of AC, EF is half the length of AB, and FD is half the length of BC. Thus, the shaded triangle, and the angles at D, E and F, are respectively equal to those at C, A and B. Furthermore, the triangles ADF, FEC, DBE and EFD are all congruent, that is identical in shape and size, and are thus all similar to triangle ABC.

acute angle an angle of less than 90°.

circles see pp. 186-87.

cos see cosine.

cosine of an angle (shortened to **cos**) the ratio of the side adjacent to the given angle to the hypotenuse. In the diagram in the entry on trigonometry (see below), $\cos\theta$ is AB/AC. (See also sine and tangent.)

hypotenuse the side of a right-angled triangle that is opposite the right angle.

line symmetry see symmetry.

plane figure a figure that lies entirely on one plane, i.e. they are two-dimensional. They includes polygons, quadrilaterals, triangles, circles and conic sections.

polygons see pp. 182-83.

polyhedra see pp. 190-91.

Pythagoras' theorem see pp. 184-85, and values for sin, cos and tan (below).

reflection (of a figure) the movement of a figure into a mirror image of its previous position. The size and shape of the figure remain unchanged.

rotation (of a figure) the movement of a figure in position in which one point of the plane - known as the centre of rotation - remains static. The size and shape of the figure remains unchanged.

rotational symmetry see symmetry.

sin see sine.

sine of an angle (shortened to **sin**) the ratio of the side opposite the given angle to the hypotenuse. The Greek letters θ and Ø are usually used to denote the angles, thus in the diagram shown in the entry on trigonometry (below) we say that the sine of θ, usually written as sin θ, is BC/AC. (See also cosine and tangent.)

solids see pp. 188-89.

symmetry the correspondence of opposite parts in size, shape and position. There are two types of symmetry - line symmetry and rotational symmetry. If the second triangle shown in the entry on the values of sin, cos and tan (see below) were to be folded along the line KL, the two halves of the triangle would fit exactly one on top of the other. The whole shape that is the triangle is therefore said to be symmetrical and the fold line KL is known as the line of symmetry. Rotational symmetry occurs when a shape is rotated about its centre point and at some position during the rotation the shape appears not have moved. The letters N and S have rotational symmetry. The letter H and the figure 8 have both line and rotational symmetry.

tan see tangent.

HYPOTENUSE

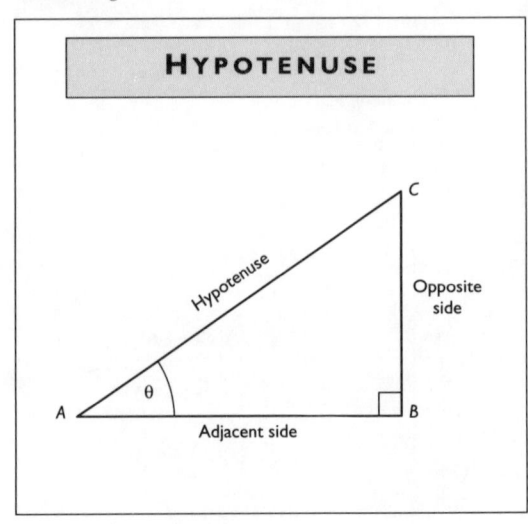

tangent of an angle (shortened to **tan**) the ratio of the opposite to the adjacent side; in the example in the diagram in the entry on trigonometry this is BC/AB. In this figure it is easy to see that $\tan\theta$ must always equal $\sin\theta/\cos\theta$. (See also cosine and sine.)

translation (of a figure) the movement of a figure from one position to another maintaining the same image.

triangle see pp. 184-85.

triangulation the determining of the distance between two points on the Earth's surface by dividing an area into connected triangles.

trigonometry the branch of mathematics that deals with the relations between the angles or the sides of triangles.

Trigonometry relies on the recognition that in a right-angled triangle the ratio of the lengths of pairs of sides depends only on the sizes of the two acute angles of the triangle. These ratios are given names - sine, cosine and tangent.

values for sin, cos and tan of 30°, 45°, and 60° can be established using Pythagoras' theorem (see pp. 184-85).

TRIANGLE DEF

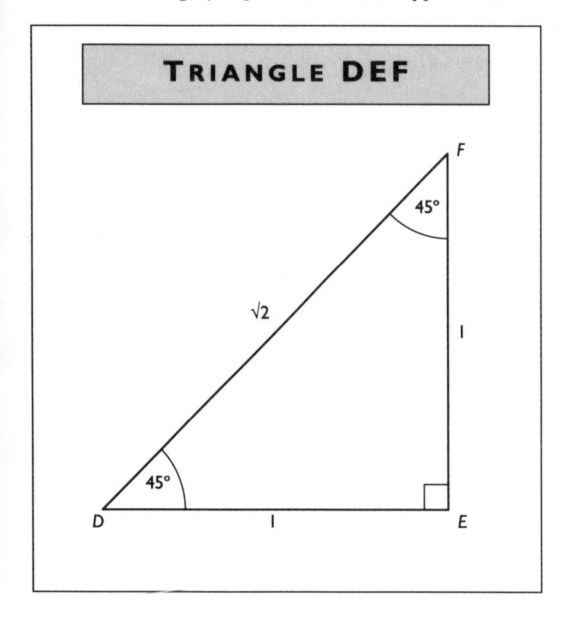

In triangle *DEF*, $DE = EF = 1$, so the angles at *D* and *F* are equal, that is they are each 45° (the internal angles of a triangle add up to 180°). Using Pythagoras' theorem (see pp. 184-85),

$$DF^2 = 1^2 + 1^2 = 2$$
so
$$DF = \sqrt{2}$$

We can therefore conclude:

$$\sin 45° = \frac{1}{\sqrt{2}}$$
$$\cos 45° = \frac{1}{\sqrt{2}}$$
$$\tan 45° = 1$$

TRIANGLE GHK

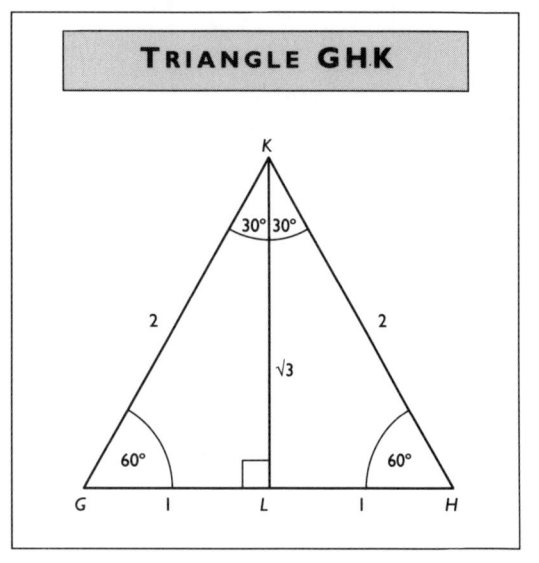

In triangle *GHK*, $GH = HK = 2$, so the angles at *G*, *H* and *K* are equal, that is they are each 60°. Using Pythagoras' theorem,

$$KL^2 + 1^2 = 2^2$$
so
$$KL = \sqrt{3}$$

We therefore have:

$$\sin 60° = \tfrac{1}{2}\sqrt{3} = \cos 30°$$
$$\cos 60° = \tfrac{1}{2} = \sin 30°$$
$$\tan 60° = \sqrt{3}$$
$$\tan 30° = \frac{1}{\sqrt{3}}$$

POLYGONS

A polygon is a closed plane figure with three or more straight sides that meet at the same number of vertices and do not intersect other than at those vertices. (A vertex is the point at which two sides of a polygon meet.) Although we tend to think of a polygon as being a many-sided figure, it can have as few sides as three (a triangle).

The sum of the interior angles =

$$\frac{(2n-4) \times 90°}{n}$$

where

n = the number of sides.

Each interior angle of a polygon =

$$\frac{(2n-4) \times 90°}{n}$$

or

$$= 180° - 360°$$

The sum of the exterior angles of any polygon = 360°, regardless of the number of sides. (An exterior angle of a polygon is the angle between one side extended and the adjacent side.)

decagon a polygon with ten sides.
dodecagon a polygon with 12 sides.
heptagon a polygon with seven sides.
hexagon a polygon with six sides.
nonagon a polygon with nine sides.
octagon a polygon with eight sides.
parallelogram a quadrilateral whose opposite sides are equal in length and parallel. If has no lines of symmetry, unless it is also a rectangle, but it does have rotational symmetry about its centre, the point where the diagonals meet.

If one angle of a parallelogram is a right angle, then all the angles are right angles, and it is a rectangle. Any parallelogram can be dissected into a rectangle by cutting a right-angled triangle off one end, and sliding it to the opposite end.

The dissection changes neither the area of the parallelogram nor the length of the sides - the area of any parallelogram is equal to the area of a rectangle with the same base and the same length.

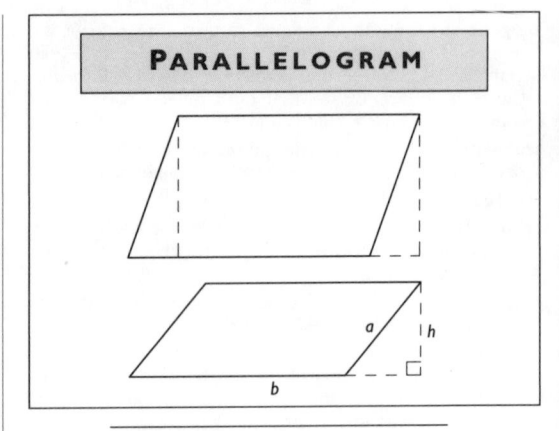

PARALLELOGRAM

Area = bh
Perimeter = $2(a+b)$

pentagon a polygon with five sides.
quadrilateral a polygon with four sides. A quadrilateral may be a rectangle, a square, a parallelogram, a rhombus or a trapezium (or trapezoid), or none of these.
rectangle a quadrilateral in which all the angles are right angles, thus the opposite sides are parallel in pairs. A rectangle may be a square (see below). A rectangle that is not a square has two lines of symmetry.

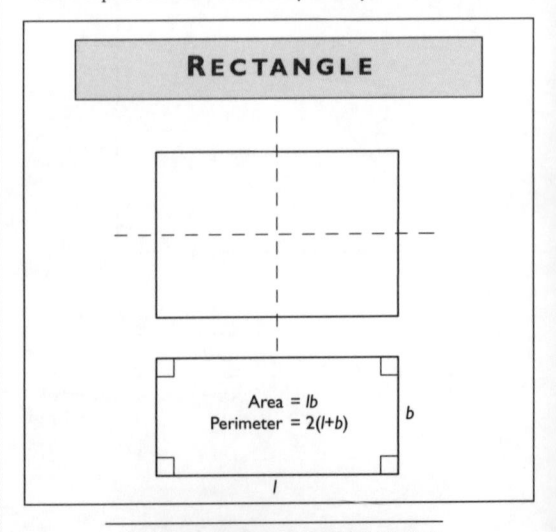

RECTANGLE

Area = lb
Perimeter = $2(l+b)$

Area = lb
Perimeter = $2(l+b)$

MATHEMATICS

rhombus a parallelogram whose sides are all equal in length. Its diagonals are both lines of symmetry, and therefore bisect each other at right angles.

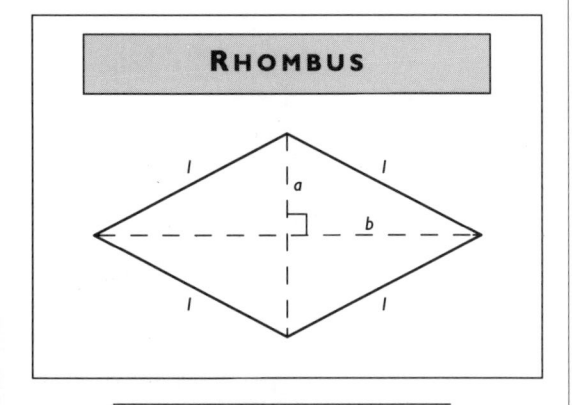

RHOMBUS

Area = $\frac{1}{2}(2a)(2b)$
Perimeter = $4l$

square a rectangle whose sides are equal. It has four lines of symmetry - both diagonals and the two lines joining the middle points of pairs of opposite sides.

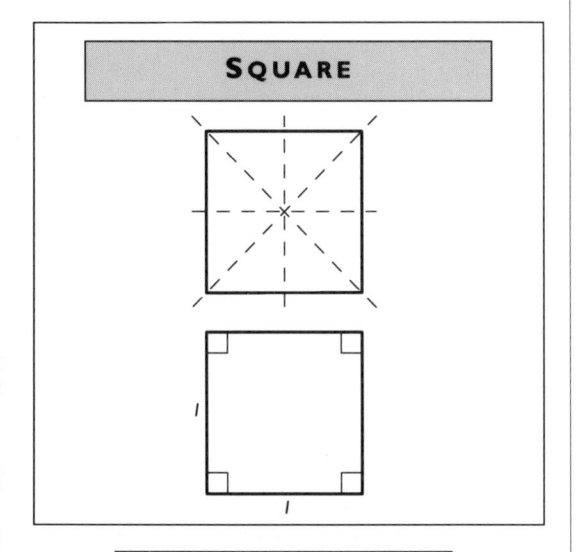

SQUARE

Area = l^2
Perimeter = $4l$

symmetry the correspondence of opposite parts in size, shape and position.

trapezium a quadrilateral with two parallel sides of unequal length. (In North America - where such a plane figure is described as a trapezoid - a trapezium is a quadrilateral with no sides parallel.)

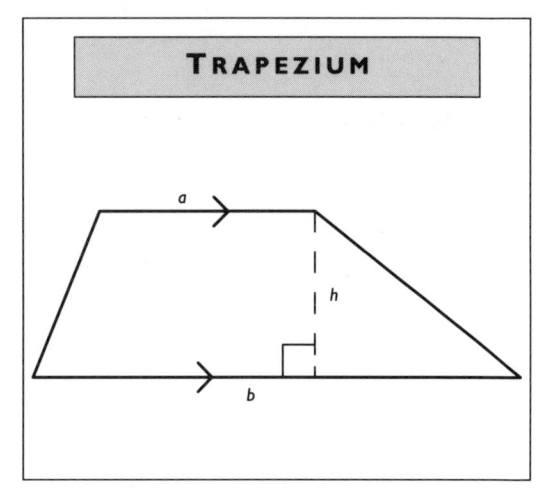

TRAPEZIUM

Area = $\frac{1}{2}(a + b)h$

To find the area three measurements have to be taken - the height between the pair of parallel sides and the length of both of the parallel sides. The area of a trapezium is equal to the height multiplied by the average length of the parallel sides; i.e.

= $\frac{1}{2}$ (the sum of the parallel side) × the perpendicular distance between them.

trapezoid (North American usage) a trapezium.
triangle a polygon with three sides. See pp. 184-85.
vertex the point where two sides of an angle intersect or the corner point of a square, cube or triangle.

TRIANGLES

A triangle is a plane figure which has three sides and three angles. It may also be defined as a three-sided polygon (see pp. 182-83).

area (of a triangle) is one half of a parallelogram with the same base and the same height (see properties of triangles; below).

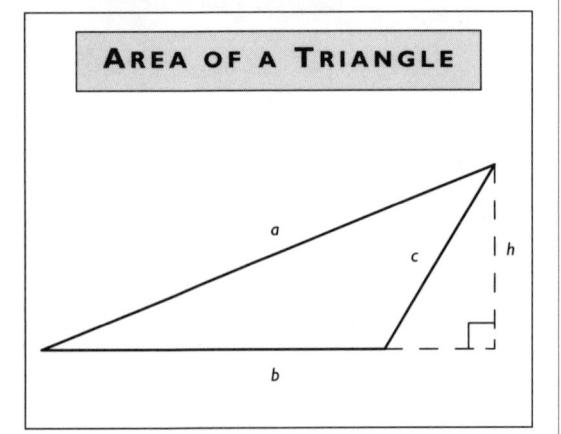

AREA OF A TRIANGLE

The area of a triangle can also be calculated from the lengths of its sides, using a formula discovered by the Greek mathematician Archimedes. If half the sum of the sides is s, then:

$$\text{Area} = \sqrt{s(s-a)(s-b)(s-c)},$$

where a, b, and c are the three sides and

$$s = \frac{a+b+c}{2}$$

axis a central line around which the parts of a figure are evenly arranged.

concurrent lines see properties of triangles (below).

congruent triangles triangles that are identical in size and shape.

diagonal a straight line between the opposite corners of a rectangle; see also pp. 182-83.

equilateral triangle a triangle which has all its sides equal. All the angles of an equilateral triangle are equal to 60°.

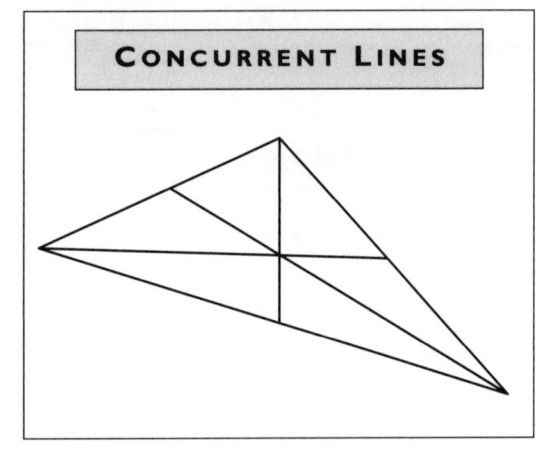

CONCURRENT LINES

hypotenuse the side of a right-angled triangle that is opposite to the right angle (see Pythagoras' theorem - below - and pp. 180-81).

integer see pp. 162-63.

isosceles triangle a triangle in which two of the sides are of equal length. An isosceles triangle has one axis of symmetry and a pair of equal angles.

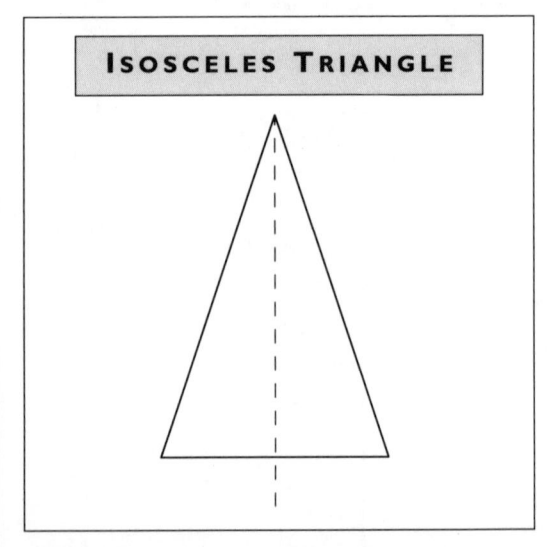

ISOSCELES TRIANGLE

medians see properties of triangles (below).

properties (of triangles) include the following curious qualities.

The three lines that join the vertices of a triangle to the middle point of the opposite sides, called the medians, meet at a point - they are said to be concurrent (see below).

The three lines that bisect the sides of a triangle at right angles also meet. The point at which they meet is the centre of the circle through the vertices of the triangle (see below).

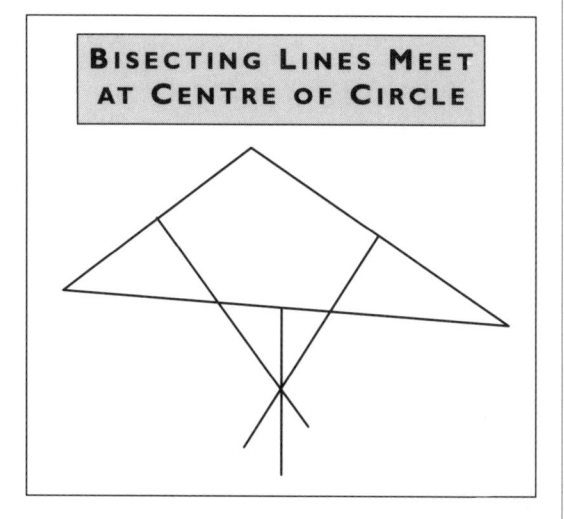

BISECTING LINES MEET AT CENTRE OF CIRCLE

Any triangle can be thought of as one half of a parallelogram (see pp. 182-83) that has been divided in two by one of its diagonals (see below).

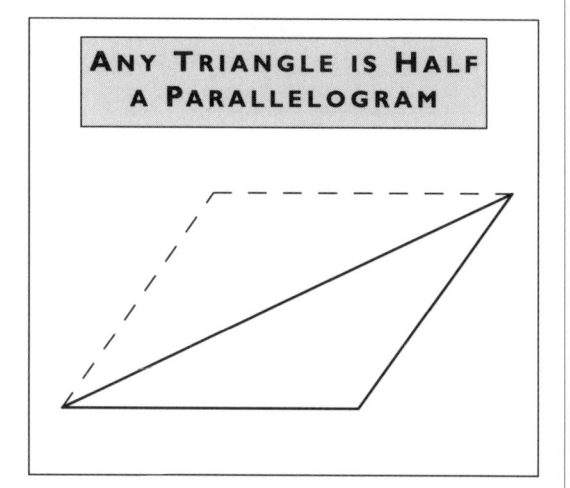

ANY TRIANGLE IS HALF A PARALLELOGRAM

Pythagoras' theorem the most-proved theorem in geometry, indeed in the whole of mathematics. (E. S. Loomis published in 1940 a collection of more than 370 different proofs of Pythagoras' theorem, and more have been discovered since.)

Pythagoras is the probable discoverer of the geometrical theorem that is named after him. (He did not, however, discover the theorem in its Euclidean form.)

The theorem states that the area of the square drawn on the hypotenuse of a right-angled triangle is equal to the sum of the squares drawn on the other two sides. (It is, however, also true that the area of any shape drawn on the hypotenuse is equal to the sum of equivalent shapes drawn on the other two sides.)

Thus, below, in the right-angled triangle ABC, the square of the hypotenuse AC is equal to the sum of the squares of the two other sides AB and BC.

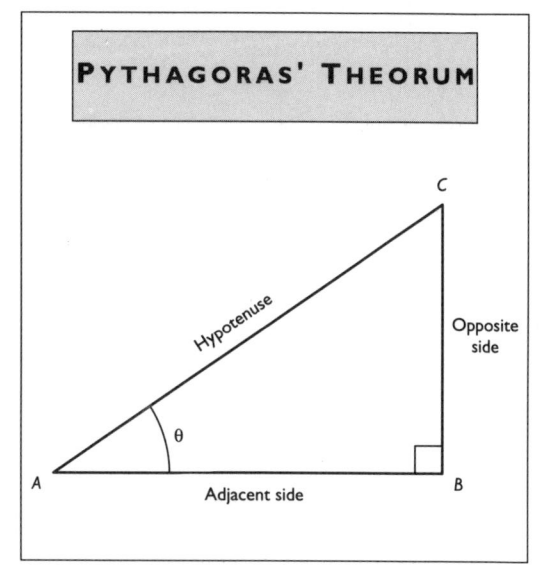

PYTHAGORAS' THEORUM

There are an infinite number of right-angled triangles whose sides are integers. Four of the smallest have the sides:

3,	4,	5
5,	12,	13
8,	15,	17
7,	24,	25.

Such whole-number sets are sometimes called Pythagorean triples.

scalene triangle a triangle which has sides of three different lengths, and has no axes of symmetry.

symmetry the correspondence of opposite parts in size, shape and position.

triples (Pythagorean) see Pythagoras' theorem.

vertex (plur. **vertices**) the point where two sides of an angle intersect or the corner point of a triangle.

CIRCLES AND OTHER CONIC SECTIONS

A circle is the path of a point that moves at a constant distance - the radius - from a fixed point (the centre of a circle). The circle is a special case of an ellipse (see below).

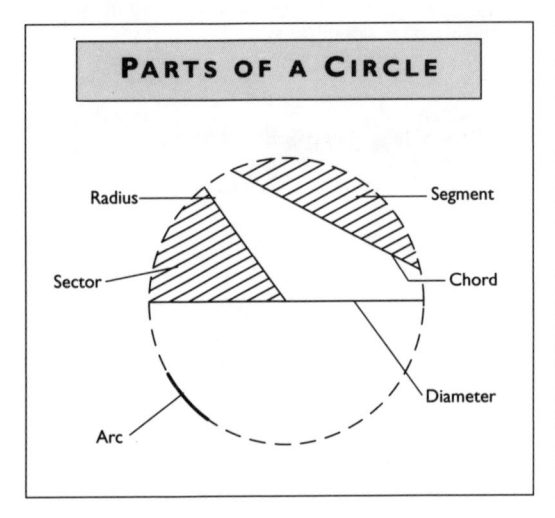

PARTS OF A CIRCLE

- Radius
- Segment
- Sector
- Chord
- Diameter
- Arc

General equation (centre at $-g, -f$)

$$x^2 + y^2 + 2gx + 2fy + c = 0$$

Basic equation (centre at origin)

$$x^2 + y^2 = r^2$$

The area of a circle =

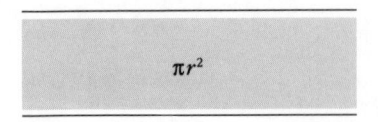

$$\pi r^2$$

arc any part of the perimeter or circumference of a circle.

chord a straight line that joins two points that are on the circumference of a circle.

circle theorems include the following:
A diameter of a circle forms a right-angle at the circumference of the circle.
A tangent to a circle is always at right angles to the radius at the point of contact.

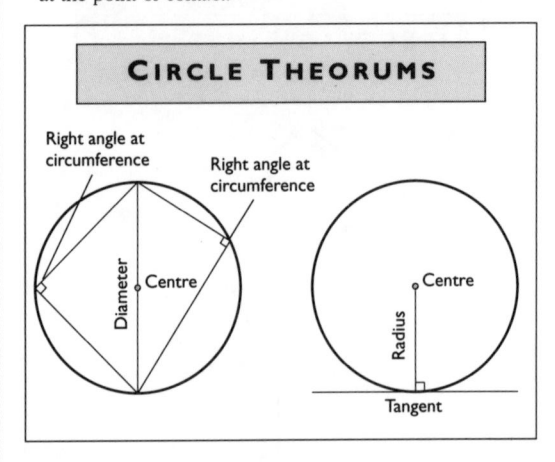

CIRCLE THEORUMS

Right angle at circumference
Right angle at circumference
Diameter
Centre
Centre
Radius
Tangent

circumference the line that forms the perimeter of a circle. The circumference of a circle =

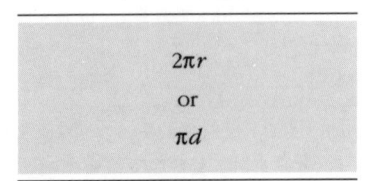

$$2\pi r$$
or
$$\pi d$$

where
r = radius, and
d = diameter.

cone a solid that has a circle as its base and curved sides that taper evenly up from that base to a point or apex. (See also Solids: areas and volumes on pp. 188-89.)

conic section curves that are formed by the intersection of a plane and a cone. An ellipse, a parabola, a hyperbola and a circle are all conic sections.

cylinder a solid figure with straight sides and a circular section; see Solids: areas and volumes (pp. 188-89).

diameter a straight line that passes through the centre of a circle (or a sphere) from one side to the other. The term diameter is also used for the length of such a line.

ellipse a closed conic section with the appearance of a flattened circle. It is formed by an inclined plane that does not intersect the base of the cone. An ellipse can also be thought of as a circle that has been stretched in one direction. The orbital path of each of the planets round the Sun is approximately an ellipse.

There are many ways to draw an ellipse. One of the simplest is to stretch a loop of thread round two pins, and hold it taut with a pencil. The path of the pencil would be an ellipse.

The area of an ellipse =

$$\pi ab$$

Basic equation (centre at the origin)

$$\left(\frac{x^2}{a^2} + \frac{y^2}{b^2}\right) = 1$$

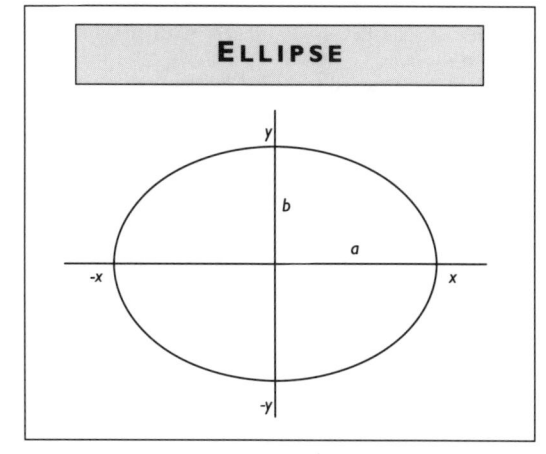

ELLIPSE

hemisphere one half of a sphere (q.v.).

hyperbola a conic section that is formed by a plane that cuts a cone making a larger angle with the base than the angle made by the side of the cone.

Basic equation (centre at the origin)

$$\left(\frac{x^2}{a^2} - \frac{y^2}{b^2}\right) = 1$$

irrational number see pp. 162-63.

parabola a conic section that is formed by the intersection of a cone by a plane parallel to its sides.

If you were to throw a ball up in the air then the path of the ball will be approximately a parabola with its axis vertical.

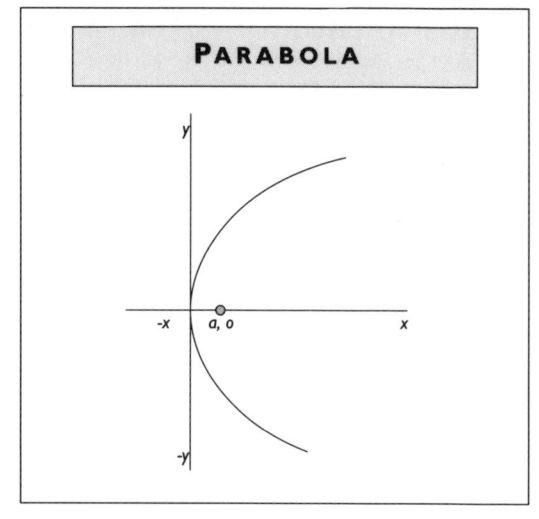

PARABOLA

Basic equation (symmetrical about the x-axis with the focus at a,O)

$$y^2 = 4ax$$

perimeter (of a circle) the circumference.

pi (which is indicated by the Greek letter π) the ratio of the circumference of a circle to its diameter (approximately 3.1411592...).

Pi was proved to be an irrational number - that is a number that cannot be written as an exact value - by Johann Heinrich Lambert (1728-77). In 1989 pi was worked out to 1 073 740 000 decimal places.

plane a surface that wholly contains a straight line connecting any two points lying on it.

radius (plur. **radii**) a straight line between the perimeter or circumference of a circle and its centre.

sector a part of a circle that is included by two radii and an arc.

segment any part of a circle (or of a sphere) that is cut off by a line (or a plane) that is not necessarily a radius, in other words a chord (q.v.).

sphere a solid figure every point of whose surface is equidistant from its centre; see also Solids: areas and volumes on pp. 188-89.

symmetry the correspondence of opposite parts in size, shape and position.

tangent a line touching but not intersecting a curve or surface at one point only. See circle theorems (above).

SOLIDS: AREAS AND VOLUMES

Solids are three-dimensional figures, i.e. they have length, breadth and depth. Solids include rectangular blocks, prisms, pyramids, tetrahedrons, cylinders, cones and spheres.

The volume of any prism is always equal to the area of the base × the perpendicular height. The volume of any pyramid is one third the base area multiplied by the perpendicular height.

apex the highest point of a triangle.

circular prism see cylinder (below).

cone a solid figure with a circular plane base, narrowing to a point or apex. If the slant height of the cone is l, the area of the curved surface of a cone =

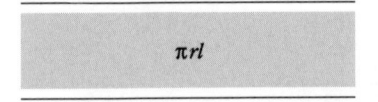

$$\pi r l$$

The volume of a cone can be calculated as if the cone were a special case of a pyramid. The volume is one third the volume of a cylinder with the same base and height.

CONE

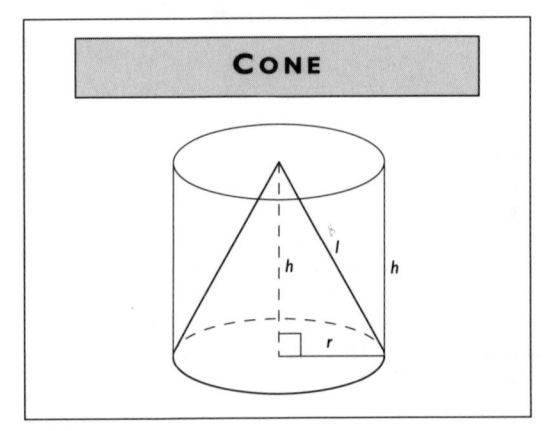

The volume of a cone =

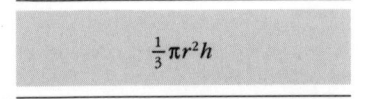

$$\frac{1}{3}\pi r^2 h$$

cuboid a solid figure, all the faces of which are rectangles.
Surface area =

$$2(lb + bh + hl)$$

Volume =

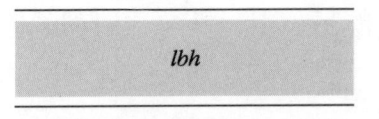

$$lbh$$

CUBOID

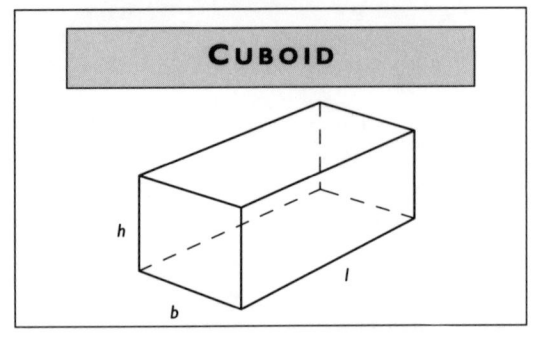

cylinder a solid figure with straight sides and a circular section.

The area of the curved surface of a cylinder =

$$2\pi r h$$

If the circles at both ends are included, then the total surface area =

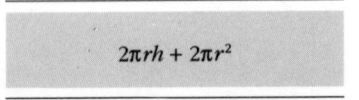

$$2\pi r h + 2\pi r^2$$

The volume of a cylinder can be found by thinking of it as a special case of a prism. The volume equals the area of the base, multiplied by the height.

The volume of a cylinder =

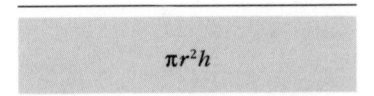

$$\pi r^2 h$$

prism a solid figure whose ends are identical polygons and whose sides are parallelograms (which could be rectangular).

The volume of a prism equals the area of either of the ends, multiplied by the perpendicular distance between the ends.

pyramid a solid figure whose base is a polygon, and whose special vertex - the apex - is joined to each vertex of the base. Therefore all its faces, apart from the base, are triangles.

MATHEMATICS

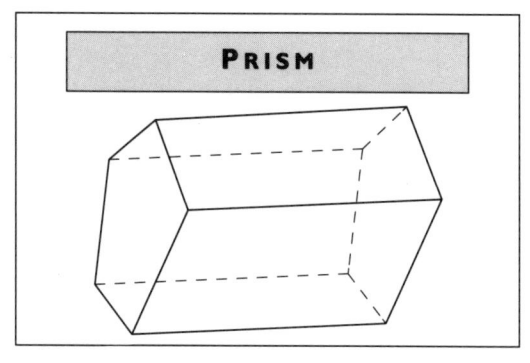

PRISM

Any pyramid can be fitted inside a prism so that the base of the pyramid is one end of the prism, and the apex of the pyramid is on the other end of the prism.

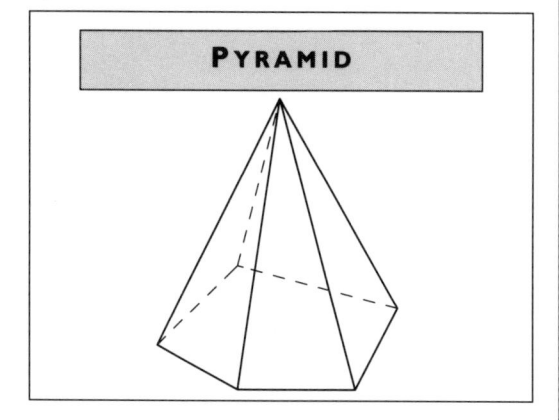

PYRAMID

The volume of a pyramid on a rectangular base =

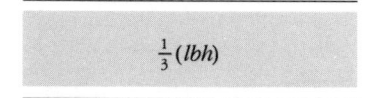

$$\frac{1}{3}(lbh)$$

rectangular block see cuboid (above).

sphere a solid figure every point of whose surface is equidistant from its centre.

Surface area =

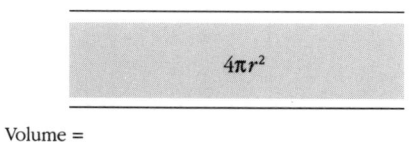

$$4\pi r^2$$

Volume =

$$\frac{4}{3}\pi r^3$$

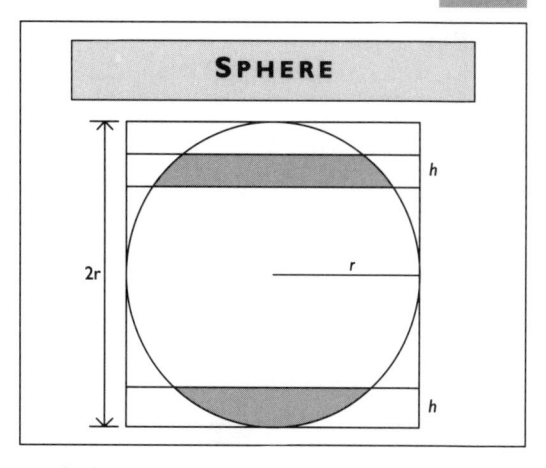

SPHERE

tetrahedron a pyramid whose base is a triangle. Any of the faces of a tetrahedron can be thought of as its base.

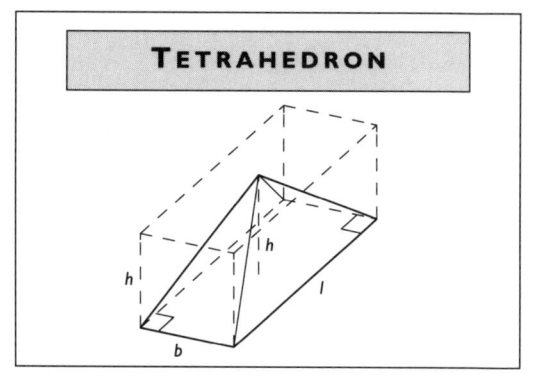

TETRAHEDRON

The volume of a tetrahedron =

$$\frac{1}{3} \text{ (the area of the triangular base} \times \text{the height)}$$

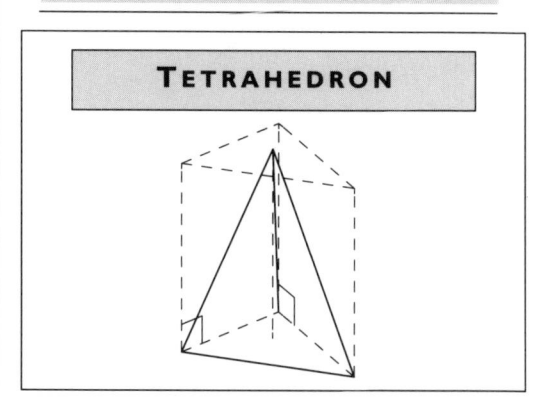

TETRAHEDRON

POLYHEDRA

A polyhedron is a solid shape with all plane faces. The faces of a regular polyhedron, or regular solid, are all identical regular polygons.

There are just five regular polyhedra - the regular tetrahedron, the cube, the regular octahedron, the regular dodecahedron and the regular icosahedron. The cube and the octahedron are dual polyhedra. The cube has six faces and eight vertices, while the octahedron has six vertices but eight faces. The regular dodecahedron and regular icosahedron are also dual polyhedra.

There are many more irregular polyhedra. The simplest to visualize have faces that are mixtures of two kinds of regular polygons. For example, the faces of the cuboctahedron are equilateral triangles and squares.

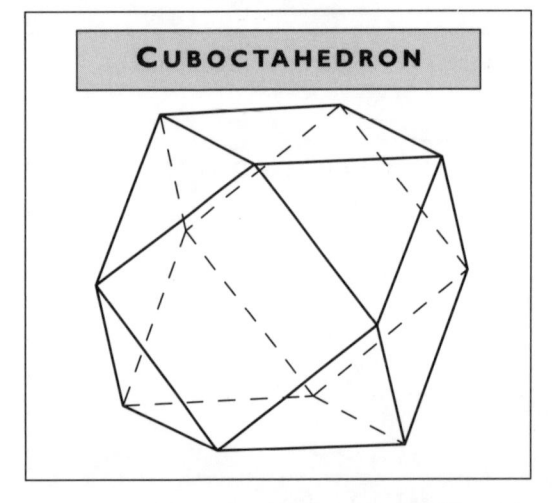

CUBOCTAHEDRON

The mathematician Euler made an interesting discovery about the relationship between the number of faces (F), vertices (V) and edges (E) of polyhedra.

The equation

$$F + V - E = 2$$

is true for all 'simple' polyhedra - the regular polyhedra listed in the glossary below.

The same relationship is true for an area divided into any number of regions (R) by boundaries or arcs (A) that join at nodes (N). Thus

$$R + N - A = 2.$$

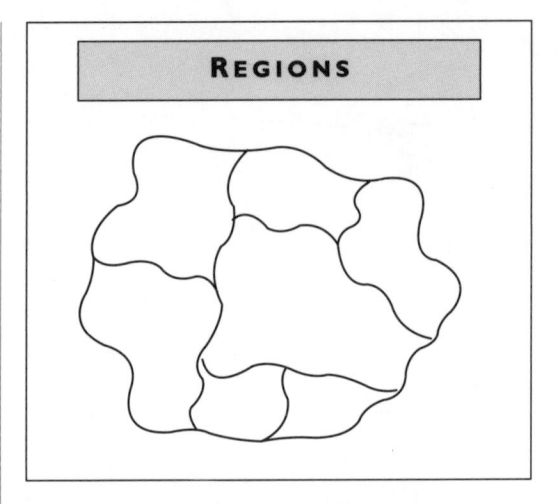

REGIONS

For the area shown below:
  R = 8 (the surrounding space counts as a region)
  N = 12
  A = 18
Thus R + N − A
  = 8 + 12 − 18
  = 2.

(It is interesting to note that for such a region, or indeed any map, no more than four colours are necessary so that no two adjoining regions have the same colour.)

cube a polyhedron which has:
  6 faces which are all squares;
  8 vertices; and
  12 edges.

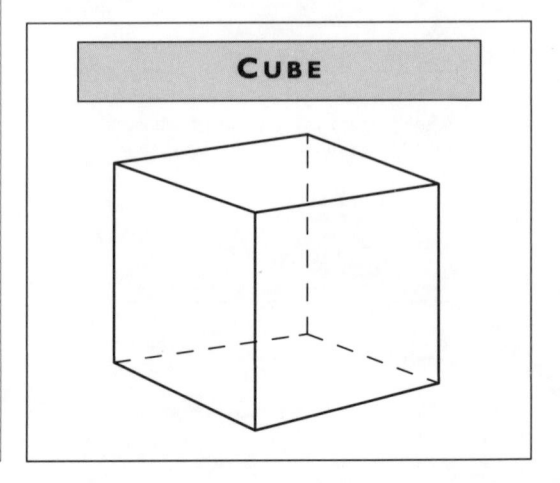

CUBE

MATHEMATICS

dodecahedron see regular dodecahedron (below).
icosahedron see regular icosahedron (below).
octahedron see regular octahedron (below).
regular dodecahedron a polyhedron which has:
   12 faces which are all regular pentagons;
   20 vertices; and
   30 edges.

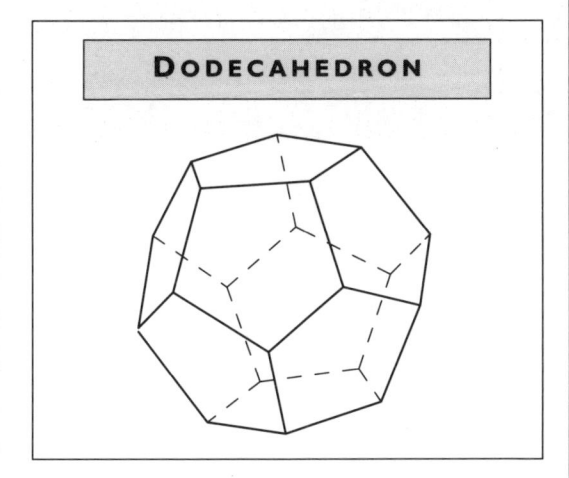

DODECAHEDRON

regular icosahedron a polyhedron which has:
   20 faces which are all equilateral triangles;
   12 vertices; and
   30 edges.

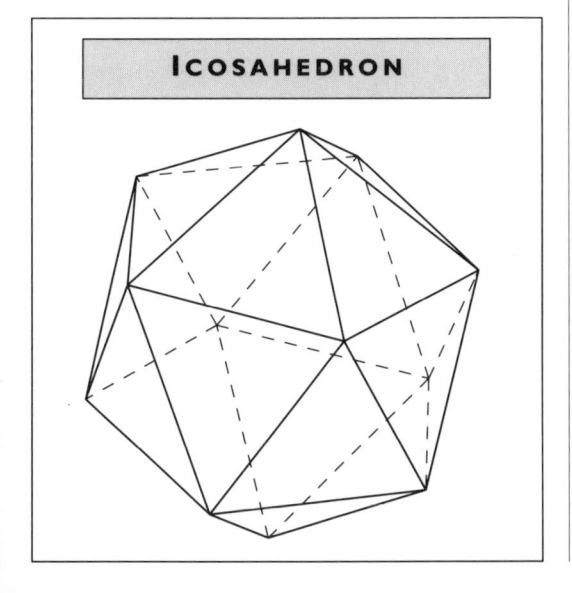

ICOSAHEDRON

regular octahedron a polyhedron which has:
   8 faces all of which are equilateral triangles;
   6 vertices; and
   12 edges.

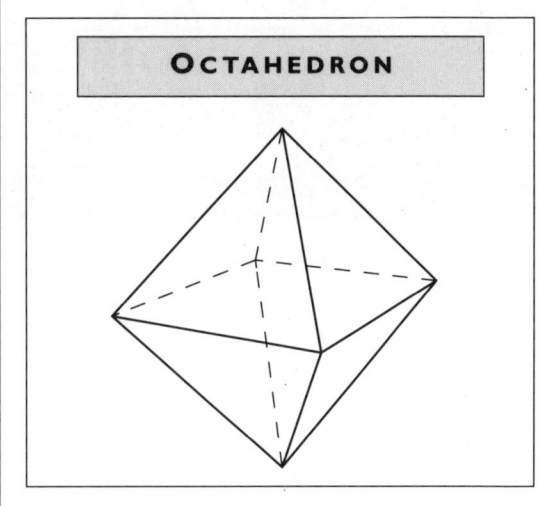

OCTAHEDRON

regular tetrahedron a polyhedron which has:
   4 faces which are all equilateral triangles;
   4 vertices; and
   4 edges.
tetrahedron see regular tetrahedron (above).

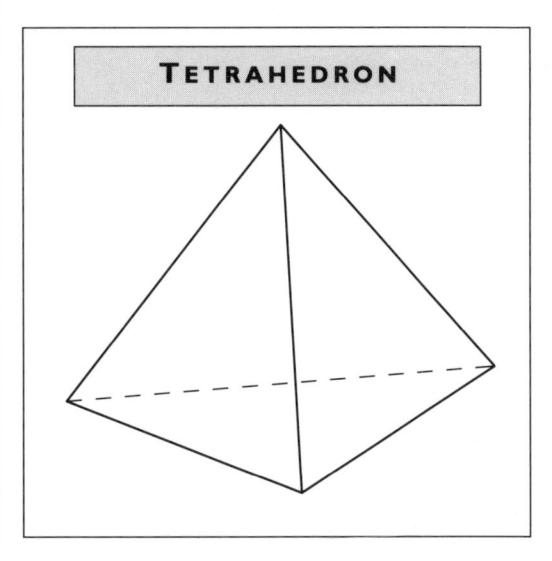

TETRAHEDRON

vertex (plur. **vertices**) a corner point of a triangle, cube or polyhedron, etc.

Famous Scientists

Abbe, Ernst (1840–1905), German physicist and professor of astronomy who was a partner in the Carl Zeiss optical works. He is best known for his research in optics.

Abbot, Charles Greely (1872–1973), US astrophysicist who is best known for his work on solar physics.

Abegg, Richard (1869–1910), German chemist whose 'rule of eight' – concerning the electric basis of linkages between atoms – was a major contribution towards the development of the modern valency theory.

Abel, Niels Henrik (1802–29), Norwegian mathematician who pioneered several branches of modern mathematics.

Abelson, Philip Haugh (1913–), US physical chemist who is best known for work on separating the isotopes of uranium.

Acheson, Edward Goodrich (1893–1971), US chemist who discovered carborundum (silicon carbide) in 1891. Acheson used an electric furnace to manufacture synthetic graphite (1899).

Adams, John Couch (1819–92), English astronomer who mathematically deduced the existence of Neptune.

Adams, Walter Sydney (1876–1956), US astronomer whose studies of stellar spectra led to the discovery of the spectroscopic method of measuring the velocities and distances of stars.

Adanson, Michel (1727–1806), French botanist whose rudimentary classification of plants predated Linnaeus.

Adrian, Lord (Edgar Douglas Adrian) (1889–1977), English physiologist who shared the Nobel Prize for Medicine in 1932 with Sir Charles Sherrington for their work regarding the nerve cell. They discovered the mechanism by which nerves carry messages to and from the brain – a discovery that increased the understanding of mental disorders.

Agassiz, Alexander E(mmanuel) R(udolphe) (1835–1910), Swiss-born US marine zoologist and oceanographer who was the son of Jean Louis Agassiz.

Agassiz, Jean Louis (1807–73), Swiss-born US naturalist who founded the Museum of Comparative Zoology at Harvard. Agassiz also made an important contribution to the study of glaciation, proving the existence of ice ages.

Agricola, Georgius (Georg Bauer; 1494–1555), German founder of the science mineralogy.

Aiken, Howard H. (1900–73), US mathematician who is credited with designing a forerunner to the digital computer.

Airy, Sir George Biddell (1801–92), English astronomer and geophysicist who measured the mass of the Earth from gravity measurements in mines.

Albert the Great, St (c. 1200–80), German bishop and philosopher who undertook 'to make intelligible' Aristotle's *Physica*.

Alcmaeon (c. 520 BC), Greek physician who is the first recorded practitioner of anatomical dissection. Alcmaeon's studies led to important advances in embryology.

Aldrovandi, Ulisse (1522–1605), Italian naturalist who attempted a multi-volume natural history.

Aleksandrov, Pavel (1896–1982), Russian mathematician who made an important contribution to topology.

Alfvén, Hannes (1908–), Swedish pioneer of plasma physics.

Alpher, Ralph Asher (1921–), US physicist who is best known for his theories concerning the origin and evolution of the universe.

Alpini, Prospero (1553–1616), Italian botanist who described sexual reproduction in plants.

Altman, Sidney (1939–), Canadian-born US chemist who shared the 1989 Nobel Chemistry Prize for his work on ribonuclease.

Alvarez, Luis (1911–), US physicist who won the 1968 Nobel Prize for Physics for the discovery of resonance particles.

Ambartsumian, Viktor (1908–), Armenian astrophysicist who devised a method of computing the mass thrown out of nova stars.

Amontons, Guillaume (1663–1705), French physicist who discovered that the temperature and pressure of gases are related.

Ampère, André Marie (1775–1836), French physicist who founded the branch of physics that he named electrodynamics and which is now known as electromagnetism. A

FAMOUS SCIENTISTS

child prodigy, he was a distinguished mathematician. Ampère was known for sudden enthusiasm for a particular area of investigation rather than for systematic research. Hearing about Oersted's discovery that an electric current affects a nearby magnet (1820), Ampère pursued and widened this area of work and (by 1824) established a mathematical and physical description of the magnetic force between two electric currents. He formulated a law of electromagnetism now known as Ampère's Law. He pioneered techniques in measuring electricity. The unit of electric current – the amp or ampere – is named after him.

Anderson, Carl David (1905–91), US physicist who discovered the positron, the first known particle of antimatter. He also discovered the first meson (1935). Anderson discovered the positron during studies of tracks of cosmic rays in a cloud chamber (1932). He shared the Nobel Prize for Physics in 1936.

Anderson, Philip Warren (1923–), US physicist who shared the 1977 Nobel Physics Prize for investigating the electronic structure of magnetic and disordered systems.

Andrews, Thomas (1813–85), Irish chemist who discovered the critical temperature of gases.

Anfinsen, Christian (1916–), US biochemist who shared the 1972 Nobel Chemistry Prize for his research on RNA.

Angström, Anders Jonas (1814–74), Swedish physicist who pioneered the science of spectroscopy. Angström undertook major research in heat conduction as well as in spectroscopy. He also discovered that hydrogen is present on the Sun. The angstrom, a unit of length used principally in measuring wavelengths of light, is named after him.

Apollonius (3rd century BC), Greek mathematician who is remembered for his treatise *Conic Sections* in which he described geometrical concepts of the cone. He was known as the Great Geometer.

Appel, Kenneth (1932–), US mathematician who solved the long-standing four-colour problem.

Appleton, Sir Edward Victor (1892–1965), English physicist who won the Nobel Prize for Physics in 1947 for his work which revealed the existence of a layer of charged particles in the ionosphere that reflects radio waves. It is now known as the Appleton layer.

Arber, Werner (1929–), Swiss microbiologist who shared the 1978 Nobel Prize for Medicine or Physiology. He is best known for his research on DNA.

Archimedes (c. 287–212 BC), Greek mathematician, philosopher and engineer, born at Syracuse in Sicily. His extensions of the work of Euclid especially concerned the surface and volume of the sphere and the study of other solid shapes. His methods anticipated the fundamentals of integral calculus. Archimedes gave the first systematic account of determining centres of gravity. He founded the science of hydrostatics and discovered the principle – now known as Archimedes' Principle – that submerged bodies displace their own volume of liquid and that their weight is diminished by an amount equal to the liquid displaced. He also invented a number of mechanical devices. Archimedes was killed when the city of Syracuse was taken by Rome.

Argand, Jean Robert (1768–1822), Swiss mathematician who revived a geometrical method of representing complex numbers. He developed what are now known as Argand diagrams, which have wide applications in mechanics and electricity as well as mathematics.

Aristarchus of Samos (3rd century BC), Greek astronomer who anticipated Copernicus in maintaining that the Earth revolves around the Sun.

Aristotle (c. 384–322 BC), Greek philosopher and scientist who reviewed the entire field of human knowledge that was known in his time. Born in Greek Macedonia he went to Plato's Athenian Academy. After Plato's death (348 or 347 BC), Aristotle travelled the Hellenic world founding academies. At the Macedonian court at Pella he tutored Alexander the Great. After 335 he resettled in Athens, devoting himself to science. His theory of causality stated that every event has four causes: material cause, the matter involved; formal cause, the way it is placed; efficient cause, which triggers the action or change; and final cause, what the event leads to.

Arrhenius, Svante August (1859–1927), Swedish chemist who developed the ionic theory of electrodes (1887) for which he was awarded the Nobel Prize for Chemistry in 1903.

Artedi, Peter (1705–35), Swedish biologist who was the virtual founder of ichthyology, the study of fishes.

Astbury, William Thomas (1889–1961), English chemist who pioneered research on protein fibres. He was also known for his studies in crystallography.

Aston, Francis William (1877–1945), English chemist and physicist who is best known for his work on atomic weights using spectographs.

Atanasoff, John Vincent (1904–), US computer pioneer and physicist.

Atiyah, Sir Michael Francis (1929–), English mathematician whose work on the mathematics of the quantum field theory has had important implications for physics.

Audubon, John James (1785–1851), Franco-Haitian ornithologist and bird artist.

Avery, Oswald Theodore (1877–1955), Canadian-born US bacteriologist whose studies of pneumococci had important implications for later DNA research.

Avicenna (Abu-Ali al-Husayn ibn-Sina; 980–1037), Uzbek-born Persian philosopher who was both a leading pioneer physician and an outstanding compiler of an encyclopedia covering medicine, mathematics and the natural sciences.

Avogadro, Amadeo (1776–1856), Italian physicist who developed the hypothesis that under the same conditions of pressure and temperature, equal volumes of different gases contain an equal number of molecules. This is now known as Avogadro's Law.

Baade, Wilhelm Heinrich Walter (1893–1960), German-born US astronomer who made important contributions to our knowledge of the age and size of galaxies.

Babbage, Charles (1792–1871), English mathematician who is regarded as the inventor of the modern computer. His 'difference machine' (1827) compiled and printed tables of logarithms from 1 to 108000. Babbage's complex 'analytical machine', which was intended to make any type of calculation, was not constructed until modern times.

Babinet, Jacques (1794–1872), French physicist who standardized light measurement. Babinet's principle states that similar diffraction patterns are produced by two complementary screens.

Bacon, Francis (1561–1626), English philosopher and statesman who was most influential in laying the foundations of the philosophy of the scientific method: he was the inspiration for the formation of the Royal Society in London (1660) and the Academy of Sciences in Paris (1666).

Bacon, Roger (c. 1214–92), English Franciscan, philosopher and scientist who was imprisoned for heresy. His scientific contributions included pioneering work on the magnifying glass and advocacy of the primacy of mathematical proof in science.

Baer, Karl von (1792–1876), Estonian-born German embryologist whose studies concerning similarities of embryos played an important role in evolutionary theory.

Baeyer, Johann Friedrich Adolph von (1835–1917), German chemist whose synthesis of indigo led to the production of many other new dyes.

Baird, John Logie (1868–1946), Scottish inventor of the television.

Balmer, Johann Jakob (1825–98), Swiss physicist who discovered the first mathematical formula to describe the wavelengths of spectral lines. This formula is basic to the development of atomic theory.

Baltimore, David (1938–), US microbiologist studied the interaction between tumor viruses and the genetic material of the cell. He shared the Nobel Prize for Physiology or Medicine in 1975. Working independently of Howard Temin he discovered that some animal cancer viruses that are composed mainly of RNA can transfer genetic information to DNA.

Banks, Sir Joseph (1744–1820), English botanist who accompanied Captain Cook on the *Endeavour*.

Banting, Sir Frederick (1891–1941), Canadian doctor who, working with Charles Best, discovered a treatment for diabetes.

Bardeen, John (1908–91), US co-inventor of the transistor.

Barkla, Charles Glover (1877–1944), English physicist who is best known for his research on X-rays. He won the 1917 Nobel Physics Prize.

Barr, Murray Llewellyn (1908–), Canadian geneticist who is best known for his discovery of what are now called Barr bodies – densely staining nuclear bodies present in the somatic cells of female mammals.

Bartholin, Thomas (1616–80), Danish physician and mathematician who described the human lymphatic system.

Basov, Nikolai (1922–), Russian physicist who developed the maser, the forerunner of the laser.

Bates, Henry Walter (1825–92), English naturalist who demonstrated the operation of natural selection through studies of animal mimicry. His work gave support to the theories of Charles Darwin. During travels in the Amazon Basin he discovered and scientifically described about 8000 insect species.

Bateson, William (1861–1926), English biologist who named the science 'genetics'. His experiments concerning plant inheritance provided evidence that was fundamental to modern studies of heredity. Bateson proposed that certain characteristics are linked rather than being inherited independently; this was later explained as gene linkage.

Bayes, Thomas (1702–61), English mathematician who was the first to study statistical inference.

Bayliss, Sir William Maddock (1860–1924), English physiologist who – with E.H. Starling – discovered the first hormone, secretin.

Beadle, George Wells (1903–89), US biochemist who proposed that specific genes control the production of specific enzymes.

Beaumont, William (1785–1853), US surgeon and physiologist whose important studies on gastric physiology were published in *Experiments and Observations on the Gastric Juice and the Physiology of Digestion* (1833).

Beckmann, Ernst Otto (1853–1923), German chemist who devised apparatus for determining freezing and boiling points, and the sensitive thermometer that is named after him.

Becquerel, Antoine Henri (1852–1908), French physicist who discovered radioactivity. After Röntgen discovered X-rays in 1896 Becquerel began work on uranium and other substances to find out whether fluorescent substances give off X-rays as well as visible light. He found a previously unknown type of radiation being emitted from uranium atoms within fluorescent uranium salts. Marie Curie – who was a colleague of Becquerel – later named this phenomenon radioactivity. Becquerel made a systematic study of the properties of this form of radiation. He shared the 1903 Nobel Physics Prize with the Curies. The unit of radioactivity – the becquerel, which corresponds to one disintegration per second – is named after him.

Behring, Emil von (1854–1917), German immunologist who – with Shibasaburo Kitasato – demonstrated immunity to tetanus and diphtheria in animals.

Beilstein, Friedrich Konrad (1838–1906), Russian-born German chemist whose *Handbook of Organic Chemistry* (1881) was a landmark in that field.

Békésy, Georg von (1899–1972), Hungarian-born US physiologist who made probably the greatest individual contribution to aural physiology.

Bell, Alexander Graham (1847–1922), Scottish-born US inventor and physicist who invented the world's first telephone (1876). His other achievements included experiments in communication, the invention of the photophone (which transmitted sound on a beam of light), medical research and studies into teaching the deaf to speak.

Bell, Sir Charles (1774–1842), Scottish anatomist whose *New Idea of Anatomy of the Brain* (1811) was a landmark in neurology. He was the first to scientifically distinguish between motor and sensory nerves.

FAMOUS SCIENTISTS

Bell, John Stuart (1928–90), Northern Irish physicist whose *On the Einstein Podolsky Rosen Paradox* (1964) was one of the most influential papers in theoretical physics of modern times.

Bell Burnell, Susan (1943–), Northern Irish astronomer who was the co-discoverer of the first pulsar.

Belon, Pierre (1517–64), French naturalist who was the first to catalogue the similarities in vertebrate skeletons.

Benzer, Seymour (1921–), US geneticist who has made a major study of mutation.

Berg, Paul (1926–), US molecular biologist who shared the 1980 Nobel Chemistry Prize. He identified the first adaptor (1956) and devised a way of inserting 'foreign' genes into bacteria.

Bergius, Friedrich (1884–1949), German chemist who converted coal dust and hydrogen into petrol.

Bernal, John Desmond (1901–71), Irish crystallographer who was also one of the pioneers of molecular biology. He is known for his work on the structure of water and for his Marxism.

Bernard, Claude (1813–78), French physiologist who made important discoveries concerning the part played by the pancreas in digestion, the glycogenic role of the liver and the way in which vasomotor nerves regulate the blood supply. Through his study of glycogen, he was the first scientist to demonstrate that the body could build up substances on its own.

Bernoulli, Daniel (1700–82), Swiss physician and mathematician who described the theory of statics and the motion of fluids in his *Hydrodynamica* (1738). He related his theories to the propulsion of ships. He formulated Bernoulli's theorem in fluid dynamics which states that the total mechanical energy of a flowing fluid – the energy associated with fluid pressure, the gravitational potential energy of elevation, and the kinetic energy of fluid motion – remains constant. He was the son of Johann Bernoulli.

Bernoulli, Jakob (1654–1705), Swiss mathematician who laid down the foundations of the calculus of variations. He studied special curves and worked on the theory of infinite series. He was the brother of Johann Bernoulli.

Bernoulli, Johann (1667–1748), Swiss mathematician who worked with his brother – Jakob Bernoulli – on calculus and the properties of curves.

Berthelot, Marcellin (1827–1907), French chemist who studied alcohols, carboxylic acids, the synthesis of hydrocarbons, and reaction velocities. He discovered the detonation wave.

Berthollet, Count Claude Louis (1748–1822), French chemist who was the first to discover that the completeness of chemical reactions depends, in part, upon the masses of the reacting substances.

Berzelius, Jöns (1779–1848), Swedish chemist who was the first to prepare an accurate list of atomic weights.

Bessel, Friedrich W(ilhelm) (1784–1846), German astronomer and mathematician who was one of the indefatigable pioneers who measured the distance to the nearest stars. In 1838 he was the first to make a successful stellar parallax measurement (as the Earth moves round its orbit the positions of the nearest stars can be observed to wobble very slightly). He is famous for his systematic study of the solutions of certain differential equations now known as Bessel Functions.

Bessemer, Sir Henry (1813–98), English metallurgist who decarbonized iron (1856), thus allowing the large-scale manufacture of steel.

Best, Charles (1899–1978), Canadian physiologist who – with Frederick Banting – discovered insulin treatment for diabetes.

Bethe, Hans Albrecht (1906–), German-born US physicist who is best known for his work on nuclear reactions.

Bichat, Marie François (1771–1802), French anatomist who founded the science of histology – the study of tissues.

Birkhoff, George David (1884–1944), US mathematician who is best known for his work in the theory of dynamic systems.

Bjerknes, Vilhelm (1862–1951), Norwegian physicist who developed weather forecasting through the identification and study of air masses and fronts.

Black, Joseph (1728–99), Scottish physicist who discovered bicarbonates and rediscovered 'fixed air' (carbon dioxide). He also developed the concept of latent heat (1761) and specific heat.

Blackett, Patrick (Baron Blackett; 1897–1974), English physicist who was the first to photograph nuclear collisions in transmutation. He also discovered the positron.

Blackman, Frederick Frost (1866–1947), English botanist who is best known for his important research into the respiration of plants.

Bloch, Felix (1905–83), Swiss-born US physicist who shared the 1952 Nobel Physics Prize. The Bloch bands are sets of discrete, closely adjacent energy levels that arise from quantum states when a nondegenerate gas condenses to a solid.

Bloembergen, Nicolaas (1920–), Dutch-born US physicist who made an important contribution to the development of the maser, the forerunner of the laser.

Bode, Johann Elert (1747–1826), German astronomer. Bode's Law is a mathematical formula relating to the distances between the planets of the solar system. The law does not hold for Pluto, the most distant planet.

Bohm, David Joseph (1917–), US physicist whose *Quantum Theory* (1951) was a landmark in that field.

Bohr, Niels (1885–1962), Danish physicist who played a major role in the development of nuclear physics for over half a century. His classic paper *On the Constitution of Atoms and Molecules* (1913) was a landmark in nuclear physics. In this paper, the Bohr theory of the atom – combining Rutherford's model with Planck's quantum theory of radiation – was able to account for the known patterns of atomic radiation seen in spectra. Bohr's theory had an important impact on the development of quantum mechanics. He was awarded the 1922 Nobel Prize for Physics.

Bok, Bart Jan (1906–83), Dutch-born US astronomer whose principal interest was the structure of the Milky Way.

Boltwood, Bertram Borden (1870–1927), US radiochemist who discovered ionium.

Boltzmann, Ludwig (1844–1906), Austrian theoretical physicist who made important advances in the theory of radiation and in statistical treatment of the behaviour of molecules in gases. The Stefan-Boltzmann Law states that the total energy radiated from a body is proportional to the fourth power of its absolute temperature, which has enormous implications for understanding such matters as loss of heat from stars and from the Earth on cloudless nights. He also developed the Boltzmann Distribution, which graphs the distribution of a collection of particles (such as molecules in gas). This work has countless applications in the understanding of processes such as the evaporation of liquids and the speed of chemical reactions. His work led to the understanding of the crucial concept of entropy, a measure of the disorder in a physical system. He was a manic-depressive and committed suicide by drowning.

Bolyai, Janos (1802–60), Hungarian mathematician who - despite being warned by his father against the attempt – investigated Euclid's parallel postulate. Bolyai eventually developed a non-Euclidean geometry, only to discover to his horror that Gauss had anticipated his work.

Boole, George (1815–64), English mathematician who despite being largely self-taught became professor of mathematics at University College, Cork. He laid the foundations of Boolean algebra, which was fundamental to the development of the digital electronic computer.

Bordet, Jules (1870–1961), Belgian bacteriologist and immunologist who won the Nobel Prize for Physiology or Medicine in 1919 for his discovery of immunity factors in blood serum. His discovery was an important step in the diagnosis and treatment of many contagious diseases.

Borelli, Giovanni (1608–79), Italian physicist and physiologist who was the first person to explain muscular movement. He suggested that muscles and bones acted in the same way as systems of levers and pulleys.

Born, Max (1882–1970), German physicist who shared the 1954 Nobel Physics Prize. His early work on crystals led to the development of the Born-Haber cycle, a theoretical cycle of reactions and changes by which the lattice energy of ionic crystals may be calculated. Born is, however, best known for his work on quantum physics, in particular statistical studies on wave functions. Born suggested that particles exist but are 'guided' by a wave. Born proposed that, at any point, the square of the amplitude indicates the probability of finding a particle there.

Bosch, Carl (1874–1940), German chemist who devised the Bosch process, by which hydrogen is obtained from water gas and superheated steam.

Bose, Sir Jagadis Chandra (1858–1937), Indian physicist and botanist who is best known for his studies of electric waves.

Bose, Salyendranath (1894–1974), Indian physicist, chemist and mathematician who developed what is now known as Bose statistics, which was the forerunner of modern quantum theory.

Bothe, Walther (1891–1957), German atomic physicist who devised the coincidence method of detecting the emission of electrons by X-rays.

Boveri, Theodor Heinrich (1862–1915), German biologist who was one of the founders of the science of cytology.

Boyer, Herbert Wayne (1936–), US pioneer in genetic engineering.

Boyle, Robert (1627–91), Anglo-Irish physicist and chemist who formulated (in *New Experiments Physio-Mechanicall, Touching the Spring of Air and its Effects* in 1660 and *Sceptical Chymist* in 1662) the law concerning the behavior of gases that now bears his name. Boyle's law states that, at a given temperature, the pressure of a gas is proportional to its volume. Boyle pioneered the concept of elements in his suggestion of a corpuscular view of matter. He also distinguished between elements and compounds.

Boys, Sir Charles Vernon (1855–1944), English physicist who determined the gravitational constant.

Bradley, James (1693–1762), English astronomer who was the third Astronomer Royal. He discovered the aberration of starlight in 1728. He noticed a very small annual oscillation in the observed positions of astronomical objects superimposed upon the expected parallax effect, which Bradley correctly ascribed to the motion of the Earth around the Sun. This was convincing confirmation of the ideas of Copernicus and Kepler. In 1747 Bradley also discovered the phenomenon of nutation, a slight nodding motion of the Earth's polar axis, which has a period of 18 years 220 days.

Bragg, Sir Lawrence (1890–1971), Australian-born British physicist whose work with his father William on the determination of crystal structures using X-rays won them the Nobel Physics Prize (1915).

Bragg, Sir William (1862–1942), English physicist who pioneered work in solid-state physics. He won – jointly with his son Lawrence – the Nobel Prize for Physics in 1915 for his studies on the determination of crystal structures.

Brahe, Tycho (1546–1601), Danish mathematician and astronomer who was one of the most successful observers of celestial bodies before the use of telescopes. His mathematical prowess in measuring and fixing the positions of the stars and the length of the year was remarkable. His measured positions of the planets were so accurate that Kepler was able to deduce that the orbits are not simple circles but ellipses, with the Sun at one focus of the ellipse.

Brattain, Walter (1902–87) US co-inventor of the transistor. He shared the 1956 Nobel Physics Prize.

Brenner, Sydney (1927–), South African-born British molecular biologist who has conducted important work on the informational code of DNA.

Brewster, Sir David (1781–1868), Scottish physicist who is best known for his research on the polarization of light.

Bridgman, Percy Williams (1882–1961), US physicist who won the 1946 Nobel Physics Prize for work on high-pressure physics and thermodynamics.

Broglie, de see de Broglie.

FAMOUS SCIENTISTS

Bronsted, Johannes Nicolaus (1879–1947), Danish physical chemist who devised the Bronsted-Lowry definition, which defines an acid as a substance with a tendency to lose a proton and a base as a substance that tends to gain a proton.

Brown, Michael Stuart (1941–), US molecular geneticist who shared the 1985 Nobel Prize for Medicine or Physiology for the discovery of low-density receptors while researching cholesterol metabolism.

Brown, Robert (1773–1858), Scottish physician and botanist, who discovered the nucleus in cells but is best known for his observation under the microscope of what came to be known as Brownian Motion, the agitation of small suspended particles (such as pollen grains in water) which we now ascribe to the buffeting effect of the random motions of the molecules in the suspending medium.

Buffon, Count Georges Louis Leclerc de (1707–88), French naturalist whose 44-volume *Histoire naturelle* was an important landmark.

Bunsen, Robert (1811–99), German chemist who developed the burner which now bears his name. With Gustav Kirchhoff he observed that, when excited in his burner, certain elements emit light of characteristic wavelengths. These studies laid the foundations for the field of spectrum analysis. Bunsen also discovered the elements caesium and rubidium.

Burbank, Luther (1849–1926), US agricultural scientist who developed the science of plant breeding.

Burbidge, Eleanor Margaret (1922–), English astrophysicist who is best known for her research on quasars.

Burnet, Sir Frank Macfarlane (1899–1985), Australian virologist who shared the 1960 Nobel Prize for Medicine or Physiology for research on immunological intolerance.

Butenandt, Adolf (1903–), German biochemist who developed the chemistry of sex hormones. Awarded the Nobel Chemistry Prize in 1939 he was forbidden to accept it by the Nazi authorities.

Calvin, Melvin (1912–), US chemist who won the 1961 Nobel Chemistry Prize for his research into the role played by chlorophyll in photosynthesis.

Camerarius, Rudolph Jacob (1665–1721), German botanist who is best known for his experiments demonstrating sexuality in plants.

Candolle, Augustin Pyrame de (1778–1841), Swiss botanist who was the first to use the word taxonomy for a classification of plants by morphology (that is their forms) rather than by their physiology (that is their functions).

Cannizzaro, Stanislao (1826–1910), Italian chemist who is best known for his studies of atomic weights.

Cannon, Annie Jump (1863–1941), US astronomer who is best known for classifying stars according to their surface temperature.

Canton, John (1718–72), English physicist who was the first to make powerful artificial magnets.

Cantor, George (1845–1918), Russian-born mathematician who spent most of his life in Germany. His most important work was on finite and infinite sets and he founded set theory. Cantor introduced the concept of transfinite numbers – numbers that are infinitely large but distinct from one another. He was also greatly interested in theology and philosophy. Cantor's last years were marred by mental illness.

Cardano, Geronimo (1501–76), Italian physician and mathematician whose *Ars Magna* played a major role in the foundation of modern algebra. He was the first person to publish solutions to cubic and quartic equations. Cardano's calculations on the theory of probability were an important development.

Carnot, Nicholas Léonard Sadi (1796–1832), French physicist who virtually founded the science of thermodynamics. His researches led him to an early form of the second law of thermodynamics. He also applied scientific principles to the working cycle and the efficiency of the steam engine.

Carothers, Wallace (1896–1937), US chemist who developed the first man-made fibre, nylon.

Cartan, Elie (1869–1951), French mathematician whose highly original ideas contributed to the development of differential geometry and differential calculus.

Carver, George Washington (1860–1943), pioneering US agricultural chemist who was born a slave in the American South.

Caspersson, Torbjörn Oskar (1910–), Swedish cytochemist who is best known for his research on chromosomes.

Cassini, Giovanni Domenico (1625–1712), Italian astronomer whose most important work was on the size of solar system.

Cauchy, Baron Augustin-Louis (1789–1857), French mathematician and physicist who developed the modern treatment of calculus and also the theory of functions. He introduced rigour to much of mathematics. As an engineer he contributed to Napoleon's preparations to invade Britain, and he twice gave up academic posts to serve the exiled Charles X.

Cavendish, Henry (1731–1810), English chemist and physicist whose discoveries included the properties of hydrogen, the composition of air, the composition of water and various properties of electricity. He also discovered the specific heat of certain substances. He measured the gravitational force of attraction between two metal balls and from this experiment he calculated the gravitational constant. Cavendish measured the density and mass of the Earth.

Cech, Thomas Robert (1947–), US chemist who has undertaken important research on RNA.

Celsius, Anders (1701–44), Swedish astronomer who suggested the scale for measuring temperature that now bears his name.

Ceulen, Ludolph van (1540–1610), Dutch mathematician who is best known for trying to find the value of *pi*. (He eventually worked it out to 35 places.)

Chadwick, Sir James (1891–1974), English physicist who won the 1935 Nobel Physics Prize for the discovery of the neutron.

Chain, Ernst (1906–79), German-born British chemist who – with Florey – isolated and purified penicillin. Chain also performed the first clinical trials of penicillin. With Fleming and Florey, he shared the 1945 Nobel Prize for Physiology or Medicine.

Chandrasekhar, Subrahmanyan (1910–), Indian-born US astrophysicist who identified the limit – now known as the Chandrasekhar limit – beyond which massive stars are unable to evolve into white dwarfs.

Chapman, Sydney (1888–1970), English mathematician and physicist who made a major contribution to the study of kinetic gases. Chapman developed the theory of thermal diffusion.

Chargaff, Erwin (1905–), Austrian-born US biochemist whose studies of nucleic acids laid the foundations for Crick and Watson's DNA research.

Charles, Jacques (1746–1823), French physicist who developed the law that now bears his name – Charles's Law relates the expansion of gas with a rise in temperature.

Chasles, Michel (1793–1880), French mathematician whose *Historical Survey of the Origin and Development of Geometric Methods* had an important role in the development of modern geometry.

Chebyshev, Pafnutii (1821–94), Russian mathematician who is best known for his contribution to the theory of the distribution of prime numbers and for his work on probability theory.

Cherenkov, Pavel (1904–), Russian physicist who noted what is now called the Cherenkov effect – the emission of blue light by water and other transparent bodies when atomic particles are passed through them at speeds greater than that of light. Cherenkov shared the 1958 Nobel Physics Prize.

Chevreul, Michel (1786–1889), French chemist who was the first to identify various fatty acids.

Chladni, Ernst (1756–1827), German physicist whose studies of the vibration of solid bodies led him to identify patterns that are now known as Chladni figures. He is regarded as one of the founders of the science of acoustics.

Chu, Paul (1941–), Chinese-born US physicist whose main studies are in superconductivity.

Clairaut, Alexis Claude (1713–65), French mathematical physicist whose main work concerned the dimensions of the Earth and the Moon.

Claisen, Ludwig (1851–1930), German chemist who developed the condensation reactions that bear his name, the Claisen-Schmidt condensation.

Claude, Georges (1870–1960), French chemist and engineer who studied inert gases. He developed a process of separating the constituent gases of air by distillation of liquefied air.

Clausius, Rudolph (1822–88), German physicist who formulated the second law of thermodynamics. He is regarded as a founder of the science of thermodynamics. Clausius formulated the concept of entropy.

Clemence, Gerald Maurice (1908–1974), US astronomer who is best known for his work on the orbital motions of celestial bodies.

Cockcroft, Sir John Douglas (1897–1967), English physicist who – with Ernest Walton – was the first to split the atom (1932). With Walton, he was awarded the Nobel Physics Prize in 1951.

Cohen, Stanley (1922–), US biochemist who shared the 1986 Nobel Prize for Medicine or Physiology for his work on growth factors and membrane receptors.

Cohn, Ferdinand Julius (1828–98), German botanist who in his *Researches on Bacteria* (1872) laid the foundations of modern bacteriology.

Compton, Arthur Holly (1892–1962), US physicist who is best known for his discovery of the Compton effect, an increased wavelength of some scattered radiation.

Conway, John Horton (1938–), English mathematician who is famous for his serious studies of mathematical recreations. Conway developed the 'Game of Life', in which an object is made up of a number of squares of an infinite chessboard. The object grows, decays and 'dies' according to very simple rules. The 'Game' can be used to simulate a general-purpose computer.

Cope, Edward Drinker (1840–97), US comparative anatomist and palaeontologist who discovered the remains of very many extinct animals.

Copernicus, Nicolaus (Mikolaj Kopernik; 1473–1543), Polish astronomer who – in *Commentariolus* (1510–14) – proposed that the Earth was in daily motion about its axis and in yearly motion about a stationary Sun. His theories had an explosive effect upon the religious and philosophical views of the time and revolutionized astronomy.

Corey, Elias James (1928–), US synthetic chemist who is responsible for over 100 first syntheses.

Cori, Charles (Carl Cori; 1896–1984), Czech-born US biochemist who – with his wife Gerty – discovered a phosphate-containing form of glucose. With Gerty (and Bernardo Houssay), he was awarded the 1947 Nobel Prize for Medicine or Physiology.

Cori, Gerty (1896–1957), Czech-born US chemist; see Charles Cori.

Coriolis, Gustave-Gaspard (1792–1843), French mathematician and engineer who formulated laws to describe moving bodies – the Coriolis force.

Cottrell, Sir Alan Howard (1919–), English physicist whose principal study is dislocations in crystals.

Coulomb, Charles-Augustin de (1736–1806), French physicist who developed the law that is now known by his name. Coulomb's law states that the force between two electrical charges is proportional to the product of the charges and inversely proportional to the square of the distance between them. Coulomb identified a similar law for magnetic poles. The coulomb, which is the unit of electric charge, is named after him.

Crick, Francis (1916–), English biophysicist who – with James Watson and Maurice Wilkins – was awarded the 1962 Nobel Prize for Physiology or Medicine for determining the molecular structure of DNA (deoxyribonucleic acid).

FAMOUS SCIENTISTS

Cronin, James Watson (1931–), US physicist who shared the 1980 Nobel Physics Prize for research into charge-conjugation and parity-inversion in the decay of neutral kaons.

Crookes, Sir William (1832–1919), English chemist and physicist who invented a cathode-ray tube which was important in the development of atomic physics. He also discovered the element thallium.

Curie, Marie Sklodowska (1867–1934), Polish-born French chemist who – with her husband Pierre – pioneered research into radioactivity. She was awarded the 1903 Nobel Physics Prize – jointly with her husband and Becquerel – and was the sole winner of the 1911 Nobel Chemistry Prize. Following Becquerel's discovery of radioactivity – which Marie Curie named – she and Pierre studied pitchblende (uranium ore) to establish whether it owed much of its radioactivity to tiny quantities of highly active impurities. During this work, the Curies discovered the elements polonium and radium. The great radioactivity of the latter confirmed the immense possibilities of the energy that could be won from atomic processes.

Curie, Pierre (1859–1906), French physicist who researched the relationship between magnetism and heat. He discovered that – above a certain critical point, which is now known as the Curie point – ferromagnetic substances lose their magnetism. He formulated Curie's law which related how easy it is to magnetize a substance to its temperature (magnetic susceptibility is *inversely* proportional to the absolute temperature). With his wife Marie, he studied radioactivity (see above).

Cuvier, Baron Georges (1769–1832), French naturalist who – in *The Animal Kingdom Arranged in Conformity with its Organization* – classified animals according to principles of comparative anatomy.

Daimler, Gottlieb (1834–1900), German engineer who (in 1882–85) developed a light, internal combustion engine.

d'Alembert, Jean Le Rond (1717–83), French philosopher and mathematician who formulated the d'Alembert principle – a generalization of Newton's third law of motion – which states that Newton's law holds not only for fixed bodies but also for those able to move.

Dalton, John (1766–1844), English chemist and physicist who developed the atomic theory of matter, defining the atom as the smallest particle of substance that can take part in a chemical reaction. He made the first proposals concerning a scale of atomic weights with hydrogen as one. His theories were published in *New System of Chemical Philosophy* (1808 and 1810). His early studies on gases led to the formulation of Dalton's law which states that the total pressure of a mixture of gases is equal to the sum of the partial pressures of the individual component gases. He also investigated colour-blindness from which he suffered and it came to be known as 'Daltonism'.

Dam, Carl Peter Henrik (1895–1976), Danish biochemist who won the 1943 Nobel Prize for Medicine or Physiology for isolating vitamin K.

Daniell, John Frederic (1790–1845), English chemist who invented the Daniell cell, an improvement on the earlier battery of Volta.

Darby, Abraham (c. 1678–1717), English metallurgist who was the first to smelt iron ore with coke successfully.

Darwin, Charles Robert (1809–82), English naturalist who proposed the modern theory of evolution in his *On the Origin of Species by Means of Natural Selection* (1859). This was based upon observations that he had made on an expedition of five years in South America and the Pacific region, in particular the Galapagos Islands. He proposed that species evolve by means of a process of natural selection involving the survival of the fittest and a gradual adaptation of animals to survive in changed circumstances or habitats. In *The Descent of Man* (1871) Darwin proposed the evolution of man from a primitive animal that was also the ancestor of the apes. His theories – which became known as Darwinism – were highly controversial and had a profound effect upon both scientific and religious opinion. His studies also included the taxonomy of barnacles, the formation of atolls, the role of earthworms in soil fertility, etc.

Davis, Raymond (1914–), US chemist and physicist who is best known for his work on neutrino emission from the Sun.

Davy, Sir Humphry (1778–1829), English chemist who investigated the electrolysis of molten salts leading to the discovery of the metallic elements sodium, potassium, calcium, barium and magnesium and the isolation of strontium. By using potassium he also discovered the element boron in borax. No individual discovered more elements but perhaps his greatest 'discovery' was Faraday who as a bookbinder's apprentice approached him after attending one of Davy's lectures at the Royal Institution. Davy became well-known to the public for his invention of the miner's safety lamp and he advised industrialists about many practical matters.

Dawkins, Richard (1941–), English ethologist who – in *The Selfish Gene* (1976) – identified the apparently altruistic behaviour of some animals as the 'selfish gene' ensuring the survival of the species.

de Beer, Sir Gavin Rylands (1899–1972), English zoologist who disproved the germ-layer theory, thus confirming that vertebrate structures are formed from the ectoderm, the outer layer of the embryo.

de Broglie, Prince Louis Victor (1892–1987), French physicist who is best known for his theory that – just as waves can behave like particles – particles can have wave-like properties. He received the 1929 Nobel Physics Prize for discovering the wave nature of the electron. He stated that an electron can behave as of it were a wave motion – now known as a de Broglie wave.

Debye, Peter (1884–1966), Dutch-born US physicist and chemist whose main work was in dipole moments. He is best known for the Debye-Hückel theory of electrolytes which states that an ion in solution tends to attract other ions of opposite charge.

Dedekind, Richard (1831–1916), German mathematician whose main work was in irrational numbers. He gave the first precise definition of an infinite set.

Dehmelt, Hans Georg (1922–), German-born US physicist who – using the Penning trap – isolated a single electron (1973).

Delbruck, Max (1907–81), German-born US pioneer in modern molecular genetics who demonstrated that phage can reproduce sexually. He shared the 1969 Nobel Prize for Medicine or Physiology.

Demerec, Milislav (1895–1966), Croat-born US geneticist whose work on *Salmonella* showed that genes controlling related functions are grouped together on the chromosome.

Desargues, Girard (1591–1661), French engineer, architect and geometrician who invented modern projective geometry. Desargues was the author of *Brouillon Project*, one of the most neglected works in the history of mathematics, lost and only rediscovered in 1845. He is remembered for Desargues' theorem on pairs of triangles in perspective.

Descartes, René (1596–1650), French philosopher, mathematician and military scientist who sought an axiomatic treatment of all knowledge, and is known for his doctrine that all knowledge can be derived from one certainty: *Cogito ergo sum* ('I think therefore I am'). His greatest contribution to mathematics was the creation of analytical geometry, which allows geometrical problems to be solved by algebra, and algebraic ideas to be expressed in geometrical imagery. This great invention appeared as the final Appendix to his *Discourses on Method*; the other Appendices deal with optics and meteorological phenomena, including the rainbow.

de Sitter, Willem (1872–1934), Dutch mathematician and astronomer who proposed what came to be known as the de Sitter universe (as opposed to the Einstein universe). The de Sitter universe was envisaged as a static universe if no matter were present.

de Vries, Hugo (1848–1935), Dutch geneticist and plant physiologist who studied heredity in plants. He mistook triploids for mutants and wrongly suggested that a new species could arise through a single sudden mutation.

Dewar, Sir James (1842–1923), Scottish physicist and chemist whose main work was in the liquefaction of gases and low temperatures.

d'Herelle, Felix (1873–1949), Canadian bacteriologist who discovered the bacteriophage, a type of virus that destroys bacteria.

Dicke, Robert Henry (1916–), US physicist who confirmed that the gravitational mass of a body is equal to its inertial mass. He also worked on the big-bang origin of the universe.

Diesel, Rudolf (1858–1913), German engineer and inventor of the internal combustion engine, which bears his name.

Diesenhofer, Johann (1943–), German chemist who shared the 1988 Nobel Chemistry Prize for research on the proteins needed in photosynthesis.

Diogenes of Apollonia (5th century BC), Greek physician and philosopher who proposed that air was fundamental to all life forms.

Diophantus of Alexandria (3rd century AD), Greek mathematician whose *Arithmetica* contains much pioneering work in algebra. The Diophantine equation – which involves up to four variables – was eventually solved in India about four centuries after his death.

Dioscorides, Pedanius (c. AD 40–c. 90), Greek physician and pharmacologist whose *De materia medica* became an important source of modern botanical terminology.

Dirac, Paul (1902–84), English mathematician and physicist (whose parents were Swiss). Dirac developed a general formulism for quantum mechanics (1926). Two years later he developed his relativistic theory to describe the properties of the electron. In *The Cosmological Constants* (1937), Dirac described large-number coincidences. He shared the 1933 Nobel Physics Prize – with Schrödinger – for introducing wave-equations in quantum mechanics.

Djerassi, Carl (1923–), Austro-Bulgarian US chemist whose main work has been on the steroid hormone progesterone.

Dobzhansky, Theodosius (1900–75), Ukrainian-born US geneticist whose studies of the fruit fly demonstrated that genetic variations within populations are far greater than had been imagined.

Dollfus, Audouin Charles (1924–), French astronomer who is best known for his studies of Saturn.

Domagk, Gerhard (1895–1964), German biochemist who won the 1939 Nobel Prize for Physiology or Medicine for the discovery of the antibacterial effects of prontosil in controlling streptococcus infections. This led to the discovery of a range of drugs that were effective for various pathogenic infections including pneumonia. A beneficiary of these discoveries was Domagk's own daughter who had become accidentally infected in the laboratory.

Donders, Frans Cornelius (1818–89), Dutch oculist whose studies of the physiology and pathology of the eye led to the discovery of scientific approaches to correct conditions including near- and farsightedness and astigmatism.

Doppler, Christian Johann (1803–53), Austrian physicist and mathematician who discovered the wave effect that now bears his name. The Doppler effect is the apparent difference between the frequency at which sound or light waves leave a source and that at which they reach an observer, caused by relative motion of the observer and the wave source.

Drake, Frank Donald (1930–), US astronomer who is best known for his radio studies of the planets.

Draper, John William (1811–82), Anglo-US chemist who founded the science of photo-chemistry. This followed from his realization that chemical reactions could be brought about when molecules absorbed light energy.

Dreyer, Johann Louis Emil (1852–1926), Danish astronomer who compiled three important catalogues of galaxies, nebulae and clusters.

Duersberg, Peter (1936–), German-born US molecular biologist who denies that AIDS is caused by HIV.

Dulbecco, Renato (1914–), Italian-born US molecular biologist and physician who introduced the concept of cell transformation in biology (a process in which special cells are mixed *in vitro* with tumour-producing viruses).

Dumas, Jean-Baptiste André (1880–84), French chemist who produced the substitution theory in chemistry. He was several times a government minister.

FAMOUS SCIENTISTS

Eddington, Sir Arthur (1882–1944), English astronomer, mathematician and physicist whose main work was in the structure, motion and evolution of stars. He made an important contribution to physics as the first English-language advocate of Einstein's theory of relativity.

Edeleman, Gerald Maurice (1929–), US biochemist whose main work elucidates the chemical structure of antibodies. He shared the 1972 Nobel Prize in Physiology or Medicine.

Edison, Thomas Alva (1847–1931), US inventor who held a remarkable total of over 1000 patents. His inventions were the product of the world's first industrial research laboratory which he established at Newark, New Jersey, USA. Concentrating in electrical technology, they included major advances in telegraphy (1874), the telephone (1877), incandescent light (1879), the phonograph (1877), celluloid film (1891), electric generating equipment and storage batteries (1901–10). In 1883 he accidentally discoverd what is now known as the Edison effect which later became the basis for the development of the electron tube.

Ehrenberg, Christian Gottfried (1795–1876), German biologist whose main work was the study of microorganisms.

Ehrlich, Paul (1854–1915), German bacteriologist who pioneered developments in haematology and immunology, and who founded the science of chemotherapy, the chemical treatment of disease. His discovery of trypan red and salvarsan made important advances in the treatment of trypanosomiasis and syphilis respectively.

Eichler, August Wilhelm (1839–87), German botanist who developed a classification comprising a plant kingdom and four divisions.

Eigen, Manfred (1927–), German chemist who developed the relaxation technique for the study of very fast chemical reactions.

Einstein, Albert (1879–1955), German-born US physicist who revolutionized physics by developing the special and general theories of relativity for which he was awarded the Nobel Physics Prize in 1921. His international impact upon physics began in 1905 when he published four research papers, each one containing a major discovery in the subject: the special theory of relativity (which he proposed to account for the constant speed of light); the theory of Brownian movement; the photon theory of light and the equation relating mass and energy. His general theory of relativity was presented in 1916 and verified in 1919. Einstein spent much of his life applying the general theory to cosmological problems, after 1933 working in the USA where he took out citizenship in 1940. On a much smaller scale Einstein never accepted the usual explanation of quantum mechanics in terms of probability and he conducted a long – and eventually unsuccessful – debate in favour of determinism. Einstein became concerned to formulate a single theory that would cover both gravitation and electrodynamics. In America he became involved in many social and charitable organizations helping Jewish refugees from Nazi Germany. In 1939 Einstein became aware – through approaches by two Hungarian scientists who wished him to approach President Franklin D. Roosevelt – that German scientists led by Hahn had discovered the fission of uranium, which opened up the possibility of enormous quantities of energy. Einstein's letter to Roosevelt led to the Manhattan Project and the development of the atom bomb.

Einthoven, Willem (1860–1927), Dutch physiologist who developed electrocardiography.

Elton, Charles Sutherland (1900–91), English ecologist who was one of the first to study animals in relation to their environment.

Enders, John F. (1897–1985), US microbiologist who – with Frederick Robbins and Thomas Weller – discovered that the polio virus could be grown in tissue other than nerve tissue. This was a major develpment that helped lead to a vaccine.

Engler, Adolf (1844–1930), German botanist who – in his monumental *The Natural Plant Families* – developed a system of plant classification that is still of importance today.

Eötvös, Baron Roland von (1848–1919), Hungarian physicist who introduced the Eötvös law, an equation that relates surface tension, temperature, density, and the relative molecular mass of a liquid.

Erlanger, Joseph (1874–1965), US neurophysicist who shared the 1944 Nobel Prize for Medicine or Physiology for his studies of the differentiated functions of nerve fibres.

Ernst, Richard Robert (1933–), Swiss chemist who is best known for his work on NMA (nuclear magnetic resonance).

Esaki, Leo (1925–), Japanese physicist who has discovered an effect he named 'tunneling', a quantum-mechanical effect in which an electron can penetrate a potential barrier through a narrow region that classical theories state it should not be able to penetrate.

Euclid (c. 3rd century BC), Greek mathematician who devised the first axiomatic treatment of geometry and studied irrational numbers. Until recent times, most elementary geometry textbooks were little more than versions of Euclid's great book *The Elements*.

Eudoxus (408–347 BC), Greek mathematician, geographer and astronomer who is known for his theory of proportion. He developed the theories of Euclid.

Euler, Leonhard (1707–83), Swiss-born mathematician who worked mainly in Berlin and St Petersburg. He was particularly famed for being able to perform complex calculations in his head, and so was able to go on working after he went blind. Euler worked in almost all branches of mathematics and made particular contributions to analytical geometry, trigonometry and calculus, and thus to the unification of mathematics. Euler was responsible for much of modern mathematical notation.

Everett, Hugo (1930–82), US physicist whose *Relative State Formulation of Quantum Mechanics* (1957) was one of the most influential works in quantum mechanics. He suggested what has become known as the 'many worlds' interpretation, establishing a universal application to quantum mechanics. He applied his universal wave function to both microscopiuc entities and macroscopic observers.

Eyring, Henry (1901–81), Mexican-born US chemist who introduced a transition-state theory to chemical kinetics.

Fabricius ab Aquapendente, Hieronymous (1537–1619), Italian anatomist whose main work established embryology as a separate science, although many of his conclusions were incorrect.

Fahrenheit, Gabriel Daniel (1686–1736), German-born physicist of Dutch ancestry who was the first person to use mercury in a thermometer. He devised for his thermometer a scale which bears his name. Fahrenheit places the freezing point of water at 32° and the boiling point at 212°.

Fairbank, William (1917–89), US physicist who announced in 1979 that he had isolated the quark, although no-one since has managed to reproduce his experiments.

Fallopius, Gabriel (Gabriello Fallopio; 1523–62), Italian anatomist who is chiefly remembered for his description of the Fallopian tubes.

Faraday, Michael (1791–1867), English physicist and chemist who made many important advances in electromagnetism. Trained as a bookbinder and bookseller, Faraday was (at 21) appointed as an assistant to Davy who recognized his potential and taught him. Having discovered electro-magnetic induction in 1821, Faraday built a primitive model. He demonstrated the continuous production of current from a conductor moving in a field – in effect, a primitive dynamo. Faraday also liquefied chlorine and isolated benzene. In 1833 he formulated quantitative laws to express magnitudes of electrolytic effects – these are now known as Faraday's laws of electrolysis. In 1831 he made observations that led to the formulation of the law of induction which now bears his name. This describes a quantitative relationship between a changing magnetic field and the electric field created by the change. Faraday also gave his name to the Faraday effect, in physics, which is the rotation of the plane of polarization (the plane of vibration) of a light beam by a magnetic field. The farad – the SI unit of electric capacitance – is named after him.

Fermat, Pierre de (1601–65), French lawyer who studied mathematics as a hobby. He contributed to the development of calculus, analytic geometry, and the study of probability (with Pascal). Fermat is regarded as the creator of the modern theory of numbers. He is most famous for his last theorem, which he claimed to have solved, without recording his proof. In 1908 a prize of 100 000 German marks was offered for a correct proof but the theorem remained unsolved until 1993, when Andrew Wiles, a British-born mathematician at Princeton University, claimed to have reached a proof.

Fermi, Enrico (1901–54), Italian-born US physicist who established the theory of beta decay in the 1930s. He discovered statistical laws obeyed by particles such as the electron. Fermi left Fascist Italy in 1938 and settled in the USA where he took citizenship. In America he researched means of producing controlled and self-sustaining nuclear fission reaction. The result – the first nuclear reactor – was built and tested in a squash court at Chicago University in 1942. At Los Alamos, New Mexico, USA, in 1943 Fermi helped to test the first atomic bomb.

Ferraris, Galileo (1847–97), Italian physicist and electrical engineer who discovered the principle of rotary magnetic field.

Feynman, Richard Phillips (1918–88), US physicist who developed the quantum approach to electromagnetic theory. A leading theoretical physicist of the modern age, Feynmarn worked with Julian Schweinger to find a method of avoiding the mathematical ambiguities that were part of the original quantum electrodynamics of Dirac and Heisenberg. As a result Feynman produced a more refined theory. He became popular for his inspirational 'Feynman Lectures' which were published and he had an engaging personality which attracted the attention of television audiences.

Fibonacci, Leonardo (also known as **Leonardo Pisano**) (c. 1170–1230), Italian mathematician who is famous for *Liber Abaci*, an account of elementary arithmetic and algebra that popularized the Hindu system of counting in Europe. It contains the problem of the breeding rabbits, which gives rise to the famous Fibonacci sequence – 1, 1, 2, 3, 5, 8, 13, etc. – in which each number is the sum of the previous two. This sequence has many uses including the design of efficient computer sorts.

Fischer, Emil (1852–1919), German chemist and biochemist whose work on sugars and peptides did much to establish biochemistry as a separate science.

Fischer, Ernst Otto (1918–), German chemist whose main work has been on inorganic complexes.

Fisher, Sir Ronald Aylmer (1890–1962), English geneticist who is best known for his major work *The Genetic Theory of Natural Selection* (1930).

Fizeau, Armand-Hippolyte-Louis (1819–96), French physicist who achieved fame for his experiments determining the speed of light. Using a powerful light on one hill and a mirror on another 8 km (5 mi) apart, he measured the time it took for the light to make the round trip from lamp to mirror and then back (1849). His ingenious experiment used a quickly rotating toothed wheel to interrupt the light beam.

Flamsteed, John (1646–1719), English astronomer whose *British Celestial Record* – published posthumously – contains the position of over 3000 stars.

Fleischmann, Martin (1927–), Czech-born British chemist who announced in 1989 that he and Stanley Pons had achieved nuclear fusion by an electrolytic method under laboratory conditions.

Fleming, Sir Alexander (1881–1955), Scottish bacteriologist who discovered that the fungus *Penicillium notatum* was able to kill bacteria (1928). Fleming's discovery of penicillin – for which he shared the 1945 Nobel Prize for Physiology or Medicine – was accidental. Some *Penicillium* had dropped by chance into a culture of bacteria that Fleming was about to dipose of; he noticed that no bacteria had grown near to the fungus. Although Fleming recognized the importance of his discovery he did not isolate the active substance.

FAMOUS SCIENTISTS

Fleming, Sir John Ambrose (1849–1945), English physicist who invented the thermionic vacuum tube.

Flerov, Georgii (1913–), Russian physicist who synthesized elements 102, 103, 104, 105, 106 and 107.

Florey, Howard Walter (Baron Florey; 1898–1968), Australian pathologist who adopted Fleming's discovery of penicillin as a treatment for diseases.

Florey, Paul John (1910–85), US chemist whose major work was on nonlinear polymers. He proposed the concept now known as the Florey temperature – the temperature for a given solution at which meaningful measurements of the properties of polymers can be made.

Flourens, Jean-Pierre (1794–1867), French anatomist who in 1824 demonstrated the main roles of different parts of the human central nervous system.

Foucault, Jean-Bernard-Léon (1819–68), French physicist who devised a method of measuring the absolute velocity of light to within 1% of its true value. By suspending a pendulum from a church dome he provided proof that the Earth rotates on its axis – the plane in which the pendulum swings twists gradually in a clockwise direction, proving that the Earth must move in the opposite direction. Foucault's inventions include the gyroscope. He also discovered the existence of eddies – now known as Foucault currents – in a copper disc moving in a strong magnetic field.

Fourier, Jean-Baptiste (1768–1830), French mathematician and physicist who in *The Analytical Theory of Heat* showed how the conduction of heat in solid bodies could be analysed in terms of infinite mathematical series, now known as Fourier Series. The Series is a theorem that governs periodic oscillation and it has wide application in many fields.

Franck, James (1882–1964), German-born US physicist who – with Gustav Hertz – won the 1925 Nobel Physics Prize for research into the excitation and ionization of atoms by electron bombardment. Although he worked on the American atomic bomb he opposed its use in warfare.

Frankland, Sir Edward (1825–99), English chemist who was a pioneer in structural chemistry. Frankland researched why atoms are linked together in the way that they are. This led him to suggest his theory of valence which states that all atoms have a certain capacity to combine with other atoms.

Franklin, Benjamin (1706–90), US inventor, diplomat, printer, publisher and scientist who played an important part in establishing American independence – he helped draft the Declaration of Independence (1776) and the United States Constitution (1787). As a scientist his most important work was in studying electricity. When flying a kite in a violent thunderstorm (1752) a spark leapt from a key he had attached to the kite and travelled down the kite string, proving that lightning is an electrical phenomenon.

Franklin, Rosalind (1920–58), English crystallographer who played an important (and still largely unacknowledged) part in the discovery of the structure of DNA. Her X-ray photograph of hydrated DNA was recognized by Watson as a helix. She worked independently of Crick and Watson and was producing a paper on the double-chain helix of DNA when they published.

Fraunhofer, Josef von (1787–1826), German physicist who discovered what are now known as the Fraunhofer lines – lines in the spectra of bright stars.

Frege, Gottlob (1848–1925), German philosopher who gave a formal definition of a cardinal number.

Fresenius, Carl (1818–97), German analytical chemist who wrote works on qualitative and quantitative analysis that became standards.

Fresnel, Augustine Jean (1788–1827), French physicist who worked on the theory that light is a transverse wave motion. He studied the laws of the interference of polarized light. He also pioneered optics. The Fresnel lens – which is used in lighthouses – is named after him.

Frisch, Karl von (1886–1982), Austrian zoologist and entomologist who is best known for his major work on the senses, communication (the 'dance') and social organization of the bee.

Frisch, Otto Robert (1904–79), Austrian-born British physicist who worked with his aunt, Lise Meitner, on the discovery of nuclear fission.

Gabor, Dennis (1900–79), Hungarian-born British physicist who pioneered holography, a method of photographically recording and reproducing three-dimensional images.

Galen (c. 131–200), Greek physician whose writing on anatomy was the standard text on the subject until the Middle Ages.

Galileo, Galilei (1564–1642), Italian astronomer, physicist and mathematician who was a pioneer in developing scientific methods of testing by systematic experimentation. He became a professor of mathematics at Pisa at the age of 25 but the popular story concerning dropping weights from the famous Leaning Tower of Pisa is a charming legend unsupported by evidence. Most of his work was concerned with the study of mechanics; Galileo was the first to apply mathematics in this field. He was also one of the first people to use a telescope and supported Copernicus' views on the nature of the universe in *Letters on the Sunspots* (1613). This led him into direct conflict with the Church which – in 1616 – declared the views of Copernicus to be blasphemous. Galileo responded with *Dialogue of the Two Great World Systems* which was widely acclaimed but for which Galileo was convicted of heresy (1633) and obliged to recant.

Gall, Franz (1758–1828), German physiologist who discovered the difference between the active (grey) and connective (white) parts of the brain.

Gallo, Robert (1937–), US physician who was one of the first people to identify HIV, the virus responsible for AIDS.

Galton, Sir Francis (1822–1911), English explorer and anthropologist who is best known for his pioneering studies into human intelligence, eugenics (a science that he may be said to have founded).

Galvani, Luigi (1737–98), Italian physicist who studied what he believed to be electricity in animal tissue. These researches resulted in the development of the voltaic pile, a type of battery giving a constant supply of electricity.

Gamov, George (1904–68), Ukrainian-born US physicist who extended Lemaître's big bang theory of the creation of the universe. Gamov – who was also interested in DNA – was a popular science writer.

Gassendi, Pierre (1592–1655), French physicist and philosopher who observed comets and attempted to measure the speed of sound. His writings had an important influence on latter physicists.

Gauss, Friedrich Carl (1777–1855), German mathematician who developed the theory of complex numbers. He was director of the astronomical observatory at Göttingen and conducted a survey, based on trigonometric techniques, of the kingdom of Hannover. He published works in many fields, including the application of mathematics to electrostatics and electrodynamics.

Gay-Lussac, Joseph (1778–1850), French chemist and physicist who pioneered investigations into the behaviour of gases. He formulated the law of combining volumes which demonstrated that gases combined with each other in proportions by volume that can be expressed as small whole numbers – for example, water is formed by hydrogen and oxygen in the ratio 2:1. Berzelius later used this law in his work to establish atomic weights. Gay-Lussac was also a leading early meteorologist.

Geiger, Hans (1882–1945), German physicist who assisted Rutherford. In 1913 Geiger developed the counter for detecting atomic particles that now bears his name.

Gelfand, Izrail (1913–), Russian mathematician whose main work has been on generalized functions.

Gell-Mann, Murray (1929–), US physicist who is best known for his work on the behaviour of mesons.

Gerhardt, Charles (1816–56), French chemist who formulated a theory of types to classify organic chemical compounds. The four types are described according to a compound's relationship to ammonia, water, hydrogen and hydrogen chloride.

Gesner, Konrad von (1516–65), Swiss physician whose impressive work *A Catalogue of Animals* (1551–58) describes large numbers of creatures. It is sometimes considered to be the first work of zoology but it is neither systematic nor analytical.

Gibbs, Josiah Willard (1839–1903), US physicist who founded the science of chemical thermodynamics. Gibbs identified the quantities – for example free energy and chemical potential – that are involved in chemical reactions and worked out their mathematical basis.

Gilbert, Walter (1932–), US molecular biologist who shared the 1980 Nobel Chemistry Prize for the chemical and biological analysis – with Frederick Sanger – of DNA. His other major study has been the lac repressor, a large protein molecule that controls the production of enzymes.

Glaser, Donald Aylmer (1926–), US physicist who developed the bubble chamber, which has become a major tool for high-energy physics experiments. He won the 1960 Nobel Physics Prize.

Glashow, Sheldon Lee (1932–), US physicist who shared the 1979 Nobel Physics Prize for an explanation of the forces that hold together elementary particles of matter.

Goddard, Robert Hutchings (1882–1945), US physicist and pioneer in the development of rockets. Although his main work was in the development of liquid fuels for rockets, Goddard also pioneered the use of liquefied gases as propellants.

Gödel, Kurt (1906–78), Austrian-born US mathematician who stunned the mathematical world in the 1930s by showing that Hilbert's dream of a general method of proving any mathematical theorem could not be realized.

Gold, Thomas (1920–), Austrian-born US astronomer who developed the steady-state theory of the expanding universe.

Goldstein, Joseph (1940–), US geneticist who has undertaken important studies on the metabolism of fats and cholesterol.

Golgi, Camillo (1843–1926), Italian cytologist who discovered the organelles now known as Golgi bodies.

Graham, Thomas (1805–69), Scottish chemist who defined diffusion. He defined substances into two categories – colloids and crystalloids – and is generally held to be the founder of colloid chemistry.

Grassman, Hermann (1809–77), German mathematician who discovered the calculus of extension, which may be regarded as an 'algebra' of geometry.

Gray, Louis Harold (1905–65), English radiologist after whom the SI unit of radiation absorbed dose – the gray – is named.

Guericke, Otto von (1602–86), German physicist who invented the first air pump. Using his invention, von Guericke studied and proved the existence of the vacuum. He also researched the importance of air in respiration and in combustion.

Haber, Fritz (1868–1934), German chemist who is best known for his discovery of an industrial process for synthesizing ammonia from nitrogen and hydrogen.

Haeckel, Ernst (1834–1919), German biologist who was an important supporter of Darwinism. He suggested that the embryological stages of an organism reflect the evolutionary history of that lifeform. Haeckel proposed that the stages of an embryo – and development of an individual lifeform – briefly, and sometimes incompletely, repeated the history of the development of the entire species.

Hahn, Otto (1879–1968), German chemist who – with Fritz Strassmann – discovered nuclear fission. From 1906 to 1938 he worked with Lise Meitner mainly studying the application of radioactive methods to solve chemical problems; during this period Hahn and Meitner discovered the element protactinium. After Meitner fled from Germany (1938), Hahn continued his studies with Fritz Strassmann. They proved that when uranium was bombarded with neutrons, one of the products formed was a much lighter radioactive form of barium. This indicated that the uranium atom had divided into two lighter atoms. Hahn sent his results abroad to Meitner. She developed an explanation for what had happened and named the process nuclear fission. Hahn won the 1944 Nobel Chemistry Prize.

FAMOUS SCIENTISTS

Haldane, J(ohn) B(urdon) S(anderson) (1892–1964), Scottish biologist who explored new fields of research in genetics and evolution. He was known for his application of mathematics to science. A born communicator, Haldane helped popularize science.

Hale, George Ellery (1868–1938), US astrophysicist who was a leading figure in the development of modern astronomy. His main work was in the area of solar spectroscopy.

Hales, Stephen (1677–1761), English cleric, chemist and botanist who was one of the first to use quantitative methods in experiments on plants and in animal physiology.

Hall, Charles Martin (1863–1914), US chemist who was the exact contemporary of the Frenchman Paul Héroult (1863–1914). Both men, in the same year (1886), independently discovered how to make aluminium commercially by electrolysing aluminium oxide in cryolite (sodium aluminium fluoride), a process which transformed aluminium from being a previously precious metal into one which could eventually become cheap enough for everyday uses.

Hall, Sir James (1811–98), English geochemist who pioneered experimental geology. He showed that crystals are formed when molten rocks cool down sufficiently slowly.

Hall, Marshall (1790–1857), English physiologist who studied the nervous system. He was the first to suggest a scientifc theory to explain reflex action.

Haller, Albrecht von (1708–77), Swiss physiologist, botanist, anatomist and poet who is regarded as the founder of experimental physiology. His experiments which involved injecting blood vessels with dyes to make their course clear produced an accurate model of the cardiovascular system.

Halley, Edmund (1656–1742), English astronomer and physicist who formulated the mathematical law relating pressure and height. His other interests in physics included the optics of the rainbow and evaporation. Halley is, however, best known for his work in astronomy. He is regarded as the founder of modern cometary study.

Hamilton, Sir William Rowan (1805–65), Irish mathematician whose studies resulted in Hamilton's equations (which relate to the positions and moments of particles), Hamilton's principle (which states that the integral with respect to time of the kinetic energy minus the potential energy of a system is a minimum), and the Hamiltonian function (whose equations are still used in quantum mechanics).

Harden, Sir Arthur (1865–1940), English biochemist who won the 1929 Nobel Chemistry Prize for his work on fermentation and enzymes.

Hardy, Godfrey Harold (1877–1947), English mathematician who solved many problems relating to prime number theory.

Harrison, Ross Granville (1870–1959), US zoologist who developed animal-tissue cultures.

Harvey, William (1578–1657), English physician who discovered the nature of the circulation of blood in the body and the function of the heart as a pump. This was described by Harvey in *On the Motion of the Heart and Blood in Animals*. Harvey was personal physician to King Charles I who gave Harvey permission to use the bodies of deer from the royal parks for experimentation.

Hawking, Stephen (1942–), English physicist who is best known for his theory of exploding black holes which draws upon both quantum mechanics and relativity theory. Hawking's *Brief History of Time* was a hugely popular exploration of space-time problems. Since 1963, Hawking has been disabled in movement and speech by a severe neuromuscular disorder.

Haworth, Sir Walter (1883–1950), English chemist who was the first person to synthesize a vitamin. He shared the 1937 Nobel Chemistry Prize.

Heaviside, Oliver (1850–1925), English physicist who predicted the existence of the ionosphere.

Hedwig, Johann (1730–99), German botanist who did major pioneering work in the study of mosses.

Heisenberg, Werner von (1901–76), German physicist who discovered how to formulate quantum mechanics in terms of matrices. Heisenberg's uncertainty principle states that certain pairs of variables describing motion cannot be measured simultaneously with accuracy because the processes involved in measurement interfere with the quantity that is being measured.

Helmholtz, Hermann von (1821–94), German scientist, mathematician and philosopher who made important discoveries in acoustics, optics, physiology and meteorology. He is best known for his statement of the law of conservation of energy, the first law of thermodynamics – energy can be converted from one form to another but cannot be destroyed.

Henry, Joseph (1797–1878), US scientist who discovered a number of important electrical properties, including self-induction. The SI unit of inductance – the henry – is named after him. Henry worked with Morse in the development of telegraphy.

Henry, William (1775–1836), English chemist and physician who formulated the law which now bears his name – Henry's law states that the amount of a gas absorbed by a liquid is in proportion to the pressure of the gas above the liquid, provided that no chemical reaction occurs.

Hero of Alexandria (7th century AD), Greek mathematician who is best known for his formula for the area of a triangle and for his method of finding a square root.

Héroult, Paul (1863–1914), French chemist who was a pioneer in electrolysis. He became famous for his discovery in 1886 of the commercial means of producing aluminium by electrolysing its oxide. (See Hall, C.M. above).

Herschel, Sir William (1738–1822), German-born British astronomer who discovered Uranus and catalogued over 2000 nebulae and 800 double stars. He was one of the first to begin to comprehend the structure of the Galaxy.

Hershey, Alfred Day (1908–), US biologist who shared the 1969 Nobel Prize for Medicine or Physiology. He is best known for an experiment in 1952 which appears to show that DNA is more involved than protein in the replication of genes.

Hertz, Gustav (1887–1975), German physicist who – with James Franck – won the 1925 Nobel Physics Prize for proving that energy can be absorbed by an atom only in definite amounts.

Hertz, Heinrich (1857–94), German physicist who discovered radio waves. Hertz established that heat and light are electromagnetic radiations. The hertz (Hz) – the unit of frequency – is named after him.

Hertzsprung, Ejnar (1873–1967), Danish astronomer who – with Henry Russell – was the independent originator of the Hertzsprung-Russell diagram (better known as the H-R diagram), which plots period-luminosity relations of stars on a numerical basis.

Hilbert, David (1862–1943), German mathematician who, in 1901, listed 23 major unsolved problems in mathematics – many of which still remain unsolved. His work contributed to the rigour and unity of modern mathematics and to the development of the theory of computability.

Hill, A(rchibald) V(ivian) (1886–1977), English biophysicist and physiologist whose main work concerned the production of heat in muscles. He shared the 1922 Nobel Prize for Physiology or Medicine.

Hipparchus (2nd century BC), Greek astronomer who is best known for his catalogue of the stars. Hipparchus was a pioneer in trigonometry.

Hippocrates (c. 400–c. 377 BC), Greek physician whose name is associated with the Hippocratic oath – which is still taken by graduating medical students – although he did not write it. Hippocrates, who is regarded as the father of medicine, complied a number of works on medicine now known as the *Hippocratic Collection*. He was the author of some of these works but relatively little is known about him for certain.

His, Wilhelm (1831–1904), Swiss-born German biologist who founded the science of histogenesis, the study of the embryonic origins of different types of animal tissue.

Hittorf, Johann Wilhelm (1824–1914), German chemist and physicist who showed how the relative speeds of ions can be calculated. Hittorf pioneered experiments on cathode rays.

Hodgkin, Sir Alan (1914–), English physiologist who shared the 1963 Nobel Prize for Physiology or Medicine for the discovery of the chemical processes that are responsible for the passage of impulses along nerve fibres.

Hodgkin, Dorothy (1910–), English chemist who is best known for her work on the structures of penicillin and vitamins.

Hooke, Robert (1635–1703), English physicist and chemist who discovered the law of elasticity that is now known by his name. Hooke's law states that the stretching of a solid body is proportional to the force applied to it – in other words, stress is proportional to strain. Hooke worked as an assistant to Robert Boyle and put forward the first rational theory of combustion.

Hooker, Sir Joseph Dalton (1817–1911), English botanist who succeeded his father (below) at Kew. His *Genera Plantarum* (1862–83) described 7569 genera and 97000 species.

Hooker, Sir William Jackson (1785–1865), English botanist who – as the first Director of Kew Gardens – built up one of the finest collections of plants in the world. Knowledge of algae, lichens and fungi was advanced by his work.

Hopkins, Sir Frederick Gowland (1861–1947), English biochemist who discovered growth stimulating vitamins. He shared the 1929 Nobel Prize for Physiology or Medicine.

Hoyle, Sir Fred (1915–), English astronomer who formulated the steady-state theory of the universe in 1948.

Hubble, Edwin Powell (1889–1953), US astronomer who is best known for his classification of galaxies. Hubble's law (discovered in 1929) states that the recessional velocity of galaxies increases proportionately with their distance from us. It is seen as a vital tool in determining the age and size of the universe.

Huggins, Sir William (1824–1910), English astronomer who studied the spectra of celestial bodies. His use of spectroscopy was a major development in cosmology.

Humboldt, Baron Alexander (1769–1859), German explorer, geographer and botanist. One of the founders of physical geography and ecology, Humboldt's work in South America established connections between plant distribution and physical conditions.

Hunter, John (1728–93), Scottish surgeon who was the founder of pathological anatomy in Britain.

Huxley, Hugh Esmor (1924–), English biologist who – with Jean Hanson – proposed the sliding filament theory of muscle contraction, which describes how muscles contract at molecular level.

Huxley, Sir Julian (1887–1975), English biologist and educator who did much to make modern biology accessible to a wider public. His studies included projects in embryology, behaviour and evolution. He was a grandson of T.H. Huxley, see below.

Huxley, T(homas) H(enry) (1825–95), English biologist who advanced Darwin's theory of evolution but who suggested that change could be 'step-like' rather than gradual.

Huygens, Christiaan (1629–95), Dutch mathematician, physicist and astronomer who formulated the wave theory of light.

Ingram, Vernon Martin (1924–), German-born British biochemist whose experiments concerning mutation of haemoglobin showed that an apparently small cause – the alteration of one amino acid in 500 – can have a profound effect. His work was a landmark in molecular biology.

Ipatieff, Vladimir (1867–1952), Russian-born US chemist who pioneered research in high-pressure catalytic reactions of hydrocarbons.

Jabir, ibn Hayyan (or **Haijan**; c. 721–815), Arabian alchemist and writer (from what is now Iran) who is regarded as one of the founders of chemistry. He is said to have isolated arsenic.

Jacob, François (1920–), French biologist who shared the 1965 Nobel Prize for Medicine or Physiology for studies of the control of bacterial enzyme production.

FAMOUS SCIENTISTS

Jeans, Sir James (1877–1946), English astronomer, mathematician and physicist who was the first person to suggest that matter is continuously created throughout the universe. He was a popular writer of books on astronomy and a well-known broadcaster.

Jenner, Edward (1749–1823), English physician who developed vaccination against smallpox.

Joliot-Curie, Frédéric (Frédéric Joliot; 1900–58), French physicist who – with his wife Irène (q.v.) – discovered artificial radioactivity.

Joliot-Curie, Irène (Irène Curie; 1896–1956), French physicist who was the daughter of Pierre and Marie Curie and the wife of Frédéric Joliot. Irène and Frédéric were jointly awarded the 1935 Nobel Prize for Chemistry for their discovery of radioactive elements that had been created artificially.

Joule, James Prescott (1818–89), English physicist who determined the relationship between heat and mechanical energy. He established that the different forms of energy – heat, mechanical energy and electrical energy – are basically the same; this led him to formulate the law of conservation of energy (the first law of thermodynamics) which states that energy can be converted from one form to another but cannot be destroyed. The joule – the SI unit of work or energy – is named after him.

Kamerlingh-Onnes, Helke (1853–1926), Dutch physicist who won the 1913 Nobel Prize for Physics for his work in low-temperature physics. He discovered superconductivity, the almost total lack of electrical resistance in certain materials when cooled to a temperature of near absolute zero. He also liquefied helium.

Kant, Immanuel (1724–1804), German philosopher who addressed the challenge of Newtonian mechanics in *Metaphysical Foundations of Natural Science* (1786). He proposed that there are two basic forces: the attractive or gravitational force and the repulsive or elastic force.

Kapitza, Pyotr (1894–1984), Russian physicist who is best known for his work in low-temperature physics.

Kekulé von Stradonitz, Friedrich (1829–96), German chemist who discovered the ring structure of benzene.

Kelvin of Largs, Baron (William Thomson; 1824–1907), Scottish engineer, physicist and mathematician who was a major influence on scientific developments in his day. He entered Glasgow University to read mathematics at the age of 10 and published two papers at 16 and 17. Among his many contributions to science are an important role in the development of the conservation law of energy, the mathematical analysis of electricity and magnetism, and major research in hydrodynamics. Thomson – who became Lord Kelvin in 1892 – is, perhaps, best known for his work on the absolute scale of temperature, which is now given in degrees Kelvin.

Kenelly, Arthur Edwin (1861–1939), British-born US electrical engineer whose work on the atmosphere correctly predicted its electrically charged layers.

Kepler, Johannes (1571–1630), German astronomer who was the first to prove that the Earth orbits the Sun rather than the other way round. He showed that the planets orbit the Sun and that the orbits are ellipses with the Sun occupying one focus of the ellipse.

Khorana, Har Gobind (1922–   ), Indian-born US biochemist who shared the 1968 Nobel Prize for Physiology or Medicine for studies of the genetic composition of the cell nucleus.

al Khwarizmi (Muhammad ibn Musa al Khwarizmi; c. 825), Arabian mathematician (from what is now Iraq) who described the Hindu system of counting. His book on algebra was influential in Europe when translated into Latin in the 12th century. The name on his treatise on algebra *Hisab al-jabr w'al-muqabalah* or 'the science of reduction and cancellation', gave us our word 'algebra', and the modern word 'algorithm' is derived from his name.

Kirchner, Athanasius (1601–80), German scientist and Jesuit who – using a powerful light, lens and a reflector – succeeded in projecting an image of a picture that had been painted onto glass onto a screen.

Kirchhoff, Gustav Robert (1824–87), German physicist who is remembered for Kirchhoff's laws which describe the currents and electric forces in electrical networks. Kirchhoff's law of radiation states that the rate of emission of energy by a body is equal to the rate at which the body absorbs energy.

Klaproth, Martin (1743–1817), German chemist who discovered uranium, zirconium and cerium.

Klein, Felix (1849–1925), German mathematician who introduced a programme for the classfication of geometry in terms of group theory. His interest in topology – the study of geometric figures that are subjected to deformations – produced the first description of what became known as a Klein bottle, which has a continuous one-sided surface.

Koch, Robert (1843–1910), German bacteriologist who discovered the tuberculosis bacillus.

Krebs, Sir Hans Adolf (1900–81), German-born British biochemist who is remembered for his discovery of the Krebs cycle (the tricarbonate acid cycle or TCA cycle), which completes the oxidation of glucose.

Kroto, Harold Walter (1939–   ), English chemist who was one of the discoverers of buckminsterfullerene.

Kurchatov, Igor (1903–60), Russian physicist who led the team that exploded the first Soviet atomic bomb.

Lagrange, Count Joseph Louis (1736–1813), Italian-born French physicist and mathematician who – in *Analytical Mechanics* (1788) – summarized research in mechanics since Newton. In astronomy he identified the Langrangian point around which three bodies (asteroids) will tend to oscillate. His work laid the foundations of the metric system.

Lamarck, Jean-Baptiste (1744–1829), French naturalist who was the first to define vertebrates and invertebrates.

Lambert, Johann Heinrich (1728–77), German mathematician who was the first to show that pi is not a rational number.

Landau, Lev (1908–68), Azeri physicist who was the first to describe ferromagnetism. He also studied the superfluid behaviour of liquid helium. He won the 1962 Nobel Physics Prize.

Landsteiner, Karl (1868–1943), Austrian pathologist who identified the blood types A, AB, B and O.

Langevin, Paul (1872–1946), French physicist who – using ultrasonic waves – developed a way to detect submerged objects. This is now known as sonar.

Laplace, Marquis Pierre-Simon de (1749–1827), French mathematician, astronomer and physicist whose five-volume *Celestial Mechanics* (1799–1825) addressed the irregularities or perturbations in the movements of the planets that seemed to contradict Newtonian laws. Laplace showed that these irregularities were periodic rather than cumulative. In *Analytical Theory of Probabilities* (1812) Laplace established probability theory on a rigorous basis.

Lavoisier, Antoine-Laurent (1743–94), French chemist whose work and discoveries laid the basis for much of modern chemistry. As well as leading a full life in public service, Lavoisier found time to conduct researches concerning the gain or loss of weight of substances that are burned or reduced with charcoal. He ascribed these differences to the absorption or loss of a substance that he later called oxygen. His theories – published in *Elementary treatise on chemistry* (1789) – described the composition of water, combustion and the chemistry of many compounds. Lavoisier sat on the committee whose findings led to the adoption of the metric system. He was executed during the French Revolution for having been a tax collector.

Lawrence, Ernest Orlando (1901–58), US physicist who invented the cyclotron, the first accelerator that was able to achieve high energies. He won the 1939 Nobel Physics Prize.

Leavitt, Henrietta (1868–1921), US astronomer whose work on variable stars led her to discover a simple method of determining very great distances when she observed that the apparent magnitude (observed brightness) of the Cepheids decreased linearly with the logarithms of the period.

Le Bel, Joseph Achille (1847–1930), French chemist who devised the graphic conventions for drawing chemical formulae – see pp. 152–55 for examples.

Le Châtelier, Henry-Louis (1850–1936), French chemist who developed what is now known as the Le Châtelier principle, which states that if a substance – or collection of substances – in a balanced state is disturbed, the system will readjust to almost neutralize the disturbance and to restore equilibrium. This principle applies not only to chemical reactions but also to physical processes.

Leclanché, Georges (1839–82), French engineer who invented an electrolytic cell to provide electric supply.

Lederberg, Joshua (1925–), US geneticist who demonstrated that bacteria possess genetic and behaviour systems. He shared the 1958 Nobel Prize for Medicine or Physiology.

Lederman, Leon Max (1922–), US physicist who shared the 1988 Nobel Physics Prize for the investigation of the decay processes that lead to neutrinos.

Leeuwenhoek, Anton van (1632–1723), Dutch biologist whose observations using a microscope have been cited as the first steps in microbiology.

Leibniz, Gottfried Wilhelm (1646–1716), German mathematician, philosopher, logician, linguist, lawyer and diplomat who invented calculus independently of Newton. Leibniz's notation was, however, superior. He was the first, in 1671, to build a calculating machine that could multiply. He was the first European mathematician to consider determinants, unaware that Kowa Seki had studied them a decade before. Leibniz also studied binary numbers and found mystical significance in the creation of all numbers out of nothing and unity.

Lenard, Philipp Eduard Anton (1862–1947), Austro-Slovak-born German physicist who came close to discovering X-rays. While researching the emission of electrons he discovered that the shorter the wavelength the faster the electron – but it was left to Einstein to realize the full significance of this. Lenard became increasingly anti-Semitic and developed a pathological hatred of Einstein. His influence helped spread a rabid nationalism in scientific circles in Nazi Germany, causing many scientists to emigrate.

Lenz, H(einrich) F(riedrich) E(mil) (1804–65), Russian physicist whose research on induced electric current led him to formulate what is now known as Lenz's law, which states that the direction of a current induced in a circuit by moving it in a magnetic field produces an effect tending to oppose the circuit's motion.

Leonardo da Vinci (1452–1519), Italian Renaissance polymath whose contributions to art, architecture and science were prodigious. In science his influence was not as great as early as he deserved because his notebooks were not published until the 19th century. These show the empirical basis of his thinking and his advocacy of scientific thought and experiment.

Levene, Phoebus Aaron Theodor (1869–1940), Russian-born US biochemist who established that nucleic acids are genuine molecules that exist independently of proteins.

Levi-Civita, Tullio (1873–1941), Italian mathematician who – with Ricci – developed absolute differential calculus. His work has important applications in many fields including relativity.

Libavius, Andreas (c. 1540–1616), German chemist whose *Alchymia* (1606) is said to be the first modern chemistry textbook.

Libby, Willard Frank (1908–80), US chemist who developed radio-carbon dating methods.

Liebig, Baron Justus von (1803–73), German chemist whose application of chemistry to biology pioneered biochemistry. He also contributed to the systematic study of organic chemistry and the development of agricultural chemistry. In 1832 he founded what became *Annalen der Chemie*, one of the most influential of all scientific journals.

Linde, Carl von (1842–1934), German engineer whose research on a continuous process for liquefying gases on a large scale led to the development of refrigeration.

FAMOUS SCIENTISTS

Lindemann, Ferdinand von (1852–1939), German mathematician who proved that pi is a transcendental number – i.e. it is not the root of any equation with rational coefficients.

Linnaeus, Carolus (Carl Linné; 1707–78), Swedish botanist whose *Systema Naturae* (1753) systematically arranged an animal, plant and mineral kingdom. Linnaeus defined classes of plants on the basis of the number and arrangement of their stamens. Although his classification has been overtaken by dramatic advances in biology his binomial nomenclature of plants – in *Species Plantarum* (1753) – remains the basic pattern for the naming of all living systems with a generic name and a specific name.

Lister, Joseph (Lord Lister; 1827–1912), English physician who introduced the principles of antiseptic surgery into medicine.

Locke, John (1632–1704), English philosopher who proposed that the macroscopic properties of matter are explained by the arrangement and rearrangement of microscopic particles.

Lockyer, Sir Joseph Norman (1836–1920), English astronomer who discovered helium.

Loeb, Jacques (1859–1924), German-born US biologist who is best known for his experimental work on reproduction without fertilization (parthenogenesis).

Lomonosov (Lommonnossoff), Mikhail (1711–65), Russian chemist, poet and grammarian who researched mass, heat and light. He is better known for developing the grammer that standardized written Russian.

Lorentz, Hendrick Antoon (1853–1928), Dutch physicist who shared the 1902 Nobel Physics Prize for the theory of electromagnetic radiation.

Loschmidt, Johann Josef (1821–95), Austrian chemist who made the first accurate estimate of the size of molecules.

Lovell, Sir Bernard (1913–), English radioastronomer whose efforts secured the Jodrell Bank telescope.

Lowry, Thomas (1874–1936), English chemist who confirmed that optical activity depends upon the wavelength of light.

Lucretius (c. 95–55 BC), Roman philosopher whose recognition of the struggle for existence anticipated Darwin by nearly 2000 years.

McClintock, Barbara (1902–93), US geneticist who, working on *Drosophila* (fruit fly), showed that gene action was connected with chromosomes.

Mach, Ernst (1838–1916), Austrian physicist whose most important work was in wave theory, optics and mechanics. Mach's principle states that inertia – which may be defined as the tendency of a body at rest to remain at rest and for a body in motion to remain in motion in the same direction – results from a relationship of that object with the other matter in the universe.

MacLaughlin, Colin (1698–1746), Scottish mathematician whose main work was in gravitation, geometry and calculus.

McMillan, Edwin Mattison (1907–), US physicist who shared the 1951 Nobel Chemistry Prize for the discovery of neptunium.

Maiman, Theodore Harold (1927–), US physicist who is best known for his pioneering work with laser beams.

Maimonides, Moses (1135–1204), Spanish-born Jewish physician and philosopher who is remembered for his medical monographs.

Malpighi, Marcello (1628–94), Italian physiologist who showed how blood reaches tissues through tiny tubes, capillaries. Malpighi is thought to be the first person to use microscopy to study animal and plant tissues and is regarded as the founder of the science of microscopic anatomy.

Mandelbrot, Benoit (1924–), Polish-Lithuanian-born US mathematician whose widespread mathematical interests are represented in *The Fractal Geometry of Nature* (1982). He is best known for the Mandelbrot set, which is constructed from simple mapping by marking dots on a complex plane.

Marconi, Giuseppe (1874–1937), Italian physicist who invented the first successful system of telegraphy. At the age of 20 he began experiments using an induction coil with a spark discharger controlled by a Morse key to send signals and a simple device to detect radio waves to receive them. He shared the 1909 Nobel Prize for Physics. In later life, Marconi studied short-wave radio communication – which became the foundation of all modern long-distance radio communication.

Markov, Andrey (1856–1922), Russian mathematician who is best known for the Markov chain, a sequence of random variables.

Martin, A(rcher) J(ohn) P(orter) (1910–), English biochemist who – with Synge – pioneered partition chromatography.

Mauchly, John William (1907–80), US computer engineer.

Maupertius, Pierre-Louis de (1698–1759), French mathematician and astronomer who formulated the principle of least action which states that all natural processes go on in such a manner that a dynamic function – 'action' – is a minimum. He also proved Newton's proposal that the Earth is a sphere which is flattened at the poles.

Maxwell, James Clerk (1831–79), Scottish physicist whose formulation of the electromagnetic theory of light – published in *Treatise on Electricity and Magnetism* (1873), which is based on the ideas of Faraday – revolutionized physics. He suggested that light is an electromagnetic vibration – this was later proved by Hertz. Maxwell also researched colour sensation and the kinetic theory of gases. 'Maxwell's demon' and Maxwell's equations of electromagnetism were formulated by, and named after, him. 'Maxwell's demon' is a hypothetical intelligence capable of detecting and responding to the movements of molecules. Maxwell's equations are four equations which – when considered together – offer a complete description of the production and interrelation of electric and magnetic fields.

Mayer, Maria Goeppert (1906–72), German-born US physicist who independently developed a theory of the structure of atomic nuclei.

Maynard Smith, John (1920–), English biologist who is best known for his influential *Theory of Evolution* (1958).

Meitner, Lise (1878–1968), Austrian-born Swedish physicist whose research, with Otto Hahn and Fritz Strassmann, led to the discovery of nuclear fission. Meitner researched with Hahn for 30 years. Together they discovered protactium, studied beta decay and the results of the nuclear bombardment of uranium. Because she was Jewish, Meitner left Nazi Germany for Sweden in 1938. Hahn and Strassmann continued to involve her in their research at a distance. She correctly described – and named – the nuclear fission that had been achieved by Hahn and Strassmann in 1938–39.

Mendel, Gregor (1822–84), Austrian Augustinian monk and botanist whose experiments with plants led him to discover the basic principles of heredity. This, in turn, led him to lay the mathematical foundations of the science of genetics. Mendel's initial experiments were with the garden pea whose characteristics and alternative differences of tallness, shortness, colour of flower, seed shape, form of pod, etc., he observed. Mendel formulated two basic laws – the law of segregation and the law of independent assortment. The first (in its modern form) states that genes are transmitted as separate and distinct units from one generation to the next. The second (in its modern form) states that the two members (alleles) of a gene pair separate during the formation of sex cells by a parent organism.

Mendeleyev, Dmitri (1834–1907), Russian chemist who discovered that, by arranging the chemical elements in order of increasing atomic weight, a curious repetition of chemical properties at regular intervals was revealed. Recognizing the underlying order that this implied, Mendeleyev discovered the Periodic Table. Mendeleyev *discovered* the Table; he did not design it. He was so confident in his discovery that he predicted the properties of missing elements – and his predictions were subsequently proved right.

Merrifield, Robert Bruce (1921–), US biochemist who won the 1984 Nobel Chemistry Prize for his work on peptides.

Meselson, Matthew Stanley (1930–), US molecular biologist who demonstrated the semiconservative nature of DNA replication.

Metchnikoff, Elie (Ilya Mechnikov; 1845–1916), Russian microbiologist and zoologist who shared the 1908 Nobel Prize for Medicine or Physiology for the discovery in animals of amoeba-like cells that engulf bacteria. This discovery revolutionized the science of immunology.

Meyer, Lothar (1830–95), German chemist who is best known for Meyer's curves, which show the periodicity of chemical elements.

Meyer, Viktor (1848–97), German chemist who devised a way of measuring the density of gases and vapours and hence the molecular masses of volatile compounds.

Michelson, A(lbert) A(braham) (1852–1931), German-born US physicist who is best known for establishing the speed of light as a fundamental constant. His other work included research in spectroscopy and meteorology. Michelson won the 1907 Nobel Physics Prize.

Miescher, Johann Friedrich (1844–95), Swiss physiologist who discovered nucleic acids (1869), which he named 'nuclein'. (The name 'nucleid acids' was not generally adopted until 1889.)

Miller, Stanley (1930–73), US biochemist who synthesized amino acids.

Millikan, Robert Andrews (1868–1953), US physicist who determined the charge on the electron (1912) and studied the photoelectric effect. He won the 1923 Nobel Physics Prize.

Milne, Edward Arthur (1896–1950), English mathematician and astrophysicist who introduced the cosmological principle, which states that the universe appears the same from wherever it is observed.

Minkowski, Hermann (1864–1909), Lithuanian-born German mathematician who won the *Grand Prix des Sciences Mechaniques* of the Paris Academy of Sciences at the age of 18. (Minkowski has previously abandoned his claims to a university prize in favour of a needy colleague.) He contributed to geometry and to the theories of numbers and of relativity. Einstein said that without the contribution of Minkowski, the general theory of relativity would not have been possible.

Minsky, Martin Lee (1927–), US computer scientist.

Monge, Gaspard (1746–1818), French mathematician who invented descriptive geometry.

Monod, Jacques (1910–76), French biochemist who shared the 1965 Nobel Prize for Medicine or Physiology for research that led him to postulate the existence of messenger-RNA.

Montagnier, Luc (1932–), French virologist who was one of the first to identify HIV.

Morgagni, Giovanni (1682–1771), Italian physician who pioneered studies in researching beyond the symptoms of diseases to discover the causes and mechanisms that produce those symptoms. He is regarded as one of the founders of pathology.

Morgan, Thomas Hunt (1866–1945), US geneticist who established the chromosome theory of heredity.

Moseley, Henry (1887–1915), English physicist whose work led to what is now known as Moseley's law, which mathematically establishes the relationship between characteristic X-rays and atomic number for each chemical element. Moseley was killed at Gallipoli in World War I.

Mössbauer, Rudolph Ludwig (1929–), German physicist who discovered the effect that now bears his name. The Mössbauer effect, which is also known as recoil-free gamma-ray resonance, is a nuclear process that allows the resonance absorption of gamma rays. Mössbauer shared the 1961 Nobel Prize for Physics.

Mueller, Erwin Wilhelm (1911–77), German-born US scientist who developed the field-emission microscope and the field-ion microscope.

Muller, Hermann Joseph (1890–1967), US geneticist who demonstrated that X-rays speed up the natural process of mutation.

Müller, Johannes (1801–58), German physiologist who correctly described the nature of sensory nerves.

Müller, Paul (1899–1965), Swiss chemist who developed DDT (dichlorodiphenyltrichloroethane).

Mulliken, Robert Sanderson (1896–1986), US physicist and chemist who – with Friedrich Hund – developed the molecular-orbital theory of chemical bonding, which states that electrons in a molecule move in the field produced by all the nuclei.

Musschenbroek, Pieter van (1692–1761), Dutch physicist who developed a device capable of storing and (under control) releasing electricity from an electrostatic machine (1745). His invention became known as the Leyden jar and was the first capacitor.

Nägeli, Karl Wilhelm von (1817–91), Swiss botanist whose pioneering work on plant cells made an important contribution to the subject.

Nambu, Yoichipo (1921–), Japanese physicist who has made important discoveries concerning the nature of quarks.

Napier, John (1550–1617), Scottish mathematician and theologian who devised logarithms as an aid to calculation. Napier's bones – a device he developed c. 1667 – was the forerunner of the slide rule.

Neher, Erwin (1944–), German biophysicist who is known for his work on the minute channels in membranes of living cells through which ions pass.

Nernst, Walter Hermann (1864–1941), German chemist who is regarded as one of the founders of modern physical chemistry and was awarded the 1920 Nobel Chemistry Prize. He formulated the third law of thermodynamics which states that, at a temperature above absolute zero, all matter tends toward random motion and all energy dissipates.

Newcomb, Simon (1835–1909), Canadian-born US astronomer and mathematician who is best known for his tables of computed locations of celestial bodies over a period of time.

Newlands, John (1838–98), English chemist who noticed that by arranging the chemical elements in order of increasing atomic weight a curious repetition of physical properties emerged. His work – summarized in Newlands' law of octaves – was ridiculed and it was left to Mendeleyev to independently bring the same observations to the discovery of the Periodic Table.

Newton, Sir Isaac (1642–1727), English mathematician and physicist who could be regarded as the founder of modern physics. By 1666, at the age of 24, he had made important discoveries in mathematics (the binomial theorem and differential calculus, which he called fluxions), optics (the theory of colours) and medicine. Newton became professor of mathematics at Cambridge, and in 1687 he published his *Philosophiae Naturalis Principia Mathematica*, generally known as the *Principia*. Through careful analysis of the available experimental data and the application of his theory he was able to explain many previously inexplicable phenomena, such as the tides and the precession of equinoxes. Using a prism, he split sunlight into the spectrum. He also invented the reflecting telescope. He is, though, chiefly remembered for his laws of motion and gravity. Newton's laws of motion state: that a body will remain at rest or travelling in a straight line at constant speed unless it is acted upon by an external force; that the resultant force exerted on a body is directly proportional to the acceleration produced by the force; and that to every action there is an equal and opposite reaction. Newton's law of gravitation states that every particle in the universe attracts every other particle with a force that is directly proportional to the product of their masses and inversely proportional to the square of the distance between them. In 1689 and 1701 he represented the University of Cambridge in Parliament. From 1703 until his death in 1727 he was President of the Royal Society. The newton, the absolute unit of force in the metre-kilogram-second system of physical units, is named after him.

Nobel, Alfred (1833–96), Swedish chemist, engineer and inventor of dynamite. He founded the Nobel Prizes.

Noether, Emily (Amalie Noether; 1882–1935), German mathematician who was described as the 'most creative abstract algebraist of modern times'. Initially she had difficulty obtaining a lectureship because she was a woman. In 1933 Noether was dismissed from her post because she was Jewish. She went to the USA where she lectured until her death.

Noguchi, Hideyo (1876–1928), Japanese bacteriologist who discovered the causative agent of syphilis.

Noyce, Robert (1927–1989), US inventor of the microchip.

Nuttall, Thomas (1786–1859), English botanist who described many of the plants of North America.

Oberth, Hermann Julius (1894–1989), Austro-Hungarian-born German rocket scientist.

Oersted, Hans Christiaan (1777–1851), Danish physicist who in 1819 discovered the magnetic effect produced by an electric current after noticing that the needle of a compass, which was close to a wire that was carrying a current, swung erratically and then came to rest at a right angle to the wire. This made Oersted realize that there was a magnetic field around the wire. This discovery was developed by Faraday and Ampère. The oersted, the unit of magnetic field strength, is named after him.

Ohm, Georg Simon (1787–1854), German physicist who discovered the law that is now named after him. Ohm's law states that the current flow through a conductor is directly proportional to the potential difference (voltage) and inversely proportional to the resistance. The ohm, the unit of electrical resistance, is named after him.

Olbers, Heinrich (1758–1840), German astronomer who is best known for Olbers' method of calculating the orbits of comets.

Oort, Jan Hendrik (1900–), Dutch astronomer whose main work has been in the structure and dynamics of the Galaxy.

Oparin, Aleksandr (1894–1980), Russian biochemist who studied the origin of life from chemical matter.

Oppenheimer, J(ulius) Robert (1904–67), US physicist who directed the Los Alamos laboratory during the development of the atomic bomb (1943–45).

Ostwald, Wilhelm Friedrich (1853–1932), Latvian-born German chemist who pioneered the development of physical chemistry as a separate branch of chemistry. He won the 1909 Nobel Chemistry Prize for his work on catalysis, whose nature he established. His other principal achievements were in chemical equilibrium and reaction velocities.

Ostwald, Wolfgang (1883–1943), Latvian-born German chemist – son of Wilhelm Ostwald (see above) – whose major work was in colloid chemistry.

Oughtred, William (1575–1660), English mathematician who invented the slide rule.

Owen, Sir Richard (1804–92), English biologist whose major work was in the study of fossil animals.

Paget, Sir James (1814–99), English physiologist and surgeon who did much to found the science of pathology.

Pallas, Peter Simon (1741–1811), German naturalist who developed new classifications of animal groups.

Pappus of Alexandria (4th century AD), Greek mathematician whose *Synagoge* is a systematic study of ancient Greek mathematics. He is best known as a geometer.

Paracelsus (Philippus Bombast von Hohenheim; 1493–1541), Swiss doctor and alchemist whose *Great Surgery Book* (1536) established the role of chemistry in medicine. He is sometimes called the Father of Chemistry.

Pardee, Arthur Beck (1921–), US biochemist who – with François Jacob and Jacques Monod – formulated the concept of a repressor molecule.

Pascal, Blaise (1623–62), French mathematician and physicist who discovered what became known as Pascal's theorem at the age of 16, when he was writing a book on conic sections. At the age of 19, to aid his father's statistical work, he invented the first calculating machines, which performed addition and subtraction. He also investigated what became known as Pascal's triangle, and helped to develop the theory of probability before abandoning mathematics in favour of theology. In physics, Pascal is remembered for the principle (or law) which bears his name and which states that in a fluid at rest in a closed container a pressure change in one part is transmitted without loss to every portion of the fluid and to the walls of the container. The pascal – the SI unit of stress or pressure – is named after him.

Pasteur, Louis (1822–95), French chemist and microbiologist who proved that living microorganisms cause disease and fermentation. Pasteur also laid the foundations of modern vaccine theory. In the 1870s, while studying anthrax, he inoculated chickens with a culture of chicken cholera. The birds survived and proved immune to subsequent inoculations of the same virus. Unwittingly, Pasteur had attenuated the virus, that is, he had weakened it to such a degree that the body's own natural defences could defeat it. In 1885 he used vaccine for the first time against rabies. As a chemist Pasteur made an important contribution to the realization that biological molecules could be left- or right-handed.

Pauli, Wolfgang (1900–58), Austrian-born US-Swiss physicist who won the 1945 Nobel Prize for Physics for his discovery of the exclusion principle, which states that in an atom no two electrons can have the same set of quantum numbers (numbers which define the exact characteristics of each electron's behaviour in respect of spatial distribution and spin).

Pauling, Linus (1901–), US chemist who applied the principles of quantum mechanics to his studies of chemical bonding. In 1939 he collected his work together in a book called *The Nature of the Chemical Bond* which was possibly the most influential chemical text of this century. In 1951 he announced that he had solved an important general structure of proteins, now known as the alpha-helix, which inspired Crick and Watson in their successful attempt to find the structure of DNA, which again proved to be helical. Pauling elucidated how oxygen binds to haemoglobin and later applied this understanding to sickle-cell anaemia. Pauling came to the attention of the media for his belief in the consumption of massive doses of vitamin C as a prophylactic. He is still scientifically active well into his nineties, recently working on superconductivity. He was awarded the 1954 Nobel Chemistry Prize and the 1962 Nobel Peace Prize for his campaign against nuclear weapons.

Pavlov, Ivan (1849–1936), Russian physiologist who is best known for developing the concept of the conditioned reflex.

Peano, Giuseppe (1858–1932), Italian mathematician who is best known for his work on the mathematical development of logic.

Pearl, Raymond (1879–1940), US zoologist who helped found biometry, the application of statistics to biology and medicine.

Pearson, Karl (1857–1936), English biometrician who introduced many concepts to mathematics including standard deviation and chi-square.

Pederson, Charles (1904–), Norwegian-born US chemist who discovered the class of compounds now known as crown ethers.

Peltier, Jean-Charles (1785–1845), French physicist who discovered the effect that now bears his name. Peltier's effect – whereby at the junction of two different metals an electric current will produce either heat or cold (depending upon the nature of the metals) – is used in temperature-measuring devices and refrigerators.

Penney, William George (1909–91), English mathematical physicist who was responsible for the design and production of the first British nuclear weapons.

Penrose, Roger (1931–), English mathematician who is also an influential theoretical physicist (in particular in the field of black holes). Penrose has developed a new cosmology based on complex geometry involving the concept of 'twistors' (massless objects which possess both linear and angular momentum in twistor space). Using these he has attempted to reconstruct the principal outlines of modern physics.

Penzias, Arno Allan (1933–), German-born US astrophysicist who – with Robert Wilson – discovered cosmic microwave background radiation, a major discovery that is considered to be evidence for the 'big bang' theory of the origins of the universe.

Perkin, Sir William (1838–1907), English chemist who discovered the first synthetic dye.

FAMOUS SCIENTISTS

Perl, Martin Lewis (1927–), US physicist who discovered the tau lepton.

Perrin, Jean (1870–1942), French physicist whose studies of the Brownian motion of suspended minute particles confirmed the atomic nature of matter. He won the 1926 Nobel Physics Prize.

Perutz, Max (1914–), Austrian-born British biochemist who shared the 1962 Nobel Chemistry Prize for work on the structure of haemoglobin.

Pfeffer, Wilhelm (1845–1925), German botanist who pioneered work in plant physiology.

Picard, Emile (1856–1941), French mathematician who worked mainly in algebraic geometry, mechanics and analysis.

Piccard, Auguste (1884–1962), Swiss-born Belgian physicist who is best known for his exploration of the upper stratosphere and the ocean depths.

Pisano, Leonardo see Fibonacci.

Planck, Max (1858–1949), German physicist who won the 1918 Nobel Prize for Physics for his discovery of the quantum theory of radiation, which states that energy from an oscillating particle is emitted not continuously but rather in discrete packets of energy called quanta. This discovery – from which quantum theory was developed – was published in *On the Theory of the Law of Energy Distribution in the Continuous Spectrum* (1900). Planck is generally regarded, with Einstein, as the co-founder of 20th-century physics. He expressed the relationship between energy emitted or absorbed by a body and the frequency of radiation mathematically as $E = nh\nu$, where E is the energy, n is the number, ν is the frequency and h is the constant of proportionality, now known as the Planck constant. Planck protested to Hitler in person concerning the harassment of Jewish scientists.

Plato (c. 428–347 BC), Greek philosopher whose influence on religion, education, politics, ethics and philosophy was profound. Plato also made important contributions to science – he questioned what accounted for the 'uniform and orderly motions' of the planets and he established the principle of the mathematical analysis of nature.

Pliny (the Elder) (AD 23–79), Roman philosopher who recognized that the Earth is round. His *Natural History* increased interest in the natural world.

Poincaré, Henri (1854–1912), French mathematician, philosopher and astronomer who greatly influenced the development of cosmology, relativity and topology. In mathematics, he is best remembered for his work on algebraic equations. A born communicator, he established a wider audience for science.

Polanyi, John Charles (1929–), Canadian chemist whose main research has been into the nature of chemical reactions.

Popov, Aleksandr (1859–1906), Russian physicist who built a primitive radio receiver in 1896.

Porter, George (1920–), English chemist who has investigated the spectroscopy of transient substances (the very short-lived intermediate species produced during a chemical reaction).

Poynting, John Henry (1852–1914), English physicist who is best known for Poynting's vector – he showed that the flow of energy at a point can be expressed by a simple formula in terms of electric and magnetic forces. Poynting is believed to have been the first to suggest the existence of the effect of radiation from the Sun that causes smaller particles that are in orbit about the Sun to be drawn into that body – the Poynting-Robertson effect.

Prandtl, Ludwig (1875–1953), German physicist who is commonly regarded as the founder of the science of aerodynamics.

Priestley, Joseph (1733–1804), English cleric, political theorist and scientist who was one of the discoverers of oxygen.

Proust, Joseph Louis (1754–1826), French chemist who, in 1799, formulated his law of definite proportions which states that all compounds contain elements in certain definite proportions.

Ptolemy, Claudius (3rd century AD), Greek mathematician, geographer and astronomer whose work laid the basis for astronomy for some 1000 years. He proposed that the Moon, Sun and planets revolved around the Earth.

Pythagoras (c. 582–500 BC), Greek philosopher and mathematician who founded a religious community at Croton in southern Italy. The Pythagorean brotherhood saw mystical significance in the idea of number. He is popularly remembered today for Pythagoras' theorem although there is no evidence that he was the originator of it.

Quételet, Adolphe (1796–1874), Belgian mathematician who pioneered the use of statistics for official purposes.

Rabi, Isidor Isaac (1899–1988), Austrian-born US physicist who pioneered atomic exploration.

Raman, Chandrasekhara Venkata (1888–1970), Indian physicist who won the 1930 Nobel Physics Prize for his discovery that when light traverses a material, some of that light changes in wavelength. This is now known as the Raman effect. Raman spectroscopy exploits the effect in order to examine the vibration of molecules and hence their structures. The use of lasers as light sources has revolutionized Raman spectroscopy and vastly increased its importance.

Ramanujan, Srinivasa Aaiyangar (1887–1920), Indian mathematician who, at the age of 16, devoted his life to mathematics after reading a book summarizing European mathematics. He had an astonishing intuition for the correct results, although many were not proved until later.

Ramsay, Sir William (1852–1916), Scottish chemist who discovered the noble gases neon, argon, krypton and xenon. He won the 1904 Nobel Chemistry Prize.

Ray, John (John Wray; 1627–1705), English naturalist whose botanical studies were of importance to the development of taxonomy.

Rayleigh, Lord (John William Strutt; 1843–1919), English physicist whose discoveries in acoustics and optics are fundamental to the theory of wave propagation in fluids. He won the 1904 Nobel Physics Prize for his isolation – with William Ramsay – of the noble gas argon.

Réamur, René-Antoine de (1683–1757), French physicist and entomologist who devised the Réamur temperature scale, isolated gastric juice and made a major contribution to the early study of insects.

Reed, Walter (1851–1902), US bacteriologist and army pathologist who proved that mosquitoes transmit yellow fever.

Regiomontanus (Johann Müller; 1436–76), German mathematician who was the leading practitioner of trigonometry in medieval Europe.

Regnault, Henri-Victor (1810–78), French chemist and physicist whose main work concerned the properties of gases.

Ricci-Curbastro, Gregorio (1853–1925), Italian mathematician who played a leading role in the discovery of the absolute differential calculus.

Richter, Burton (1931–), US physicist who shared the 1976 Nobel Physics Prize. In 1974, Richter and his colleagues created and detected a heavy elementary particle which they called the psi.

Riemann, Bernhard (1826–66), German mathematician who was a major influence on analysis and geometry. His work on the geometry of space had important implications for modern theoretical physics.

Roberts, Richard (1943–), English molecular biologist who discovered the phenomenon of split genes. (Philip Sharp made the same discovery independently and they were jointly awarded the 1993 Nobel Prize for Medicine or Physiology.)

Romer, Alfred Sherwood (1894–1973), US palaeontologist who is best known for his theories concerning the evolution of vertebrate animals.

Romer, Ole (1644–1710), Danish astronomer who demonstrated conclusively – from his observations of Jupiter's moons – that light travels at finite speed.

Röntgen, Wilhelm Conrad (1845–1923), German physicist who won the 1901 Nobel Prize for Physics for his discovery, in 1895, of X-rays. (The true nature of X-rays was not, however, established until 1912.)

Rose, William Cumming (1887–1984), US biochemist who is best known for his research into amino acids.

Ross, Sir Ronald (1857–1932), English pathologist who won the 1902 Nobel Prize for Physiology or Medicine for his work on malaria.

Rowland, Sherry (1927–), US chemist who – with Mario Molina – have pioneered studies of the increase of CFCs in the upper atmosphere.

Rumford, Count (Sir Benjamin Thompson; 1753–1814), US-born British physicist who was the first to suggest that heat is a form of energy. Observing the boring of cannon barrels, he noted the great heat given off by the drill and suggested that – as the drill bit into the metal – its mechanical energy was translated into heat. Rumford – who was a British spy during the American War of Independence – acquired his German title through service as a government minister in Bavaria.

Russell, Lord Bertrand (1872–1970), English philosopher and mathematician who did much of the basic work on mathematical logic and the foundations of mathematics. He found the paradox, now named after him, in the theory proposed by the German logician Gottlob Frege (1848–1925), and went on to develop the whole of arithmetic in terms of pure logic. He was jailed for his pacifist activities in World War I. Russell was awarded the 1950 Nobel Prize for Literature.

Rutherford, Ernest (Baron Rutherford; 1871–1937), New Zealand physicist who is acknowledged as the founder of modern atomic theory. His theory of the scattering of alpha particles (1910) led him to suggest that the atom consists of a positively charged nucleus which is surrounded by orbiting planetary electrons. This model is known as the Rutherford electron. Rutherford was the first person to split the atom.

Rydberg, Johannes Robert (1854–1919), Swedish physicist who proposed that the periodicity of an atom is the result of the structure of its atom. The Rydberg constant is a formula giving the frequency of lines in the spectral series of elements.

Sabin, Albert (1906–), Polish-born US microbiologist who developed oral polio vaccine.

Sachs, Julius von (1832–97), German botanist whose researches into transpiration and photosynthesis greatly advanced the knowledge of plant physiology.

Sagan, Carl Edward (1934–), US astronomer who has researched the physics and chemistry of the atmospheres and surfaces of planets.

Sakharov, Andrei (1921–89), Russian physicist who made a major contribution to the development of Soviet nuclear weapons and who offered evidence for the existence of quarks. He is best known, however, as a campaigner for civil rights in the former Soviet Union.

Salk, Jonas Edward (1914–), US physician who developed a vaccine for polio.

Sarich, Vincent (1934–), US biochemist who has worked on methods of dating the genetic relationship between humans and apes.

Sauveur, Albert (1863–1939), Belgian-born US metallurgist who laid the foundations of physical metallurgy.

Scheele, Karl Wilhelm (1742–86), Swedish chemist who discovered oxygen (1771) and chlorine (1774). As Scheele did not publish his results until 1777, Priestley – who had discovered oxygen in the meantime – was credited with the discovery.

Schmidt, Bernhard Voldemar (1879–1935), Estonian telescope maker whose Schmidt telescope is common to modern observatories.

Schmidt, Maarten (1929–), Dutch-born US astronomer who is best known for his research on quasars.

Schrödinger, Edwin (1887–1961), Austrian theoretical physicist who – with Paul Dirac – made an important contribution to the wave theory of matter. He shared – with Dirac – the 1933 Nobel Physics Prize.

FAMOUS SCIENTISTS

Schuster, Sir Arthur (1851–1934), German-born British physicist who was the first to show that an electric current is conducted by ions.

Schwann, Theodor (1810–82), German physiologist who was one of the first to propose the cell theory in biology.

Schwarzschild, Karl (1873–1916), German astronomer who established the photographic magnitude of 3500 stars.

Schwinger, Julian Seymour (1918–), US physicist who combined the electromagnetic theory with quantum mechanics to found the science of quantum electrodynamics.

Seaborg, Glenn Theodore (1912–), US nuclear chemist who is associated with the discovery or first isolation of elements 93–98, 101 and 102.

Seki, Kowa (1642–1708), Japanese mathematician who invented a form of calculus, and who used determinants before Liebniz. Seki suggest that an equation on the nth degree has in general n roots. He also gave 355/113 as an approximation of pi, used positive and negative numbers and algebraic quantities, and studied magic squares.

Shapley, Harlow (1885–1972), US astronomer whose main work was on the structure of the Galaxy. He played a major role in the development of the Harvard Observatory.

Sharp, Phillip (1944–), US molecular biologist who shared – with Roberts – the 1993 Nobel Prize for Medicine or Physiology for his discovery of split genes. (See also Richard Roberts.)

Sherrington, Sir Charles (1861–1952), English neurologist who shared – with Adrian – the 1932 Nobel Prize for Physiology or Medicine for his research into the integrated nervous system (the motor-nerve system) of higher animals.

Shockley, William (1910–), US physicist who – with Brattain – developed the transistor.

Sidgwick, Nevil (1873–1952), English chemist who developed the electronic theory of chemical bonding.

Siemens, Charles William (Carl Wilhelm Siemens; 1823–83), German-born British engineer who made important advances in electric telegraphy.

Siemens, Werner von (1816–92), German electrical engineer who was one of the founders of telegraphy. The SI unit of electric conductance – the siemens – is named after him.

Skolem, Thoralf Albert (1887–1963), Norwegian mathematician who is best known for his construction of the Skolen paradox.

Slipher, Vesto Melvin (1875–1969), US astronomer who is best known for determining the radial velocity of spiral nebulae.

Sloane, Sir Hans (1660–1753), Anglo-Irish physician and naturalist who built up and catalogued an impressive botanical collection.

Smith, Michael (1932–), English-born Canadian biochemist who introduced the concept of site specific mutagenesis into molecular biology. He shared the 1993 Nobel Chemistry Prize.

Smith, William (1769–1839), English geologist who founded the science of stratigraphy.

Snell, Willebrord (1591–1626), Dutch mathematician and physicist who is best known for Snell's law in optics, which formulates the relationship between the path taken by a ray of light in crossing the boundary of separation between two contacting surfaces and the refractive index of each.

Soddy, Frederick (1877–1966), English chemist who won the 1921 Nobel Chemistry Prize for his work on the origins and nature of isotopes.

Somerville, Mary (1780–1872), Scottish astronomer, physicist and geographer whose *Mechanism of the Heavens* (1831) was an important landmark.

Spallanzani, Lazzaro (1729–99), Italian biologist who studied animal reproduction, mammalian bodily functions and microscopic life in nutrient cultures.

Sperry, Roger Wolcott (1913–), US neurobiologist whose principal work has been on the hemispheres of the brain.

Stahl, Georg Ernst (1660–1734), German physician and chemist who developed the phlogiston theory, which attempted the first systematic account of chemical transformation, but without, of course, any knowledge of the part played by atoms and elements. He defined phlogiston as the property of being combustible which is lost in combustion.

Stanley, Wendell (1904–71), US biochemist who shared the 1946 Nobel Prize for Chemistry with John Northrop for the crystallization of viruses.

Starling, Ernest Henry (1866–1927), English physiologist who researched the mechanical controls on heart functions. He also studied digestion and coined the word 'hormone'.

Stefan, Josef (1835–93), Austrian physicist who is best known for his work on heat radiation.

Stephenson, George (1781–1848), English engineer who developed the first successful steam engines.

Stern, Otto (1888–1969), German-born US physicist who won the 1943 Nobel Physics Prize for developing the molecular beam as a tool for studying the characteristics of molecules.

Stevin, Simon (1548–1620), Flemish mathematician who was influential in establishing the use of decimal fractions. He discovered the triangle of forces, a basic theorem in statics.

Stokes, Sir George Gabriel (1819–1903), Anglo-Irish physicist and mathematician who is best known for Stokes's law, which gives the force resisting motion of a spherical body through a viscous fluid.

Stoney, George Johnstone (1826–1911), Irish physicist who introduced the term 'electron'.

Sturgeon, William (1783–1850), English physicist who devised the first electromagnet.

Svedberg, Theodor (1884–1971), Swedish chemist who introduced the ultracentrifuge to investigate the molecular weights of very large molecules.

Swammerdam, Jan (1637–80), Dutch entomologist whose pioneering studies of insects under the microscope resulted in the first detailed descriptions of many species.

Swan, Sir Joseph Wilson (1828–1914), English physicist and chemist who produced an early electric light bulb.

Sydenham, Thomas (1624–89), English physician who founded the science of epidemiology.

Sylvester, James Joseph (1814–97), English mathematician who developed the theory of algebraic invariants, algebraic-equation coefficients that are not changed when the coordinate axes are rotated.

Szent-Györgi, Albert von (1893–1986), Hungarian-born US biochemist who is best known for his work on the biochemistry of muscular contractions.

Szilard, Leo (1898–1964), Hungarian-born US physicist who was one of the first people to realize the significance of nuclear fission.

Talbot, William Henry Fox (1800–77), English chemist and linguist who developed an early camera.

Tatum, Edward L(awrie) (1909–75), US biochemist who shared the 1958 Nobel Prize for Physiology or Medicine with George Beadle for pioneering work in molecular genetics.

Taylor, Joseph Hooton (1941–), US astrophysicist who shared the 1993 Nobel Physics Prize with Russell Hulse for the discovery of a new type of pulsar.

Teller, Edward (1908–), Hungarian-born US physicist who was the main scientific force behind the development of the US hydrogen bomb.

Tesla, Nikola (1856–1943), Croat-born US electrical engineer who discovered the rotating magnetic field, which became the basis of most AC machinery. The tesla – the SI unit of magnetic flux density – is named after him.

Theorell, Axel (1903–82), Swedish biochemist who won the 1955 Nobel Prize for Physiology or Medicine for research into enzymes.

Thompson, Benjamin see Rumford.

Thomson, Sir Wyville (1830–82), Scottish naturalist who was a pioneer in studying and describing the life forms of the ocean depths.

Thomson, Sir George Paget (1892–1975), English physicist who shared the 1937 Nobel Prize for Physics for demonstrating that electrons undergo diffraction.

Thomson, Sir Joseph John (1856–1940), English physicist whose discovery of the electron revolutionized understanding of atomic structure. He won the 1906 Nobel Physics Prize.

Tiselius, Arne (1902–71), Swedish biochemist who won the 1948 Nobel Chemistry Prize for his work in electrophoresis.

Tomonaga, Shin'ichiro (1906–79), Japanese physicist who shared the 1965 Nobel Physics Prize for his development – with others – of changes that made the quantum theory of mechanics fully consistent with the quantum theory of relativity.

Torricelli, Evangelista (1608–47), Italian mathematician and physicist who invented the barometer.

Townes, Charles Hard (1915–), US physicist who is best known for his pioneering work on the maser, the forerunner of the laser.

Tsui, Lap-Chee (1950–), Chinese-born Canadian geneticist who located the cystic fibrosis gene.

Turing, Alan (1912–54), English mathematician who pioneered computer theory. Turing discovered that it is impossible in general to predict if or when the Turing machine – a universal automatic machine capable of mathematical problem-solving which he designed – would stop.

Tyndall, John (1820–93), Anglo-Irish physicist who was a popular author of scientific books. He described the Tyndall effect, the scattering of light by particles of matter in its path.

Urey, Harold Clayton (1893–1981), US chemist who discovered deuterium (heavy hydrogen).

Van Allen, James Alfred (1914–), US physicist who is best known for atmospheric studies. He identified zones in the Earth's magnetic field where high-speed charged particles are found – the Van Allen belts.

Van de Graaff, Robert Jemison (1901–67), US physicist who developed the Van de Graaff generator and the Van de Graaff accelerator.

van der Waals, Johannes Diderik (1837–1923), Dutch physicist who formulated the van der Waals equation, which describes the behaviour of real gases.

Van Maanen, Adriaan (1884–1946), Dutch-born US astronomer who measured the minute changes in position of astronomical bodies over a period of time.

Van't Hoff, Jacobus Henricus (1852–1911), Dutch chemist who created stereochemistry, the study of the three-dimensional structure of organic compounds. He won the 1901 Nobel Chemistry Prize.

Varmus, Harold (1939–), US microbiologist whose major research has been into oncogenes (genes with a cancer-causing capacity).

Vauquelin, Nicolas-Louis (1763–1829), French chemist who discovered chromium and beryllium.

Vavilov, Nikolai (1887–1943), Russian plant geneticist who made an important contribution to the study of botanical populations.

Veblen, Oswald (1880–1960), US mathematician whose contributions to differential equations and topology had implications in atomic physics.

Vesalius, Andrea (Andreas van Wesel; 1514–64), Flemish physician who was one of the first to dissect human corpses. His *On the Fabric of the Human Body* (1543) describes his anatomical discoveries. Vesalius was the first to suggest that thought and personality were based in the nervous system rather than in the heart.

Virchow, Rudolf (1821–1902), German pathologist who was the principal founder of cellular pathology.

Volta, Count Alessandro (1745–1827), Italian physicist who invented the battery cell – the 'voltaic pile' made from copper and zinc discs which were separated by cardboard soaked in salt solution. He also discovered methane gas. The unit of electromagnetic force, the volt, is named after him.

FAMOUS SCIENTISTS

von Braun, Wernher (1912–77), German-born US rocket engineer who developed the V-2 and who played a major role in US space exploration.

von Karman, Theodore (1881–1963), Hungarian-born US aerodynamicist who discovered Karman vortices, the alternating vortices found behind obstacles placed in moving fluids.

von Neumann, John (Johann von Neumann; 1903–57), Hungarian-born US mathematician who is best known for his theory of games, which has economic applications.

Waddington, Conrad Hal (1905–75), English geneticist whose *Principles of Embryology* (1956) became a standard.

Waksman, Selman (1888–1973), Ukrainian-born US microbiologist who isolated antibiotics – a word that he coined – from micro-organisms. He won the 1952 Nobel Prize for Physiology or Medicine.

Wallace, Alfred Russell (1823–1913), Welsh naturalist who is best known for developing a theory of evolution independently of Darwin. He coined the phrase 'survival of the fittest'. Wallace's Line – which he delineated – geographically divides the flora and fauna of Asia from that of Australasia.

Watson, James (1928–), US geneticist who – with Francis Crick – discovered the molecular structure of DNA. He shared the 1962 Nobel Prize for Physiology or Medicine with Crick and Maurice Wilkins.

Watson-Watt, Sir Robert Alexander (1892–1973), Scottish physicist who played a major role in the development of radar.

Watt, James (1736–1819), Scottish inventor who is best known for his steam engine. The watt – the unit of power in the SI system – is named after him.

Weber, Wilhelm Eduard (1804–91), German physicist who investigated magnetism. In 1833 he devised the electromagnetic telegraph. The SI unit of magnetic flux – the weber – is named after him.

Weierstrass, Karl (1815–97), German mathematician who was a pioneer in the modern theory of functions.

Weinberg, Steven (1933–), US physicist whose main research has concerned leptons and bosons.

Weismann, August (1834–1914), German biologist who made a major contribution to modern genetic theory by his 'germ plasm' theory, a forerunner to the DNA theory.

Werner, Alfred (1866–1919), Swiss chemist whose influential *New Ideas on Inorganic Chemistry* (1911) won him the 1913 Nobel Chemistry Prize. Werner distinguished between a primary and secondary valence of a metal.

Weyl, Hermann (1885–1955), German mathematician whose interests were particularly wide. He is remembered for his work on group theory.

Wheatstone, Sir Charles (1802–75), English physicist famous for his 'Wheatstone Bridge' which accurately measures electrical resistance and for his work in telegraphy. (Wheatstone had many talents: he patented an early design of the harmonica in 1829 and in the same year the first concertina.)

White, Gilbert (1720–93), English cleric and naturalist whose *Natural History of Selborne* (1789) was a landmark in nature study.

Whitehead, Alfred (1861–1947), English mathematician who collaborated with Russell on *Principia Mathematica* (1910–14), a major work on logic and mathematics.

Wiener, Norbert (1894–1964), US mathematician who founded the science of cybernetics.

Wilkins, Maurice (1916–), New Zealand-born British biophysicist who researched the molecular model of DNA with Francis Crick and James Watson, and shared with them the 1962 Nobel Prize for Physiology or Medicine.

Williamson, Alexander William (1824–1904), English chemist whose work on ethers and alcohols clarified molecular structure.

Wilson, Kenneth (1936–), American physicist who won the 1982 Nobel Physics Prize for developing a theory that explains the behaviour of different substances under pressure and temperature.

Witten, Edward (1951–), US physicist who has developed string theory, which attempts to unify general relativity with quantum mechanics.

Wöhler, Friedrich (1800–82), German chemist who was the first person to synthesize an organic chemical compound (urea) from an inorganic chemical compound (1828). This disproved the theory that living organisms could only be formed in living systems.

Wolfram, Stephen (1959–), English physicist who is best known for his work on cellular automata.

Wollaston, William Hyde (1766–1828), English physicist and chemist who discovered palladium and rhodium.

Woodward, R(obert) B(urns) (1917–79), US chemist who synthesized complex organic substances including quinine, cholesterol and vitamin B. He won the 1965 Nobel Chemistry Prize.

Wu, Chien-Shiun (1912–), Chinese-born US physicist who discovered the mechanism of beta disintegration. She also proved that the electron and the positron have opposite polarity.

Young, Thomas (1773–1829), English physicist who established the principle of interference of light and thus confirmed the wave theory of light. Young calculated the approximate wavelengths of the seven colours and he was the first person to define the word 'energy' scientifically. He was also interested in Egyptology and helped decipher the Rosetta Stone.

Yukawa, Hideki (1907–81), Japanese physicist who predicted the existence of the meson. He won the 1949 Nobel Physics Prize.

Zermelo, Ernst (1871–1953), German mathematician who was a pioneer of axiomatic set theory.

Zernike, Frits (1888–1966), Dutch physicist who won the 1953 Nobel Physics Prize for his invention of the phase-contrast microscope.

Ziegler, Karl (1898–1973), German chemist who introduced stereospecific polymers.

The Nobel Prizes

THE NOBEL PRIZES

The six Nobel Prizes (five before the addition of a prize for economic science in 1969) are awarded annually from a fund established under the terms of the will of Alfred Bernhard Nobel (1833–96), a Swedish chemist, industrialist and engineer who made his fortune through the discovery of dynamite. His interest in science and literature – and his pacifism – determined the fields in which the first five prizes were awarded: physics, chemistry, medicine or physiology, literature and peace. Only the three scientific Nobel Prizes are detailed below.

Nobel stipulated certain organizations in Sweden and Norway as the awarding bodies for specific prizes.

Chemistry – the Royal Swedish Academy of Sciences, Stockholm, Sweden;

Medicine or Physiology – the Royal Caroline Medico-Chirurgical Institute, Stockholm, Sweden;

Physics – the Royal Swedish Academy of Sciences, Stockholm, Sweden.

Nobel Prizewinners in Chemistry

1901 Jacobus Van't Hoff, Dutch; for laws of chemical dynamics and osmotic pressure.

1902 Emil Fischer, German; for work on sugar and purine syntheses.

1903 Svante Arrhenius, Swedish; for his theory of electrolytic dissociation.

1904 Sir William Ramsay, Scottish; for the discovery and periodic system classification of inert gas elements.

1905 Adolf von Baeyer, German; for work on organic dyes and hydroaromatic compounds.

1906 Henri Moissan, French; for the Moissan furnace, and the isolation of fluorine.

1907 Eduard Buchner, German; for the discovery of non-cellular fermentation.

1908 Ernest Rutherford, New Zealand; for his description of atomic structure and the chemistry of radioactive substance.

1909 Wilhelm Ostwald, Latvian-born German; for pioneering catalysis, chemical equilibrium and reaction velocity work.

1910 Otto Wallach, German; for pioneering work on alicyclic combinations.

1911 Marie Curie, Polish-born French naturalized citizen; for the discovery of radium and polonium and the isolation of radium.

1912 Victor Grignard, French; for Grignard reagents.

Paul Sabatier, French; for his method of hydrogenating compounds.

1913 Alfred Werner, Swiss naturalized citizen; for his work on the linkage of atoms in molecules.

1914 Theodore Richards, US; for the precise determination of atomic weights of many elements.

1915 Richard Willstätter, German; for pioneer research on plant pigments, especially chlorophyll.

1916 No award.

1917 No award.

1918 Fritz Haber, German; for the synthesis of ammonia.

1919 No award.

1920 Walther Nernst, German; for work in thermo-chemistry.

1921 Frederick Soddy, English; for research on radioactive materials, and the occurrence and nature of isotopes.

1922 Francis Aston, English; for work on mass spectrography, and on the whole number rule.

1923 Fritz Pregl, Austrian; for his method of microanalysis of organic substances.

1924 No award.

1925 Richard Zsigmondy, German; for the elucidation of the heterogeneous nature of colloidal solutions.

1926 Theodor Svedberg, Swedish; for work on disperse systems.

1927 Heinrich Wieland, German; for research into the constitution of bile acids.

1928 Adolf Windaus, German; for work on the constitution of sterols and their connection with vitamins.

THE NOBEL PRIZES

1929 Sir Arthur Harden, English, and Hans von Euler-Chelpin, German-born Swedish naturalized citizen; for studies on sugar fermentation and the enzymes involved in the process.

1930 Hans Fischer, German; for chlorophyll research, and the discovery of haemoglobin in the blood.

1931 Karl Bosch, German, and Friedrich Bergius, German; for the invention and development of high-pressure methods.

1932 Irving Langmuir, US; for furthering understanding of surface chemistry.

1933 No award.

1934 Harold Urey, US; for the discovery of heavy hydrogen.

1935 Frédéric Joliot-Curie, French, and Irène Joliot-Curie, French; for the synthesis of new radioactive elements.

1936 Peter Debye, Dutch-born US naturalized citizen; for work on dipole moments and the diffraction of X-rays and electrons in gases.

1937 Sir Walter Haworth, English; for research into carbohydrate and vitamin C.

Paul Karrer, Swiss; for research into carotenoid, flavin, and vitamins.

1938 Richard Kuhn, German; for carotenoid and vitamin research (the award was declined as Hitler forbade Germans to accept Nobel prizes).

1939 Adolf Butenandt, German; for work on sex hormones (the award was declined as Hitler forbade Germans to accept Nobel prizes).

Leopold Ruicka, Croat-born Swiss citizen: for research on steroid hormones.

1940 No award.

1941 No award.

1942 No award.

1943 George von Hevesy, Hungarian; for the use of isotopes as tracers in research.

1944 Otto Hahn, German; for the discovery of the fusion of heavy nuclei.

1945 Arturri Virtanen, Finnish; for the invention of fodder preservation method.

1946 James Sumner, US; for the discovery of enzyme crystallization.

John Northrop, US, and Wendell Stanley, US; for the preparation of pure enzymes and virus proteins.

1947 Sir Robert Robinson, English; for research on alkaloids and plant biology.

1948 Arne Tiselius, Swedish; for electrophoretic and adsorption analysis research, and for work on serum proteins.

1949 William Giauque, US; for work on the behaviour of substances at very low temperatures.

1950 Otto Diels, German, and Kurt Alder, German; for the discovery and development of diene synthesis.

1951 Edwin McMillan, US, and Glenn Seaborg, US; for the discovery of and research on trans-uranium elements.

1952 Archer Martin, English, and Richard Synge, English; for the development of partition chromatography.

1953 Hermann Staudinger, German; for work on macromolecules.

1954 Linus Pauling, US; for studies on the nature of the chemical bond.

1955 Vincent Du Vigneaud, US; for pioneer work on the synthesis of a polypeptide hormone.

1956 Nikolay Semyonov, Russian, and Sir Cyril Hinshelwood, English; for work on the kinetics of chemical reactions.

1957 Sir Alexander Todd, Scottish; for work on nucleotides and nucleotide coenzymes.

1958 Frederick Sanger, English; for determining the structure of the insulin molecule.

1959 Jaroslav Heyrovsky, Czech; for the discovery and developed polarography.

1960 Willard Libby, US; for the development of radio-carbon dating.

1961 Melvin Calvin, US; for studies the chemical stages that occur in photosynthesis.

1962 John C. Kendrew, English, and Max F. Perutz, British (Austrian-born); for determining the structure of haemoproteins.

1963 Giulio Natta, Italian, and Karl Ziegler, German; for research into the structure and synthesis of plastics polymers.

1964 Dorothy M. C. Hodgkin, English; for determining the structure of compounds essential in combating pernicious anaemia.

1965 Robert B. Woodward, US; for synthesizing sterols, chlorophyll, etc. (previously produced only by living things).

1966 Robert S. Mulliken, US; for investigations into chemical bonds and the electronic structure of molecules.

1967 Manfred Eigen, German, Ronald G. W. Norrish, English, and George Porter, English; for studies on extremely fast chemical reactions.

1968 Lars Onsager, Norwegian-born US naturalized citizen; for his theory of the thermodynamics of irreversible processes.

1969 Derek H. R. Barton, English, and Odd Hasell, Norwegian; for determining the actual three-dimensional shape of certain organic compounds.

1970 Luis F. Leloir, French-born Argentinian naturalized citizen; for the discovery of sugar nucleotides and their role in carbohydrate biosynthesis.

1971 Gerhard Herzberg, German-born Canadian citizen; for research into the structure of molecules.

1972 Christian B. Anfinsen, US, Stanford Moore, US, and William H. Stein, US; for contributions to the fundamentals of enzyme chemistry.

1973 Ernst Fischer, German, and Geoffrey Wilkinson, English; for work in organometallic chemistry.

1974 Paul J. Flory, US; for studies on long-chain molecules.

1975 John W. Cornforth, Australian-born British citizen, and Vladimir Prelog, Bosnian-born Swiss; for work on stereochemistry.

1976 William N. Lipscomb, US; for work on the structure of boranes.

1977 Ilya Prigogine, Russian-born Belgian; for work in advanced thermodynamics.

1978 Peter D. Mitchell, English; for his theory of energy transfer processes in biological systems.

1979 Herbert C. Brown, English-born US, and Georg Wittig, German; for the introduction of boron and phosphorus compounds in the synthesis of organic compounds.

1980 Paul Berg, US; for pioneer work in the preparation of a hybrid DNA.

Walter Gilbert, US, and Frederick Sanger, English; for chemical and biological analysis of the structure of DNA.

1981 Fukui Kenichi, Japanese, and Roald Hoffmann, US naturalized citizen; for orbital symmetry interpretation of chemical reactions.

1982 Aaron Klug, Lithuanian-born South African (a British naturalized citizen); for determining the structure of some biologically active substances.

1983 Henry Taube, Canadian; for studies into electron transfer reactions.

1984 Bruce Merrifield, US; for formulating a method of polypeptide synthesis.

1985 Herbert A. Hauptman, US, and Jerome Karle, US; for the development of means of mapping the chemical structure of small molecules.

1986 Dudley R. Herschbach, US, Yuan T. Lee, Taiwanese-born US naturalized citizen, and John C. Polanyi, Canadian; for the introduction of methods for analysing basic chemical reactions.

1987 Donald J. Cram, US, Charles J. Pedersen, Norwegian-born US, and Jean-Marie Lehn, French; for developing molecules that could link with other molecules.

1988 Johann Deisenhofer, German, Robert Huber, German, and Hartmut Michel, German; for studies into the structure of the proteins needed in photosynthesis.

1989 Tom Cech, US, and Sidney Altman, Canadian-born US naturalized citizen; for establishing that RNA catalyses biochemical reactions.

1990 Elias Corey, US; for work on synthesizing chemical compounds based on natural substances.

1991 Richard R. Ernst, Swiss; for refining the technology of nuclear magnetice resonance imaging (NMR and MRI).

1992 Rudolph A. Marcus, US; for mathematical analysis of the cause and effect of electrons jumping from one molecule to another.

1993 Kary B. Mullis, US; for the invention of his PCR method.

Michael Smith, English-born Canadian; for contributions to oligonucleotide-based, site-directed mutagenesis

Nobel Prizewinners in Physics

1901 Wilhelm Röntgen, German; for the discovery of X-rays.

1902 Hendrik Antoon Lorentz, Dutch, and Pieter Zeeman, Dutch; for investigating the influences of magnetism on radiation.

1903 Antoine-Henri Becquerel, French; for the discovery of spontaneous radioactivity.

Pierre Curie, French, and Marie Curie, Polish-born French naturalized citizen; for investigating radiation phenomena (inspired by Becquerel's discovery).

1904 Lord Rayleigh (John William Strutt), English; for the discovery of argon, an unreactive gas in the atmosphere.

1905 Philipp von Lenard, Austro-Slovak-born German; for research on cathode rays.

1906 Sir Joseph J. Thomson, English; for investigating electrical conductivity of gases.

1907 A. A. Michelson, German-born US naturalized citizen; for establishing the speed of light as a constant, and other spectroscopic and meteorological investigations.

1908 Gabriel Lippmann, Luxembourg-born French; for the photographic reproduction of colours.

1909 Guglielmo Marconi, Italian, and Karl Braun, German; for the development of wireless telegraphy.

1910 Johannes van der Waals, Dutch; for investigating the relationships between the states of gases and liquids.

1911 Wilhelm Wien, German; for investigating the laws governing heat radiation.

1912 Nils Gustav Dalén, Swedish; for the invention of automatic regulators for lighting buoys and beacons.

1913 Helke Kamerlingh Onnes, Dutch; for studies into the properties of matter at low temperatures and producing liquid helium.

1914 Max von Laue, German; for achieving diffraction of X-rays using crystals.

1915 Sir William Bragg, English, and Sir Lawrence Bragg, English; for the analysis of crystal structure using X-rays.

1916 No award.

1917 Charles Barkla, English; for the discovery of characteristics of X-radiation of elements.

1918 Max Plank, German; for the formulation of the first quantum theory.

1919 Johannes Stark, German; for the discovery of the Doppler effect in positive ion rays and the division of spectral lines when the source of light is subjected to strong electric force fields.

1920 Charles Guillaume, Swiss; for discovering anomalies in alloys.

1921 Albert Einstein, German-born US naturalized citizen; for elucidating theories fundamental to theoretical physics.

1922 Niels Bohr, Danish; for investigations into atomic structure and radiation.

1923 Robert Millikan, US; for work on elementary electric charge and the photoelectric effect.

THE NOBEL PRIZES

1924 Karl Siegbahn, Swedish; for work on X-ray spectroscopy.

1925 James Franck, German, and Gustav Hertz, German; for the definition of the laws governing the impact of an electron upon an atom.

1926 Jean-Baptiste Perrin, French; for work on the discontinuous structure of matter.

1927 Arthur Holly Compton, US; for the discovery of wavelength change in diffused X-rays.

Charles Wilson, Scottish; for the invention of the cloud chamber, which made visible the paths of electrically-charged particles.

1928 Sir Owen Richardson, English; for the discovery of Richardson's Law, which concerns the electron emissions by hot metals.

1929 Prince Louis de Broglie, French; for the discovery of the wave nature of electrons.

1930 Sir Chandrasekhra Raman, Indian; for work on light diffusion and the discovery of the Raman effect.

1931 No award.

1932 Werner Heisenberg, German, for formulating the indeterminacy principle of quantum mechanics.

1933 Paul Dirac, English (of Swiss parentage), and Erwin Schrödinger, Austrian; for the introduction of wave-equations in quantum mechanics.

1934 No award.

1935 Sir James Chadwick, English; for the discovery of the neutron.

1936 Victor Hess, Austrian; for discovering cosmic radiation.

Carl Andersen, US; for the discovery of the positron and the meson.

1937 Clinton Davisson, US, and Sir George Thomson, English; for demonstrating the interference phenomenon in crystals irradiated by electrons.

1938 Enrico Fermi, Italian-born US; for the discovery of radioactive elements produced by neutron irradiation.

1939 Ernest Lawrence, US; for the invention of the cyclotron.

1940 No award.

1941 No award.

1942 No award.

1943 Otto Stern, German-born US naturalized citizen; for the discovery of the magnetic moment of the proton.

1944 Isodor Isaac Rabi, Austrian-born US naturalized citizen; for the resonance method for observing the magnetic properties of atomic nuclei.

1945 Wolfgang Pauli, Austrian-born US-Swiss citizen; for the discovery of the exclusion principle.

1946 Percy Bridgman, US; for discoveries in high-pressure physics.

1947 Sir Edward Appleton, English; for the discovery of the Appleton Layer in the upper atmosphere.

1948 Patrick Blackett, English; for discoveries in nuclear physics and cosmic radiation.

1949 Hudeki Yukawa, Japanese; for predicting the existence of mesons.

1950 Cecil Powell, English; for the development of the photographic method of studying nuclear processes and for discoveries about mesons.

1951 Sir John Cockcroft, English, and Ernest Walton, Irish; for pioneering the use of accelerated particles to study atomic nuclei.

1952 Felix Bloch, Swiss-born US naturalized citizen, and Edward Purcell, US; for discovering nuclear magnetic resonance in solids.

1953 Frits Zernike, Dutch; for the phase-contrast microscopy method.

1954 Max Born, German-born British naturalized citizen; for statistical studies on wave functions.

Walther Bothe, German; for his invention of coincidence method.

1955 Willis Lamb, Jr., US; for discoveries in the hydrogen spectrum.

Polykarp Kusch, German-born US naturalized citizen; for measuring the magnetic moment of the electron.

1956 William Shockley, English-born US naturalized citizen, John Bardeen, US, and Walther Brattain, US; for investigating semi-conductors and for the discovery of the transistor effect.

1957 Tsung-Dao Lee, Chinese-born US naturalized citizen, and Chen Ning Yang, Chinese-born US naturalized citizen; for the discovery of violations of the principle of parity.

1958 Pavel A. Cherenkov, Russian, Ilya M. Frank, Russian, and Igor Y. Tamm, Russian; for investigating the effects produced by high-energy particles – the Cherenkov effect.

1959 Emilio Segrè, Italian-born US naturalized citizen, and Owen Chamberlain, US; for confirmation of the existence of the antiproton.

1960 Donald Glaser, US; for the development of the bubble chamber (the device that enables the tracks of ionizing particles to be photographed).

1961 Robert Hofstadter, US; for determining the shape and size of atomic nucleons.

Rudolf Mössbauer, German; for the discovery of the Mössbauer effect (the emission of gamma rays from certain crystal substances).

1962 Lev D. Landau, Azeri; for contributions to the understanding of condensed states of matter.

1963 Johannes H. D. Jensen, German, and Maria Goeppert-Mayer, Polish-German-born US naturalized citizen; for developing the shell model theory of the structure of atomic nuclei.

Eugene Paul Wigner, Hungarian-born USA naturalized citizen; for work on principles governing the interaction of protons and neutrons in the nucleus.

1964 Charles H. Townes, US, Nikolay G. Basov, Russian, and Aleksandr M. Prokhorov, Russian; for studies in quantum electronics leading to the construction of instruments based on maser-laser principles.

1965 Julian S. Schwinger, US, Richard P. Feynman, US, and Tomonaga Shin'ichiro, Japanese; for work on basic principles of quantum electrodynamics.

1966 Alfred Kastler, French; for work on optical methods for studying Hertzian resonances in atoms.

1967 Hans A. Bethe, German-born US naturalized citizen; for discoveries concerning the energy production of stars.

1968 Luis W. Alvarez, US; for the discovery of resonance states as part of work with elementary particles.

1969 Murray Gell-Mann, US; for the classification of elementary particles and their interactions.

1970 Hannes Alfvén, Swedish, and Louis Néel, French; for work on magneto-hydrodynamics and antiferromagnetism and ferrimagnetism.

1971 Dennis Gabor, Hungarian-born British naturalized citizen; for the invention of holography.

1972 John Bardeen, US, Leon N. Cooper, US, and John R. Schrieffer, US; for the development of the theory of superconductivity.

1973 Leo Esaki, Japanese, Ivar Giaever, Norwegian-born US naturalized citizen and Brian Josephson, Welsh; for tunnelling in semiconductors and superconductors.

1974 Sir Martin Ryle, English, and Antony Hewish, English; for work in radio astronomy.

1975 Aage Bohr, Danish, Ben R. Mottelson, US-born Danish naturalized citizen, and L. James Rainwater, US; for pioneering the understanding of the atomic nucleus that paved the way for nuclear fusion.

1976 Burton Richter, US, and Samuel C. C. Ting, US; for the discovery of a new class of elementary particles (psi, or J).

1977 Philip W. Anderson, US, Sir Neville Mott, English, and John H. Van Vleck, US; for contributions to the understanding of the behaviour of electrons in magnetic, non-crystalline solids.

1978 Pyotr L. Kapitsa, Russian; for the invention of the helium liquefier, and for its applications.

Arno A. Penzias, German-born US naturalized citizen and Robert W. Wilson, US; for the discovery of cosmic microwave background radiation (which supported the big-bang theory).

1979 Sheldon Glashow, US, Abdus Salam, Pakistani, and Steven Weinberg, US; for establishing the analogy between electromagnetism and the 'weak' interactions of subatomic particles.

1980 James W. Cronin, US, and Val L. Fitch, US; for work on the simultaneous violation of both charge-conjugation and parity-inversion.

1981 Kai M. Siegbahn, Swedish, and Nicolaas Bloembergen, Dutch-born US naturalized citizen; for work on electron spectroscopy for chemical analysis.

Arthur L. Schalow, US; for applications of lasers in spectroscopy.

1982 Kenneth G. Wilson, US; for his analysis of continuous phase transitions.

1983 Subrahmanyan Chandrasekhar, Indian-born US naturalized citizen, and William A. Fowler, US; for contributions to understanding the evolution and devolution of stars.

1984 Carlo Rubbia, Italian, and Simon van der Meer, Dutch; for the discovery of subatomic particles, supporting the electro-weak theory.

1985 Klaus von Klitzing, German; for the discovery of the Hall effect, permitting exact measurements of electrical resistance.

1986 Ernst Ruska, German, Gerd Binnig, German, and Heinrich Rohrer, Swiss; for the development of special electron microscopes.

1987 J. Georg Bednorz, German, and K. Alex Müller, Swiss; for the discovery of new superconducting materials.

1988 Lwon Lederman, US, Melvin Schwartz, US, and Jack Steinberger, German-born US naturalized citizen; for research into subatomic particles.

1989 Norman Ramsey, US; for the development of the separated field method.

Hans Dehmelt, German-born US naturalized citizen, and Wolfgang Paul, German; for the development and exploitation of the ion trap.

1990 Richard E. Taylor, Canadian, Jerome Friedman, US, and Henry Kendall, US; for proving the existence of the quark.

1991 Pierre-Gilles de Gennes, French; for studies in changes in liquid crystals.

1992 George Charpak, Polish-born French naturalized citizen; for devising an electronic detector that reads trajectories of sub-atomic particles.

1993 Russell A. Hulse, US, and Joseph H. Taylor, US; for the discovery of a new type of pulsar.

Nobel Prizewinners in Physiology or Medicine

1901 Emil von Behring, German; for his work in serum therapy.

1902 Sir Ronald Ross, English; for the discovery of how malaria enters an organism.

1903 Niels R. Finsen, Danish; for work on light radiation treatment of skin diseases.

1904 Ivan Pavlov, Russian; for work on the physiology of digestion.

1905 Robert Koch, German; for tuberculosis research.

1906 Camillo Golgi, Italian, and S. Ramón y Cajal, Spanish; for studies of the structure of the nervous system.

1907 Alphonse Laveran, French; for his discovery of the role of protozoa in diseases.

1908 Paul Ehrlich, German, and Elie Metchnikoff, Russian; for immunity systems research.

THE NOBEL PRIZES

1909 Emil Kocher, Swiss; for work on the physiology, pathology and surgery of the thyroid gland.

1910 Albrecht Kossel, German; for cellular chemistry research.

1911 Allvar Gullstrand, Swedish; for his work on the dioptics of the eye.

1912 Alexis Carrel, French; for studies of vascular suture and transplantation of organs.

1913 Charles Richet, French; for anaphylaxis research.

1914 Robert Bárány, Austro-Hungarian; for studies on the vestibular apparatus of the inner ear.

1915 No award.

1916 No award.

1917 No award.

1918 No award.

1919 Jules Bordet, Belgian; for his studies of the immunity system.

1920 August Krogh, Danish; for the discovery of the capillary motor-regulating mechanism.

1921 No award.

1922 Archibald Hill, English; for studies of heat production in muscles.

Otto Meyerhof, German-born US naturalized citizen; for work on the metabolism of lactic acid in muscles.

1923 Sir Frederick Banting, Canadian, and John James R. Macleod, Scottish; for the discovery of insulin.

1924 Willem Einthoven, Dutch; for the discovery of electrocardiogram mechanism.

1925 No award.

1926 Johannes Fibiger, Danish; for cancer research.

1927 Julius Wagner-Jauregg, Austrian; for malaria inoculation in dementia paralytica.

1928 Charles Nicolle, French; for typhus research.

1929 Christiaan Eijkman, Dutch; for his discovery of anti-neuritic vitamin.

Sir Frederick Hopkins, English; for his discovery of growth stimulating vitamins.

1930 Karl Landsteiner, Austrian-born US naturalized citizen; for his work in the grouping of human blood.

1931 Otto Warburg, German; for the discovery of the nature and action of a respiratory enzyme.

1932 Edgar D. Adrian (Lord Adrian), English, and Sir Charles Sherrington, English; for studies on the function of neurons.

1933 Thomas Hunt Morgan, US; for work on the role of chromosomes in transmission of heredity.

1934 George R. Minot, US, William P. Murphy, US, and George H. Whipple, US; for work on liver therapy to treat anaemia.

1935 Hans Spemann, German; for work on organization in embryos.

1936 Sir Henry Dale, English, and Otto Loewi, German-born US naturalized citizen; for work on the chemical transmission of nerve impulses.

1937 Albert von Szent-Györgyi, Hungarian-born US naturalized citizen; for studies on biological combustion.

1938 Corneille Heymans, Belgian; for research on the role of sinus and aortic mechanisms in respiration regulation.

1939 Gerhard Domagk, German (who declined the award as Hitler refused to allow Germans to accept Nobel Prizes); for work on the antibacterial effect of prontosil.

1940 No award.

1941 No award.

1942 No award.

1943 Henrik Dam, Danish; for the discovery of vitamin K.

Edward A. Doisy, US; for the discovery of the chemical nature of vitamin K.

1944 Joseph Erlanger, US, and Herbert S. Gasser, US; for studies of the differentiated functions of nerve fibres.

1945 Sir Alexander Fleming, Scottish, Ernst Boris Chain, German-born British naturalized citizen, and Howard Florey (Lord Florey), Australian: for the discovery of penicillin and its curative value.

1946 Hermann J. Muller, US; for the production of mutations by X-ray irradiation.

1947 Carl F. Cori and Gerty Cori, both Czech-born US naturalized citizens; for their discovery of the catalytic conversion of glycogen.

Bernardo Houssay, Argentinian: for research on the pituitary hormone function in sugar metabolism.

1948 Paul Müller, Swiss; for work on the properties of DDT.

1949 Walter Rudolf Hess, Swiss; for the discovery of the function of the midbrain.

António Egas Moniz, Portuguese; for work on the therapeutic value of leucotomy in psychoses.

1950 Philip S. Hench, US, and Edward Kendall, US, and Tadeusz Reichstein, Polish-born Swiss naturalized citizen; for adrenal cortex hormones research.

1951 Max Theiler, South African-born US naturalized citizen; for yellow fever research.

1952 Selman A. Waksman, Ukrainian-born US naturalized citizen; for the discovery of streptomycin.

1953 Fritz A. Lipman, German-born US naturalized citizen, and Sir Hans Krebs, German-born British naturalized citizen; for the discovery of coenzyme, a citric acid cycle in metabolism of carbohydrates.

1954 John F. Enders, US, Thomas H. Weller, US, and Frederick Robbins, US; for work on tissue culture of poliomyelitis viruses.

1955 Axel Hugo Theorell, Swedish; for work on the nature and mode of action of oxidation enzymes.

1956 Werner Forssmann, German, Dickinson Richards, US, and André F. Cournand, French-born US naturalized citizen; for work on heart catheterization and circulatory changes.

1957 Daniel Bovet, Swiss-born Italian naturalized citizen; production of synthetic curare.

1958 George W. Beadle, US, and Edward L. Tatum, US; for genetic regulation of chemical processes.

Joshua Lederberg, US: for work on genetic recombination.

1959 Severo Ochoa, Spanish-born US naturalized citizen, and Arthur Kornberg, US; for the production of artificial nucleic acids.

1960 Sir Frank MacFarlane Burnet, Australian, and Sir Peter B. Medawar, British (of Anglo-Lebanese parentage); for research into acquired immunity in tissue transplants.

1961 Georg von Békésy, Hungarian-born US naturalized citizen; for research on the functions of the inner ear.

1962 Francis Crick, English, James D. Watson, US, and Maurice Wilkins, New Zealand-born British citizen; for the discovery of the molecular structure of DNA.

1963 Sir John Eccles, Australian, Sir Alan Lloyd Hodgkin, English, and Sir Andrew Huxley, English; for work on the transmission of nerve impulses along a nerve fibre.

1964 Konrad Bloch, Swiss-born US naturalized citizen, and Feodor Lynen, German; for research into cholesterol and fatty acid metabolism.

1965 François Jacob, French, Jacques Monod, French, and André Lwoff, French; for research into regulatory activities of body cells.

1966 Charles B. Huggins, Canadian-born US naturalized citizen, and Francis Peyton Rous, US; for cancer research.

1967 Haldan Keffer Hartline, US, George Wald, US, and Ragner A. Granit, Finnish-born Swedish citizen; for research on the chemical and physiological visual processes in the eye.

1968 Robert W. Holley, US, H. Gobind Khorana, Indian-born US naturalized citizen, and Marshall W. Nirenberg, American; for research into deciphering the genetic code.

1969 Max Delbrück, German-born US naturalized citizen, Alfred D. Hershey, US, and Salvador E. Luria, Italian-born US naturalized citizen; for research into viruses and viral diseases.

1970 Julius Axelrod, US, Sir Bernard Katz, German-born British citizen, and Ulf von Euler, Swedish; for work on the chemistry of nerve transmission.

1971 Earl W. Sutherland, US; for studies of the action of hormones.

1972 Gerald M. Edelman, US, and Rodney Porter, English; for research into the chemical structure of antibodies.

1973 Karl von Frisch, Austrian, and Konrad Lorenz, Austrian, and Nikolaas Tinbergen, Dutch; for research on animal behaviour patterns.

1974 Albert Claude, Luxembourg-born US naturalized citizen, Christian R. de Duve, Belgian, and George E. Palade, Romanian-born US naturalized citizen; for work on the structural and functional organization of cells.

1975 Renato Dulbecco, Italian-born US naturalized citizen, Howard M. Temin, US, and David Baltimore, US; for research on the interactions between tumour viruses and the genetic material of the cell.

1976 Baruch S. Blumberg, US, and Daniel Carleton Gajdusek, US; for studies on the origin and spread of infectious diseases.

1977 Rosalyn S. Yalow, US, Roger Guillemin, French-born US naturalized citizen, and Andrew Schally, Polish-born US naturalized citizen; for the development of radio immunoassay and research on pituitary hormones.

1978 Werner Arber, Swiss, Daniel Nathans, US, and Hamilton O. Smith, US; for the discovery and application of enzymes that fragment DNA.

1979 Allan M. Cormack, South African-born US naturalized citizen, and Sir Godfrey N. Hounsfield, English; for the development of computerized axial tomography scanning.

1980 Baruj Benacerraf, Venezuelan-born US naturalized citizen, George D. Snell, US, and Jean Dausset, French; for research on genetic control of the immune response to foreign substances.

1981 Roger W. Sperry, US; for studies of the functions of the celebral hemispheres.

Torsten N. Wiesel, Swedish, and David H. Hubel, Canadian-born US naturalized citizen; for work on visual information processing by the brain.

1982 Sune K. Bergström, Swedish, Bengt I. Samuelsson, Swedish, and Sir John R. Vane, English; for work on the biochemistry and physiology of prostaglandins.

1983 Barbara McClintock, US; for the discovery of mobile plant genes which affect heredity.

1984 Niels K. Jerne, British/Danish, Georges J. F. Köhler, German, and César Milstein, Argentine-born British citizen; for the technique for producing monoclonal antibodies.

1985 Michael S. Brown, US, and Joseph L. Goldstein, US; for the discovery of cell receptors involved in cholesterol metabolism.

1986 Stanley Cohen, US, and Rita Levi-Montalcini, Italian; for the discovery of chemical agents that help regulate cell growth.

1987 Susumu Tonegawa, Japanese; for research into genetic aspects of antibodies.

1988 Sir James W. Black, Scottish, Gertrude Elion, US, and George H. Hitchings, US; for the development of new classes of drugs.

1989 Harold Varmus, US, and Michael Bishop, US; for cancer research.

1990 Joseph Murray, US, and E. Donnall Thomas, US; for transplant surgery.

1991 Erwin Neher, German, and Bert Sakmann, German; for research in cell biology, particularly the understanding of disease mechanisms.

1992 Edmond Fischer, US, and Edwin Krebs, US; for the discovery of a cellular regulatory mechanism used to control a variety of metabolic processes.

1993 Richard Roberts, English, and Phillip Sharp, US; for their discovery of split genes.

Weights and Measures

Measurement – in terms of length, weight or capacity – involves comparison. The measurement of any physical quantity entails comparing it with an agreed and clearly defined standard. The result is expressed in terms of agreed units. Each measurement is expressed in terms of the appropriate unit preceded by a number which is the ratio of the measured quantity of that unit.

Crude measurements probably date back to prehistory. The first – units of weight and length – were based upon parts of the human body. The average pace of a man was a common unit in many ancient civilizations. The length of the human thumb was another widely used measure – in England, it was the precursor of the inch. The length of ploughs and of other agricultural implements were also frequently used as early units of measurement. As civilization and trade developed the need for standardization grew. Units were fixed by local tradition or by national rulers, and many different (though sometimes related) systems developed.

acceleration, derived SI unit of see metre per second squared.

ampere the SI base unit of electric current, defined as that constant current which, if maintained in two straight parallel conductors of infinite length of negligible circular cross-section, and placed 1 metre apart in a vacuum, would produce between these conductors a force equal to 2×10^{-7} newtons per metre of length; symbol A. The ampere is also defined as the unit of magnetomotive force.

angular acceleration, derived SI unit of see radian per second squared.

angular momentum, derived SI unit of see kilogram metre squared per second.

angular velocity, derived SI unit of see radian per second.

apothecaries' system see Imperial system (below).

area, derived SI unit of see square metre.

area, metric units of are given as:

> 100 sq millimetres = 1 sq centimetre
> 100 sq centimetres = 1 sq decimetre
> 100 sq decimetres = 1 sq metre
> 100 sq metres = 1 are
> 100 ares = 1 hectare
> 1000 sq metres = 1 hectare
> 100 hectares = 1 sq kilometre

avoirdupois system see Imperial system (below).

base units (SI) are the metre, kilogram, second, ampere, kelvin, candela and mole.

becquerel the SI derived unit of radiation activity; symbol Bq.

candela the SI base unit of luminous intensity, defined as the luminous intensity, in a given direction, of a source that emits monochromatic radiation of frequency 540×10^{12} Hz and has a radiant intensity in that direction of (1/683) watts per steradian; symbol cd.

capacity, metric units of are given as follows:

> 10 millilitres = 1 centilitre
> 10 centilitres = 1 decilitre
> 10 decilitres = 1 litre
> 1000 millilitres = 1 litre
> 1 litre = 1 cu decimetre
> 10 litres = 1 dekalitre
> 10 dekalitres = 1 hectolitre
> 10 hectolitres = 1 kilolitre
> 1 kilolitre = 1 cu metre

Celsius see degree Celsius.

Centigrade see degree Celsius.

coulomb the SI derived unit of electric charge; symbol C.

cubic metre the derived SI unit of volume; symbol m^3.

degree Celsius the derived SI unit of temperature; symbol °C. The Celsius temperature scale is, with the kelvin scale, the most commonly used temperature scale in science. In a meteorological context Celsius is the most widely used temperature scale – the Fahrenheit scale is used in North America, in a few other English-speaking countries and is (decreasingly) quoted alongside Celsius in the United Kingdom (where it is sometimes wrongly referred to as Centigrade).

The Celsius scale may be said to have been devised in concept in 1743 by J. P. Christen (1683–1755), but is referred to by its present name because it took its present form through the work of the Swedish astronomer Anders Celsius (1701–44).

See also degree Fahrenheit and kelvin, and the box on temperature comparisons.

degree Fahrenheit a unit of temperature that has been largely replaced in most contexts by the degree Celsius and kelvin. The Fahrenheit scale is named after Gabriel Daniel Fahrenheit (1686–1736), a German-born physicist of Dutch parentage.

See also degree Celsius and kelvin, and the box on temperature comparisons.

density, derived SI unit of see kilogram per cubic metre.

derived units (SI) see metric system.

dynamic viscosity, derived SI unit of see newton second per metre squared.

electric capacitance, derived SI unit of see farad.

electric charge, derived SI unit of see coulomb.

electric conductance, derived SI unit of see siemens.

electric resistance, derived SI unit of see ohm.

electromotive force, derived SI unit of see volt.

energy, derived SI unit of see joule.

farad the SI derived unit of electric capacitance; symbol F.

force, derived SI unit of see newton.

frequency, derived SI unit of see hertz.

gray the derived SI unit of radiation absorbed dose; symbol Gy.

heat capacity, derived SI unit of see joule per kelvin.

henry the SI derived unit of inductance; symbol H.

hertz the SI derived unit of frequency; symbol Hz.

illumination, derived SI unit of see lux.

Imperial conversion see box.

Imperial system the traditional British system of units, which is related to the US Customary Units.

The two basic units are the yard (the unit of length) and the pound (the unit of mass). Subdivisions and multiples of these units are traditional in origin and do not follow the logical tenfold stages of the metric system.

The Imperial System is complicated by the existence of three different systems of measurement of weight. The avoirdupois system is the most widely used. The troy system is used to measure precious metals, while the apothecaries' system uses the same units as the troy system but with certain differences of name.

The use of the metric system was legalized in the United Kingdom in 1897. The intention to switch to the metric system 'within ten years' was declared in 1965 by the President of the Board of Trade, although in 1976 the government decided not to proceed with the second reading of the Weights and Measures (Metrication) Act. However, since 1965 the metric system has replaced the Imperial System for many purposes, although loose fruit and vegetables continue to be sold by the pound, the pint and the dram will remain as the unit of capacity for alcohol, and the mile will not be replaced as the standard unit of length over long distances.

The principal units were defined by the Weights and Measures Act (1963) as follows:

yard (symbol yd) is equal to 0.9144 metre;

pound (symbol lb) is equal to 0.45359237 kilogram;

gallon (symbol gal) is the space occupied by 10 pounds weight of distilled water of density 0.998859 gram per millilitre weighed in air of density 0.001217 gram per millilitre against weights of density 8.136 gram per millilitre.

inductance, derived SI unit of see henry.

joule the SI base unit of work, energy, and quantity of heat; symbol J.

joule per kelvin the SI derived unit of heat capacity; symbol JK^{-1}.

joule per kilogram the SI derived unit of specific latent heat; symbol Jkg^{-1}.

joule per kilogram kelvin the SI derived unit of heat capacity; symbol Jkg^{-1}K^{-1}.

kelvin the SI base unit of thermodynamic temperature (a degree of temperature measured on the kelvin scale), defined as the fraction $\frac{1}{273.15}$ of the thermodynamic temperature of the triple point of water. The triple point of water is the point where water, ice and water vapour are in equilibrium; symbol K. The kelvin scale is most commonly employed in any non-meteorological context. See degree Celsius and degree Fahrenheit, and the box on temperature comparisons.

kilogram the SI base unit of mass, defined as the mass of the international prototype of the kilogram, which is in the custody of the Bureau International des Poids et Mésures (BIPM) at Sèvres near Paris, France; symbol kg.

kilogram metre per second the SI derived unit of momentum; symbol kgms^{-1}.

kilogram metre squared the SI derived unit of moment of inertia; symbol kgm^2.

kilogram metre squared per second the SI derived unit of angular momentum; symbol kgm^{2}s^{-1}.

kilogram per cubic metre the SI derived unit of density; symbol kgm^{-3}.

kinematic viscosity, derived SI unit of see metre squared per second.

length, metric units of are given as follows:

10 ångström = 1 nanometre
1000 nanometres = 1 micrometre
1000 micrometres = 1 millimetre
10 millimetres = 1 centimetre
10 centimetres = 1 decimetre
1000 millimetres = 1 metre
100 centimetres = 1 metre
10 decimetres = 1 metre
10 metres = 1 dekametre
10 dekametres = 1 hectometre
10 hectometres = 1 kilometre
1000 metres = 1 kilometre
1000 kilometres = 1 megametre

lumen the SI derived unit of luminous flux; symbol lm.

luminous flux, derived SI unit of see lumen.

lux the SI derived unit of illumination; symbol lx.

magnetic flux, derived SI unit of see weber.

magnetic flux density, derived SI unit of see tesla.

magnetomotive force, derived SI unit of see ampere.

mass see weight.

metre the SI base unit of length, defined as the length of a path travelled by light in a vacuum during a time interval of 1/299 792 458 of a second; symbol m.

metre per second the SI derived unit of velocity; symbol ms^{-1}.

metre per second squared the SI derived unit of acceleration; symbol ms^{-2}.

metre squared per second the SI derived unit of kinematic viscosity; symbol $m^2 s^{-1}$.

metric conversion see box.

metric system was adopted in Revolutionary France in 1799 to replace the existing traditional illogical units. It was based upon a natural physical unit to ensure that it should be unchanging. The unit selected was 1/10000000 of a quadrant of a great circle of the Earth, measured around the poles of the meridian that passed through Paris. This unit – equivalent to 39.37003 inches in the British Imperial system – was called the metre (from Greek metron, 'measure').

Several other metric units are derived from the metre. The gram – the unit of weight – is one cubic centimetre of water at its maximum density, while the litre – the unit of capacity – is one cubic decimetre. Prefixes – from Danish, Latin and Greek – are commonly used for multiples of ten from atto ($\times 10^{-18}$) to exa ($\times 10^{18}$).

In 1875 an international conference established the International Bureau of Weights and Measures and founded a permanent laboratory at Sèvres, near Paris, where international standards of the metric units are kept and metrological research is undertaken. The prototype metre was an archive standard rather than an actual measurement upon the ground. In 1983 the metre was redefined as the length of the path travelled by light in vacuum during a time interval of 1/299 792 458 of a second.

The metric system is centred on a small number of base units (q.v.). These relate to the fundamental standards of length, mass and time, together with a few others to extend the system to a wider range of physical measurements, e.g. to electrical and optical quantities. There are also two geometrical units that are sometimes referred to as supplementary units (q.v.). These few base units can be combined to form a large number of derived units. For example, units of area, velocity and acceleration are formed from units of length and time. Thus very many different kinds of measurement can be made and recorded employing very few base units.

See the metric units of length, area, weight (mass), volume and capacity. See also SI units.

metrology the science of measurement.

mole the SI base unit of substance, defined as the amount of substance of a system that contains as many elementary entities as there are atoms in 0.012 kilogram of carbon-12; symbol mol.

moment of inertia, derived SI unit of kilogram metre squared.

momentum, derived SI unit of see kilogram metre per second.

newton the SI derived unit of force; symbol N.

newton per metre the SI derived unit of surface tension; symbol Nm^{-1}.

newton second per metre squared the SI derived unit of dynamic viscosity; symbol Nsm^{-2}.

ohm the SI derived unit of electric resistance; symbol Ω.

pascal the SI derived unit of pressure and stress; symbol Pa.

per degree Celsius an SI derived unit of thermal coefficient of linear expansion; symbol $°C^{-1}$. See also per kelvin.

per kelvin an SI derived unit of thermal coefficient of linear expansion; symbol K^{-1}. See also per degree Celsius.

potential difference, derived SI unit of see volt.

power, derived SI unit of see watt.

pressure, derived SI unit of see pascal.

quantity of heat, derived SI unit of see joule.

radian the SI supplementary unit of plane angle – the plane angle between two radii of a circle that cut off on the circumference an arc equal in length to the radius; symbol rad.

radian per second the derived SI unit of angular velocity; symbol $rad\ s^{-1}$.

radian per second squared the derived SI unit of angular acceleration; symbol $rad\ s^{-2}$.

radiation absorbed dose, derived SI unit of see gray.

radiation activity, derived SI unit of see becquerel.

second the SI base unit of time, defined as the duration of 9 192 631 770 periods of the radiation corresponding to the transition between the two hyperfine levels of the ground state of the caesium-133 atom; symbol s.

siemens the SI derived unit of electric conductance; symbol S.

SI multiples and submultiples of both base and derived units are used. These are decimal multiples and submultiples, for example 1 kilogram is divided into 1000 milligrams.

The submultiples are:

$\times 10^{-24}$, prefix yocto-, symbol y

$\times 10^{-21}$, prefix zepto-, symbol z

$\times 10^{-18}$, prefix atto-, symbol y

$\times 10^{-15}$, prefix femto-, symbol f

$\times 10^{-12}$, prefix pico-, symbol p

$\times 10^{-9}$, prefix nano-, symbol n

$\times 10^{-6}$, prefix micro-, symbol m

$\times 10^{-3}$, prefix milli-, symbol m

$\times 10^{-2}$, prefix centi-, symbol c

$\times 10^{-1}$, prefix deci-, symbol d

The multiples are:

$\times 10$, prefix deca-, symbol da

$\times 10^2$, prefix hecto-, symbol h

$\times 10^3$, prefix kilo-, symbol k

$\times 10^6$, prefix mega-, symbol M

$\times 10^9$, prefix giga-, symbol G

$\times 10^{12}$, prefix tera-, symbol T

$\times 10^{15}$, prefix peta-, symbol P

$\times 10^{18}$, prefix exa-, symbol E

$\times 10^{21}$, prefix zetta-, symbol Z

$\times 10^{24}$, prefix yotta-, symbol Y

SI units the 'standard units' of the metric system (see above).

A number of systems of units based upon the metric system have been in use. Initially the cgs system – based upon the centimetre for length, the gram for mass and the second for time – was widespread. It has, however, largely been replaced by the mks system in which the fundamental units are the metre for length, the kilogram for mass and the second for time. The mks system is central to the Système International d'Unités, which was adopted by the 11th General Conference on Weights and Measures in 1960. The SI units are now employed for all scientific and most technical purposes, and are in general use for most other purposes in the majority of countries.

See base, supplementary and derived SI units. See also metric system (above).

specific heat capacity, derived SI unit of see joule per kilogram kelvin.

specific latent heat, derived SI unit of see joule per kilogram.

square metre the SI derived unit of area; symbol m^2.

steradian the SI supplementary unit of solid angle – the solid angle that, having its vertex in the centre of a sphere, cuts off an area of the surface of the sphere equal to that of a square having sides of length equal to the radius of the sphere; symbol sr.

stress, derived SI unit of see pascal.

supplementary units (SI) see radian and steradian.

surface tension, derived SI unit of see newton per metre.

tesla the SI derived unit of magnetic flux density; symbol T.

temperature, derived SI unit of see degree Celsius.

thermal coefficient of linear expansion, derived SI unit of see per degree Celsius and per kelvin.

thermal conductivity, derived SI unit of see watt per metre degree Celsius.

troy system see Imperial system (above).

US Customary Units are related to the (British) Imperial system. Some units of the Imperial system have fallen into disuse in North America. The yard, for example, is only encountered in sport. The differences between English and American units make conversion difficult; for example, a ton in Britain is a unit of mass equivalent to 2240 pounds (or 1016.046909kg), while a ton in the USA and Canada is equivalent to 2000 pounds (or 907.184kg). There are also considerable differences between the English and American gallon.

velocity, derived SI unit of see metre per second.

volt the SI derived unit of electromotive force and potential difference; symbol V.

volume, derived SI unit of see cubic metre.

volume, metric units of are given as follows:

    1000 cu millimetres = 1 cu centimetre
    1000 cu centimetres = 1 cu decimetre
    1000 cu decimetres = 1 cu metre
    1000 cu metres = 1 cu dekametre

watt the SI derived unit of power; symbol W.

watt per metre degree Celsius the SI derived unit of thermal conductivity; symbol $Wm^{-1}°C^{-1}$.

weber the SI derived unit of magnetic flux; symbol Wb.

weight (mass), metric units of are given as follows:

    1000 milligrams = 1 gram
    10 grams = 1 dekagram
    10 dekagrams = 1 hectogram
    10 hectograms = 1 kilogram
    100 kilograms = 1 quintal
    1000 kilograms = 1 tonne

work, derived SI unit of see joule.

MATHEMATICAL SYMBOLS

The following symbols are those most commonly encountered:

=	equal to
≠	not equal to
≡	identically equal to
>	greater than
<	less than
≯	not greater than
≮	not less than
≥	equal to or greater than
≤	equal to or less than
≅	approximately equal to
+	plus
−	minus
±	plus or minus
×	multiplication (times)
÷	divided by
( ) [ ] { }	brackets, square brackets, enveloping
∥	parallel
∦	not parallel
#	numbers to follow (US usage)
%	per cent
‰	per mille (thousand)
∝	varies with
∞	infinity
$r!$ or $\angle r$	factorial r
√	square root
$^n\sqrt{}$	nth root
r^n	r to the power of n
Δ	triangle, finite difference or increment
~	difference
Σ	summation
∫	integration sign
°	degree
′	minute
″	second
→	approximate time; tends to
∴	therefore
∵	because
⇒	implies that
⇐	is implied by
⇔	is equivalent to

WEIGHTS AND MEASURES

METRIC AND IMPERIAL CONVERSIONS (* = exact)

Column One	Equivalent	Column Two	To convert Col.2 to Col.1 multiply by	To convert Col.1 to Col.2 multiply by
Length				
inch (in)	–	centimetre (cm)	0.393 700 78	2.54*
foot (ft)	12 in	metre	3.280 840	0.3048*
yard (yd)	3 ft	metre	1.093 61	0.9144*
mile	1760 yd	kilometre (km)	0.621 371 1	1.609 344*
ångström unit (Å)	10^{-10} m	nanometre	10	10^{-1}
Area				
square inch	–	square centimetre	0.155 00	6.4516*
square foot	144 sq in	square metre	10.763 9	0.092 903*
square yard	9 sq ft	square metre	1.195 99	0.836 127*
acre	4840 sq yd	hectare (ha) ($10^4 m^2$)	2.471 05	0.404 686*
square mile	640 acres	square kilometre	0.386 10	2.589 988*
Volume				
cubic inch	–	cubic centimetre	0.061 024	16.387 1*
cubic foot	1728 cu in	cubic metre	35.314 67	0.028 317*
cubic yard	27 cu ft	cubic metre	1.307 95	0.764 555*
Capacity				
litre	100 centilitres	cubic centimetre or millilitre	0.001*	1000*
pint	4 gills	litre	1.759 753	0.568 261
UK gallon	8 pints or 277.4in^3	litre	0.219 969	4.546 092
fluid ounce	0.05 pint	millilitre	0.035 195	28.413 074
Velocity				
feet per second (ft/s)	–	metres per second	3.280 840	0.3048
miles per hour (mph)	–	kilometres per hour	0.621 371	1.609 344
Acceleration				
foot per second per second	–	metres per second per second	3.280 840	0.3048*
Mass				
ounce (avoirdupois)	16 drams	gram	0.035 274 0	28.349 523 125
pound (avoirdupois)	16 ounces	kilogram	2.204 62*	0.453 592 37*
stone	14 pounds	kilogram	0.157 473 04	6.350 293 18*
hundredweight (cwt)	112 pounds	kilogram	0.019 684 1	50.802 345 44*
ton (long)	2240 pounds	tonne (= 1000kg)	0.984 206 5	1.016 046 908 8
Density				
pounds per cubic inch	–	grams per cubic centimetre	0.036 127 2	27.6799
pounds per cubic foot	–	kilograms per cubic metre	0.062 428 0	16.0185
Force				
poundal (pdl)	–	newton	7.233 01	0.138 255
pound-force (lbf)	–	newton	0.224 809	4.448 22
tons-force	–	kilonewton (kN)	0.100 361	9.964 02
kilogram-force (kgf)	–	newton	0.101 972	9.806 65
Energy (Work, Heat)				
erg	10^{-7} joule	joule	10^7	10^{-7}
horse-power (hp) (550ft/lbf/s)	–	kilowatt (kW)	1.341 02	0.745 700
therm	–	mega joule (MJ)	0.009 478 17	105.506
kilowatt hour (kWh)	–	mega joule (MJ)	0.277 778	3.6
calorie (international)	–	joule	0.238 846*	4.1868*
British thermal unit (Btu)	–	kilo-joule (kJ)	0.947 817	1.055 06
Pressure, Stress				
millibar (mbar or mb)	1000 dynes/cm^2	Pa	0.01*	100*
pounds per square inch (psi)	–	Pa	0.000 145 038	6894.76
pounds per square inch (psi)	–	kilogram-force per cm^2	14.223 3	0.070 307 0

TEMPERATURE COMPARIONS

A quick conversion of Celsius to Fahrenheit, and vice versa, may be obtained as follows:
To convert °C to °F, multiply the °C reading by 9, divide by 5 and add 32.
To convert °F to °C, subtract 32 from the °F reading and multiply by 5, divide by 9.
Temperature comparisons may be obtained in the following tables, which compare points on the Celsius and Fahrenheit (and, in the first table, kelvin) scales.

(1) Absolute zero = −273.15°C = −459.67°F = 0K
(2) Zero Fahrenheit = −17.8°C = −0.0°F = 255.35K
(3) Freezing point of water = 0.0°C = 32.0°F = 273.15K
(4) Triple point of water = 0.01°C = 32.0°F = 273.16K
(5) Normal human blood temperature = 36.9°C = 98.4°F = 310.05K
(6) Boiling point of water (at standard pressure) = 100.0°C = 212.0°F = 373.15K

−40°C	= −40°F	−9°C	= 16°F	22°C	= 72°F	53°C	= 127°F
−39°C	= −38°F	−8°C	= 18°F	23°C	= 73°F	54°C	= 129°F
−38°C	= −36°F	−7°C	= 19°F	24°C	= 75°F	55°C	= 131°F
−37°C	= −35°F	−6°C	= 21°F	25°C	= 77°F	56°C	= 133°F
−36°C	= −33°F	−5°C	= 23°F	26°C	= 79°F	57°C	= 135°F
−35°C	= −31°F	−4°C	= 25°F	27°C	= 81°F	58°C	= 136°F
−34°C	= −29°F	−3°C	= 27°F	28°C	= 82°F	59°C	= 138°F
−33°C	= −27°F	−2°C	= 28°F	29°C	= 84°F	60°C	= 140°F
−32°C	= −26°F	−1°C	= 30°F	30°C	= 86°F	61°C	= 142°F
−31°C	= −24°F	0°C	= 32°F	31°C	= 88°F	62°C	= 144°F
−30°C	= −22°F	1°C	= 34°F	32°C	= 90°F	63°C	= 145°F
−29°C	= −20°F	2°C	= 36°F	33°C	= 91°F	64°C	= 147°F
−28°C	= −18°F	3°C	= 37°F	34°C	= 93°F	65°C	= 149°F
−27°C	= −17°F	4°C	= 39°F	35°C	= 95°F	66°C	= 151°F
−26°C	= −15°F	5°C	= 41°F	36°C	= 97°F	67°C	= 153°F
−25°C	= −13°F	6°C	= 43°F	37°C	= 99°F	68°C	= 154°F
−24°C	= −11°F	7°C	= 45°F	38°C	= 100°F	69°C	= 156°F
−23°C	= −9°F	8°C	= 46°F	39°C	= 102°F	70°C	= 158°F
−22°C	= −8°F	9°C	= 48°F	40°C	= 104°F	71°C	= 160°F
−21°C	= −6°F	10°C	= 50°F	41°C	= 106°F	72°C	= 162°F
−20°C	= −4°F	11°C	= 52°F	42°C	= 108°F	73°C	= 163°F
−19°C	= −2°F	12°C	= 54°F	43°C	= 109°F	74°C	= 165°F
−18°C	= 0°F	13°C	= 55°F	44°C	= 111°F	75°C	= 167°F
−17°C	= 1°F	14°C	= 57°F	45°C	= 113°F	76°C	= 169°F
−16°C	= 3°F	15°C	= 59°F	46°C	= 115°F	77°C	= 171°F
−15°C	= 5°F	16°C	= 61°F	47°C	= 117°F	78°C	= 172°F
−14°C	= 7°F	17°C	= 63°F	48°C	= 118°F	79°C	= 174°F
−13°C	= 9°F	18°C	= 64°F	49°C	= 120°F	80°C	= 176°F
−12°C	= 10°F	19°C	= 66°F	50°C	= 122°F	81°C	= 178°F
−11°C	= 12°F	20°C	= 68°F	51°C	= 124°F	82°C	= 180°F
−10°C	= 14°F	21°C	= 70°F	52°C	= 126°F	83°C	= 180°F

Index/Factfinder

a posteriori probability 168
a priori probability 168
aardvark 54, 58, 59
aardwolf 54, 56
Abbe, Ernst 192
Abbot, Charles Grely 192
abdomen 42, 43, 45, 64, 65
Abegg, Richard 192
Abel, Niels Henrik 192
Abelson, Philip Haugh 192
aberration 106
abies 27
abomasum the fourth chamber of a ruminant's stomach.
abscission the shedding of a leaf, fruit, flower, etc., by a plant.
absolute zero 87
absorption the taking up of water, solutes and other substances by both active and passive mechanisms. Also, the taking up of radiant energy (from the sun) by pigments in plants.
AC see alternating current
acacia 28
Acari 45
acceleration 74
acceleration due to gravity 74
acceleration of free fall 74, 75
acceleration, derived unit of 225
Accipitridae 52, 53
accumulator 114
Aceraceae 28
acetone 152
acetyl choline a neurotransmitter chemical found in the nervous systems of both vertebrates and invertebrates.
acetyl glucose 41
acetylcholine 68, 69
achene a simple one-seeded indehiscent dry fruit.
Acheson, Edward Goodrich 192
Achilles tendon the tendon in the heel, linking the muscles in the calf to the heel bone.
achromatic lens 106
acid 144, 147, 159
acid rain 144, 150
acidity 147
acoelomate an animal lacking any form of coelom.
acoustics 7, 100-03, 102-03

acrosome 72
ACTH the usual name for adrenocorticotrophic hormone.
actinides 128-29
actinium 128, 134
actinopterygian 46
Actinopterygii 46
active transport the transport of substances across a membrane - e.g. cell membrane - against a concentration gradient.
acute angle 180
acyclic skeleton 152, 153
Adams, John Couch 192
Adams, Walter Sydney 192
Adanson, Michel 192
adaptation 70
adaptations for life in water 73
adaptive evolution 10
adaptor 195
addition 162
addition polymerization 156
addition rule of probability 168
adenine 14, 15
adenoid a lymph tissue at the back of the nose.
ADH the usual name for antidiuretic hormone.
adipose tissue a fatty tissue occurring under the skin in mammals.
adrenalin a hormone secreted in many groups of higher animals. It is said to prepare the body for 'fight and flight'.
adrenals the adrenal glands.
Adrian, Edgar Douglas 192
Adrian, Lord 192
adventitious organs organs that arise in unexpected sites, e.g. leaves that grow roots.
aerial root a root that appears above soil level, usually hanging down in moist air.
aerobe an organism that can live only in the presence of oxygen.
aerodynamics 213, 217
afferent nerves 68
Afinsen, Christian 193
African camel 55
agamid 50
Agamidae 50
agaric 24
Agaricales 24
Agassiz, Alexander 192

Agassiz, Jean Louis 192
Agave 28
age of the reptiles 50
Agnatha 46
agnathan 46
agouti 54
Agricola, Georgius 192
AIDS 19, 200, 203
Aiken, Howard H. 192
air 64, 65, 197
air bladder 46, 47
air sac 64, 65
Airy, Sir George Biddell 192
alanine 152
albatross 52
Albert the Great, St 192
alchemy 124
Alcmaeon 192
alcohol 21, 217, 152
alcomax 110
aldehyde 152
alder 28
aldosterone a hormone produced by the adrenals. It affects the kidneys, regulating the excretion of salt.
Aldrovandi, Ulisse 192
Aleksandrov, Pavel 192
Alfven, Hannes 192
algae 18, 19, 20, 206
algebra 7, 165, 197, 200, 207, 211, 213, 216
alimentary canal 65, 66
aliphatic compound 152
Alismataceae 28
Alismatidae 28
alkali 144
alkali metals 130, 146, 147-48
alkaline earth metals 147-48
alkalinity 147-48
alkane 152, 154
alkene 152
allele 14
Alliaceae 28
alligator 50
allotrope 139, 140
allowed orbit 120
alloy 110, 118, 125, 161
alni 110

Alouatia 56
alpaca **54**, 56
alpha decay 120
alpha particle 120, 214
Alpher, Ralph Asher 192
Alpini, Prospero 192
alternating current 116, 117
alternation of generations 25, 37
alternator 116
Altman, Sidney 192
aluminium 91, 118, 129, **131**, 142, 146, 205, 205
aluminium oxide 146
Alvarez, Luis 192
alveolus **64**, 65
AM see amplitude modulation
Amaryllidaceae 28
Ambartsumian, Viktor 192
americium 129, 135
amide 152, 153
amide linkage **156**, 158
amine 153
amino acid 12, 13, **14**, 15, 20, 21, 22, 66, 152, **153**, 156, 160, 206, 214
ammonia 136, **139**, 140, **150**, 152, 204
ammonium 22, 142
ammonium chloride 114, 136
amnion 50
amoeba 18, 19
Amontons, Guillaume 192
ampere **110**, 115, 193, 225
Ampère's Law 193
Ampère, André Marie 112, **192**
amphibian 46, **48-49**, 64, 72
amphibians (respiration in) 64
amphipod 44
Amphipoda 44
Amphisbaenia 50
amphisbaenian **50**, 51
amplitude **96**, 98, **100**, **104**, 108
amplitude modulation 100
ampullae of Lorenzini 70
anaerobe an organism that can live in the absence of free oxygen.
anaerobic respiration a number of processes by which chemical energy is obtained from various substrates without the use of free oxygen.
anal fin 46
Anatidae 52, 53
anatomy 7, 60-69, 203, 206, 216
Anderson, Carl David 193
Andrews, Philip Warren 193
Andrews, Thomas 193
androdioecious male and hermaphrodite flowers occurring on separate plants.
androecium the male component of a flower, consisting of several stamens.
andromonoecious male and hermaphrodite flowers carried on the same plant.
anemone 28
anemophily pollination by wind.
aneroid barometer 84
angiosperm 28
Angiospermatophyta 20
angle 182, 184, 185
angle of incidence **98**, 104

angle of reflection 98
angle of refraction 98
angstrom 193
Angstrom, Anders Jonas 193
angular acceleration, derived unit of 225
angular momentum, derived unit of 225
angular velocity, derived unit of 225
Animalia 16
animals **35**, 36-73
anion **139**, 141
ankle 60
annelid **39**, 62
Annelida 39
annual 28
annuli 25
anode 109, 114, **116**, 145
Anseriformes 52
ant 42, 54
Antarctic floral region 20
anteater **54**, 56, 57
antelope **54**, 55, 56, 58, 59
antenna 42
anther 28
antheridia 25
Anthocerotae 25
Anthozoa 37
antibiotics 217
antimony 129, **133**, 146
antineutrino 120
antinode **100**, 103
antiparticle 120
antiquark **120**, 122
antiserum serum extracted from the blood of an animal that is immune to a specific microorganism, e.g. hepatitis.
antitoxin an antibody that neutralizes a specific toxin or antigen.
antler 55, 56, 57
Anura 48
anuran 48
anus 66
anvil 60
aorta the large blood vessel that carries blood from the heart to the body. It is found in the higher four-limbed animals.
ape **54**, 56, 57, 58
apex 188
aphid 42, 43
Apiaceae 28
Apodiformes **52**, 53
apollination the transfer of pollen from the male to the female parts in seed plants, i.e. from the anthers to the stigma.
Apollonius 193
apomixis asexual reproduction.
apothecaries' system 225
Appel, Kenneth 193
appendix 66
apple 28
Appleton layer 193
Appleton, Sir Edward Victor 193
apterygote 42
Aquifoliaceae 28
arachnid 41, **45**
Arachnida 45

Araldite resin **157**, 160
Aranea 45
Arber, Werner 193
arc **186**, 187
archaebacteria 18
archegonia 25
Archimedes 184, **193**
Archimedes' principle **84**, **90**, 193
Ardeidae 53
area (of circle) 186
area (of triangles) 184
area, derived unit of 225
area, metric units of 225
Arecaceae 28
Argand diagram 193
Argand, Jean Robert 193
Argasidae 45
argon 114, 124, 129, **131**, **138**, 213
Aristarchus of Samos 193
Aristotle 16, **193**
Aristotle's lantern 38
arm **60**, 61
armadillo **54**, 56
armature 116
aromatic compound 153
Arrhenius, Svante August **193**, 218
arsenic 129, **132**, 146, 206
Artedi, Peter 193
arteriole a small artery linking an artery to the capillaries.
artery 62, 63
arthropod 41, **42-43**, 62, 63, 64, 65
artichoke 66
articulation the movement of one part of a skeleton over another, often at a joint.
artiodactyl **54**, 55, 56, 57, 58, 59
Ascomycota 24
ascorbic acid 66
ascus 24
asexual reproduction 25, 36, 38, **72**
ash 28
asparagus 28
aspen 28
ass **55**, 56, 58
assimilation the incorporation of the simple molecules resulting from digestion into more complex molecules.
Astacus 44
astatine 129, **134**, **149**
Astbury, William Thomas 193
Asteraceae 28
Asteridae 28
Asteroidea 38
Aston, Francis William 193
Atanasoff, John Vincent 193
Atiyah, Sir Michael Francis 193
atlas vertebra the first vertebra; it allows free movement of the head.
atmosphere 65
atmospheric pressure 84
atom 94, 120, 121, 122, 123, 124, **128**, 137, 139, 140, 141, 142, 144, 146, 152, 153, 154, 195, 198, 199, 203, 209, 213, 214, 216
atomic bomb 207, 211
atomic number **128-29**, 138

INDEX/FACTFINDER

atomic particles 120-23
atomic weight 130, 131-35, 193, 195, 197, 199, 204
ATP adenosine triphosphate, a nucleotide occurring in all plants.
atrium 62, 63
attraction 110
Audubon, John James 193
auricle a small projection from the base of a leaf or petal.
auroch 55
Australian faunal region 35
Australian floral region 20
autonomic function 68
autonomic nervous system the vertebrate nervous system concerned with controlling bodily functions.
autotrophic organism 20
autotrophic protoctists 18
auxin a variety of plant growth hormones that promote the elongation of shoots and roots.
Avery, Oswald T 14, 193
Aves 52
Avicenna 192
Avogadro's Law 193
Avogadro, Amadeo 193
avoirdupois system 225
axil the upper angle formed where the leaf or a similar organ joins the stem.
axis (of graph) 170, 184
axis vertebra the second vertebra; its articulation with the atlas vertebra allows rotational movement of the head.
axolotl 48, 49
axon 68, 69
aye-aye 55, 58

Baade, Wilhelm Heinrich Walter 193
Babbage, Charles 194
Babinet's principle 194
Babinet, Jacques 194
baboon 55, 57
Bacillariophyta 18
bacillus 18
backbone 60, 61
Bacon, Francis 194
Bacon, Roger 194
bacteria 12, 14, 16, 17, 18, 19, 20, 21, 42, 54, 66, 157, 195, 198, 201, 202, 207, 208, 210
bacteriophage 18
bacteroid 20, 21
Bactrian camel 55
badger 55, 57
Baer, Karl von 194
Baeyer, Johann Friedrich Adolph von 194
Baird, John Logie 194
Bakelite 157, 160
balance 60
balanced chemical equation 144
baleen 73
ball and socket joint 60
Balmer, Johann Jakob 194
Baltimore, David 194
bamboo 28, 58
bandicoot 55, 57, 58
Banks, Sir Joseph 194

banteng 55
Banting, Sir Frederick 194
bar chart 177
bar magnet 110
barb a hair-like structure attached to the shaft of a feather.
Barbary ape 55, 57
barbet 53
barbule one of the 'teeth' on the barb of a feather. The barbules interlock, linking the barbs together.
Bardeen, John 194
barium 128, 133, 147, 161, 204
bark 20, 21, 28
Barkla, Charles Glover 194
barnacle 44
barometer 84
Barr bodies 194
Barr, Murray Llewellyn 194
Bartholin, Thomas 194
baryon 120, 122
basalisk 50
base (triangle) 184
base 119, 144
base units 225
basic 144
basidia 24
Basidiomycota 24
basidiospore 24
Basov, Nikolai 194
bat 55, 56
Bates, Henry Walter 194
Bateson, William 194
Bathyergidae 57
batteries 114-15
Bayes, Thomas 194
Bayliss, Sir William Maddock 194
Beadle, George Wells 194
beak 51, 52
beam (light) 104
bean 21, 28, 67
bear 55, 59, 73
beat frequency 98
Beaumont, William 194
beaver 55
Beckmann thermometer 194
Beckmann, Ernst Otto 194
becquerel 194, 225
Becquerel, Antoine Henri 194
bee 42, 43, 72, 203
bee-eater 52
beech 28
beetle 42, 43
begonia 28
Behring, Emil von 194
Beilstein, Friedrich Konrad 194
Bekesy, Georg von 194
bel 102
Bell Burnell, Susan 195
Bell, Alexander Graham 102, 194
Bell, John Stuart 195
Bell, Sir Charles 194
bellflower 28, 29
Belon, Pierre 195

benzene 160, 207
benzene ring 153
Benzer, Seymour 195
Berg, Paul 195
Bergius, Friedrich 195
berkelium 129, 135
Bernal, John Desmond 195
Bernard, Claude 195
Bernouilli, Daniel 195
Bernouilli, Jakob 195
Bernouilli, Johann 195
berry 28
Berthelot, Marcellin 195
Berthollet, Count Claude Louis 195
beryllium 128, 131, 147, 148, 216
Berzelius, Jons 195
Bessel Functions 195
Bessel, Friedrich 195
Bessemer, Sir Henry 195
Best, Charles 195
beta decay 120, 202, 217
Bethe, Hans Albrecht 195
Betulaceae 28
Bichat, Marie Francois 195
bichir 47
biennial 28
big-bang theory 200, 204, 212
bilateral symmetry the arrangement of the body and organs of an animal in only one plane of symmetry.
bile 66, 67
bill 52, 53
binding energy 120
binomial 16, 167
binomial theorem 167
biochemistry 7, 12-15, 208
bioengineering 8
biological control the control of pests by making use of their natural predators.
biological sciences 8, 10-73
biomechanics 8
biometry 212
bionics 8
biophysics 8, 12-15
Biot, Jean-Baptiste 112
biotin 66
bipolar transistor 119
birch 28
bird 52-53, 54, 57, 63, 65, 72
bird of prey 52
birds (respiration in) 64
Birkhoff, George David 195
birth canal 57, 72
bismuth 129, 134
bison 55, 59
bittern 53
bivalent 153
bivalve 40
Bivalvia 40
Bjerknes, Vilhelm 195
black hole 205, 212
Black, Joseph 195
blackcurrant 67
Blackett, Patrick 195

Blackman, Frederick Frost 195
bladder a fluid- or gas-filled sac, often taken to mean the muscular sac into which urine drains from the kidneys.
blade 20
blastula the early stage in the development of an animal embryo after fertilization.
bleaching properties 149
Bloch Bands 195
Bloch, Felix 195
Bloembergen, Nicolaas 195
blood 46, **62**, 63, 64, 65, 68, 205
blood cell 12, 65
blood fluid 65
blood pressure 62
blood system (of insects) 42
blood vessel 62, 63, 65, 67
bloodstream 66
bloom (algae) a noticeable increase in the numbers of a species in the plankton.
blubber 73
blue-green algae 18
blue-green bacteria 18
bluebell 28
boa 50
bobcat 55
Bode's Law 195
Bode, Johann Elert 195
body growth 68, 69
bog myrtle 21
Bohm, David Joseph 195
Bohr, Niels 123, 195
boiling point 89, 138
Bok, Bart Jans 195
bolete 24
Boltwood, Bertram Borden 196
Boltzmann, Ludwig 196
Bolyai, Janos 196
bone 60
bone marrow 67
bongo 55
bony fish 46, 64
Boole, George 196
Boolean algebra 196
borage 28
Bordet, Jules 196
Boreal floral region 20
Borelli, Giovanni 196
Born, Max 196
Born-Haber cycle 196
boron 129, 131
Bos 55
Bosch process 196
Bosch, Carl 196
Bose statistics 196
Bose, Salyendranath 196
Bose, Sir Jagadis Chandra 196
Bose-Einstein particle 120
boson 120, 122, 217
botany 8, 20-23, 25-33
Bothe, Walther 196
bottom quark 120
Boveri, Theodor Heinrich 196
bovid 54, **55**, 56

Bovinae **55**, 56
bowel 67
bowfin 47
Boyer, Herbert Wayne 196
Boyle's Law 92, 196
Boyle, Robert 196
Boys, Sir Charles Vernon 196
brackets 165
bract a small leaflike structure that subtends a flower or inflorescence.
bract scale 27
bracteole a small bract, typically on a flower stalk.
Bradley, James 196
Bradypodidae **55**, 59
Bragg, Sir Lawrence 196
Bragg, Sir William 196
Brahe, Tycho 196
brain 58, 61, **68**, 215
branchiopod 44
Branchiopoda 44
Brassicaceae 28
Brattain, Walter 196
breastbone **60**, 61
Brenner, Sydney 196
Brewster, Sir David 196
Bridgman, Percy William 196
brine shrimp 44
bristletail 42
bristleworm 39
brittlestar 38
broadleaved tree 28
broccoli 66
bromeliad 28
bromine 129, **132**, 142, 143, **149**, 152
bronchi **64**, 65
bronchiole **64**, 65
Bronsted, Johannes Nicolaus 197
Bronsted-Lowry base 144
Bronsted-Lowry definition **144**, 197
bronze 146
broom 28
Brown, Michael Stuart 197
Brown, Robert 197
Brownian motion 197
browser 55
Bryophyta 20, 21, 25
bryophyte 25
buccal cavity the mouth cavity.
Bucerotiformes 52, 53
buckminsterfullerene **139**, 153, 207
bud 28
budding 24, 37, 72
buffalo 55
Buffon, Count Georges Louis Leclerc de 197
Bufo bufo 48
Bufonidae 48
bulb 28
bulbil a small bulb found on an aerial bud. It functions as a means of vegetative propagation.
bulk modulus model of elasticity 82
Bunsen burner 197
Bunsen, Robert 197
buoyancy force 84

Burbank, Luther 197
Burbidge, Eleanor Margaret 197
Burnet, Sir Frank Macfarlane 197
bush 28
bushbuck 55
butane 153
butanol 153
Butenandt, Adolf 197
buttercup 28
butterfly 42, 43
buttonquail 53
butene 153
buzzard 52
byssi 40

cabbage 28
cactus 28
cadmium 129, **133**
caecilian 48, 49
caecum 66
caesium 128, 130, **133**, 146, 147
calcaneus 60
Calcarea 36
calcite 38
calcitonin a hormone produced by the thyroid. It lowers the concentration of calcium in the blood.
calcium 67, 128, **131**, 142, **147**, 148
calcium carbonate 144
calcium fluoride 146
calcium hydroxide 144, 148
calcium oxide 144, 148
calculus 8, **170-73**, 172-73, 195, 197, 202, 208, 211, 214, 215
californium 129, **135**
calorie 87
Calvin, Melvin 197
calyx **28**, 38
Cambarus 44
cambium 20, 21, 29
camel 54, **55**
camelid **55**, 57, 59
camellia 29
camera 216
Camerarius, Rudolph Jacob 197
campanula 29
campion 29
Cancer 44
cancer an uncontrolled cell growth, in which the body's usual checks and controls are absent for some reason.
Candolle, Augustin Pyrame de 197
canid **55**, 56, 57, 59
Canidae 55
canine teeth 54, **55**, 57
Canis 57, 59
canker a plant disease in which there is an area of necrosis which becomes surrounded by layers of callus tissue.
Cannizzaro, Stanislao 130, **197**
Cannon, Annie Jump 197
Canton, John 197
Cantor, George 197
capacity, units of 225
capillary **62**, 65, 209

INDEX/FACTFINDER

capillary action the process in which the effect of surface tension on a liquid in a fine tube causes that liquid to rise. The supply of water throughout a plant is largely the result of capillary action.
capitate 60
Caprifoliaceae 29
capuchin 55, 57
capybara 55
carapace 44, 50, 51
carbohydrate 12, 20, 21, 22, 23, 66, 67, 139
carbon 12, 14, 20, 22, 66, 114, 129, **131**, **139**, 140, 142, 152, **153**, 154, 155, 156, 157
carbon dating 120
carbon dioxide 22, 42, 62, 64, 124, **139**, 142, 147, 148, **150**, 154, 195
carbon monoxide 150
Carboniferous period 50
carboxyl acids 153
carboxyl groups 14
Cardano, Geronimo 197
cardiac pertaining to the heart.
cardinal number 203
caribou 55, 56
carnassial teeth 57
Carnivora 73
carnivore 40, **55**, 57, 59, 73
Carnot engine 87
Carnot, Nicholas Leonard Sadi 197
Carothers, Wallace 197
carpal 60
carpel 29
carpus 60, 61
carrier waves 100, 101
carrot 29
Cartan, Elie 197
Cartesian coordinates 170
cartilage 46
cartilaginous fish 46
Carver, George Washington 197
Caryophyllidae 29
Caspersson, Torbjorn Oskar 197
Cassini, Giovanni Domenico 197
Castoridae 55
cat **55**, 56, 57, 59
catalysis 212
catalyst 124, 125, **144**
catastrophism 10
caterpillar 41, 42
catfish 47
cathode **109**, 114, 115, 116, 145
cathode rays 109
cathode-ray oscilloscope 109
cathode-ray tube **109**, 110
cation **109**, 139, 141, **144**, 145
catkin 29
cattle 54, **55**, 59
Cauchy, Baron Augustin-Louis 197
caudal fin 46
Caudata 48, 49
Cavendish, Henry 197
Cavia 55
cavy 54, **55**, 57, 58
Cebus 55
Cech, Thomas Robert 197

cell biology 8
cell structure 13
cells (physics) 114-15
cells 12-13, 215
cellular pathology 216
cellulose 20, 22, **66**, 156, **157**, 159
Celsius (degree) 225
Celsius scale 89, 197
Celsius, Anders **197**, 225
Celsius-Fahrenheit conversions 230
Centigrade 225
centipede 41
central nervous system 68
centre of mass 79, 80
centrifugal force 74, 125
centrifugation 125
centriole 12
centripetal acceleration 74, 75
Cephalophinae 56
cephalopod 40
Cephalopoda 40
cephalothorax 44, 45
Ceratocystis ulmi 24
cereals 29, 67
cerebellum 68
cerebral hemisphere 68
cerebrospinal fluid the fluid, derived from blood, which surrounds, and is found within cavities of, the brain and spinal cord.
cerebrum 68
cerium 128, **133**
cervical either pertaining to the cervix of the womb or pertaining to the neck of the womb.
cervical vertebrae 60, 61
Cervidae 55, 56
cervix 72
Cestoda 39
cetacean 73
Ceulen, Ludolph van 197
CFC 214
Chadwick, Sir James 197
chaetae 39
Chain, Ernst 198
chalk 40
chameleon 50
Chameleonidae 50
Chandrasekhar limit 198
Chandrasekhar, Subrahmanyan 198
change of subjects 165
change of subjects 165
Chapman, Sydney 198
Chargaff, Erwin 198
charge carrier 117, 118, **119**
Charles's law 92, 198
Charles, Jacques 198
charmed quark 120
Chasles, Michel 198
Chebyshev, Pafnutii 198
cheetah 55
chela pincers found in arthropods.
Chelonia 50
chelonian 51
chemical bonds 137, **139-143**, 211, 212, 215
chemical equation 144

chemical formulae 208
chemical reactions 144-45, 213
chemistry 8, 124-61
Chemistry, Nobel Prizes 218-20
chemoreception 70
chemotherapy 201
Chenopodiaceae 29
Cherenkov effect 198
cherry 29
chest 64, 65
Chevreul, Michel 198
chicken mite 45
chimaera 46
chimpanzee **55**, 56
chinchilla 55
chipmunk **55**, 59
chiral 154
Chiroptera 55
chitin 41, 42, 44
Chladni, Ernst 198
chloride 67
chlorine 129, **131**, 142, 147, **149**, 152, 158, 214
chloroethene 157
chlorophyll 20, 22, 197
chloroplast 12, 13, 18, **20**, 22
chloroxybacteria 18
cholesterol 217
chondrichthyan 46
Chondrichthyes 46
chondrocytes 46
chord 186
Chordata 17
chordates 17
chorion 50
Christen, J.P. 225
chromaticity 106
chromatograph 126
chromatography 125, 209
chromatophore a cell containing pigment that is involved in colour changes.
chromium 128, **132**, 216
chromoplast 12
chromosome 12, 13, **14**, 15, 72, 197
chrysalis the pupal form of butterflies and moths.
chrysanthemum 29
Chu, Paul 198
chyme 66
Ciconiidae 53
Ciconliformes 52
cilia 18, **39**, 40
ciliate any part of the plant fringed with hairs.
circle 180, 186-87
circle theorems 186
circuits 114-15
circular motion 74, 174
circular prism 188
circulatory systems 62-63
circumference 186
Cirripedia 44
citrus fruit 29, 67
civet **55**, 59
clade 16
cladistics 16
cladogram 16

235

Clairaut, Alexis Claude 198
Claisen, Ludwig 198
Claisen-Schmidt condensation 198
clam 40
clasper 42, 43
class 16
Claude, Georges 198
Clausius, Rudolph 198
clavicle 60
claw 52, 55
cleistogamy the self-pollination of flowers which do not open to reveal the reproductive organs, thus preventing cross-pollination.
clematis 29
Clemence, Gerald Maurice 198
climbing plants 29
clitellum 39
cloaca 50
closed circulatory system 63, 62-63
closed system 87
clotting (of blood) 67
cloven hoof 55
clover 29
club moss 20, 21, 25
club root a fungal disease in which roots become swollen and malformed, causing wilting, yellowing and stunting.
Clupea 46
Cnidaria 37
cnidarians 37
coati 59
coatl 55
cobalt 128, 132
cobra 50
coccus 18
coccyx 60, 61
Cockcroft, Sir John Douglas 198
cockle 40
cocoon a protective covering around the eggs or larvae of many invertebrates.
cod 46
codon 14, 15
coefficient 167, 172
coefficient of friction 82
coelacanth 46
coelenterates see cnidarians
coelenteron the body cavity in lower animals that functions as a digestive cavity. There is one opening, and the cavity itself is lined with two layers of cells.
coelom a body cavity in higher animals.
coenocarpium fruit that includes ovaries, floral parts and receptacles of a number of flowers on a fleshy axis.
coevolution 10
Cohen, Stanley 198
Cohn, Ferdinand Julius 198
coinage metals 146
cold receptor 70
cold-blooded 46, 50, 51
Coliiformes 52
collar bone 60
collector 119
collembolans 42
collenchyma long cells with thickened primary cell walls; a supporting tissue.

collies 52
colloid 139, 204, 212
colobus 55, 57
colon 66
colour charge 120
colubrid 50, 51
Colubridae 50
Columbiformes 52, 53
column graph 178
comet 205, 211
Commeliniceae 29
Commelinidae 29
common cold 19
common European frog 48
common European toad 48
complement 169
complex number 204
component of a force 174
Compositae 29
compound (chemical) 124, 125, 139, 142, 147, 154
compound eye 42, 70
Compton effect 198
Compton, Arthur Holly 198
computer 192, 194, 209, 210, 216
concentration (chemical) 124, 125
concentration of particles 136
concurrent triangle 184
condensation 127
condensation polymerization 157
conduction (heat) 90, 117, 146
conductivity (electrical) 117, 146
conductor 114, 115, 117-18
condyloid joint 60
cone (plant) 27
cone 186, 187, 188
configuration (chemical) 154
congruent triangle 184
conic sections 186-87, 193, 212
conifer 10, 20, 27
Coniferophyta 20
conjugation 18
Connochaetes 59
conservation of energy 87, 205, 207
conservation of momentum 75, 174
constrictor 50, 51
constructive interference 98, 99
contour feather 52
contractile root a specialized thickened root that pulls a rhizome, bulb, corm, etc., down into the soil.
contractile vacuole 18
convection (heat) 90
convergent evolution 10
convex 106
convolutions 68
convulvulus 29
Conway, John Horton 198
coordinates 170-71
Cope, Edward Drinker 198
copepod 44
Copepoda 44
Copernicus, Nicolaus 198
copper 114, 129, 132, 142, 146, 161, 216

copper chloride 145
copper oxide 118
copper sulphate 144
copulation 72
Coraciiformes 53
coracoid 60
coral 37
coral snake 50
Coraliformes 52
coralline sponge 36
cordate heart-shaped, e.g. leaves.
Corey, Elias James 198
Cori, Charles 198
Cori, Gerty 198
Coriolis force 198
Coriolis, Gustave-Gaspard 198
corm 29
cormorant 52
cornea 70
corolla tube the fusion of the edges of the petals.
corona a crown-like leafy outgrowth of a corolla tube.
coronary arteries the arteries providing the blood supply to the heart muscle.
corrosion 146
corymb a flat-topped cluster of flowers on lateral stalks of different lengths.
cosine 180
cosmic rays 112
costal pertaining to the ribs.
Cottrell, Sir Alan Howard 198
cotyledon 29
coulomb 110, 225
Coulomb's law 76, 110, 198
Coulomb, Charles Augustin de 110, 198
couple 81
covalent bonding 139, 141, 143, 154
covering cell 36
coyote 55
crab 41, 44
Craciformes 52
cracking 154
crane 52
Crangon 44
cranial kinesis 50, 51
cranium the skull of vertebrates.
crayfish 44
creeping fern 25
Crick, Francis 14, 198
cricket 43
Crinoidea 38
critical mass 120
crocodile 50
Crocodilia 50
Crocodilidae 50
crocus 29
Cronin, James Watson 199
Crookes, Sir William 199
crop 52
cross-link 160
cross-pollination 29
Crotolidae 50, 51
crown ethers 212
Cruciferae 29

INDEX/FACTFINDER

Crustacea **44**
crustacean 41, **44**, 72
cryolite 146, 205
cryptophyte any plant with perennating buds below ground or water.
crystal 198
crystal lattice **139**, 141, 146, 196
crystal structure 196
crystallization 125
crystalloid **139**, 204
ctenidia 40
cube 183, **190**
cubic metre 225
cubic number 162
cuboctahedron **190**
cuboid 60, **188**
cuckoo 52
Cuculiformes 52
cud 54
cultivar 29
curassow 52
Curie point 199
Curie's law 199
Curie, Marie Sklodowska **199**
Curie, Pierre **199**
curium 129, 135
curve 172
curved mirror 106
curved space-time 93
cuticle 25, 41, 42, 44
cutin a waterproof substance that forms the waxy cuticle.
cuttlefish 40
Cuvier, Baron Georges 10, **199**
cyanobacteria 18
cyanocobalamin 66
cybernetics 217
cycad 27
cyclic skeleton 153, **154**
cyclotron **120**, 208
cylinder **186**, **188**
cyst 18
cystic fibrosis 216
cytology 8
cytoplasm **12**, 13, 23
cytosine **14**, 15

d'Alembert principle 199
d'Alembert, Jean le Rond **199**
d'Herelle, Felix **200**
d-block elements 146
d-sub level **128**
Dacron 156
daffodil 29
Daimler, Gottlieb **199**
dairy product 66, 67
daisy 29
Dalton, John **120**, **199**
Daltonism 199
Dam, Carl Peter Henrik **199**
dandelion 29
Daniell cell 199
Daniell, John Frederic **199**
Darby, Abraham **199**

dark reactions part of the photosynthetic process that is not light dependent.
Darwin's frog 48
Darwin, Charles **10**, **11**, 16, **199**
Darwin, Erasmus 10
Darwinism 10-11, 199, 204
dasurid **56**, 59
Dasuridae 56
Dasypodidae 54
Davis, Raymond **199**
Davy, Sir Humphrey **199**
Dawkins, Richard **199**
DC see direct current
DDT 211
de Beer, Sir Gavin Rylands **199**
de Broglie wave 199
de Broglie, Prince Louis Victor **199**
de Sitter universe 200
de Sitter, Willem **200**
de Vries, Hugo **200**
Debye, Peter **199**
Debye-Huckel theory 199
decagon **182**
decantation 125
decapod 44
Decapoda 44
decibel 102
deciduous 29
decimal 215
decimal point **164**
decimals **164**
decreasing velocity 75
decumbent a stem that lies along the ground.
Dedekind, Richard **199**
Dedelphidae 56
deer 54, 55, **56**, 57, 59
defaecation the discharge of waste from the body through the anus.
definite growth the maximum size beyond which the plant can grow no more.
deformation (physics) 82
dehiscence the bursting open of certain plant organs at maturity - especially reproductive structures - to release their contents.
Dehmelt, Hans Georg **199**
Delbruck, Max **200**
Delphidae 58
Delphinidae 73
Demerec, Miloslav **200**
Demospongia 36
dendrite **68**, 69
denominator **164**
density 85
density, derived unit of **226**
dentate a leaf margin that is toothed.
dentine the main bulk of a tooth. It is served by blood vessels and covered with enamel.
deoxyribonucleic acid see DNA
deoxyribose 14
dependent variable 172
depolymerization 157
derivative 172
derived units **225-228**
Dermanyssidae 45
dermis 47

Desargues, Girard **200**
Descartes, Rene 170, **200**
descent with modification 10
destructive interference **98**, **99**
detonation wave 195
deuterium 216
Deuteromycota 24
deviation 125
Dewar, Sir James **200**
diabetes 194
diaccharides **66**
diagonal **184**
diameter **186**
diamond **139**, 153
diaphragm 54, **64**
diastole the phase of the heart beat in which the heart muscle is relaxed.
diatom **18**, 19
dichogamy anthers and stima maturing at different times on the same plant, thus reducing the chance of self-fertilization.
Dicke, Robert Henry **200**
dicliny the male and female reproductive parts in different flowers.
dicotyledon 29
Diesel, Rudolf **200**
Diesenhofer, Johann **200**
difference (mathematics) **162**
difference machine 194
differential calculus 172
differential coefficient 172
differential variable 172
differentiation 172, **173**
diffraction **100-101**
diffusion **136**, 204
digestion (in birds) 52
digestion (insects) 42
digestion 12, **66-67**, 68
digestive system **66-67**
digestive tract 66
digit 53, 58, **60**
digital clocks 118
digital signals 108
dilatation a process of widening, either by means of one of the body's reflexes or by mechanical means.
Dillenidae 29
dinitrogen monoxide **150**
dinosaur **50**, 52
diode, thermionic **116**
dioecious the male and female reproductive organs on different individuals, making cross-fertilization necessary and ensuring genetic variation.
Diogenes of Apollonia **200**
Dioscorides, Pedanius **200**
diploid **14**, 72
diplurans 42
Dirac, Paul **200**
direct current **116**, 117
direct proportion **166**
disaccharide 66
disc a cartilaginous pad between each vertebra, acting as a shock absorber and imparting flexibility to the spinal column as a whole.
disjoint sets 169

dispersion 104, 106
displacement 75, 77
disproportionate reaction 144
distillate 126
distillation 126, 127, 152
diver 52
division (biology) 16
division (mathematics) 162
Djerassi, Carl 200
DNA 10, 12, 13, 14-15, 18, 193, 198, 203, 204, 205, 210, 212, 217
Dobzhansky, Theodosius 200
dock 29
dodecagon 182
dodecahedron 190, 191
dog 55, 56
Dollfus, Andouin Charles 200
dolphin 73
dominant gene 15
Donders, Frans Cornelius 200
donkey 55, 56
dopamine 68, 69
Doppler effect 102, 200
Doppler shift 102
Doppler, Christiaan Johann 102, 200
dormancy an inactive phase of seeds, spores and buds, often in order to survive adverse conditions.
dorsal fin 46
double bond 139, 156, 157
double circulatory system 63
double flower a flower with more than the usual number of petals.
double helix 14, 15
Douglas fir 27
dove 52
down 52
down quark 120
dragonfly 42, 43
drain 119
Drake, Frank Donald 200
Draper, John William 200
Dreyer, Johann Louis Emil 200
drift velocity 117
dromedary 55, 56
drupe 29
dual nature of light 93
dual polyhedra 190
duck 52
duck-billed platypus 56, 57
Duersberg, Peter 200
dugong 73
duiker 55, 56
Dulbecco, Renato 200
Dumas, Jean-Baptiste Andre 200
duodenum 66, 67
dye 125, 144, 212
dynamic viscosity, derived unit of 226
dynamo 114, 116
dynamo generator 112
dynamo theory 112, 113
dysprosium 129, 133

eagle 52
ear 60, 61, 70, 73, 102
eardrum 102
earthworm 39, 57
earwig 42
ecdysis 41
echidna 56, 57, 59
echinoderm 38, 62
Echinoidea 38
echo 100, 101
echolocation 56, 102
ecology 8
ectotherm 46, 50
Eddington, Sir Arthur 201
Edeleman, Gerald Maurice 201
edentate 54, 56, 57, 59
edge 190, 191
Edison, Thomas Alva 201
eel 47
eelworm 39
efferent nerves 68
efficiency 86
egg 12, 39, 40, 43, 46, 48, 50, 57, 59, 67, 72
egg yolk 66, 67
egret 53
Ehrenberg, Christian Gottfried 201
Ehrlich, Paul 201
Eichler, August Wilhelm 201
Eigen, Manfred 201
eight-fold way 120, 122
Einstein, Albert 10, 93, 95, 123, 201
einsteinium 129, 135
Einthoven, Willem 201
eland 55, 56
elapid 50, 51
Elapidae 50
elastic force 97
elasticity 82-83, 206
elder 29
electric capacitance, derived unit of 226
electric charge 110-11, 112-13, 114, 130, 198
electric charge, derived unit of 226
electric conductance, derived unit of 226
electric conductor 115
electric current 110, 112-13, 114, 115, 116, 117, 122, 123, 211, 212, 215
electric field 110, 112
electric force 110
electric motor 116
electric resistance, derived unit of 226
electric switch 115
electrical impulses 69
electrical potential energy 110
electrical resistance 88, 211
electricity 110-11, 112-13, 114-15, 116-17, 119, 193, 197, 203, 211, 216
electro-weak force 75, 112
electrode 114, 193
electrodynamics 192
electrolyis 145, 47, 199, 205
electrolyte 114
electromagnet 215
electromagnetic force 75, 112
electromagnetic force of repulsion 123
electromagnetic induction 112
electromagnetic radiation 97, 112

electromagnetic spectrum 96, 112
electromagnetic waves 112
electromagnetism 112-13, 192, 202, 209
electromotive force 114
electromotive force, derived unit of 226
electron 109, 110, 113, 114, 116, 117, 118, 120, 122, 123, 129, 130, 140, 141, 142, 145, 146, 147, 149, 154, 199, 200, 203, 210, 212, 215, 216, 217
electron conduction 117
electron emission 116
electron gun 109
electron structures 143
electronic configuration 120
electronics 117
electronvolt 120
electrophilic 154
electrophoresis 216
electrovalent bonding 139, 141
electroweak force 121
elements 120, 122, 124, 125, 128-35, 139, 141, 142, 144, 146, 154, 199, 203, 207, 210, 211, 215
elephant 56, 58
elephant shrew 56
Elephantidae 56
elk 56, 57, 59
ellipse 186, 187
elm 29
Elton, Charles Sutherland 201
embryo 50, 194, 204, 206
embryology 192, 202, 217
EMF see electromotive force
emission of particles 120
emitter 119
empirical formula 154
empirical probability 168
empty set 169
Emydidae 50, 51
enantiomer 152, 154
enantioners 104
Enders, John 201
endocarp the innermost layer of the pericarp of an angiosperm fruit, outside the seeds. It can sometimes be woody.
endocrine cell 68
endocrine gland 68
endocrine system 68-69
endocytosis 12
endoplasmic reticulum 12
Endopterygota 42
endopterygote 42, 43
endoscopy 108
endoskeleton a rigid and often articulated structure that lies within the body tissues. It provides support and shape, and often sites of attachment for muscles.
endosperm the storage tissue in seeds of angiosperms.
energy 66, 87, 117-18, 120, 121, 130, 141, 144, 207, 214
energy of particles 136
energy production 68
energy, derived unit of 226
Engler, Adolf 201
entomophily pollination by insects.

INDEX/FACTFINDER

entropy 198
enzyme 12, 66, 67, 72, 125, 157, 194, 205, 206, 216
Eotvos law 201
Eotvos, Baron Roland von 201
epicalyx a calyx-like extra ring of floral appendages below the calyx, resembling a ring of sepals.
epicotyl the apical end of the axis of an embryo, immediately above the cotyledon or cotyledons. It grows into the stem.
epidemiology 216
epidermis 42, 47
epiglottis a cartilaginous flap that closes off the windpipe of mammals during the swallowing reflex.
epigyny the floral parts found above the ovary.
epiphyte 29
epiphytic fern 25
equation 165
Equidae 55, 56, 58, 59
equilateral triangle 184
equilibrium 79, 80, 144, 198, 208,
equilibrium temperature 87
equiprobable (of outcomes) 168
equivalent fraction 164
erbium 129, 133
Ericaceae 29
Erinaceidae 56, 57
Eriophydae 45
Erlanger, Joseph 201
Ernst, Richard Robert 201
erythrocyte 62, 63
Esaki, Leo 201
essential amino acid 66
ester linkage 157, 158
ethane 154
ethanethiol 154
ethanoic acid 145, 153
ethanol 152, 154
ethene 154, 157
ether 217
Ethiopian faunal region 35
ethmoid 61
ethology 8
ethylene 152
eucalyptus 29
Euclid 185, 201
Eudoxus 201
eukaryote cell 12
eukaryotes 14, 15, 16, 18, 20
Euler, Leonhard 190, 201
Euphorbiaceae 29
European wall lizard 50
europium 129, 133
Eustachian tube a passage leading from the back of the nose to the middle ear.
evaporation 126, 137
even number 162
Everett, Hugo 201
evergreen 30
evolution 10-11, 16, 41, 199, 205, 206, 210, 214, 217
exchange of gases 48, 62, 63, 64
exclusion principle 212
excretion the elimination of waste chemicals from the body. This is not the same as defaecation.

excurrent opening 36
exocarp the outermost layer of an angiosperm fruit, usually forming a skin.
exocrine gland a gland, found in vertebrates, in which the secretion is carried down a duct to the site of activity.
exocytosis 12
Exopterygota 42
exopterygotes 42, 43
exoskeleton 41, 42, 60
experimental physiology 205
exponent 162
expression 165
external fertilization 72
external gills 64
external laser beam 108
extraction of metals 146
eye 53, 70, 200
eyelid 49
Eyring, Henry 202

f-sub level 129
Fabaceae 30
Fabricius ab Aquapendente, Hieronymous 202
face 190, 191
factor 162
factorization of expression 165
faeces 66
Fagaceae 30
Fahrenheit (degree) 226
Fahrenheit, Gabriel Daniel 202, 226
Fahrenheit-Celsius conversions 230
Fairbank, William 202
fairy shrimp 44
Fallopian tube 72, 202
Fallopius, Gabriel 202
fallow deer 56
family 16
fan worm 39
fang 45, 50, 51
farad 226
Faraday, Michael 112, 116, 202
fat 66, 139, 157, 158
faunal regions 35
feather 52, 53
feather star 38
feature detector 70
feet 58
Felidae 55, 56
Felis 16
female gamete 32
femur 60
Fermat, Pierre de 202
fermentation the anaerobic respiration of glucose and other organic substrates to obtain energy.
Fermi, Enrico 202
Fermi-Dirac particle 121
fermion 121, 122
fermium 129, 135
fern 20, 25-26
Ferraris, Galileo 202
fertilization 72
fetus see foetus
Feynman, Richard Phillips 202

Fibonacci, Leonardo 202
fibre optics 108
fibrin a protein produced in the blood during the clotting process.
fibula 60
field-effect transistor 119
filament 114, 115, 116
Filicinophyta 20, 25
filoplume hairlike feathers scattered over the surface of a bird.
filtration 125, 126
fin 46
finch 52
finger 55, 60
fir 27
first law of thermodynamics 87, 89
Fischer, Emil 202, 218
Fischer, Ernst Otto 202
fish 46-47, 63, 64, 65, 67
fish liver oil 67
Fisher, Sir Ronald Aylmer 202
fission (nuclear) 121
fission (reproduction) 37
fix 150
Fizeau, Armand-Hippolyte-Louis 202
flagella 18, 36, 72
flamingo 52
Flamsteed, John 202
flatfish 46
flatworm 39, 62
flavour (quarks) 121, 123
Fleischmann, Martin 202
Fleming, Sir Alexander 202
Fleming, Sir John Ambrose 203
Flerov, Georgii 203
flight 52
flipper 53
floral region 20
floret a small flower.
Florey, Howard Walter 203
Florey, Paul John 203
Flourens, Jean-Pierre 203
flow of heat 87
flower 30
flowering plants 10, 20, 28-34
fluids 84-85
fluke 39
fluorine 129, 130, 131, 140, 141, 142, 149
fly 42
FM see frequency modulation
focal length 106
foetus 60
folic acid 66
follicle a small sac or cavity.
food poisoning 19
foot 60
force 74-78, 86, 174
force multiplier 86
force, derived unit of 226
forearm the part of the arm between the elbow and the wrist.
forebrain 68, 68
forelimb 60
forget-me-not 30

239

formaldehyde 154
formula 165
fossils 10, 11
Foucault, Jean-Bernard-Leon 203
Fourier Series 203
Fourier, Jean-Baptiste 203
fox 55, 56, 57
fraction (chemical) 124, 126
fraction (mathematics) 164, 166
fractional distillation 126
fragmentation asexual reproduction in which the parent splits into two or more pieces which develop into new individuals.
frame of reference 93
francium 128, 134, 147
Franck, James 203
Frankland, Sir Edward 203
Franklin, Benjamin 203
Franklin, Rosalind 203
Fraunhofer, Josef von 203
freezing point (water) 89
Frege, Gottlob 203
frequency 96, 98, 102, 106, 168
frequency chart 179
frequency distribution diagram 178
frequency modulation 101
Fresenius, Carl 203
Fresnel, Augustin Jean 203
friction 82-83
frigate bird 52
Frisch, Karl von 203
frog 48, 49, 72
frond 25
fructose 66, 67
fruit 30, 66
fruit bat 55
fruit fly 200, 209
fulcrum 81
fullerene 139
function 170, 172, 197, 217
functional group 154
fundamental forces 75
Fungi (kingdom) 16
Fungi imperfecti 24
fungi 16, 24, 206
funicle 30
fur 56, 57
furcula 52, 53
fuse 114
fusion (nuclear) 121
fusion 90

Gabor, Dennis 203
Gadidae 46
gadolinium 129, 133
galactose 66
galaxy 193
Galbuliformes 52
Galen 203
Galileo Galilei 75, 203
gall bladder 66
gall mite 45
Gall, Franz 203
Galliformes 52

gallium 122, 129, 132
gallium arsenide 118
Gallo, Robert 203
Galton, Sir Francis 203
Galvani, Luigi 203
gamete 15, 25, 72
gametophyte 26
Gamma rays 113
gamma decay 121
gamopetalous petals fused along their margins forming a corolla tube.
gamosepalous sepals that are fused to form a tubular calyx.
Gamov, George 204
ganglia 42, 68
gannet 52
Gapra 56
gar 47
garlic 30
gas chromatography 126, 127
gas discharge lamp 114, 115
gases 92, 124, 130, 136, 137, 139, 192, 193, 196, 200, 208, 210, 214, 216
Gassendi, Pierre 204
gastric pertaining to the stomach.
gastrointestinal tract 66, 67
gastropod 40
Gastropoda 40
gate 119
gaur 55
Gauss, Friedrich Carl 204
Gay-Lussac, Joseph 204
Gazella 56
gazelle 54, 56
gecko 50
Geiger counter 204
Geiger, Hans 204
Gekkonidae 50
Gelfand, Izrail 204
Gell-Mann, Murray 204
gemma a multicellular structure for vegetative reproduction found on some mosses and liverworts.
gemmule 36
gene 10, 12, 14, 15, 194, 199, 200, 202, 214, 215, 216
gene pool 10
general relativity theory 93
generator 114, 116
genet 56, 59
genetic drift 10
genetics 8, 14-15, 205, 214, 216
genitalia 72
genome 15
genotype 15
gentian 30
genus 16
geometry 8, 180-81, 182-91, 196, 197, 198, 200, 201, 202, 207, 209, 210, 212, 212, 213, 214
geranium 32
gerbil 56
Gerhardt, Charles 204
germanium 122, 129, 132, 146
germination the changes undergone by a reproductive body, e.g. zygote, spore, pollen, grain, seed, before and during the first signs of growth.

Gesner, Konrad von 204
gestation the time between conception, i.e. fertilization, and birth.
giant molecule 139
giant structure 141
gibbon 56
Gibbs, Josiah Willard 204
Gilbert, Walter 204
gill 40, 46, 64
gill cavity 46
gill raker 46
gill slit 46
ginger 30
ginkgo 27
giraffe 56, 58
Giraffidae 58
gizzard 52
glabrous a plant surface that has no hairs.
glaciation 192
gladiolus 30
gland 68
glandular system 68
Glaser, Donald Aylmer 204
Glashow, Sheldon Lee 204
glass sponge 36
glaucous plant surfaces with a waxy blue-grey bloom.
glider 58
glottis the opening of the larynx into the pharynx.
glucose 20, 22, 23, 66, 67, 198
glume bracts subtending each spikelet in the flowers of grasses.
glycerin 157
glycerol 157
glycine 69
glycogen 22, 66, 157
glycolysis 12
glyoxysome 12
gnu 56, 59
goat 55, 56
Goddard, Robert Hutchings 204
Godel, Kurt 204
gold 129, 134
Gold, Thomas 204
Goldstein, Joseph 204
Golgi apparatus 12
Golgi body 204
Golgi, Camillo 204
gonad 72
goose 52
gorilla 56
gorse 30
gradient (of graph) 171
Graham, Thomas 204
Graminae 30
grape vine 30
grapefruit 30
graph 172
graphic formula 172
graphite 115, 139, 140, 141, 153, 192
graphs 170-71
grasses 30
grasshopper 42, 43
Grassman, Hermann 204

INDEX/FACTFINDER

gravitation 75
gravitational acceleration 75
gravitational constant 75, 196
gravitational force 75, 76
gravity 75, 76, 79, 93, 192
gray 204, 226
Gray, Louis Harold 204
grazer 56
grebe 52
greenhouse effect 150
grey matter the region of the vertebrate brain that contains nerve cell bodies and synapses.
Grossulariaceae 30
group position (elements) 129-30
group theory 217
grouped data 178
grouse 52
growth hormone 68
growth ring secondary xylem produced in a growing period in the stems and roots of many plants.
Gruidae 52
Gruiformes 52
guanaco 55, 56
guanine 15
guard cell 20, 23
Guericke, Otto von 204
guinea pig 55
gull 52
gullet 66
gustation 70
gut 67
Gymnophiona 48
gymnosperm 27
gynandrous (flowers) stamens and styles united in a single structure.
gynodioecious plants that bear female and hermaphrodite flowers on separate individuals.
gynoecium the female part of the angiosperm flower, consisting of one or more carpels.
gynomonoecious plants that bear female and hermaphrodite flowers on the same individual.

Haber, Fritz 204
hadron 120, 121, 122
Haeckel, Ernst 204
haem the basis of many respiratory pigments, e.g. haemoglobin, as it can combine reversibly with oxygen.
haemocoel 42, 63
haemocyanin 64, 65
haemoglobin 64, 65, 66, 213
haemolymph 42
hafnium 128, 134
hagfish 46
Hahn, Otto 204
hahnium the provisional name for element 105, unnilpentium
hair 56, 73
Haldane, J.B.S. 205
Hale, George Ellery 205
Hales, Stephen 205
half-life 122
halide 147, 149
Hall, Charles Martin 205
Hall, Marshall 205

Hall, Sir James 205
Haller, Albrecht von 205
Halley, Edmund 205
hallux a vestigial digit on the inside of the rear limbs of most higher terrestrial vertebrates.
halogen 130, 149
halogen 153
halophyte any plant that can live in soil with a high salt concentration.
halothane an anaesthetic gas.
haltere a modified wing of flies that provides information on stability in flight.
Hamamelidaceae 30
Hamamelidae 30
hamate 60
Hamilton's equations 205
Hamilton's principle 205
Hamilton, Sir William Rowan 205
Hamiltonian function 205
hammer 60
hamstring muscles the muscles at the back of the thigh that flex the knee.
hand 58, 60
haploid 15
hard seed a seed with a hard coat that is impervious to water.
Harden, Sir Arthur 205
hardwood 30
Hardy, Godfrey Harold 205
hare 56, 57, 58
harmonics 102, 103
Harrison, Ross Granvile 205
harvest mite 45
harvestman 45
Harvey, William 205
hassium the provisional name for element 108, unniloctium
hastate a leaf shaped like a three-lobed spear.
haustorium an organ produced by a parasite to absorb nutrients from the host plant.
haw the nictitating membrane found in reptiles, birds and some domestic animals - e.g. horse and cat - that can be drawn upwards across the eye.
hawk 52
Haworth, Sir Walter 205
Hawking, Stephen 205
hazel 30
HCF 162
head (insects) 42, 43
head 54, 60
heart 61, 62, 63, 205
heart disease 66
heartwood the central part of secondary xylem in some woody plants. It is derived from the sapwood that has deteriorated with age.
heat 87, 88, 89, 90-91, 199, 207, 214
heat 144
heat 158, 160
heat capacity, derived unit of 226
heat transfer 80
heather 30
heating (by electricity) 114
Heaviside, Oliver 205
hedgehog 56, 57

Hedwig, Johann 205
Heisenberg uncertainty principle 74, 93, 95, 205
Heisenberg, Werner von 95, 205
helium 121, 129, 130, 131, 138, 208, 209
Helmholtz, Baron Hermann von 87, 205
hemicellulose a carbohydrate found in plant cell walls, often in association with cellulose.
hemicryptophyte any plant with perennating buds just below the soil surface.
hemisphere 187
hemlock spruce 27
henry 205, 226
Henry, Joseph 112, 205
Henry, William 205
hepatic pertaining to the liver.
Hepaticae 26
heptagon 182
herb 30
herbaceous perennial 30
herbivore 30, 40, 56, 57, 58, 59
heredity 14, 194, 200, 210
hermaphrodite 30, 72
Hero of Alexandria 205
heron 52, 53
Heroult, Paul 205
Herpestes 57
herring 46
Herschel, Sir William 205
Hershey, Alfred Day 205
hertz 96, 97, 102, 206, 226
Hertz, Gustav 206
Hertz, Heinrich 97, 102, 112, 206
Hertzsprung, Ejnar 206
Hertzsprung-Russell diagram 206
hesperidium any berry with a leathery epicarp, e.g. citrus fruit.
heterocercal any fish whose vertebral column extends into the tail fin.
heterosporous condition 26
heterotrophic protoctists 18
heterozygous cells 15
heterozygous gene 15
Hexactinellida 36
hexagon 182
hibernation a period of time during winter when an animal becomes inactive and the basal metabolic rate drops, thus conserving energy.
hidden heat 90
high-density polythene 158
high-temperature superconductors 161
highest common factor 162
Hilbert, David 206
hilium a scar on the seed coat at the point of abscission.
Hill, A.V. 206
hindbrain 68, 68
hinge joint 60
hip 60
Hipparchus 206
Hippocastanaceae 30
Hippocrates 206
hippopotamus 54, 56
Hirudinea 39
His, Wilhelm 206
histogenesis 206

histogram 177, 178
histology 195
Hittorf, Johann Wilhelm 206
Hodgkin, Dorothy 206
Hodgkin, Sir Alan 206
hole 118, 119
holly 30
holmium 129, 133
holograms 108
holography 108, 203
Holothuroidea 38
Homarus 44
hominid a primate in the family Hominidae, including early and modern man.
Hominidae 56
Hominoidea 54, 55
Homo sapiens 54, 58
Homo sapiens sapiens 56
homogamy the maturation of anthers and stigmas at the same time.
homoiothermic 54, 56, 64
homologous chromosome 14-15
homologous series 154
homologous structures 10
homosporous condition 26
homozygous cells 15
honey 66
honey guide dots or lines on petals that guide pollinating insects to the nectaries.
honeysuckle 30
hood 50
hoof 58, 59
hoofed mammal 73
Hooke's law 82, 206
Hooke, Robert 82, 206
Hooker, Sir Joseph Dalton 206
Hooker, Sir William Jackson 206
hookworm 39
hoopoe 53
Hopkins, Sir Frederick Gowland 206
horizontal component of velocity 75
hormone 66, 68, 72, 157, 160, 194, 197, 200, 215
horn 51, 55
hornbill 52, 53
hornwort 26
horse 56, 58, 59
horse chestnut 30
horsetail 20, 23, 26
howler 56, 57
Hoyle, Sir Fred 206
HPLC 127
Hubble, Edwin Powell 206
Huggins, Sir William 206
human being 56
human circulatory system 62-63
human digestive system 66-67
human respiratory system 64-65
human skeleton 60-61
Humboldt, Baron Alexander 206
humerus 60
hummingbird 53
Hunter, John 206
Huxley, Hugh Esmor 206
Huxley, Sir Julian 206

Huxley, T.H. 206
Huygens' construction 101
Huygens' principle 101
Huygens, Christiaan 101, 206
Hyaenidae 56
hybrid an individual plant produced by genetically distinct parents.
hydra 37, 64
Hydrachnidae 45
hydrangea 30
hydrocarbon 152, 154, 206
hydrochloric acid 66, 145
hydrogen 15, 20, 22, 23, 66, 114, 123, 124, 128, 130, 131, 140, 141, 142, 143, 144, 145, 147, 148, 150, 153, 193, 197, 204
hydrogen bomb 121, 216
hydrogen bond 141
hydrogen chloride 144, 157
hydrogen oxide 124
hydrolysis 157
Hydrophiidae 50, 51
hydrophily pollination by water transport of pollen grains.
hydrostatics 85
hydroxide compounds 147
hydroxides 148
Hydrozoa 37
hyena 54, 56
Hylobatidae 56
hyoid 60
hyperbola 187
hypha 24
hypocotl that part of the stem between the cotyledons and the radicle in the embryo.
hypogyny the floral parts inserted below the ovary.
hypotenuse 180, 184, 185
hypothalamus 69
Hyracoidea 56
hyrax 56, 58
hyroxide 142

ibex 56
ibis 52, 53
ice 141
icosahedron 190, 191
ideal gas 87, 92
ideal gas scale 87
identity matrix 176
ideograph 178
iguanadon 50
iguanid 50
Iguanidae 50
ileum 66
ilium 60
illumination, derived unit of 226
imago 43
immersion heater 91
immunity the natural or acquired resistance of a body to invading, i.e. 'foreign', chemicals.
immunization the production of immunity to a specific disease.
immunology 197, 210
impact of elastic bodies 174
impala 54, 56
imperfect elasticity 82

Imperial and metric conversions 229
Imperial system 226
implantation the attachment of a vertebrate fertilized ovum to the wall of the uterus.
improper fraction 164
impulse 69b
incident ray 104
incisor 55, 56
increasing velocity 75
increment 172
incubation period the time it takes between infection by a disease-carrying microorganism and the production of symptoms of the disease.
incurrent pore 36
incus 54, 60
indehiscent 30
independent variable 172
indeterminism 93
index 162, 165
indicator 144
indirect proportion 166
indium 129, 133
inductance, derived unit of 226
induction 116
inequality 165
inequation 165
inert gas 114, 130, 138, 198
inertia 76
inertial frame 93
inferior nasal conchae 61
infinitesimal 172
inflorescence 30
infrared waves 113
infrasonic frequency 102
Ingram, Vernon Martin 206
inheritance of acquired characteristics 10
inheritance, theories of 14
initiator 157
inorganic chemistry 8, 121-51, 156-61
insect 41, 42-43, 65, 72, 216
Insecta 42-43
Insectivora 57
insectivore 56, 57, 59
insects (respiration in) 64
instantaneous velocity 76
instar the form adopted by an insect between moults.
insulator 118
insulin 68, 69, 157, 160
integer 162, 184, 185
integument a protective envelope around the ovule of seed plants.
intelligence 54, 68, 203
intensity (of light) 108
intercept 170
intercostal muscles the muscles between the ribs.
intercourse 72
interference (light) 108
interference 93, 95
intermediate cuneiform 60
intermolecular forces 141
internal energy 87, 92
internal fertilization 72
interneurone 68, 69

INDEX/FACTFINDER

intersection (sets) 169
interstitial fluid 62, 63
interval 172
intestine 66
intromittent organ 72
invertebrate 40, 49, 50, **58**, 65, 72, 207
iodide 149
iodine 129, **133**, 142, **149**
ion 12, 23, 139, 140, **141**, **144**, 146, 206, 211, 215
ionic bonding 140, **141**
ionic compounds **142**, 149
ionic salt 147
ionic theory of electrodes 193
ionium 196
ionosphere 193
Ipatieff, Vladimir **206**
Iridaceae 30
iridium 76, 128, **134**
iris (plant) 30
iris 71
iron 67, 110, 128, **132**, 142, 146, 195, 199
irrational number **162**, 187, 199
irregular polyhedra 190, 191
ischium 60
islets of Langerhans the groups of cells in the pancreas responsible for the production of insulin.
isolating mechanisms 11
isomer 154
isometric transformation 176
isopod 44
Isopoda 44
isosceles triangle 184
isotope 121, **122**, 144, 192, 215
itch mite 45
ivy 30
Ixodidae 45

Jabir, ibn Hayyan **206**
jacamar 52
jackal 55, **57**
Jacob, Francois **206**
Jacobsen's organ 71
jaw 55, 60
jawless fish 46
Jeans, Sir James **207**
jejunum **66**, 67
jellyfish 37, 44
Jenner, Edward **207**
joint 54, **60**
Joliot-Curie, Frederic **207**
Joliot-Curie, Irene **207**
joule 87, **90, 226**
Joule, James 87, **207**
Joyce, James 122
Juglandaceae 30
junction transistor 119
Jupiter 214
Jurassic period 51

kalium 147
Kamerlingh-Onnes, Heike **207**
kangaroo **57**, 59
Kant, Immanuel **207**
Kapitza, Pyotr **207**

Karman vortices 217
Kekule von Stradonitz, Friedrich **207**
kelvin 87, 89, **90, 226**
Kelvin, Lord 87, **207**
Kenelly, Arthur Edwin **207**
Kepler, Johannes **207**
keratin 51, **52**
ketone **154, 155**
Kevlar 156, **158**
Khorana, Har Gobind **207**
Khwarizmi, al **207**
kidney one of two excretory organs found in vertebrates.
kilocalorie 87
kilogram **76, 226**
kinematic equations 76, **77**
kinematics 76
kinetic energy 88, 89, 92, 114, **136**, 137, **174**
kinetic frictional force **82**
kinetic theory of gases 92
kingdoms, of living organisms **16,** 209
kingfisher 52, **53**
kinkajou **57**, 59
Kirchhoff's laws 207
Kirchhoff, Gustav Robert **207**
Kirchner, Athanasius **207**
kite 52, **53**
Klaproth, Martin **207**
Klein, Felix **207**
koala 57
Koch, Robert **207**
Komodo dragon 50
kouprey 55
krait 50
Krebs cycle 207
Krebs, Sir Hans Adolf **207**
krill **44**
Kroto, Harold Walter **207**
krypton 129, 130, **132, 138,** 213
Kurchatov, Igor **207**

labellum the distinct lower three petals of an orchid.
labrum the upper 'lip' in insects that helps in feeding.
Lacerta muralis 50
lacertid **50**
Lacertidae **50**
lack of reactivity 138
lacrimal 60
lactation milk production. One of the chief characteristics of mammals.
lacteal lymph vessels in vertebrates, involved in the absorption of digested fats in the intestine.
lactose **66**
lagomorph **57,** 58
Lagrange, Count Joseph Louis **207**
Lamarck, Jean-Baptiste 10, **207**
Lambert, Johann Heinrich 187, **207**
lamina 20
laminar flow **85**
lamprey 46
lanceolate (leaves) narrow; tapering at both ends.
Landau, Levi **208**

Landsteiner, Karl **208**
Langevin, Paul **208**
Langrangian point 207
language 68
langur 57
lanthanides **128-29**
lanthanum 128, **133,** 161
Laplace, Marquis Pierre-Simon de **208**
larch 27
large intestine **66**
large molecule 142
Laridae **52**
Laris 27
larva 36, 38, 40, 42, **43**
larynx 65
laser **108,** 94, 195, 209
latent heat **90**
lateral cuneiform 60
lateral-line system **46**
Lauraceae 30
laurel 30
Lavoisier, Antoine-Laurent **208**
law of averages 168
law of constant composition **144**
law of large numbers 168
Lawrence, Ernest Orlando **208**
lawrencium 129, **135**
laws of thermodynamics **88, 89**
laws of uniformly accelerated motion **76**
LCM **162, 164**
Le Bel, Joseph Achille **208**
Le Chatelier principle **208**
Le Chatelier, Henry-Louis **208**
lead 129, **134,** 142
leaf **20,** 22, 23
leaf base the point of attachment of a leaf to the stem.
Leavitt, Henrietta **208**
Leclanche, Georges **208**
Lederberg, Joshua **208**
Lederman, Leon Max **208**
leech 39
Leeuwenhoek, Anton van **208**
leg **60,** 62
leghaemoglobin **20,** 21
legume **20,** 21, **30**
Leibniz, Gottfried Wilhelm **208**
lemma the lower of a pair of bracts beneath each flower in a grass.
lemming **57**
Lemmus 57
lemon 30
lemur **57,** 58
Lenard, Philipp Eduard Anton **208**
lens 71, **106-07**
lenticel a small pore containing loose cells in the periderm of plants. Gaseous exchange takes place through the lenticel.
Lenz's law **208**
Lenz, H.F.E. **208**
Leonardo da Vinci **208**
leopard 16, **57,** 58
Leporidae **57,** 59, 121
lepton **122,** 213, 217

243

Lepus 56
lettuce 30
leucocyte 62, 63
leucoplast 12, 13
Levene, Phoebus Aaron Theodor 208
lever 86
Levi-Civita, Tullio 208
Libavius, Andreas 208
Libby, Willard Frank 208
lice insects which infest the hair of the body.
lichen 24, 26, 206
Liebig, Baron Justus von 208
life cycle the various stages an organism passes through, from fertilized egg in one generation to fertilized egg in the next generation.
ligament a band of fibre holding two bones together at a vertebrate joint.
light 93, 94, **104**, 196, 197, 202, 203, 206, 209, 210, 213, 216, 217
light bulb **114**, 115
light reactions 21, 22
light-emitting diode 118
lignin 21, 23, 30
like terms 165
likelihood ratio 168
Liliaceae 30
Liliidae 30
Liliopsida 32
lily 32
limb 73
lime 32
limiting frictional force 82
limonene 152, **154**
limpet 40
Linde, Carl von 208
Lindemann, Ferdinand von 209
linden 32
line symmetry 180
linear (leaves) flat and parallel-sided leaves.
linear accelerator 122
linear motion 76
lines of force **110**, 111
Linnaean system 16
Linnaeus, Carl 16, 209
lion 16, **57**
lipid 12, **157**, **158**
lipoprotein an association of lipid and protein usually found in plant cell membranes.
liquid 137
liquid crystals 161
Lister, Joseph 209
lithium 128, **131**, 142, 146, **147**, 148
lithium chloride 148
lithium hydroxide 148
litmus 144
liver 66, **67**
liverwort 20, 21, 26
lizard **50**, 51, 72
llama 55, 56, **57**
lobster 44
Locke, John 209
Lockyer, Sir Joseph Norman 209
locomotion (of snakes) 51
Loeb, Jacques 209

Lomonosov, Mikhail 209
longitudinal wave **96, 97**, 101
Lorentz, Hendrick Loorentz 209
loris **57**, 58
Lorisidae 57
Loschmidt, Johann Josef 209
loudness 102, 103
Lovell, Sir Bernard 209
low-density polythene 157
lowest common denominator 164
lowest common multiple 162
lowest term 164
Lowry, Thomas 209
Lucretius 209
lugworm 39
lumbar pertaining to the lower back.
lumbar vertebrae 61
lumen 226
luminous flux, derived unit of 226
lung 48, **61**, **64**, 65, 66
lung books 65
lungfish 46
lungless salamander 48
lupin 32
lutetium 129, **134**
lux 226
Lycopodophtya 20, 21
Lycopodophyta 26
Lyell, Sir Charles 11
lymph a fluid from the blood which leaks out of the capillaries, bathes the tissues and returns to the blood system via the lymphatic system.
lymphatic system 62, **63**, 194
lynx 57
lyosome 12

Macaca 57
macaque **57**, 59
Mach's principle 209
Mach, Ernst 209
machines 86
mackerel 46, 47
MacLaughlin, Colin 209
Macrochiroptera 55
macromolecule 142
macronutrient a chemical element required by a plant in relatively large amounts.
macrophyllous leaf 26
Macropodidae 57
Macropodus 57
Macroscelidae 56
magnesium 128, **131**, 141, 142, 146, **147**, 148
magnesium oxide 145, 148
magnet **110-11**, 112, 197
magnetic field **110-11**, 112, **113**, 116, 216, 216
magnetic flux density 216
magnetic flux density, derived unit of 227
magnetic flux, derived unit 226
magnetic meridian 111
magnetism **110-11**, **112-13**, 199, 207, 211, 217
magnetomotive force, derived unit 227
magnolia 32
Magnoliopsida 32
maidenhair tree 27

Maiman, Theodore Harold 209
Maimodes, Moses 209
maize 22
malaria 214
malic acid **21**, 23
malleus 54, **60**
Malphigian tubules 42
Malpighi, Marcello 209
maltose 66, **67**
mamba 51
mammal **54-59**, 63, 64, 65, 72, **73**
Mammalia 16
mammary gland 54, 68
man, classification of 17
manatee 73
Mandelbrot, Benoit 209
mandible **60**, 61
manganese 128, **132**, 146
Manidae 58
manometer 85
mantles folds of skin in molluscs which secrete the shell, if present, and protect the gills.
manubrium 61
map 166
maple 32
mara 57
Marconi, Giuseppe 209
marigold 32
marine mammal 57, **73**
marjoram 32
Markov chain 209
Markov, Andrey 209
marmoset 57
marmot 57
Marmota 57
marsupial 11, 56, **57**, 58, 59
marsupium a pouch possessed by female marsupials.
Martin, A.J.P. 209
maser 194, 195, 216
mass **76**, **78**, **79**, 93, 227
mass defect 122
mass of particles 136
mastication the process of chewing food.
mathematical symbols 228
mathematics 8, **162-91**
matrices 176
matter **137**, 204, 207, 209, 211, 213, 214, 216
Mauchly, John William 209
Maupertius, Pierre-Louis de 209
maxilla 61
Maxwell's demon 209
Maxwell's equations 209
Maxwell's theory 112
Maxwell, James Clerk 94, 112, 209
Mayer, Maria Goeppert 209
Maynard Smith, John 210
McClintock, Barbara 209
McMillan, Edwin Mattison 209
meatus the passage leading from the outer ear to the eardrum.
mechanical advantage 86
mechanical strength (metals) 146
mechanical waves 97
mechanics 8, **74-85**, **174-75**

INDEX/FACTFINDER

medial cuneiform 60
median 184
median eye a third eye in the top of the head found in many invertebrates.
Medicine or Physiology, Nobel Prizes 222-24
medulla oblongata 68, **69**
medulla the central part of an organ.
medusa 37
Megalonychidae **57**, 59
meiosis 24, **26**
Meitner, Lise 210
meitnerium the provisional name for element 109, unnilennium
melamine 158
melanin a pigment found in skin, hair, etc.
melting point 131-35, **138**, 139, 141, 142, 146, **149**
member (sets) **169**
membrane 12-13
memory 68
Mendel's laws of inheritance 14-15
Mendel, Gregor 11, **14**, 15, 210
mendelevium 129, **135**
Mendeleyev, Dmitri 122, **130**, 210
Mendelism 11
meninges the membranes surrounding the central nervous system of vertebrates and the spaces within it.
mercury 84, 88, 114, 129, **134**
meristem **21**, 32
Meropidae 52
Merrifield, Robert Bruce 210
Meselson, Matthew Stanley 210
mesocarp the middle layer of the pericarp of an angiosperm fruit - absent in some species.
mesoglea 37
meson **122**, 193, 217
mesophyte any plant with no adaptations to environmental extremes.
mesosome 18
messenger-RNA 15
metabolism the sum total of the chemical reactions in the body by which nutrients are converted to energy, tissues are renewed, replaced and regenerated and waste products are broken down.
metacarpal 60
metal oxide 144
metalloids **146**
metals 124, **139**, 141, **146**, 147, 161, 212, 214, 217
metameric segmentation the division of the body into a number of similar segments along its length.
metamorphosis 42, **43**, 48, **49**
metatarsal 60
metatarsus 60
Metatheria 57
Metchnikoff, Elie 210
methanal 152, **155**
methane 139, **140**, **150**, **155**
methanol 152, **155**
methoxymethane 154
methyl methacrylate 158
metre **227**
metric and Imperial conversions **229**
metric system **227**
metrology **227**
Meyer, Lothar 210

Meyer, Viktor 210
Michelson, A.A. 210
microbiology 208
microchip 211
Microchiroptera 55
microflora small plants found in a given area.
micronutrient a chemical element required in small quantities, i.e. a trace element.
microorganism 212, 217
microphyllous leaf **26**
microscope **106**, 210, 216
Microtus 59
microwaves 113
midbrain 68, **69**
midrib the vein running down the middle of a leaf.
midwife frog 48
Miescher, Johann Friedrich 210
migration 49
mildew a fungal disease of plants in which the fungus is seen on the plant surface.
milk 54, **57**, 66, **67**
milk teeth 54
Miller, Stanley 210
Millikan, Robert Andrews 210
millipede **41**
Milne, Edward Arthur 210
mimicry the ability of one animal to resemble another, usually for protection.
mimosa **32**
mineral salts 66, **67**
mineralogy 192
minerals **124**
Minkowski, Hermann 210
Minsky, Martin Lee 210
mint **32**
mirrors **106-07**
miscarriage an accidental abortion.
miscible 155
mite **41**, 45
mitochondria **13**, 23
mitosis 15, **21**
mitral valve the heart valve of the higher vertebrates.
mixed number **164**
mixture **124**, 125-27
model **166**
modulation **100-101**
molar 54
molar quantity **144**
mole (chemistry) **144**, 145
mole (biology) **57**, 59
mole (measures) **227**
mole concept **144**
mole cricket **43**
mole rat 57
molecular formula **155**
molecule 41, **124**, 140, 143, 144, **215**
molecule, large **142**, **215**
molecules, small 141, **150-51**
mollusc **40**, 46, 62, 63, 64
Mollusca **40**
molybdenum 128, **132**
moment 81, **86**
moment of inertia, derived unit of **227**

momentum **76**, **174**
momentum, derived unit of **227**
Monera **16**, **18**
Monge, Gaspard 210
mongoose **57**, 59
monitor 51
monkey 55, 56, **57**, 58
monoatomic molecule **138**
monochromatic beam **104**
monocotyledon **32**
Monod, Jacques 210
monoecious female and male reproductive parts in separate floral structures on the same plant.
monomer **157**, **158**, **159**
monophyletic group **16**
monopodial branching the condition in which secondary shoots or branches arise behind the main growing tip and remain subsidiary to the main stem.
monosaccharide 66, **67**
monotreme 54, 56, **57**, 58, 59
Montagnier, Jacques 210
Moon, the **75**, 198
moonrat 56, **57**
moose 56, **57**
Morgagni, Giovanni 210
Morgan, Thomas Hunt 210
morphology that branch of biology concerned with the form and structure of organisms.
Moschidae 57
Moseley's law 210
Moseley, Henry 210
moss 20, **21**, **26**
Mossbauer effect 210
Mossbauer, Rudolph Ludwig 210
moth **42**
motile organism **18**, 24
motion **74-78**
motor nerve 68, **69**
mould a fungus that produces a velvety growth on the surface of its host.
moulting **41**, 44, **51**
mouse **57**, 59
mouth 54, 65, **67**
mucous membrane a surface membrane that secretes mucus.
mucus a slimy protective secretion that does not dissolve in water.
Mueller, Erwin Wilhelm 210
Muller, Hermann Joseph 210
Muller, Johannes 211
Muller, Paul 211
Mulliken, Robert Sanderson 211
multicellular **13**
multiple **162**
multiple fruit fleshy fruit incorporating the ovaries of many flowers and derived from a complete inflorescence.
multiplication **163**
mumetal **111**
Muridae 56, **57**, 59
Musci **26**
muscle a contractile tissue that produces movement in invertebrates and vertebrates.
muscle cell 12

245

muscular contractions 216
mushroom 24
musk deer 57
Musophagiformes 53
Musschenbroek, Pieter van 211
mussel 40
mustard 32
mustelid 55, 57, 58, 59
Mustelidae 57
mutagenesis 15, 215
mutation 10, 11, 15, 195
MVK 155
mycelium 24
mycoplasm 18
Mycoplasma 18
mycorrhizal 24
myelin sheath a membranous sheath around nerve fibres.
Myrataceae 32
Myremecobiidae 57
Myremecophagidae 54, 57
myriapod 41
Mysticeti 73
Mytilidae 40
myxomycetes 18

n-type 117, 118
n-type doping 118
n-type semiconductor 118
Nageli, Karl Wilhelm von 211
nail 58
Naja 50
Nambu, Yoichipo 211
Napier's bones 211
Napier, John 211
nasal cavity the cavity in the head of vertebrates containing the organs of smell.
nastic movement a plant response caused by an external stimulus.
natural classification 16
natural compounds 152-55
natural frequency 101
natural number 163
natural polymer 158
natural selection 10, 11
navicular 60
Nearctic faunal region 35
nectar 32
nectary 32
negative charge 110, 111
negative electrode 115
negative gradient 171
negative index 165
negative terminal 114
Neher, Erwin 211
nematocysts 37
Nematoda 39
nematode 39
neodymium 128, 133
neon 114, 129, 130, 131, 138, 141, 213
neon lamp 115
neotony 48, 49
Neotropical faunal region 35
Neotropical floral region 21

nephridium an excretory organ in invertebrates.
Neptune 192
neptunium 128, 135, 209
Nernst, Walter Hermann 211
nerve cell 69, 70, 192
nerve fibre 201, 206
nerve net 69
nervous system 68, 69, 203, 205, 215
nettles 32
neural system 68, 69
neurology 194
neuromodulator 69
neurone 69, 70
neurosecretory cell 69
neurotransmitters 69
neutral point 111
neutralization 145
neutrino 122, 199, 208
neutron 120, 121, 122, 123, 130, 197
Newcomb, Simon 211
Newlands, John 130, 211
newt 48, 49
newton 77, 78, 227
newton metre 88
Newton's first law of motion 78
Newton's law of gravitation 76, 211
Newton's law of gravitational force 75
Newton's laws of motion 77, 78, 174, 211
Newton's second law of motion 74, 78
Newton's third law of motion 78, 113
Newton, Sir Isaac 10, 74, 76, 93, 113, 211
Newtonian mechanics 74-78
newtonian fluid 85
niacin 67
nickel 129, 132
nicotinic acid 67
nictating membrane 49
nidiculous any bird that hatches in an undeveloped state and is unable to fend for itself.
nidifugous any bird that hatches in a well-developed state and is soon able to fend for itself.
nielsbohrium the provisional name for element 107, unnilseptium
nightjar 53
nightshade 32
niobium 128, 132, 161
nit the egg of a louse.
nitrate 21, 22, 142
nitrogen 20, 21, 124, 129, 131, 140, 141, 143, 151, 153, 158, 204
nitrogen dioxide 151
nitrogen fixation 21
nitrogen monoxide 150-51
NMA 201
Nobel Prizes 218-24
Nobel, Alfred 211
nobelium 129, 135
noble gas 114, 124, 130, 138, 140, 141, 147, 213
nocturnal birds 52
node 101, 103
Noether, Emily 211
Noguchi, Hideyo 211
nonagon 182
noradrenaline 69

normal (optics) 104
normal contact force 82
normal waves 99
north pole 110, 111, 113
north-seeking pole 111
nostril 57
notochord a form of primitive cartilaginous spinal column.
nova star 192
Noyce, Robert 211
nuclear accelerator 122, 123
nuclear fission 121, 203, 204, 210, 216
nuclear fusion 121, 202
nuclear particle 122
nuclear physics 8, 120-23
nuclear power 121
nuclear reaction 122
nuclear reactor 121
nuclear structure 122
nuclear weapons 212
nucleic acid 12, 210
nucleocapsid 18, 19
nucleoid 13
nucleon 120, 122
nucleophilic 154, 155
nucleotide 14, 15
nucleotide base 15
nucleus (cell) 13
nucleus 120, 122, 123, 197, 207
nudibranch 40
Nudibranchia 40
null set 169
numbat 57
number patterns 167
numbers 162-63
numerator 164
numerical data 177
nutrients 62
nutrition 87
Nuttall, Thomas 211
nylon 156, 157, 158
nymph an immature form of some insects, in which the wings and reproductive organs are not fully developed.
Nymphaeaceae 32

oak 32
Oberth, Hermann Julius 211
object beam 108
objective 106
occelot 57
occipital 61
occiput 60
ocellus 43, 71
Ochotonidae 57, 58
octagon 182
octahedron 190, 191
octane 155
octocorals 37
Octopoda 40
octopus 40
odd number 163
odd-toed ungulates 58
odds 168

INDEX/FACTFINDER

Odobenidae 73
Odontoceti 73
Oersted, Hans Christiaan 112, 211
oesophagus 65, 66, **67**
oestradiol 68, **69**
oestrogen 72
oestrous cycle the reproductive cycle of female mammals.
oestrus the period of the oestrous cycle.
ohm 115, 118, 211, **227**
Ohm's law **115**, 211
Ohm, Georg Simon 115, **211**
oil, crude 125
oils 124
okapi 58
Olbers, Heinrich 211
Oleaceae 32
olfaction 71
olfactory organs organs of smell.
Oligochaeta 39
olive 32
omasum the third chamber of a ruminant's stomach.
omega-particle 123
omnivore 58
Onchychopora 41
oncogene 216
onion 32
Onnes, Johannes Kammerlingh 118, 161
ontogeny the changes that occur during the life cycle of an organism.
Oomycota 18
Ooort, Jan Hendrik 211
Oparin, Aleksandr 211
open circulatory system 63
operculum the muscular flap covering the gills in bony fish.
Ophidia 51
Ophiophagus 50
Ophiuroidea 38
Opiliones 45
opossum 57, **58**
Oppenheimer, J. Robert 211
optic chiasma the point at which the optic nerves cross over between the vertebrate eyes and brain.
optic nerve the nerve connecting the vertebrate eye with the brain.
optical activity 104
optical isomers 104
optical reflection and refraction **104-05**
optics 8, **104-08**, 215
orang-utan 58
orange 32
orbit **122**, 123
orchid 32
order 16
ordered pair (of numbers) 171
organ (human) 60
organelle 12, **13**
organic chemistry 8, **152-55**
Oriental faunal region 35
origin (of graph) 170, **171**
Ornithischia 51
ornithophily pollination by birds.
orthodox classification 16

oryx 54, **58**
oscillatory change 97
osculum 36
osmium 128, **134**
osmosis 21, **23**
osprey 52
ossicle 54, **60**
ossification the process by which bone is formed from cartilaginous or other tissue.
Osteichthyes 46
Ostreidae 40
ostrich 53
Ostwald, Wilhelm Friedrich 212
Ostwald, Wolfgang 212
Otariidae 73
otter 57
Oughtred, William 212
ovary 32, 69, **72**
oviparous any female animal that lays eggs within which the embryo develops.
ovipositor 43
Ovis 59
ovoviviparous 51
ovulation the release of an ovum from the ovary.
ovule 32
ovuliferous scale 27
ovum 12, **32**, 36, 72
Owen, Sir Richard 212
owl 52, **53**
oxidation 130, **145**
oxidation state 130
oxides 1146, 47, 148
oxone 140
oxonium oxide 144
oxygen 20, 18, 21, 22, 23, 42, 46, 62, 63, 64, **65**, 66, 73, 124, 129, **131**, 140, 141, 142, **143**, 144, 145, 146, 147, 148 , **151**, 153, **158**, 208, 213, 214
oxytocin a hormone produced by the pituitary gland. It stimulates contractions of the uterus during labour.
oyster 40
ozone 151

p sub-level 130
p-n-p transistor 119
p-type 117, **118**
p-type doping 118
p-type semiconductor 118
paddlefish 47
paddleworm 39
Paeoniaceae 32
Paget, Sir James 212
Palaearctic faunal region 35
Palaemon 44
Palaeotropical floral region 21
palate the roof of the mouth.
palea the upper bract of the pair found beneath each floret in a grass inflorescence.
palladium 129, **133**, 217
Pallas, Peter Simon 212
palm 32
Palmae 32
palp 43
Pan 55
pancreas 67, **69**, 195

panda (giant) 58
pangolin 58
panicle an inflorescence in which the flowers are formed on stalks arising spirally or alternately from the main stem.
pansy 32
panther 57, **58**
Papaveraceae 32
Papio 55
Pappus of Alexandria 212
parabola **187**
parabronchi 65
Paracelsus 212
paraffin 148
parallel connection 115
parallel evolution 11
parallelogram **182**
paramagnetic 151
Paramecium 18
parapodia 39
parasite 10, 18, **19**, 40, 41
parasympathetic nervous system part of the autonomic nervous system in vertebrates.
parathyroids a group of small endocrine glands associated with the thyroid gland.
Pardee, Arthur Reid 212
parietal 61
parietal bone **60**
parrot 53
parsley 32
parthenocarpy the production of a fruit without the process of fertilization.
parthenogenesis 72
particle (Newtonian) 78
particle size 124
particle, nuclear 120, 121, **122**, 123
particles (behaviour of) 93
particles (movement of) 123
partridge 52, **53**
pascal **85**, **227**
Pascal's theorem 212
Pascal's triangle **167**
Pascal, Blaise 167, **212**
Passeriformes 53
passerine bird 52, **53**
Pasteur, Louis 212
patella the kneecap.
pathology 210, 212
Pauli, Wolfgang 212
Pauling, Linus 212
Pavlov, Ivan 212
pea 14, **32**, 67
Peano, Giuseppe 212
peanut 32, 66
pear 32
pearl 40
Pearl, Raymond 212
Pearson, Karl 212
peccary 54, **58**, 59
pectoral fin 46
pectoral girdle 60
pectoral pertaining to the chest.
Pederson, Charles 212
pedicel a stalk attaching flowers to the main stem of the inflorescence.

pedicellarium 38
pelagic any animal inhabiting the open waters of the sea or a lake.
Pelecanidae 53
pelican 52, 53
Peltier's effect 212
Peltier, Jean-Charles 212
pelvic fins the rear pair of lateral fins in fishes.
pelvic girdle 60, 61
pelvic patch 49
pelvic seat 49
pelvis 60
penguin 52, 53
penicillin 198, 202, 203, 206
penis 72
Penney, William George 212
Penrose, Roger 212
pentadactyl limb the characteristic limb of the vertebrates, with five digits or 'fingers'.
pentagon 182
Penzias, Arno Allan 212
peony 32
pepo any berry with a hard exterior.
pepsin 66, 67
peptic ulcer an ulcer of the stomach or duodenum.
peptide 67, 202, 210
peptide 152
peptide 160
Peramelidae 58
percentage 164
percentages 164
perception 70-71
perching bird 53
perennating organ 32
perennial 32
perfect elasticity 82
perfect number 163
perfect plasticity 82
perianth the protective structure encircling the reproductive parts, consisting of the calyx and corolla or a ring of petals.
pericarp the wall of the fruit, derived from the ovary wall.
periderm the protective secondary tissue replacing the epidermis as the outer cellular layer of stems and roots.
perimeter 187
period (waves) 96, 97
periodic oscillation 97
periodic table 128-30, 138, 139, 146, 147, 210
peripheral nervous system 69
Perissiodactyla 58
perissodactyl 55, 58, 59
peristalsis 62, 63
peritoneum the membrane lining the abdominal cavity.
periwinkle 40
Perkin, Sir William 212
Perl, Martin Lewis 213
Perrin, Jean 213
perspex 158
Perutz, Max 213
petal 32
petiole 21
petrel 52

Petrogale 58, 59
Pfeffer, Wilhelm 213
pH scale 147
phalange 60
phalanger 58
phalanx 60
pharynx 64, 65
phase (of light) 108
phase change 90
phase speed 97
Phasianidae 52, 53
phasic response 71
phasitonic response 71
pheasant 52, 53
phenetic classification 17
phenol 21
phenylethene 158
pheromone 71
phloem 20, 21, 26, 32
phlogiston 215
phlox 32
Phocidae 73
Phocoenidae 73
Phoenicopteridae 52
Pholidota 58
phosphate 67, 142
phosphorus 129, 131, 153
photic zone the surface waters of lakes and seas, in which light penetrates and which is inhabited by plankton.
photodiode 118
photon 93, 95, 108, 118, 201
photoperiodism 21
photopigment 71
photoreception 71
photoreceptor 70
photorespiration respiration that occurs in plants in the light.
photosynthesis 18, 20, 21, 22, 23, 200, 214
phototransistor 118
phototropism 22
photovoltaic effect 118
phycobiont an algal partner in a lichen.
phyllody the transformation of parts of a flower into leaflike structures.
phylogentic classification 17
phylum 16, 17
physical chemistry 8, 136-60
physics 8, 74-123
Physics, Nobel Prizes 220-22
Physiology or Medicine, Nobel Prizes 222-24
physiology 8, 18-73
phytogeographical region 22
Phytophthora infestans 18
pi 187, 197, 207, 209
Picard, Emile 213
Piccard, Auguste 213
Picea 27
Picidae 53
Piciformes 53
pictogram 177, 178
pictograph 178
pie chart 177, 178
pie graph 178

pig 54, 58, 59
pigeon 52, 53
pika 57, 58
pileus the cap of a mushroom or toadstool.
pinacocytes 36
pine 27
pink 32
pinna (in plants) a first-order leaflet in a compound leaf or (in mammals) the outermost part of the outer ear in some mammals.
pinniped 73
Pinus 27
pipe corals 37
Pisano, Leonardo see Fibonacci
pisiform 60
pistil a single carpel or group of carpels.
pit viper 51
pitch 53, 102-03
pith an area of parenchyma in the centre of many plant stems.
pituitary gland 68, 69
pivotal joint 54, 60
placenta the series of membranes within the uterus of viviparous animals that nourishes the developing foetus.
planarian 39
Planck constant 123
Planck's constant 94
Planck, Max 10, 94, 123, 213
plane (tree) 32
plane 171, 186, 187
plane figure 180
plane mirror 106
plane wavefront 99
planet 120, 195, 200, 208, 214
plankton 46, 73
plantae 16, 20, 21, 23
plants 16, 20-23, 67, 192, 205, 209, 211, 213, 214
plasma 62, 63, 65
plasma physics 192
plasmamembrane 12, 13
plasmid 18
plastic 142, 158
plastic flow 83
plastid 13
plastron 51
Platanaceae 32
plate joint 60
platelet a small particle found in blood. Platelets are involved in the clotting mechanism.
platinum 76, 129, 134
Plato 213
Platyhelminthes 39
platypus 56, 58
Plethodontidae 48, 49
pleura a double membrane surrounding the lungs.
Pleuronectiformes 46
plexus a network of nerve cells.
Pliny (the Elder) 213
plum 33
plumage 53
plumule 33
plutonium 121, 128, 135, 146
Poaceae 33
Podicipedidae 52

INDEX/FACTFINDER

poikilothermic **46, 51**
Poincare, Henri **213**
point-contact transistor **119**
points of incidence **104**
Polanyi, John Charles **213**
polarizing **114, 115**
Polemoniaceae **33**
polio **201, 214**
pollen **33**
pollen sac the chambers on the anther in which pollen is formed.
pollen tube **33**
pollex the inner digit on the forelimbs of the higher vertebrates. It is often vestigial, or may be adapted for a variety of purposes.
pollination **33**
pollinator **33**
pollutant **136**
polonium **129, 134, 199**
polyamide **156, 158**
Polychaeta **39**
polychaetes **39**
polychloroethene **158**
polyester **158, 160**
polyethene **156, 158**
polygon **180, 182-83**
Polygonaceae **33**
polyhedra **180, 190-91**
polymer **13, 15, 21, 22, 67, 142, 152, 156-60, 217**
polymerization **156-60**
polymethyl methacrylate **159**
polyp **37**
polypeptide **13, 22, 67, 159**
polyphenylethene **159**
polypropene **159**
polysaccharide **12, 13, 20, 22, 23, 66, 67, 159**
polystyrene **159**
polytetrafluoroethene **159**
polythene **158, 159, 160**
polyunsaturated animal or vegetable fats consisting of long carbon chains with many double bonds.
pome a fleshy pseudocarp in which tissues develop from the receptacle and enclose the true fruit.
pondweed **33**
poplar **33**
Popov, Aleksandr **213**
poppy **33**
porcupine **58**
pore cell **36**
Porifera **36**
porocyte **36**
porpoise **73**
Porter, George **213**
Portuguese man-of-war **37**
positive charge **110, 111**
positive electrode **115**
positive gradient **171**
positive hole **117, 118**
positive terminal **114**
positron **122, 195, 217**
possum **57, 58**
Potamogetonaceae **33**
potassium **23, 128, 131, 142, 144, 145, 147, 148**
potassium chloride **147, 148**

potassium hydroxide **144, 145, 148**
potassium manganate **136**
potato **33, 67**
potato blight **18**
potential difference **111, 114, 115**
potential difference, derived unit of **227**
potential energy **88, 174**
power (electric) **115**
power (mathematics) **163**
power (multiplication and division) **165**
power, derived unit of **227**
Poyning, John Henry **213**
Poynting's vector **213**
prairie dog **58, 59**
Prandtl, Ludwig **213**
praseodymium **128, 133**
prawn **44**
precipitation **127**
precipitation reactions **145**
preening **52, 53**
premolar **54**
Presbytis **57**
pressure **84-85, 92, 156**
pressure, derived unit of **227**
Priestley, Joseph **213**
primary growth the increase in size as a result of cell division at apical meristems.
primary meristem **21, 22**
primary phloem **22**
primary xylem **22**
primate **16, 55, 57, 57, 58**
prime number **163, 198**
primitive ungulates **58**
primrose **33**
primula **33**
principal focus **106**
principle of indeterminism **95**
principle of least action **209**
prism (solid) **188, 189**
prism **104, 211**
probability **167, 168, 198, 202**
probability scale **168**
proboscide **58**
proboscidean **58, 73**
Procellaridae **52**
Procolobus **55**
Procyon **59**
Procyonidae **55, 57, 59**
product (chemistry) **145**
product (mathematics) **163**
progesterone a hormone produced in the ovaries. It acts on the uterus to prepare to receive fertilized ovum.
projectiles **174**
Prokaryota **16, 18**
prokaryote **12, 16, 18**
prokaryote cell **13**
prolactin a hormone produced by the pituitary that stimulates the breasts to secrete milk.
promethium **128, 133, 146**
pronghorn **59**
propane **155**
propanol **155**
propanone **155**

propene **155, 160**
proper fraction **164**
properties of lenses **106-07**
properties of triangles **184-85**
proportion **166**
prostate gland a gland found only in males. It secretes part of the seminal fluid into the urethra.
protactinium **128, 134, 204**
protein **12, 13, 14, 15, 21, 22, 40, 41, 52, 66, 67, 125, 139, 142, 152, 160, 193, 200, 204, 208**
protist **19**
Protista **16, 18, 19**
Protoctista **16, 18, 19**
proton **23, 112, 120, 121, 122, 123, 128, 129, 130, 145**
protoplasm **13**
Prototheria **59**
prototherian **59**
protozoa **18, 19**
proturans **42**
Proust, Joseph Louis **213**
Przewalski's horse **56**
pseudocarp any fruit consisting of tissues other than those derived from the gynoecium.
pseudopodium **19**
Pseudotsuga **27**
psi **214**
Psittaciformes **53**
Pterygota **43**
pterygotes **43**
PTFE **160**
Ptolemy, Claudius **213**
pubis **60**
puerperal pertaining to childbirth.
puffbird **52**
pulley **86**
pulmonary pertaining to the lungs.
Pulmonata **40**
pulsar **195, 216**
pulses **67**
pupa **42, 43**
pupil **70, 71**
PVC **157, 158, 160**
pygmy hippopotamus **56**
pyloric sphincter **67**
pyramid **188, 189**
pyridoxine **67**
Pythagoras **185, 213**
Pythagoras' theorem **180, 185, 213**
python **51**

quadrilateral **182**
quadruple bond **142**
quanta **94, 122, 213**
quantity of heat, derived unit of **227**
quantity of motion **78**
quantum **94, 122**
quantum field **193**
quantum mechanics **8, 93-95, 123, 200, 201, 205, 212, 215, 217**
quantum theory **93-95, 123, 195, 196, 213, 216**
quark **122, 123, 202, 211, 214**
quartz **141**
quasar **197, 214**

Quetelet, Adolphe 213
quill 52, 53
quinine 217
quotient 163

rabbit 57, 59
Rabi, Isidor Isaac 213
raceme an inflorescence in which flowers are formed on individual pedicels on the main axis.
rachis the shaft of a feather.
racoon 59
radar 217
radial symmetry the form of symmetry found in sedentary animals.
radian 83, 227
radiant energy 123
radiation 90, 121, 123, 194, 196, 204
radiation absorbed dose, derived unit of 227
radiation activity, derived unit of 227
radical 142, 155, 160
radio 114, 209, 213
radio transmission 100
radio waves 100, 113
radio-carbon dating 208
radioactive decay 121
radioactivity 123, 199, 207
radium 128, 134, 147, 199
radius (mathematics) 186, 187
radius 60
radon 129, 134, 138
radula 40
ragworm 39
rail 52
rainbow 104
Raman effect 213
Raman spectroscopy 213
Raman, Chandrasekhara Venkata 213
Ramanujan, Srinivasa 213
Ramsay, Sir William 213, 218
Rana temporaria 48, 49
Ranidae 48, 49
Ranunculaceae 33
rarefaction 97, 103
raspberry 33, 66
rat 59
rate of diffusion 136
rate of reaction 124
ratio 166
rational number 163, 164
ray (fish) 46
ray (light) 104
Ray, John 16, 213
ray-finned fish 47
Rayleigh, Lord 213
reactant 145
reaction equilibrium 145
reaction techniques 127
reactions, of halogens 149
reactive metals 148
real image 107
Reamur temperature scale 214
Reamur, Rene-Antoine de 214
rear focal point 107
receptacle 33

recessive gene 15
rectangle 182
rectangular numbers 167
rectifier 118
rectum 66, 67
recurring 164
red blood cell 63, 67
red deer 56
redox reaction 145
reduction 145
reed 33
Reed, Walter 214
reference beam 108
reflection (of a figure) 180
reflection 98-99, 104
reflection of sound 101
reflex 212
reflex action 69
reflux 127
refraction 98-99, 105
refraction of sound 103
refractive index 98, 105
regeneration the regrowth or replacement of tissues and body parts lost owing to injury.
Regiomontanus 214
regions (mathematics) 190
Regnault, Henri-Victor de 214
regular dodecahedron 191
regular icosahedron 191
regular octahedron 191
regular polyhedra 190, 191
regular tetrahedron 191
reindeer 55, 56
relative complement 169
relative frequency 168
relative molecular mass 145
relative velocity 175
relativity 93-95, 123, 201, 210, 216, 217
relativity theory 93-95
remainder 163
renal pertaining to the kidneys.
replication 15
repressor molecule 212
reproduction (echinoderms) 38
reproduction (insects) 43
reproduction 68, 72
reptile 50-51, 57, 65
repulsion 111
resistance 114, 115
resistivity 118
resonance particles 192
resonant frequency 101
respiration 64-65, 67, 68
respiration in air 65
respiration in water 65
respiration, animals 22
respiration, insects 43
respiration, plants 22, 23, 195
respiration, reptiles 65
respiratory pigment 62, 63, 65
respiratory systems 64-65
rest energy 119
retia mirabilia 73
retina 71

retinol 67
retrosynthetic analysis 155
retrovirus 19
Reynold's number 85
rhea 53
Rheidae 53
rhenium 128, 134
rhesus monkey 57, 59
rheumatic fever 19
rhinoceros 58, 59
Rhinocerotidae 58, 59
Rhizobium 20, 21, 23
rhizoid 26
rhizome 33
rhodium 128, 133, 217
rhododendron 33
rhombus 183
Rhynchocephalia 51
rib 61
riboflavin 67
ribonucleic acid see RNA
ribosome 13, 15
Ricci-Curbastro, Gregorio 214
Richter, Burton 214
Riemann, Bernhard 214
right angle 182, 183, 184
rigidity modulus 83
RMM 145
RMS 117
RNA 12, 13, 15, 19, 192, 193, 194, 210
Roberts, Richard 214
rod (eyes) a light-sensitive cell in the retina of vertebrate eyes. or (plants) a section of a plant - usually underground - that is involved with fixing the plant in position and absorbing water and nutrients.
rodent 54, 55, 56, 57, 58, 59
roe deer 56
roller 52
Romer, Alfred Sherwood 214
Romer, Ole 214
Rontgen, Wilhelm Conrad 214
root 26, 33
root mean square values see RMS
Rosaceae 33
rose 33
Rose, William Cumming 214
Rosidae 33
rotation (of a figure) 180
rotation 79-81
rotational symmetry 180
roundworm 39
Rowland, Sherry 214
rubber 82
rubidium 128, 132, 146, 147, 148
rumen 54
Rumford, Count 214
ruminant 54, 55, 59
runner a creeping stem arising from an axillary bud, giving rise to new plants at the nodes.
Russell's paradox 214
Russell, Lord Bertrand 214
Russulales 24
rust 24

INDEX/FACTFINDER

Rutaceae **33**
ruthenium **128, 132**
Rutherford electron **214**
Rutherford model **120**
Rutherford, Ernest **120, 123, 214**
rutherfordium   the provisional name for element 104, unnilquadium
Rydberg constant **214**
Rydberg, Johannes Robert **214**

s sub-level **130**
Sabin, Albert **214**
Saccharomyces **24**
Sachs, Julius von **214**
sacral vertebrae **61**
sacrum **60, 61**
saddle joint **60**
Sagan, Carl Edward **214**
Sakharov, Andrei **214**
salamander **48, 49**
Salamandridae **49**
Saliceae **33**
salinity **103**
saliva   a secretion of mucus and enzymes which moistens the food and starts off the process of digestion.
salivary gland **68**
Salk, Jonas Edward **214**
Salmonella **19**
salt **139, 145, 148**
samarium **128, 133**
sand hopper **44**
sandgrouse **52**
sap   a liquid containing mineral salts and sugars dissolved in water, found in xylem and phloem vessels.
saprophyte **18, 19**
saprotroph **19, 24**
sapwood   the outer functional part of the secondary xylem.
Sarcina **19**
Sarcoptidae **45**
Sarich, Vincent **214**
saturated hydrocarbon **156**
Saturn **200**
Saurischia **51**
Sauveur, Albert **214**
Savart, Felix **112**
saxifrage **33**
scalar quantity **77**
scale (mathematics) **166**
scale **58**
scalene triangle **185**
scales **47**
scaly anteater **58**
scandium **128, 132**
scape   a leafless stem of a solitary flower or inflorescence.
scapula **60, 61**
scarlet fever **19**
scent **33**
scent gland **57**
Scheele, Karl Wilhelm **214**
schizocarp   a dry fruit formed from two or more one-seeded carpels that divide into one-seeded units when mature.

Schmidt, Bernhard Voldemar **214**
Schmidt, Maarten **214**
Schrodinger, Edwin **214**
Schuster, Sir Arthur **215**
Schwann, Theodor **215**
Schwinger, Julian Seymour **215**
Scincidae **51**
scion   a shoot or bud cut from one plant and grafted or budded on to another.
Sciuridae **59**
sclerenchyma   a strengthening tissue composed of dead cells.
Sclerospongia **36**
scolex   the head of a tapeworm, which anchors it to the intestinal wall of the host.
scoliosis   the curvature of the spine sideways.
scorpion **41, 45**
screen **109**
scrotal sac **72**
Scyphozoa **37**
sea anemone **37**
sea cow **73**
sea cucumber **38**
sea fan **37**
sea lily **38**
sea pen **37**
sea slug **40**
sea snake **51**
sea squirt **17**
sea urchin **38**
sea whip **37**
Seaborg, Glenn Theodor **215**
seaborgium   the provisional name for element 106, unnilhexium
seal **73**
seaweed   a group of large algae found in the littoral zone and floating freely in the sea.
sebum   a greasy material produced by the sebaceous glands in the skin of mammals. It greases the hair and protects the skin.
second **227**
second law of thermodynamics **88, 89**
secondary growth   the increase in diameter of a plant organ as a result of cell division in the cambium.
secondary meristem **21, 23**
secondary palate **50, 54**
secondary phloem **23**
secondary xylem **23**
secretary bird **52**
secretin **194**
sector **187**
seed **33**
seed dispersal **33**
seedling   a young plant.
segment **187**
segmented worm **39**
Seki, Kowa **215**
Selachii **47**
selenium **129, 132**
self-pollination   the transfer of pollen to the stigma of the same flower or flowers on the same plant.
semiconductor **117, 118**
semiconductor diode **118**

seminal roots   roots growing from the base of the stem and taking over from the radicle during early seedling growth.
sensation **71**
sense organ **71**
senses **70-71**
sensilla **71**
sensory nerves **69**
sensory perception **71**
sensory signals **71**
sepal **34**
separation of mixtures **125-27**
Sepia **40**
series connection **115**
serotonin **69**
serpent star **38**
Serpentes **51**
serum   a straw-coloured fluid that separates from the blood as it clots.
sessile **36**
sets **169, 197, 217**
sex cell **15, 72**
sexual reproduction **12, 25, 36, 40, 72, 192**
Shapley, Harlow **215**
shark **46, 47, 63**
Sharp, Phillip **215**
shear modulus **83**
shearwater **52**
sheep **54, 55, 59**
shell (electron) **123, 141**
shell **40, 51**
Sherrington, Sir Charles **215**
Shockley, William **215**
shoulder **60**
shoulder blade **61**
shoulder girdle **60, 61**
shrew **57, 59**
shrimp **44**
shrub **23, 34**
Shwarzschild, Karl **215**
SI multiples and submultiples **227**
SI units **227, 228**
sibling   one of a number of offspring of the same two parents.
Sidgwick, Nevil **215**
siemens **215**
siemens **227**
Siemens, Charles William **215**
Siemens, Werner von **215**
sika deer **56**
silica **140**
silicon **129, 131**
silicon dioxide **141**
silicula   a broad dry dehiscent fruit developed from the fusion of two carpels.
silver **115, 129, 133, 142**
silver nitrate **145**
simple distillation **126**
simple harmonic motion **175**
sine **180**
sinew   a tendon or ligament.
singing organ (birds) **53**
single circulatory system **63**
singular matrix **176**

sink 88
sinus a hollow cavity opening off a passageway, e.g. the nasal sinuses opening off the nose.
Siphonophora 37
siphonophore 37
siren 49, 58
sirenian 73
Sirenidae 49
skeleton (human) 60-61, 195
skeleton (sponges) 36
skin 49, 67
skink 51
Skolem, Thoralf Albert 215
Skolen paradox 215
skull 60, 61
skunk 57, 59
slaked lime 148
slater 44
sliding friction 83
slime mould 19
Slipher, Vesto Melvin 215
Sloane, Sir Hans 215
sloth 55, 56, 57, 59
sloughing 51
slug 40, 49
small bowel 67
small intestine 67
small molecules 150-51
smell 71
smelting 146
Smith, Michael 215
Smith, William 215
snail 40, 49, 51
snake 50, 51
Snell's law 98, 105, 215
Snell, Willebrord 98, 215
snowdrop 34
Soddy, Frederick 215
sodium 67, 114, 128, 131, 141, 142, 144, 145, 146, 147, 148
sodium chloride 139, 145, 147, 148
sodium fluoride 141
sodium hydrogen sulphate 145
sodium hydroxide 144, 145, 147
sodium oxide 148
sodium sulphate 145
soil the surface layer of the Earth's crust, consisting of water, air, living organisms, dead and decaying organisms and mineral particles.
solar cell 118
solar constant 118
solar energy 118
solar heating 118
solenoid 110, 111
solid (state) 137
solids (area and volume) 188-89
solids 180
solute 125
solution 124, 125, 126
solvent 125
somatic cell 15
Somerville, Mary 215
sonar 208
songbird 53

sonic boom 103
Soricidae 59
sound 56, 100-101, 102-03
sound signal 101
sound wave 70, 97, 101, 103
source 119
South African floral region 23
south pole 110, 111
south-seeking pole 111
soybean 21
space exploration 217
Spallanzani, Lazzaro 215
special relativity 94-95
speciation 11
species 16, 17, 72
specific heat 90-91
specific heat capacity, derived unit 228
specific latent heat of fusion 91
specific latent heat of vaporization 91
specific latent heat, derived unit 228
spectroscopy 193, 205, 213
spectrum 71, 123, 105
speed 77
speed of light 105
speed of propagation 97
Spencer, Herbert 11
sperm 43, 72
spermatazoa 72
spermatophore 72
Sperry, Roger Wolcott 215
Spheniscidae 53
sphenoid 61
Sphenophyta 20, 23, 26
sphere 187, 189
spherical wavefront 99
sphincter a ring of muscle around an opening to a hollow or tubular organ.
spicule 36
spider 45, 65
spider crab 44
spider mite 45
spinach 34, 66
spinal column 61
spinal cord 69
spinneret 45
spiny anteater 56, 59
spiracle 43, 65
spirillium 19
spirochaetae 19
spleen an organ at the top of the abdominal cavity, responsible for white blood cell production, breakdown of red blood cells and some control of immunity.
split genes 215
sponge 36, 62, 64
spongin 36
spoonbill 52
sporangia 24, 26
spore 26
spore-bearing plants 25-26
sporophyte 26
sporopollenin 26
springtail 42
spruce 27

Squamata 51
square 183, 190
square metre 228
square number 163, 167
square root 163, 205
squid 40
squirrel 57, 58, 59
stable elements 128
Stahl, Georg Ernst 215
stalk 21, 23, 34
stamen 34
standing waves 100-101
Stanley, Wendell 215
stapes 54, 60, 61
star 192, 197, 202, 203, 205, 206, 208, 215
starch 22, 23, 66, 67, 142, 156, 158, 159, 160
starfish 38
Starling, Ernest Henry 215
states of matter 137
static equilibrium 81
static friction 83
statics 79-81, 195
stationary waves 101
statistics 8, 177-79
statistics 167, 177-79
steam engine 215, 217
steel 146, 195
Stefan, Josef 215
stele the vascular cylinder responsible for transport of water and solutes in the stems and roots of vascular plants.
stellar spectrum 102
stem 23, 34
stenosis a constriction or narrowing of a tube, e.g. part of the digestive tract.
Stephenson, George 215
stereochemistry 216
Stern, Otto 215
sternebrae 61
sternum 60, 61
Stevin, Simon 215
stigma 34
stimulated emission of light 108
sting 43
stirrup 61
stoat 59
stoichiometry 145
Stoke's law 215
Stokes, Sir George Gabriel 215
stolon a long branch that bends over and touches the ground, at which point a new plant may develop.
stoma see stomata
stomach 54, 66, 67
stomata 21, 22, 23, 26
Stoney, George Johnstone 215
stork 52, 53
straight line equation 171
strain 83
strange quark 123
stratigraphy 215
strawberry 34
Streptococcus 19
stress (physics) 83

INDEX/FACTFINDER

stress, derived unit of **228**
stridulation the production of sound by insects, usually by rubbing body parts together.
Strigiformes **53**
strong acid/base **145**
strong nuclear force 112, 122, **123**
strontium 128, **132, 148**
structural formulae **155**
Struthionidae **53**
Struthioniformes **53**
Strutt, John William **213**
sturgeon **47**
Sturgeon, William **215**
style **34**
subatomic particles **120-23**, 128, 129, 130
subcutaneous beneath the skin.
sublimation **127**
subphylum **17**
subset **169**
substitution reaction **155**
substrate molecules upon which enzymes act.
subtraction **163**
succulent plants **22**
succus entericus digestive secretions of the walls of the small intestine in vertebrates.
sucker a shoot that develops from the roots and forms its own root system.
sucrose 21, 22, **23**, 66, 67
sugar 12, 14, 15, 20, 21, **23**, 41, **160**, 202
Suidae 58, **59**
sulfur see sulphur
Sulidae **52**
sulphate **142**
sulphur (sulfur) 129, **131**, 140, 142, 146, 153
sulphur dioxide **151**
sulphur trioxide 144, **151**
sulphuric acid **144**
sum **163**
Sun, the 66, 186, 196, 198, 207
sunstar **38**
superconductivity **118**, **161**, 198, 207
superimposition principle **98**
supplementary units **228**
surface tension, derived unit of **228**
survival of the fittest **11**
Svedberg, Theodor **215**
Swammerdam, Jan **216**
swan 52, **53**
Swan, Sir Joseph Wilson **216**
swift 52, **53**
swim bladder **47**
swimming (fish) **47**
swimming **73**
sycamore **34**
Sydenham, Thomas **216**
Sylvester, James Joseph **216**
symbiont **19, 23**
symbiosis 18, **19, 23, 24**
symbols **167**
symmetry **180**, 182, **183, 184, 185**, 187
sympathetic nervous system part of the vertebrate autonomic nervous system.
synapse **69**
syncarpous a gynoecium with fused carpels.

synchotron **123**
synchrocyclotron **123**
syngenesious an androecium with fused anthers.
synthesizing **160**
synthetic products **161**
synthetic zeolites **161**
syphilis **211**
syrinx **53**
Szent-Gyorgi, Albert von **216**
Szilard, Leo **216**

table of values **171**
table sugar **66**
Tachyglossidae **59**
tadpole 48, **49**, 64
tail 48, **49**, 53, 60
Talbot, William Henry Fox **216**
tally chart 177, 178, **179**
Talpidae 57, **59**
talus **60**
tamarind **57**
tangent **181**, 186, **187**
tanning **41**
tantalum 128, **134**
tapeworm **39**
tapir 58, **59**
Tapiridae **58**
taproot **34**
tarnishing **146**
tarpan 56, **59**
tarsal **61**
tarsier **57**
tarsus 60, **61**
Tasmanian devil **59**
taste **71**
taste bud **71**
Tatum, Edward **216**
taxonomy **8**, **16-17**, 197, 213
Tayassuidae 58, **59**
Taylor, Joseph Hooton **216**
tears **60**
teat 54, **57**
technetium 128, **132**, 146
teeth 54, 56, **57**
teiid **51**
Teiidae **51**
telecommunications **108**
telegraphy 201, 205, 209, 215
teleost **47**
Teleostei **47**
telephone **194**, 201
telescope **107**, 196, 214
television 109, **114**
television set **109**
television tube **109**
Teller, Edward **216**
tellurium 129, **133**
telson the tail appendage found in some arthropods.
temperature 46, 51, 54, 70, 87, **88**, 90, 91, 92, 114, 115, 118, 124, 136, 137, 156, 161, 197, 200
temperature coefficient **115**
temperature comparisons **230**
temperature reception **71**
temperature, derived unit of **228**

temporal bone **61**
tendon a fibrous band connecting a muscle to a bone.
tendril **34**
tenrec 57, **59**
tentacles **37**, 40
terbium 129, **133**
termite 54, **56**
terrapin **51**
terylene **157, 160**
tesla **216**, **228**
Tesla, Nikola **216**
test **38**
testa **34**
testes 69, **72**
testosterone 68, **69**
Testudinidae **51**
tetrafluoroethene **159, 160**
tetrahedron **189, 190, 191**
Tetranychidae **45**
tetrapod any vertebrate having four limbs.
thallium 129, **134**
thallus a plant body undifferentiated into leaves, stem and roots.
Theaceae **34**
Theorell, Axel **216**
therian **59**
thermal coefficient of linear expansion, derived unit of **228**
thermal conductivity, derived unit of **228**
thermal energy **88**
thermal motion **116**
thermionic emission **117**
thermionic triode **117**
thermistor **118**
thermodynamic temperature scale **89**
thermodynamics **87-89**, 196, 197, 198, 204, 205, 211
thermodynamics, laws of **89**
thermometer 88, **89**, 194, 202
thermoplastics **160**
thermoreception **71**
thermosets **160**
thermosetting plastics **160**
thermosoftening plastics **160**
thiamin **67**
thigh **60**
thiols **155**
thistle **34**
Thompson, Sir Benjamin **214**
Thompson, William **87**
Thomson, Sir George Paget **216**
Thomson, Sir Joseph John **216**
thoracic vertebrae **61**
thorax (insects) **43**
thorium 120, 128, **134**
thorn a modified reduced branch forming a pointed woody structure. It has a vascular structure within it.
threadworm **39**
three-toed sloth 55, **59**
threshold intensity **102**
threshold level **71**
throat 60, **65**
thrush **53**

thulium 129, 134
thumb 60, 61
thyme 34
thymine 15
thyroid gland 69
thyroxine 68, 69
thysanurans 42
tibia 60, 61
tick 45
ticonal 111
tiger 16, 59
Tiliaceae 34
timbre 53
time 94, 95
time dilation 95
time/distance graph 170
tin 129, 133, 146
Tinaformes 53
tinamous 53
Tiselius, Arne 216
tissue a collection of cells, usually of the same type, specialized to perform a particular function.
titanium 128, 132, 144, 145, 146, 161
toad 48, 49
toadstool 24
tobacco 34
tobacco mosaic virus 19
toe 52, 58, 60, 61
tomato 34, 66
Tomonaga, Shin'ichiro 216
tones 103
tongue 49, 60
tonic response 71
tonsil a lymph tissue at the back of the mouth.
top quark 123
topology 192, 207, 216
topshell 40
torque 81
Torricelli, Evangelista 216
tortoise 51
total internal reflection 105
Townes, Charles Hard 216
toxin a poisonous substance, usually produced by bacteria.
trachea 43, 53, 64, 65
tracheid 23
tracheophyte 23
tradescantia 34
transcription 15
transfer (of electrons) 145
transfer RNA 15
transformation 165
transformation matrix 176
transformer 117
transistor 119, 194, 215
transition metals 146
translation (of a figure) 181
transmutation 10, 11
transpiration 23, 214
transverse waves 97, 101
trapezium 60, 183
trapezoid 60, 183
travelling wave 97
tree 23, 34

tree ferns 26
tree shrew 59
Trematoda 39
trematode 39
Treshkiornithidae 53
triangle 167, 181, 182, 183, 184-85, 190, 200, 205
triangle of velocities 175
triangulation 181
Triassic period 51
trifoliate a compound leaf with three leaflets.
trigonometry 8, 180-81, 182-91, 206
triode 117
triple covalent bond 142
triple point 89, 91
triples 185
triplet 14, 15
triploid 200
Trochilidae 53
Trochiliformes 53
trogon 53
Trogoniformes 53
Trombidiidae 45
troy system 228
trunk 56, 58, 59
Tsuga 27
Tsui, Lap-Chee 216
tuatara 51
tube feet 38
tuber 34
Tubulidentata 54, 58, 59
tulip 34
tumour a group of cells that starts to divide without the usual checks and controls imposed by the body. It may be malignant or benign.
tungsten 114, 128, 134
tunneling 201
tunny 47
Tupalidae 59
turaco 53
Turbellaria 39
turbellarian 39
turbine 116
turbulence 85
turgor the pressure of cell contents on cell walls, swelling them out as the cell takes in water by osmosis.
Turing machine 216
Turing, Alan 216
Turniciformes 53
turning effect 86
turnip 34
twistor 212
two-toed sloth 57, 59
tympanic membrane 102
Tyndall effect 216
Tyndall, John 216

Ulmaceae 34
ulna 60, 61
ultrasonic frequency 103
ultrasonic scanning 103
ultrasound sound waves at a frequency well above the range of human hearing, used to provide an image of internal structures.
ultraviolet waves 113

umbel an inflorescence in which flowers are borne on undivided stalks that arise from the main stem.
umbilical cord the connection between the embryo and the placenta in pregnant mammals.
uncertainty principle 95
ungulate 56, 73, 58, 59
unicellular 13
uniform acceleration 77
uniform acceleration equations 175
uniform velocity 77, 78
uniformitarianism 11
uniformly accelerated motion, laws of 77
unipolar transistor 119
unnilennium (provisionally named meitnerium) 128, 135
unnilhexium (provisionally named seaborgium) 128, 135
unniloctium (provisionally named hassium) 128, 135
unnilpentium (provisionally named hahnium) 128, 135
unnilquadium (provisionally named rutherfordium) 128, 135
unnilseptium (provisionally named nielsbohrium) 128, 135
unsaturated fat 67
unsaturated hydrocarbon 156
up quark 123
Upupliformes 53
uranium 120, 121, 128, 134, 146, 192, 194, 199, 201, 204
Uranus 205
urea 217
ureter the tube leading from the kidney to the bladder.
urethra the tube leading from the bladder to the exterior.
Urey, Harold Clayton 216
uric acid the waste product of birds and some other animals.
urine liquid produced in the kidney, containing waste products such as urea or uric acid.
Urodela 48
urodele 49
Ursidae 55, 59
Urticaceae 34
US Customary Units 228
uterus 72
uvula the soft projection hanging down at the back of the mouth.

V-2 217
vaccination 207
vaccine the dead or harmless microorganisms used in a vaccination.
vacuole 13
vacuum 84, 112, 204
vacuum filtration 126
vagina 72
valency 142, 192, 203, 217
values for sin, cos and tan 181
valve 40, 62, 63
Van Allan belts 216
Van Allan, James Alfred 216
Van de Graaff accelerator 216
Van de Graaff generator 216

INDEX/FACTFINDER

Van de Graaff, Robert Jemison 216
Van Maanen, Adriaan 216
van der Waals force 142, 143
van der Waals, Johannes Diderik 216
Van't Hoff, Jacobus Henricus 216, 218
vanadium 128, 132
vanarid 51
Vanaridae 51
vaporization 91
vapour 137
Varmus, Harold 216
varying amplitude 98
vascular bundle a strand of primary vascular tissue, consisting largely of xylem and phloem.
vascular pertaining to blood vessels.
vascular plant 23
vascular system 23
vascular tissue 26
Vauquelin, Nicolas-Louis 216
Vavilov, Nikolai 216
Veblen, Oswald 216
vector 78, 175
vegetable 34
vegetable oil 67
vegetative reproduction asexual reproduction in which specialized multicellular organs are formed and detached from the parent, generating new individuals.
vein 63
veliger 40
velocity 74, 77, 174
velocity of separation 175
velocity of sound 103
velocity ratio 86
velocity, derived unit of 228
velvet worm 41
venation (in insects) the arrangement of veins in an insect's wing or (in plants) the pattern of veins in a leaf.
venom 45, 50, 51
ventilation 65
ventral the surface of an animal furthest away from the notochord or spinal column.
ventricle 62, 63
venule a small blood vessel linking a vein to the capillary network.
Venus 40
vermiform appendix 66
vernalization 23
vertebra 51, 54, 55, 60, 61
vertebral column 51
vertebral rib 61
Vertebrata 17
vertebrate 17, 46, 54, 60, 62, 63, 64, 65, 69, 199, 207
vertebrates (respiration in) 65
vertex 182, 183, 185, 190, 191
vertical component of velocity 78
Vesalius, Andrea 216
vesicle 13
vestigial structures 11
vetch 34
vibrio 19
Victrex PEEK 156, 160
vicuna 59

villus a projection from a body surface - usually designed to increase the surface area of a tissue.
violet 34
viper 51
Viperidae 51
viral replication 19
Virchow, Rudolf 216
viroid 19
virus 18, 19, 200, 201, 215
viscosity 85
visible waves 113
Vitaceae 34
vitamin 67
vitamin A 67
vitamin B 217
vitamin B1 67
vitamin B12 66, 67
vitamin B2 67
vitamin B6 67
vitamin C 67
vitamin D 67
vitamin E 67
vitamin K 67, 199
vitamins 66, 205, 206
viverrid 56, 57, 59
Viverridae 59
viviparous any animal in which embryos develop within and are nourished by the mother.
vivipary young plants forming at the axils of flowers, or the germination of seeds on the parent plant before release.
vocal cords elastic fibres in the larynx that produce sounds in vertebrates.
vole 59
volt 110, 111, 114, 115, 228
Volta, Count Alessandro 110, 114, 216
voltage 114, 115, 117
voltaic pile 114, 115, 203, 216
volume (metric units of) 228
volume, derived unit of 228
Vombatidae 59
vomer 61
vomeronasal response 71
von Braun, Werner von 217
von Karman, Theodore 217
von Neumann, John 217
vulgar fraction 164
vulture 53
vulva 72

W particle 123
Waddington, Conrad Hal 217
wader 52
Waksman, Selman 217
wallaby 57, 58, 59
Wallace's line 217
Wallace, Alfred Russell 217
wallflower 34
walnut 34
walrus 73
wapiti 56, 59
warm receptor 71
warm-blooded 56, 64, 65
wart a small tumour of the outer layer of the skin caused by a virus.

wasp 42, 43
waste products 62, 64, 66
water 46, 47, 48, 64, 65, 66, 73, 84, 89, 90, 91, 124, 136, 140, 141, 142, 144, 145, 146, 147, 148, 151, 154, 157, 159, 195, 197
water boatman 43
water fern 26
water flea 44
water lily 34
water loss 23
water mite 45
water plantain 34
water vapour 65, 124, 148
water wave 97
waterbuck 54, 59
Watson, James 14, 217
Watson-Watt, Sir Robert Alexander 217
watt 115, 217, 228
Watt, James 115, 217
wave attenuation 96, 97
wave effect 200
wave motion 96
wave nature 199
wave particle duality 95
wave theory of light 217
wave types 96-97
wave-particle duality 123
wavefront 99
wavelength 96, 97, 213
waves 96-97, 98-99
weak acid/base 145
weak force 75, 112, 123, 143
weasel 57, 73, 59
webbed feet 52, 58
weber 217, 228
Weber, Wilhelm Eduard 217
Weierstrass, Karl 217
weight 78
weight 228
Weinberg, Steven 217
Weismann, August 217
Werner, Alfred 217
western hemlock 27
Weyl, Hermann 217
whale 56, 73
whalebone 73
wheatgerm 67
Wheatstone bridge 217
Wheatstone, Sir Charles 217
whisk fern 26
white blood cell 63
white matter the region of the vertebrate central nervous system consisting of nerve cell fibres.
White, Gilbert 217
Whitehead, Alfred 217
whole number 163
Wiener, Norbert 217
wild boar 58
wildebeest 54, 59
Wilkins, Maurice 217
Williamson, Alexander William 217
willow 34
Wilson, Kenneth 217
wind pollination 34

windpipe 65
wing 43, 52, 53
winged insect 42, 43
wingless insect 42, 43
wisent 55, 59
wishbone 52
Witten, Edward 217
Wohler, Friedrich 217
wolf 55, 59
Wolfram, Stephen 217
Wollaston, William Hyde 217
womb 72
wombat 57, 59
wood 34
woodlouse 44
woodpecker 52, 53
woody perennial 34
work 89
work 175

worm 39, 49, 50, 51, 64
wrist 60, 61
Wu, Chien-Shiun 217

x-axis 171
x-particles 112
X-rays 113, 194, 196, 208, 210, 214
xenon 129, 130, 133, 138, 213
xeromorphic 27
xiphisternum 61
Xiphosura 45
xylem 20, 21, 23, 26, 34

y-axis 171
yak 55, 59
yeast 24, 67
yolk 50
Young's modulus of elasticity 83
Young, Thomas 101, 217

ytterbium 129, 134
yttrium 128, 132, 161
Yukawa, Hideki 217

Z particle 123
zebra 56, 58, 59
zeolite 161
Zermelo, Ernst 217
Zernike, Frits 217
zero force 81
Ziegler, Karl 217
zinc 114, 115, 129, 132, 142, 146, 216
Zingiberaceae 34
zirconium 128, 132
zoology 8, 34-73
zygodactylous bird 52, 53
Zygomycetes 24
zygospore 24
zygote 26, 72